BIOLOGICAL NUTRIENT REMOVAL (BNR) OPERATION IN WASTEWATER TREATMENT PLANTS

Prepared by the Biological Nutrient Removal (BNR) Operation in Wastewater Treatment Plants Task Force of the Water Environment Federation and the American Society of Civil Engineers/Environmental and Water Resources Institute

Jeanette A. Brown, P.E., DEE, *Co-Chair*
Carl M. Koch, P.E., DEE, Ph.D., *Co-Chair*

James L. Barnard, P. Eng., Ph.D.
Mario Benisch
Stephen A. Black, P. Eng., M.A.Sc.
Bob Bower
William C. Boyle, P.E., DEE, Ph.D.
Rhodes R. Copithorn
Glen T. Daigger
Christine deBarbadillo, P.E.
Paul A. Dombrowski, P.E., DEE
Alex Ekster, P.E., DEE, Ph.D.
Ufuk G. Erdal, Ph.D.
Zeynep K. Erdal, Ph.D.

John R. Harrison
Joseph A. Husband
Samuel S. Jeyanayagam, P.E., DEE, Ph.D.
Philip R. Kiser
Edmund A. Kobylinski, P.E.
Curtis I. Kunihiro, P.E., DEE
Troy A. Larson
Neil Massart, P.E.
Krishna R. Pagilla, P.E., Ph.D.
Barry Rabinowitz, P. Eng., Ph.D.
Keith A. Radick
Andrew R. Shaw
Troy Stinson, P.E.
Cindy Wallis-Lage, P.E.
Richard Watson

Under the Direction of the Municipal Subcommittee of the Technical Practice Committee

2005

Water Environment Federation
601 Wythe Street
Alexandria, VA 22314-1994 USA
http://www.wef.org
and
American Society of Civil Engineers/Environmental and Water Resources Institute
1801 Alexander Bell Drive
Reston, VA 20191-4400
http://www.asce.org

BIOLOGICAL NUTRIENT REMOVAL (BNR) OPERATION IN WASTEWATER TREATMENT PLANTS

WEF Manual of Practice No. 29

ASCE/EWRI Manuals and Reports on Engineering Practice No. 109

Prepared by the Biological Nutrient Removal (BNR) Operation in Wastewater Treatment Plants Task Force of the Water Environment Federation and the American Society of Civil Engineers/ Environmental and Water Resources Institute

Water Environment Federation
and
American Society of Civil Engineers/
Environmental and Water Resources Institute

WEF Press

McGraw-Hill

New York Chicago San Francisco Lisbon London Madrid
Mexico City Milan New Delhi San Juan Seoul
Singapore Sydney Toronto

The **McGraw·Hill** Companies

Cataloging-in-Publication Data is on file with the Library of Congress.

Copyright © 2006 by the Water Environment Federation and the American Society of Civil Engineers/Environmental and Water Resources Institute. Permission to copy must be obtained from both WEF and ASCE/EWRI.
All Rights Reserved. Except as permitted under the United States Copyright Act of 1976, no part of this publication may be reproduced or distributed in any form or by any means, or stored in a data base or retrieval system, without the prior written permission of WEF and ASCE/EWRI.

Water Environment Research, WEF, and *WEFTEC* are registered trademarks of the Water Environment Federation.
American Society of Civil Engineers, ASCE, Environmental and Water Resources Institute, and EWRI are registered trademarks of the American Society of Civil Engineers.

1 2 3 4 5 6 7 8 9 0   DOC/DOC   0 1 0 9 8 7 6 5

ISBN 0-07-146415-8

The sponsoring editor for this book was Larry S. Hager and the production supervisor was Pamela A. Pelton. It was set in Palatino by Lone Wolf Enterprises, Ltd. The art director for the cover was Anthony Landi.

Printed and bound by RR Donnelley.

 This book is printed on recycled, acid-free paper containing a minimum of 50% recycled, de-inked fiber.

McGraw-Hill books are available at special quantity discounts to use as premiums and sales promotions, or for use in corporate training programs. For more information, please write to the Director of Special Sales, McGraw-Hill Professional, Two Penn Plaza, New York, NY 10121-2298. Or contact your local bookstore.

IMPORTANT NOTICE

The material presented in this publication has been prepared in accordance with generally recognized engineering principles and practices and is for general information only. This information should not be used without first securing competent advice with respect to its suitability for any general or specific application.

The contents of this publication are not intended to be a standard of the Water Environment Federation (WEF) or the American Society of Civil Engineers (ASCE)/Environmental and Water Resources Institute (EWRI) and are not intended for use as a reference in purchase specifications, contracts, regulations, statutes, or any other legal document.

No reference made in this publication to any specific method, product, process, or service constitutes or implies an endorsement, recommendation, or warranty thereof by WEF or ASCE/EWRI.

WEF and ASCE/EWRI make no representation or warranty of any kind, whether expressed or implied, concerning the accuracy, product, or process discussed in this publication and assumes no liability.

Anyone using this information assumes all liability arising from such use, including but not limited to infringement of any patent or patents.

Water Environment Federation

Improving Water Quality for 75 Years

Founded in 1928, the Water Environment Federation (WEF) is a not-for-profit technical and educational organization with members from varied disciplines who work toward the WEF vision of preservation and enhancement of the global water environment. The WEF network includes water quality professionals from 79 Member Associations in over 30 countries.

For information on membership, publications, and conferences, contact

Water Environment Federation
601 Wythe Street
Alexandria, VA 22314-1994 USA
(703) 684-2400
http://www.wef.org

American Society of Civil Engineers/Environmental and Water Resources Institute

A Better World by Design

Founded in 1852, the American Society of Civil Engineers (ASCE) represents more than 133,000 members of the civil engineering profession worldwide, and is America's oldest national engineering society. Created in 1999, the Environmental & Water Resources Institute (EWRI) is an Institute of the American Society of Civil Engineers. EWRI services are designed to complement ASCE's traditional civil engineering base and to attract new categories of members (non-civil engineer allied professionals) who seek to enhance their professional and technical development.

For more information on membership, publications, and conferences, contact

ASCE/EWRI
1801 Alexander Bell Drive
Reston, VA 20191-4400
703-295-6000
http://www.asce.org

Manuals of Practice of the Water Environment Federation

The WEF Technical Practice Committee (formerly the Committee on Sewage and Industrial Wastes Practice of the Federation of Sewage and Industrial Wastes Associations) was created by the Federation Board of Control on October 11, 1941. The primary function of the Committee is to originate and produce, through appropriate subcommittees, special publications dealing with technical aspects of the broad interests of the Federation. These publications are intended to provide background information through a review of technical practices and detailed procedures that research and experience have shown to be functional and practical.

Water Environment Federation Technical Practice
Committee Control Group

G. T. Daigger, Vice Chair
B. G. Jones, Vice Chair

S. Biesterfeld
R. Fernandez
L. Ford
Z. Li
M. D. Moore
M. D. Nelson
A. B. Pincince
S. Rangarajan
J. D. Reece
E. P. Rothstein
A. T. Sandy
J. Witherspoon

Manuals and Reports on Engineering Practice

(As developed by the ASCE Technical Procedures Committee, July 1930, and revised March 1935, February 1962, and April 1982)

A manual or report in this series consists of an orderly presentation of facts on a particular subject, supplemented by an analysis of limitations and applications of these facts. It contains information useful to the average engineer in his everyday work, rather than the findings that may be useful only occasionally or rarely. It is not in any sense a "standard," however; nor is it so elementary or so conclusive as to provide a "rule of thumb" for nonengineers.

Furthermore, material in this series, in distinction from a paper (which expressed only one person's observations or opinions), is the work of a committee or group selected to assemble and express information on a specific topic. As often as practicable the committee is under the direction of one or more of the Technical Divisions and Councils, and the product evolved has been subjected to review by the Executive Committee of the Division or Council. As a step in the process of this review, proposed manuscripts are often brought before the members of the Technical Divisions and Councils for comment, which may serve as the basis for improvement. When published, each work shows the names of the committees by which it was compiled and indicates clearly the several processes through which it has passed in review, in order that its merit may be definitely understood.

In February 1962 (and revised in April 1982) the Board of Direction voted to establish:

A series entitled "Manuals and Reports on Engineering Practice," to include the Manuals published and authorized to date, future Manuals of Professional Practice, and Reports on Engineering Practice. All such Manual or Report material of the Society would have been refereed in a manner approved by the Board Committee on Publications and would be bound, with applicable discussion, in books similar to past Manuals. Numbering would be consecutive and would be a continuation of present Manual numbers. In some cases of reports of joint committees, bypassing of Journal publications may be authorized.

Contents

Preface xliii

Chapter 1 Introduction 1

Chapter 2 Overall Process Considerations

Introduction 6
Nutrient Sources 8
    Sources of Nitrogen 8
    Source of Phosphorus 9
Effects of Nutrients on Receiving Waters 12
    Eutrophication 12
    Ammonia Toxicity 12
    Nitrate in Groundwater 13
Wastewater Characteristics 13
    Carbonaceous Materials 13
    Nitrogen 16
    Phosphorus 20
    Solids 21
    Temperature 23
    pH 24
    Alkalinity 24
Variations in Flows and Loads 25
Effect of Recycle Flows 27

Review of Recycle Flows 27
Management of Return Flows 28
Effect of Effluent Permit Requirements 29
Technology-Based Permits 29
Monthly Average. 29
Annual Average. 30
Seasonal Permit 30
Water-Quality-Based Permit 30
References 31

Chapter 3 Nitrification and Denitrification

Introduction 35
Wastewater Characteristics 35
Assimilation 35
Hydrolysis and Ammonification 35
Nitrifier Growth Rate 37
Nitrification 37
Process Fundamentals 37
Stoichiometry 37
Nitrification Kinetics 38
Biomass Growth and Ammonium Use 38
Wastewater Temperature. 41
Dissolved Oxygen Concentration 42
pH and Alkalinity 42
Inhibition 43
Flow and Load Variations 43
Suspended-Growth Systems 43
General. 43
Determining the Target $SRT_{aerobic}$ 50
Example 3.1—Single Sludge Suspended-Growth Nitrification 51
Single Sludge Systems. 52
Separate Sludge Systems. 53

Attached Growth Systems	55
General	55
Trickling Filter	57
Rotating Biological Contactors	62
Biological Aerated Filter	65
Coupled Systems	67
Denitrification	68
Process Fundamentals	68
Stoichiometry	70
Denitrification Kinetics—Biomass Growth and Nitrate Use	71
Example 3.2—Single Sludge Suspended-Growth Postdenitrification	73
Carbon Augmentation	74
Separate-Stage Denitrification	75
Suspended-Growth Systems	75
Attached-Growth Systems	76
Denitrification Filter	76
Moving Bed Biofilm Reactor	77
Combined Nitrification and Denitrification Systems	78
Basic Considerations	78
Suspended-Growth Systems	78
Wuhrmann Process	79
Modified Ludzack-Ettinger Process	79
Bardenpho Process (Four-Stage)	82
Sequencing Batch Reactors	84
Cyclically Aerated Activated Sludge	86
Oxidation Ditch Processes	86
Countercurrent Aeration	87
Hybrid Systems	87
Introduction	87
Integrated Fixed-Film Activated Sludge	87
Descriptions of Integrated Fixed-Film Activated Sludge Processes	88
Rope-Type Media	88

Sponge-Type Media 90
Plastic Media 90
Rotating Biological Contactors 92
Operational Issues 92
Rope-Type Media 92
Media Location 92
GROWTH 92
KINETIC RATE 92
Worms 93
Media Breakage 93
Adequate Dissolved Oxygen Level 94
Mixing 94
Access To Diffusers 94
Odor 94
Sponge Media 94
Screen Clogging 94
Sinking Sponges 95
Loss of Sponges 95
Taking Tank Out of Service 96
Loss of Solids 96
Air Distribution System 96
Plastic Media 96
Startup Procedures 96
Growth 97
Worms 97
Media Breakage 97
Media Mixing 97
Accumulation of Growth 97
SCREEN CLOGGING 97
FOAMING 98
TAKING TANK OUT OF SERVICE 98
Membrane Bioreactor 98

Secondary Clarification 100
   Suspended Growth 100
      Denitrification 101
      Flocculation Problem 1 101
      Flocculation Problem 2 101
      High Sludge Blanket 101
      Hydraulic Problem 102
   Attached Growth 102
References 102

Chapter 4 Enhanced Biological Phosphorus Removal

Introduction and Basic Theory of Enhanced Biological
Phosphorus Removal 106
Basic Enhanced Biological Phosphorus Removal Theory 107
Basic Enhanced Biological Phosphorus Removal Design Principles 108
Operational Parameters 113
   Influent Composition and Chemical-Oxygen-Demand-to-
   Phosphorus Ratio 113
   Solids Retention Time and Hydraulic Retention Time 117
   Temperature 122
   Recycle Flows 127
      Internal Recycles 127
      Plant Recycles 129
Types of Enhanced Biological Phosphorus Removal Systems 129
   Suspended-Growth Systems 130
      Anaerobic/Oxic and Anaerobic/Anoxic/Oxic Configurations 130
      Modified Bardenpho Configuration 132
      University of Cape Town, Modified University of Cape Town, and Virginia
      Initiative Process Configurations 133
      Johannesburg Configuration 135
      Oxidation Ditches, BioDenipho, and VT2 Configurations 136
   Hybrid Systems 138
      PhoStrip Configuration 138

  Biological Chemical Phosphorus and Nitrogen Removal Configuration 139
Process Control Methodologies 140
 Influent Carbon Augmentation 140
  Volatile Fatty Acid Addition..140
  Pre-Fermentation..140
 Solids Separation and Sludge Processing 143
 Chemical Polishing and Effluent Filtration 144
Case Studies 146
 The Lethbridge Wastewater Treatment Plant, Alberta, Canada 146
 Durham, Tigard, Oregon, Clean Water Services 148
 Unique New Designs 149
  McAlpine Creek Wastewater Management Facility of Charlotte, North Carolina 150
  Traverse City Regional Wastewater Treatment Plant, Traverse City, Michigan ..150
References 153

Chapter 5   Combined Nutrient Removal Systems

Combined Nitrogen and Phosphorus Removal Processes 162
Flow Sheets for Combined Nutrient Removal 162
 The Five-Stage Bardenpho Process 165
 Phoredox (A^2/O) Process 166
 The University of Cape Town and Virginia Initiative Processes 166
 Modified University of Cape Town Process 168
 Johannesburg and Modified Johannesburg Processes 168
 Westbank Process 169
 The Orange Water and Sewer Authority Process 171
 Phosphorus Removal Combined with Channel-Type Systems 171
 Cyclical Nitrogen and Phosphorus Removal Systems 173
 General Remarks about the Various Process Configurations 174
Interaction of Nitrates and Phosphorus in Biological Nutrient Removal
 Plants 178
Process Control Methodologies 182
 Effect of Oxygen 182

    Temperature Effects 183
    pH Effects 184
    Sufficient Dissolved Oxygen in the Aeration Zone 185
    Chemical Oxygen Demand to Total Kjeldahl Nitrogen Ratio 186
    Selection of Aeration Device 187
    Clarifier Selection 187
Effect of Chemical Phosphorus Removal on Biological Nutrient Removal Systems 189
    Primary Clarifiers 189
    Secondary Clarifiers 190
    Tertiary Filters 190
Process Selection for Combined Nitrogen and Phosphorus Removal 192
    Effluent Requirements 192
    Phosphate Removal but No Nitrification 192
    Phosphate Removal with Nitrification but No Denitrification 193
    Phosphate Removal with Nitrification Only in Summer 194
    High-Percentage Nitrogen and Phosphate Removal 195
Benefits from Converting to Biological-Nutrient-Removal-Type Operation 196
    Reliable Operation 196
    Restoring Alkalinity 198
    Improving the Alpha Factor 199
    Improving Sludge Settleability 199
Troubleshooting Biological Nutrient Removal Plants 202
    Plant not Designed for Nitrification but Nitrification in Summer Causes Problems 202
        Problem 203
        Correction 203
        Problem 203
        Correction 203
        Problem 203
        Correction 203

Plant Designed for Nitrification but No Denitrification ... 203
Problem ... 203
Correction ... 203
Problem ... 203
Correction ... 203
Problem ... 203
Correction ... 204
Problem ... 204
Correction ... 204

Plant Designed for Nitrification and Denitrification ... 204
Problem ... 204
Correction ... 204
Problem ... 204
Correction ... 205
Problem ... 205
Correction ... 205

Plant Designed for Phosphorus Removal Only ... 206
Problem ... 206
Correction ... 206
Problem ... 206
Correction ... 206
Problem ... 207
Correction ... 207

Plant Designed for Ammonia and Phosphorus Removal ... 207
Problem ... 207
Correction ... 207
Problem ... 207
Correction ... 207
Problem ... 207
Correction ... 208
Problem ... 208
Correction ... 208

  Problem 208
  Correction 208
  Problem 208
  Correction 208
  Problem 208
  Correction 208
 Retrofitting Plants for Nutrient Removal 208
 Return Activated Sludge and Internal Recycle Rates 212
 Minimizing the Adverse Effect of Storm Flows .213
 Foam Control 214
 Waste Sludge and Return Stream Management .215
Summary 216
Case Studies 216
 City of Bowie Wastewater Treatment Plant (Bowie, Maryland) 216
  Facility Design 216
  Effluent Limits. 216
  Wastewater Characteristics. 217
  Performance. 217
 Potsdam Wastewater Treatment Plant (Germany) 217
  Facility Design 217
  Performance. 217
 Goldsboro Water Reclamation Facility, North Carolina 217
  Facility Design 217
  Effluent Limits. 218
  Wastewater Characteristics. 218
  Performance. 218
 South Cary Water Reclamation Facility, North Carolina 218
  Facility Design 218
  Effluent Limits. 218
  Wastewater Characteristics. 219
  Performance. 219
 North Cary Water Reclamation Facility, North Carolina 219

    Facility Design 219
    Effluent Limits 219
    Wastewater Characteristics 220
    Performance 220
 Wilson Hominy Creek Wastewater Management Facility, North Carolina 220
    Facility Design 220
    Effluent Limits 220
    Wastewater Characteristics 220
    Performance 221
 Greenville Utilities Commission Wastewater Treatment Plant, North Carolina 221
    Facility Design 221
    Effluent Limits 221
    Wastewater Characteristics 221
    Performance 222
 Virginia Initiative Plant, Norfolk 222
    Facility Design 222
    Effluent Limits 222
    Wastewater Characteristics 222
    Performance 222
References 223

Chapter 6 Models for Nutrient Removal

Introduction 227
History and Development of Models for Biological Nutrient Removal ... 227
Description of Models 228
  Mechanistic Models 228
  Simulators 228
Use of Simulators for Plant Operation 230
  Ease of Use 230
  Developing Data for Model Input 230
  Using Simulators to Troubleshoot 231

Reference 232

Chapter 7 Sludge Bulking and Foaming

Introduction 233
Microscopic Examination 234
Filamentous Bulking 236
Process Control for Filamentous Bulking Problems 239
    Dissolved Oxygen 239
    Nutrient Balance 240
    pH 241
    Chemical Addition 241
        Chlorination 241
        Biological Selectors 242
        Nonfilamentous Bulking 243
    Troubleshooting Sludge Bulking Problems 243
    Foaming Problems and Solutions 243
        Stiff White Foam 247
        Excessive Brown Foam and Dark Tan Foam 247
        Dark Brown Foam 247
        Very Dark or Black Foam 247
        Filamentous Foaming 247
Conclusion 248
References 248

Chapter 8 Chemical Addition and Chemical Feed Control

Introduction 256
Carbon Supplementation for Denitrification 257
    Methanol Addition 257
        Methanol Addition to Activated Sludge Biological Nutrient Removal Processes . 259
        Methanol Addition to Tertiary Denitrification Processes 263
        Methanol Feed Control 265
            Manual Control 265
            Flow-Paced Control 265

  Feed-Forward Control 265
  Feed-Forward and Feedback with Effluent Concentration Control ...266
 Alternate Carbon Sources for Denitrification 266
 Case Studies 268
  Havelock Wastewater Treatment Plant, Havelock, North Carolina 268
  Long Creek Wastewater Treatment Plant, Gastonia, North Carolina 269
Volatile Fatty Acid Supplementation for Biological Phosphorus Removal 271
 Acetic Acid 273
 Alternate Chemical Volatile Fatty Acid Sources 274
 Case Study: McDowell Creek Wastewater Treatment Plant,
  Charlotte, North Carolina 274
  Effect of Dewatering Filtrate 275
  Optimization of Chemical Dosages 275
Alkalinity Supplementation 277
 Alkalinity 277
 Alkalinity Supplementation 279
  Sodium Hydroxide 279
  Calcium Hydroxide 281
  Quicklime 281
  Magnesium Hydroxide 282
  Sodium Carbonate 282
  Sodium Bicarbonate 282
 Alkalinity Considerations 282
  Usable Alkalinity 282
  Volatile Fatty Acids and Other Alkalinity 283
  Alkalinity Measurement 284
  High-Purity-Oxygen Activated Sludge 284
  Phosphorus Precipitation 285
 Practical Examples 285
  Example 1 285
   Step 1 285

Step 2	285
Example 2	286
Step 1	286
Step 2	286
Example 3	287
Step 1	287
Step 2	287
Phosphorus Precipitation	288
Iron Compound Chemical Addition	290
Aluminum Compound Chemical Addition	297
Lime Addition	303
Other Options for Chemical Precipitation of Phosphorus	305
Chemical Feed Control	306
Case Study: Northwest Cobb Water Reclamation Facility, Cobb County, Georgia	306
Chemical Feed System Design and Operational Considerations	309
References	311

Chapter 9 Sludge Fermentation

Overview of Fermentation Processes	314
Function and Relationship to Biological Nutrient Removal Process	314
Hydrolysis	317
Acidogenesis	317
Acetogenesis	318
Methanogenesis	318
Primary Sludge Fermentation	319
Return Activated Sludge Fermentation	320
Primary Sludge Fermenter Configurations	322
Activated Primary Sedimentation Tanks	322
Complete-Mix Fermenter	326
Single-Stage Static Fermenter	327
Two-Stage Complete-Mix/Thickener Fermenter	328

Unified Fermentation and Thickening Process	329
Primary Sludge Fermentation Equipment Considerations	329
Sludge Collector Drives	329
Primary Sludge Pumping	330
Fermented Sludge Pumping	330
Sludge Grinders or Screens	330
Fermentate Pumping	330
Mixers	331
Scum Removal	331
Odor Control and Covers	331
Corrosion and Protective Coatings	332
Instrumentation	332
Flow Measurement	332
Oxidation–Reduction Potential	332
Level Measurement	333
Sludge Density Meters	333
pH Meters	333
Headspace Monitoring	333
Return Activated Sludge Fermentation	333
Configuration	333
Equipment	334
Control Parameters	335
Case Studies	337
Kelowna Wastewater Treatment Plant, Canada	337
Kalispell, Montana	338
South Cary Water Reclamation Facility, North Carolina	341
Plant Description	341
Fermentation Process Description	341
Operating Parameters	343
References	345

Chapter 10 Solids Handling and Processing

Introduction 350
Issues and Concerns 351
    Influent Load Variations 351
    Influent Amenability to Biological Nutrient Removal 351
    Mean Cell Residence Time 352
    Struvite Formation 352
Sludge Production 353
Nutrient Release 353
    Release Mechanisms 353
    Sources of Secondary Release 354
        Primary Clarification 354
        Final Clarification 354
        Thickening 354
        Stabilization 355
        Dewatering 355
Estimating Recycle Loads 357
Eliminating or Minimizing
Recycle Loads 360
Sidestream Management and Treatment 362
    Sidestream Management Alternatives 365
    Recycle Equalization and Semitreatment 366
        Equalization 366
        Solids Removal 366
        Aeration 366
        Operational Issues 367
    Sidestream Treatment 367
        Nitrogen Removal 367
        Stand-Alone Sidestream Treatment and Full-Centrate Nitrification 368
        Operational Issues with Separate Recycle Nitrification Process 370
            Influent Solids 370

      Struvite 371
      Alkalinity Feed 371
      Aeration Efficiency 372
      Reactor Configuration 372
      Foaming 372
  Separate Recycle Treatment—Nitrogen Removal
  (Nitrification and Denitrification) 372
  Single Reactor System for High Activity Ammonium Removal over Nitrite 372
      ANAMMOX 373
      Ammonia Stripping 374
Combination Sidestream Treatment and Biological Nutrient
Removal Process—Return Activated Sludge Reaeration 374
Formation of Struvite and Other Precipitates 376
  Struvite Chemistry 377
  Biological Nutrient Removal and Struvite 380
  Areas Most Susceptible to Struvite Formation 382
  Struvite Control Alternatives 387
    Phosphate Precipitating Agents 388
    Dilution Water 389
    Cleaning Loops 389
    Hydroblasting 389
    Equipment and Pipe Lining Selection 389
    Magnetic and Ultrasonic Treatment 389
    Lagoon Flushing 391
    Controlled Struvite Crystallization (Phosphorus Recovery) ... 391
    Facility Design 391
    Process Design 392
Case Studies 395
Conclusion 395
References 396

Chapter 11 Laboratory Analyses

Nitrogen 400
   Types 400
   Sampling and Storage 400
   Analyses Methods 401
      Ammonia-Nitrogen. 401
         Colorimetric Methods 401
         Titrimetric Methods 402
         Ion Selective Method 402
         Ion Chromatography 403
      Nitrite-Nitrogen 404
         Colorimetric Method 404
         Ion Chromatography 404
      Nitrate-Nitrogen. 404
         Nitrate Electrode Screening Method 405
         Chromotropic Acid Method 406
         Ion Chromatography for Nitrite and Nitrate .407
      Organic Nitrogen 408
      Kjeldahl Method 409
         Storage of Samples 409
         Interference 410
Phosphorus 411
   Types 411
   Sampling and Storage 411
   Analyses Methods 412
      Digestion Methods for Total Phosphorus Analyses 412
      Methods for Orthophosphate Analysis 413
         Vanadomolydophosphoric Acid Colorimetric Method 413
         Ascorbic Acid Method 413
         Ion Chromatography 413

Short-Chain Volatile Fatty Acid Analysis 415
Analytical Methods for Short-Chain Volatile Fatty Acid Measurement ...416
References 417

Chapter 12 Optimization and Troubleshooting Techniques

Process Evaluation 422
Sampling and Testing 422
    Sampling Locations and Techniques 422
        Sampling Plan... 422
        Sample Handling.. 423
            In Situ Sampling 424
            Grab Sampling 424
            Interval Sampling 427
            Composite Sampling 428
        Sample Location 428
    Mixed Liquor Suspended Solids, Mixed Liquor Volatile Suspended Solids, Return Activated Sludge, and Waste Activated Sludge 430
        What 430
        Where... 431
        Why 431
        When 432
        How 432
Settleability and Sludge Volume Index 432
    What 432
    Where... 433
    Why 433
    When 433
    How 433
pH 434
    What 434
    Where... 434

Why	434
When	434
How	434

Alkalinity ... 435
What ... 435
Where ... 435
Why ... 435
When ... 435
How ... 436

Temperature ... 436
What ... 436
Where ... 436
Why ... 436
When ... 437
How ... 437

Dissolved Oxygen ... 437
What ... 437
Why ... 437
Where and When ... 438
How ... 438

Oxidation–Reduction Potential ... 438
What ... 438
Where ... 439
Why ... 440
When ... 440
How ... 440

Ammonia and Total Kjeldahl Nitrogen ... 440
What ... 440
Where ... 441
Why ... 441
When ... 441

How	442
Nitrite-Nitrogen	442
What	442
Where	442
Why	442
When	442
How	442
Nitrate-Nitrogen	442
What	442
Where	442
Why	443
When	443
How	443
Total Phosphorus	443
What	443
When and Where	443
Why	443
How	444
Orthophosphorus	444
What	444
When and Where	444
Why	444
How	445
Chemical Oxygen Demand	445
What	445
Where	446
Why	446
When	446
How	446
Volatile Fatty Acids	448
What	448

Where..448
Why..449
When...449
How..449
Soluble Biochemical Oxygen Demand.............................450
What...450
Where..450
Why..450
When...451
How..451
Nitrification Test..451
What...451
Where..451
Why..451
When...452
How..452
Denitrification Test..453
What...453
Where..453
Why..453
When...453
How..454
Biological Phosphorus Removal Potential Test 454
What...454
Where..454
Why..455
When...455
How..455
    Method 455
    Preservation 455
    Analyze 455
    Equipment and Supplies 456

Reagents and Chemicals	456
Principle	456
Preparation of Samples	457
Considerations	457
Procedure	458
Test Evaluation	459

Microbiological Activity 460
What 460
Where 461
Why 461
When 462
How 462

Data Analysis and Interpretation 462
Nitrification 462
Chemical Environment 463
Solids Retention Time 463
Performance Indicators 463
Denitrification 463
Chemical Environment 464
Solids Retention Time 464
Performance Indicators 464
Biological Phosphorus Removal 464
Chemical Environment 464
Solids Retention Time 465
Performance Indicators 465

Optimization and Troubleshooting Guides 465
Overview 465
Optimization and Troubleshooting Guide Format 466
Indicator and Observations 466
Probable Cause 466
Check or Monitor 466

	Solutions 466
	References 468

Optimization and Troubleshooting Guides 468
Case Studies 468
   Wolf Treatment Plant, Shawano, Wisconsin 468
   City of Stevens Point, Wisconsin 491
   City of Dodgeville, Wisconsin 491
   Eastern Water Reclamation Facility, Orange County, Florida ... 495
   Wastewater Treatment Plant, Stamford, Connecticut 497
References 499

Chapter 13 Instrumentation and Automated Process Control

Introduction 505
Online Analyzers 506
   General Considerations 506
      Meters Reproducibility and Accuracy. 506
      Instrument Maintenance. 507
Specific Analyzers 508
   Basic Instruments 508
      Total Suspended Solids Meters 509
         Measuring Method 509
         Accuracy and Repeatability 509
         Installation 509
         Maintenance Requirements 510
         Application 510
      Dissolved Oxygen Measurement 510
         Introduction 510
         Membrane Technology 510
         Accuracy and Repeatability 511
         Maintenance 511
         Zullig Technology 511

    Fluorescent and Luminescent Dissolved Oxygen 512
   pH Measurement.. 512
    Principles of Operation 513
    Accuracy and Repeatability 514
    In-Tank Or Open-Channel Installation 514
    Flow-Through Installation 514
    Maintenance Requirements 515
   Oxidation–Reduction Potential 515
    Introduction 515
    Principle of Operation 515
    Accuracy and Repeatability 515
    Installation 516
    Maintenance Requirements 516
    Application 516
  Advanced Instruments 516
   Ammonia and Ammonium 517
    Measurement 517
    Accuracy and Repeatability 518
   Nitrate and Nitrite 518
    Principle of Operation 518
    Accuracy and Repeatability 519
   Phosphorus and Orthophosphate 519
    Orthophosphate 519
    Total Phosphorus 519
   Installation of Ammonia and Nutrient Analyzers 520
    Maintenance Requirements 520
    Applications 521
Process Parameters for Optimization and Automatic Control 521
  General Considerations 521
  Selecting Optimum Set Points 523
  Basic Automatic Control 523

  Excess Sludge Flow Control 523
    Improved Calculation Methods 526
    Selecting Sludge Age Target 526
    Maintaining Optimum Sludge Age 527
  Case Studies. 528
    Oxnard Trickling Filter Solids Contact Activated Sludge System 528
    Toronto Main Wastewater Treatment Plant 528
    Santa Clara/San Jose Water Pollution Control Plant 529
  Dissolved Oxygen Control for Biological Nutrient Removal Plants 532
    The Need for Good Dissolved Oxygen Control 532
    Challenges 533
    Control Strategies 534
  Control of Chemical Addition 536
Advanced Control 537
  Ammonia Control 537
  Control of Denitrification 537
  Respirometry 538
  Intermittent Aeration 539
  Sequencing Batch Reactors 540
  COST Model for Control Strategy Development 541
Supervisory Control and Data Acquisition System Requirements 541
  General Considerations 541
  Supervisory Control and Data Acquisition Functions 541
  Continuous Process Control 542
    Programmable Logic Controller and Logic Program 542
    Programmable Logic Controller Programming Software 543
  Data Acquisition 543
    Input/Output Software 543
    Data Highways and Ethernet Communications 543
    Redundancy 544
    Network Mapping and Monitoring 544

Supervisory Control 544
    Supervisory Control and Data Acquisition Engine (Core) 544
    Supervisory Control and Data Acquisition Database 545
    Graphics 545
    Real-Time Trending 545
    Data Integrity 546
    Data Distribution 546
Distributed Alarming 546
Historical Collection 547
    Historian Software 547
    Historical Trend Charts 547
    Historical Metric Charts 548
Information Systems 548
    Reporting 548
    Automated Alarm Notification 548
    Thin Client Software 549
    Server Emulation Sessions 549
Security 549
Comprehensive Supervisory Control and Data Acquisition System
Summary 550
References 551

Glossary 557

Index 585

LIST OF TABLES

Table		Page
2.1	Typical raw wastewater characteristics	7
2.2	Distribution of nitrogen sources in Chesapeake Bay and Long Island Sound	9
2.3	Forms of nitrogen and their definitions	10
2.4	Forms and typical concentration of phosphates in U.S. wastewater	11
2.5	Ratio of load to the average annual load	25
3.1	Elemental composition of bacterial cells	36
3.2	Key nitrification relationships	39
3.3	Organic compounds reported as inhibitory to nitrification	44
3.4	Historical classification of trickling filters	59
3.5	Comparative physical properties of trickling filter media	60
3.6	Rotating biological contactor manufacturer recommended staging	66
3.7	Electron donors and acceptors for carbon oxidation, nitrification, and denitrification	69
3.8	Key nitrification relationships	71
3.9	Monitoring requirements for MLE process	82
3.10	Monitoring requirements for four-stage Bardenpho process	83
4.1	Major events observed in anaerobic and aerobic zones of a EBPR plant	108
4.2	Volatile fatty acids typically found in fermented wastewater	112
5.1	Summary of BNR process zones	164
5.2	Typical design parameters for commonly used biological nitrogen and phosphorus removal processes	176
5.3	Alkalinity consumed or produced by certain processes	185
5.4	Nitrogen and phosphorus removal process selection	197
6.1	Examples of BNR computer software	230
7.1	Filament types as indicators of conditions causing activated sludge bulking	237
7.2	Troubleshooting guide for bulking sludge	244
7.3	Troubleshooting guide for foaming sludge	248
8.1	Properties of 100% methanol	258
8.2	Effect of organic substrate on enhanced biological phosphorus removal	272
8.3	Properties of acetic acid	274
8.4	Comparison of mass use per calcium carbonate equivalent	280
8.5	Mole and weight ratios for iron addition	291
8.6	Chemical properties	294
8.7	Mole and weight ratios for phosphorus removal using aluminum compounds	298
8.8	Solubility of aluminum phosphate	298
8.9	Chemical properties	301

8.10	Northwest Cobb WRF chemical dosages for phosphorus removal	308
9.1	Sidestream RAS fermentation zone sampling parameters	336
9.2	Summary of operating parameters at the South Cary Water Reclamation Facility	344
10.1	Recycle loads from BNR sludge stabilization processes and their effects	356
10.2	Biochemical oxygen demand and TSS levels in sludge processing sidestreams	358
10.3	Separate centrate nitrification design criteria	368
10.4	Sludge treatment strategies implemented at various BNR facilities	394
11.1	Sampling and sample locations for nitrogen species in an MLE process	400
11.2	Required sample volume for titrimetric method as a function of ammonia-nitrogen concentration in sample	403
11.3	Standard solution concentrations and volumes for standard curve generation	406
11.4	Volume of concentrated nitrite and nitrate solutions and deionized water needed to prepare 100-mL standard solutions in the range 0.5 to 20 mg/L	409
11.5	Required sample size for various organic nitrogen concentrations	410
11.6	Sample types and locations for an anaerobic/oxic EBPR system	412
12.1	Example of sampling plan	425
12.2	Effects of organic substrate on EBPR	448
12.3	Optimization/troubleshooting guide 1: loadings	469
12.4	Optimization/troubleshooting guide 2: aeration/mixing—diffused aeration	475
12.5	Optimization/troubleshooting guide 3: aeration/mixing—mechanical aeration	478
12.6	Optimization/troubleshooting guide 4: biomass inventory	481
12.7	Optimization/troubleshooting guide 5: clarifier operation	483
12.8	Optimization/troubleshooting guide 6: internal recycle	485
12.9	Optimization/troubleshooting guide 7: pH/alkalinity	486
12.10	Optimization/troubleshooting guide 8: toxicity	487
12.11	Optimization/troubleshooting guide 9: sudden loss of chemical phosphorus removal	488
12.12	Optimization/troubleshooting guide 10: gradual loss of chemical phosphorus removal	489
12.13	Control parameters of the Wolf Treatment Plant, Shawano, Wisconsin	492
12.14	Effect of operational adjustments on effluent phosphorus levels	495
12.15	Primary effluent characteristics	500
12.16	Average effluent nitrogen data, March to August 1990	500
13.1	Considerations to be taken into account during selection of set points	524
13.2	Average percent deviation from 24-hour moving average	530

LIST OF FIGURES

Figure   **Page**

Figure	Title	Page
2.1	Forms of BOD	14
2.2	Fractionation of COD	15
2.3	Nitrogen transformations	16
2.4	Forms of nitrogen	18
2.5	Solids fractionation	22
2.6	Hourly variation in flow and strength of municipal wastewater	26
3.1	Influence of ammonia concentration on nitrification	40
3.2	Influence of temperature on nitrification	42
3.3	Daily versus seven-day moving average SRT	52
3.4	Nitrification aerobic SRT requirements	53
3.5	Single sludge nitrification	54
3.6	Separate sludge nitrification	54
3.7	Biofilm schematic	56
3.8	Trickling filter schematic	57
3.9	Trickling filter recirculation layouts	61
3.10	Trickling filter nitrification	62
3.11	Rotating biological contactor schematic	63
3.12	Rotating biological contactor nitrification rates	64
3.13	Rotating biological contactor temperature correction	65
3.14	Biological aerated filter	67
3.15	Roughing filter/activated sludge process	68
3.16	Deep bed denitrification filter	77
3.17	Wuhrmann process	79
3.18	Ludzack–Ettinger process	80
3.19	Modified Ludzack-Ettinger process	81
3.20	Nitrified recycle rate	81
3.21	Four-stage Bardenpho process	82
3.22	Sequencing batch reactor process cycle	85
3.23	Sequencing batch reactor cycle timeline—2 basin system	85
3.24	Oxidation ditch reactor	86
3.25	Examples of rope-type media	89
3.26	Examples of sponge-type media	91
3.27	Examples of plastic media	91

List of Figures

4.1 Typical concentration patterns observed in a generic EBPR system 109
4.2 Typical phosphate and cation profiles observed in batch experiments conducted on EBPR biomass 110
4.3 Typical release and uptake of magnesium and phosphate observed in an EBPR system 111
4.4 Typical release and uptake of potassium and phosphate observed in an EBPR system. 111
4.5 Observed COD:TP ratio effect on mixed liquor PAO enrichment and phosphorus storage at different EBPR plants 114
4.6 Effect of influent COD:TP ratio on phosphorus release and uptake 115
4.7 Effect of influent COD:TP ratio on PHA storage and phosphorus uptake 115
4.8 Data collected using pilot-scale EBPR plants fed with VFA at various feed COD:P ratios 116
4.9 Effect of EBPR biomass observed yield and SRT on mixed liquor volatile suspended solids phosphorus content 118
4.10 Effect of anaerobic HRT on system phosphorus removal performance 120
4.11 Data collected during batch tests performed on an enriched EBPR sludge 121
4.12 Effect of acclimation on cold-temperature performance of enriched EBPR populations 123
4.13 Comparison of two aerobic sludges obtained from 5 and 20°C 123
4.14 Effect of feed quality and nutrient limitation on EBPR aerobic washout SRT 124
4.15 Effect of system SRT on phosphorus removal at cold temperatures 125
4.16 Phosphorus mass balance performed for phosphorus in a pilot-scale, VIP-type EBPR system 128
4.17 Summary of processes developed for phosphorus removal 130
4.18 A/O and A^2/O process configurations 131
4.19 Modified Bardenpho process configuration 132
4.20 The UCT and modified UCT process configurations 133
4.21 The VIP process configuration 134
4.22 The JHB process configuration 134
4.23 Oxidation ditch designs for nitrogen and phosphorus removal 135
4.24 VT2 process configuration 136
4.25 BioDenipho process configuration and operation cycles. 137
4.26 BioDenipho trio process configuration and operation cycles 138
4.27 PhoStrip process configuration 139
4.28 BCFS process configuration 139
4.29 Recycle streams generated at typical wastewater treatment plant solids handling facilities 143

4.30	Contribution of the effluent TSS to the total phosphorus in the effluent for different phosphorus contents in the MLSS (assuming that the VSS/TSS is 75%)	144
4.31	Step bio-P reactor configuration at Lethbridge wastewater treatment plant, Alberta, Canada	145
4.32	Monthly average total phosphorus concentration in the plant effluent and ammonia concentration in the aerobic stage of the last tank of the Lethbridge wastewater treatment plant	147
4.33	McAlpine Creek Wastewater Management Facility of Charlotte, North Carolina, liquid process train	149
4.34	Process additions and improvements designed for the Traverse City Wastewater Treatment Plant	151
5.1	Five-stage (modified) Bardenpho process	165
5.2	Three-stage Phoredox (A^2/O) process	167
5.3	The UCT and VIP process configurations	167
5.4	Modified UCT process	169
5.5	Johannesburg and modified Johannesburg processes	170
5.6	Westbank process	171
5.7	Orange Water and Sewer Authority process	172
5.8	Example of SBR for nitrogen and phosphorus removal	174
5.9	Primary and secondary release in the anaerobic zone	179
5.10	Profile of soluble phosphorus through Rooiwal, South Africa, plant	180
5.11	Secondary release of phosphorus in the second anoxic zone	181
5.12	Relative life cycle cost for removal of phosphorus to low levels	191
5.13	Flow diagram for phosphorus removal with occasional nitrification	193
5.14	Alternative way to prevent nitrates to the anaerobic zone in high-rate plants with occasional nitrification	193
5.15	Removal of nitrogen in BNR plant in the United States	196
5.16	Relationship between SRT, BPR, and nitrification	200
5.17	Effect of under-aeration on the SVI	201
5.18	The SVI at start of nitrification in summer in Bonnybrook, Calgary, Alberta, Canada	202
5.19	Typical diurnal pattern of nitrogen in influent	205
5.20	Improvement to deal with nitrates in RAS	209
5.21	Phosphorus removal at Reedy Creek, Florida, by switching off aerators	210
5.22	Conversion of high-rate plant to partial nitrogen removal at Tallman Island, New York	211
5.23	Step feed approach for partial nitrification in summer	211
5.24	Surface of anoxic zones in Southwest plant, Jacksonville, Florida	215

List of Figures

7.1	Filamentous bacteria	234
7.2	Stalked ciliates	235
7.3	Rotifers	235
7.4	Troubleshooting guide on settleability tests	238
7.5	Sludge bulking in clarifiers	246
8.1	Methanol addition to first anoxic zone	260
8.2	Methanol addition to post-anoxic zone	261
8.3	Methanol addition to denitrification filters	263
8.4	Methanol addition to MBBR	264
8.5	Havelock WWTP schematic	268
8.6	Long Creek WWTP schematic	270
8.7	VFA addition for biological phosphorus removal	273
8.8	McDowell Creek WWTP effluent phosphorus (2001 through 2003)	276
8.9	Chemical phosphorus removal dosing locations	289
8.10	Ratio of Iron (Fe^{3+}) dose to phosphorus removed as a function of residual soluble orthophosphate concentration	293
8.11	Ratio of Aluminum (Al^{3+}) dose to phosphorus removed as a function of residual orthophosphate concentration	300
8.12	Northwest Cobb WRF phosphorus profile	307
8.13	Northwest Cobb WRF effluent phosphorus (2000 through 2003)	310
9.1	A typical WWTP operation schematic	315
9.2	Four phases of anaerobic digestion	318
9.3	Activated primary sedimentation tanks	323
9.4	Complete-mix fermenter	324
9.5	Single-stage static fermenter	324
9.6	Two-stage fermenter/thickener	325
9.7	Unified fermentation and thickening fermenter	325
9.8	Schematic of sidestream biological phosphorus removal process with RAS fermentation	334
9.9	South Cary Water Reclamation Facility sidestream RAS fermentation zone	335
9.10	Kelowna static fermenters	337
9.11	Kalispell complete-mix fermenter and thickener	340
9.12	Plan view of South Cary Water Reclamation Facility BNR system	342
9.13	Biological nutrient removal process configurations with sidestream RAS fermentation: (a) MLE with sidestreams, (b) three-stage BNR, (c) Bardenpho A, and (d) Bardenpho B	343
10.1	Partitioning of flow and particulate and soluble fractions thickening and dewatering	359
10.2	Primary sludge management options	361

10.3	Waste activated sludge management options	361
10.4	Influent loading to a BNR system	364
10.5	In-Nitri sidestream nitrification process	370
10.6	Flow scheme of Prague wastewater treatment plant	375
10.7	Separate centrate nitrification in four-pass system	376
10.8	Struvite deposit from a 0.3-m (12-in.) lagoon decant line, which broke loose and got caught in a check valve at Columbia Boulevard Wastewater Treatment Plant, Portland, Oregon	377
10.9	Struvite deposits found in 75-mm (3-in.) magnetic flowmeter on dewatering centrate line	378
10.10	Image of struvite deposit in 25-mm (1-in.) pipe to a streaming current meter, which controlled dewatering polymer feed. Inaccurate measurement by the streaming current meter, as a result of the deposits, can result in polymer overdosing or reduced dewatering performance	379
10.11	Close-up of recovered struvite deposit	380
10.12	Photograph of recovered struvite crystals from anaerobic digester at Durham Advanced Wastewater Treatment Plant, Tigard, Oregon	381
10.13	Struvite deposits in a 75-m (3-in.), rubber-lined, 90-deg elbow; the buildup occurred during a two-month material testing period	382
10.14	Struvite solution product versus pH	383
10.15	Struvite deposits on a belt filter press	384
10.16	Photograph captured during a camera inspection image of a 10-km (6-mile) sludge transfer line at Eugene Springfield Water Pollution Control Facility (Eugene, Oregon). The photograph shows some struvite deposits on the pipe surface and a layer of struvite grit	386
10.17	Struvite solution product versus pH	387
10.18	Example of struvite crystals found in digested sludge at Eugene Springfield Water Pollution Control Facility	388
10.19	Surface of pipe linings magnified 5000×	390
10.20	Struvite control through process design: decision support chart	393
11.1	A typical anion separation in ion chromatography	415
11.2	An example of fatty acid run via BP21	417
12.1	Flow schematic example	423
12.2	Illustration of an in situ measurement. The parameter measurement is conducted in the tank or channel of interest	426
12.3	Illustration of a grab sample. Grab samples can be useful when parameters are time-sensitive and are best when taken from well-mixed or homogeneous solutions	427
12.4	Illustration of an interval sample	429
12.5	Illustration of a time-composite sample	429

12.6	Illustration of a flow-composite sample	430
12.7	Relative ORP readings	439
12.8	Division of the total influent COD in municipal wastewater	445
12.9	Examples of possible outcomes of BPR potential test	460
12.10	Decision tree for using optimization/troubleshooting guides	467
12.11	Version of the UCT process at Wolf Treatment Plant, Shawano, Wisconsin	490
12.12	Stevens Point, Wisconsin, effluent phosphorus concentrations during initial startup	493
12.13	City of Dodgeville, Wisconsin, A/O process with RAS denitrification	494
12.14	Schematic flow diagram of EWRF	496
12.15	Clarifier effluent quality and MLSS in the two trains at EWRF	498
12.16	Stamford, Connecticut, process for nitrogen removal	500
13.1	Influent ammonia concentration with and without ammonia equalization	522
13.2	Automatic waste control system schematic	528
13.3	Oxnard trickling filter–solids contact activated sludge system improvements in sludge settleability	529
13.4	Effect of automatic control on MLSS	530
13.5	The effect of different methods of sludge wasting on SRT data	531

Preface

Nutrient removal is being required at many plants throughout the United States, Europe, and Asia. Virtually all of the plants use biological processes for nitrogen, phosphorus, and/or ammonia removal. Operating a biological nutrient removal (BNR) process is not simple and requires a high level of operator involvement and knowledge. Recognizing this need, the Water Environment Federation jointly with the Environmental and Water Resources Institute of the American Society of Civil Engineers developed this manual to help those disciplines associated with the operation of biological nutrient facilities better understand the process and the way that process should be controlled and operated. Furthermore, the information in this manual can be applied to any BNR plant, large or small, any where in the world.

The purpose of this manual is to give the reader an understanding of the theory behind these processes and design requirements for the various types of processes currently used. Most importantly, this manual will give guidance to operational personnel on the most accepted process control parameters to optimize the performance of this process and troubleshoot it.

The manual is written for plant managers and operators but it will be useful to consulting engineers and regulatory agency staff. Moreover, it can be used as a training document, both by trainers and college professors, to ensure that personnel operating and designing these processes will understand the requirements needed to develop and operate a highly efficient BNR facility. A separate study guide, titled *Biological Nutrient Removal Operation Study Guide*, contains more than 100 detailed problems and solutions, an acronym list, conversion factors (metric to U.S. customary and U.S. customary to metric), and a glossary. The study guide will further this manual's use as a training tool or can be used for self study (available at www.wef.org and www.asce.org).

This manual was produced under the direction of Jeanette A. Brown, P.E., DEE, and Carl M. Koch, P.E., DEE, Ph.D., *Co-Chairs*.

Principal authors of the publication are

James L. Barnard, P. Eng., Ph.D.   (5)
Jeanette A. Brown, P.E., DEE   (1, 6, 7)

Rhodes R. Copithorn	(2)
Christine deBarbadillo, P.E.	(8, 9)
Paul A. Dombrowski, P.E., DEE	(3)
Alex Ekster, P.E., DEE, Ph.D.	(13)
Ufuk G. Erdal, Ph.D.	(2, 11)
Zeynep K. Erdal, Ph.D.	(4)
Samuel S. Jeyanayagam, P.E., DEE, Ph.D.	(10)
Curtis I. Kunihiro, P.E., DEE	(12)
Krishna R. Pagilla, P.E., Ph.D.	(5)
Cindy Wallis-Lage, P.E.	(8, 9)

Additional chapter content was provided by the following individuals: Mario Benisch (10), Rhodes R. Copithorn (3), Zeynep K. Erdal, Ph.D. (9), Joseph A. Husband (10), Philip R. Kiser (13), Edmund A. Kobylinski, P.E. (8), Troy A. Larson (12), Neil Massart, P.E. (8), Barry Rabinowitz, P. Eng., Ph.D. (9), Bob Rutemiller (13), Andrew R. Shaw (13), and Troy Stinson, P.E. (12).

Additional review was provided by Joni Emrick.

Authors' and reviewers' efforts were supported by the following organizations:

Associated Engineering, Calgary, Alberta, Canada
Black & Veatch, Kansas City, Missouri
CH2M HILL, Santa Ana, California; Englewood, Colorado; Burnaby, British
    Columbia, Canada; Toronto, Ontario, Canada
City of Kalispell, Montana
Ekster and Associates, Fremont, California
Floyd Browne Group, Marion, Ohio
Hach Company, Loveland, Colorado
HDR Engineering, Inc., Portland, Oregon, Bellevue, Washington
Illinois Institute of Technology, Chicago, Illinois
Jacobs Engineering Group, Inc., Pasadena, California, Orlando, Florida
Kennedy Jenks Consultants, Portland, Oregon
Malcolm Pirnie, Inc., White Plains, New York; Columbus, Ohio
Stamford Water Pollution Control Authority, Stamford, Connecticut
Stearns & Wheler, Bowie, Maryland
Strand Associates, Inc., Madison, Wisconsin
Tighe & Bond, Inc., Westfield, Massachusetts
University of Wisconsin–Madison, Wisconsin

Chapter 1

Introduction

Many wastewater treatment plants throughout North America, Europe, and Asia are required to remove nitrogen and phosphorus. Though effective nitrogen removal relies on biological processes, phosphorus removal may use a biological process or a combination of a biological process plus chemical precipitation.

Operation of biological nutrient removal (BNR) facilities requires considerable operator involvement and knowledge. To optimize the process, additional sampling, analysis, and monitoring beyond that required for biochemical oxygen demand removal is necessary. Moreover, successful operation of BNR facilities requires understanding of process control and troubleshooting techniques.

The manual was written to give plant managers and operators an understanding of theory and typical design requirements for processes currently used for nitrogen and phosphorus removal. Most importantly, it gives guidance on process control and troubleshooting methodologies, thus assisting with optimizing process performance and solving operational problems. Consulting engineers, professors, and regulatory agency staff will also find this manual useful not only to broaden their understanding of BNR processes but also for training operators and other professionals.

The chapters are organized in such a way that the reader develops an understanding of the theory before reading about process control parameters and requirements and troubleshooting methodology.

Chapter 2 describes overall process considerations for BNR processes. The focus is on the sources and types of inorganic and organic carbon, nitrogen, and phosphorus that compose typical wastewater. There is also a detailed discussion on wastewater characteristics and the effect of excessive levels of nutrients on the environment.

Chapter 3 contains a detailed explanation of process fundamentals associated with nitrification and denitrification. There are descriptions of suspended-growth and attached growth processes and microbiology, stoichiometry, and kinetics of each system. Included are process control parameters and effects of flow and load variations on process stability. There is also a discussion on typical carbon sources, such as methanol and other alternative sources. In addition, there is a discussion on various types of integrated fixed-film systems, including equipment requirements and performance.

Chapter 4 is dedicated to enhanced biological phosphorus removal. Among the subjects discussed in this chapter are the basic theory of phosphorus removal; types of system, such as suspended growth and hybrid and coupled system; operational parameters; and process control methodologies.

Chapter 5 is dedicated to descriptions of combined nutrient removal systems, where both nitrogen and phosphorus are removed biologically. Described in this chapter are some patented processes, such as the Phoredox (A^2O) process, Bardenpho process, UCT (VIP) process, and Modified UCT process. There is a discussion on process control methodologies, operational parameters, effect of chemical phosphorus removal on BNR systems, and limits for simultaneous nitrogen and phosphorus removal.

Chapter 6 explores various computer models presently used for design and control of BNR processes. There is a discussion on the history and development of the processes and a description of design versus simulators. There is a discussion on how simulators can help operators understand plant operations and optimize the process.

Chapter 7 concerns solids separation problems caused by filamentous sludge bulking and foaming, which are operating problems that occur with BNR processes. There is a discussion about the causes of filamentous bulking and foaming, how to identify the problem (visually and microscopically), some of the control strategies, and effects of solids handling side streams.

Chapter 8 focuses on chemical addition and chemical feed control, including carbon supplementation for denitrification using methanol or alternative carbon sources. There is also a discussion on volatile fatty acid supplementation for biological phosphorus removal. Lastly, there is a discussion on alkalinity supplementation.

Chapter 9 describes primary sludge fermentation and the way it can enhance biological phosphorus or nitrogen removal processes. Included is a description of primary sludge fermentation and types of fermenters, such as activated primary tanks, static fermenters, complete-mix fermenters, and two-staged fermenters.

Chapter 10 describes solids handling and processing. It includes a detailed discussion on the effects of recycle streams on various processes. There is a discussion on how to estimate sludge production and recycle loads and the sources of secondary release of nutrients from various other processes within the treatment train, such as primary clarifiers, bioreactors, secondary clarifiers, thickening, stabilization, and dewatering. There is also a discussion on side-stream treatment of various process streams, such as digester supernatant.

Chapter 11 outlines the various types of laboratory analysis required for BNR process control and optimization. Included is a discussion on sampling, preservation, and storage of samples; which species must be determined; and the types of analytical methods available.

Chapter 12 details various optimization and troubleshooting techniques. Such things as process evaluation and data interpretation are discussed, and guides for optimization and troubleshooting of BNR plants are provided. There is also a series of case studies to help describe various troubleshooting techniques.

The final chapter, Chapter 13, describes instrumentation and automated process control. There is a detailed discussion on various types of inline analyzers, such as total suspended solids meters and dissolved oxygen meters; oxidation–reduction potential; nitrate; and ammonia. The chapter ends with a detailed description of supervisory control and data acquisition systems.

Furthermore, troubleshooting guides have been included where appropriate for easy reference by the operators. In addition, there are case studies that exemplify various aspects of nutrient removal.

Chapter 2

Overall Process Considerations

Introduction	6	pH	24
Nutrient Sources	8	Alkalinity	24
Sources of Nitrogen	8	Variations in Flows and Loads	25
Source of Phosphorus	9	Effect of Recycle Flows	27
Effects of Nutrients on Receiving Waters	12	Review of Recycle Flows	27
		Management of Return Flows	28
Eutrophication	12	Effect of Effluent Permit Requirements	29
Ammonia Toxicity	12		
Nitrate in Groundwater	13	Technology-Based Permits	29
Wastewater Characteristics	13	*Monthly Average*	29
Carbonaceous Materials	13	*Annual Average*	30
Nitrogen	16	*Seasonal Permit*	30
Phosphorus	20	Water-Quality-Based Permit	30
Solids	21	References	31
Temperature	23		

INTRODUCTION

Wastewater treatment is much like an industry where the final product is well-defined and, in fact, highly regulated, and where the quality and quantity of the raw materials used to produce that product are uncontrollable. This is not a very easy position for the treatment plant operator. Fortunately, wastewater characteristics are fairly predictable, at least over a range of flows and loads, and treatment plants are generally designed to operate effectively over that range of influent flow and load conditions. An understanding of the various wastewater characteristics and their typical ranges will assist the operator in maintaining effective treatment and in troubleshooting the process should problems develop.

This section will focus on the sources and types of organic and inorganic forms of carbon, nitrogen, and phosphorus that comprise a typical wastewater. A summary of typical raw wastewater characteristics is presented in Table 2.1. The effects of excessive levels of nutrients on the environment will also be reviewed.

Wastewater characteristics are influenced by a number of factors, including water usage, type of collection system (combined versus separate), infiltration and inflow, use of garbage grinders, and the presence of industrial sources of wastewater. Each of the main wastewater constituents can be divided into biodegradable and nonbiodegradable fractions and further subdivided into soluble and particulate forms. This approach to differentiating between fractions is significant because the form of the substrate (soluble, particulate, biodegradable, etc.) directly affects how the substrate is processed in a wastewater treatment plant. This differentiation is also very important in computer modeling of a wastewater treatment process because modeling is concerned with how these materials behave and interact in an activated sludge process.

The concentration of the various forms of each wastewater constituent will change as the wastewater flows through each unit process. An activated sludge process with primary clarification, for example, will be subjected to a somewhat different wastewater then will an oxidation ditch without primary clarifiers. Also, the relative concentration of various constituents will change. For example, as settling removes organic solids in a primary clarifier, the ratio of biochemical oxygen demand (BOD) to total phosphorus (TP) may decrease.

Another important factor that affects the influent wastewater characteristics to a biological nutrient removal process is the recycle flow from unit processes such as sludge thickening, dewatering and stabilization, and from filter backwash operations. This will be discussed further in the Effect of Recycle Flows section.

TABLE 2.1 Typical raw wastewater characteristics.

Contaminants	Units[a]	Concentration[b]		
		Low-strength	Medium-strength	High-strength
Solids, total	mg/L	390	720	1230
Dissolved, total	mg/L	270	500	860
Fixed	mg/L	160	300	520
Volatile	mg/L	110	300	340
Suspended solids, total	mg/L	120	210	400
Fixed	mg/L	25	50	85
Volatile	mg/L	95	160	315
Settleable solids	mL/L	5	10	20
Biochemical oxygen demand, 5-d, 20°C (BOD_5, 20°C)	mg/L	110	190	350
Total organic carbon	mg/L	80	140	260
Chemical oxygen demand	mg/L	250	430	800
Nitrogen (total as N)	mg/L	20	40	70
Organic	mg/L	8	15	25
Free ammonia	mg/L	12	25	45
Nitrites	mg/L	0	0	0
Nitrates	mg/L	0	0	0
Phosphorus (total as P)	mg/L	4	7	12
Organic	mg/L	1	2	4
Inorganic	mg/L	3	5	10
Chlorides[c]	mg/L	30	50	90
Sulfate[c]	mg/L	20	30	50
Oil and grease	mg/L	50	90	100
Volatile organic compounds	mg/L	<100	100 to 400	>400
Total coliform	No./100 mL	10^6 to 10^8	10^7 to 10^9	10^7 to 10^{10}
Fecal coliform	No./100 mL	10^3 to 10^5	10^4 to 10^6	10^5 to 10^8
Cryptosporidum oocysts	No./100 mL	10^{-1} to 10^0	10^{-1} to 10^1	10^{-1} to 10^2
Giardia lamblia cysts	No./100 mL	10^{-1} to 10^1	10^{-1} to 10^2	10^{-1} to 10^3

[a] mg/L = g/m^3.
[b] Low-strength is based on an approximate wastewater flowrate of 750 L/cap·d (200 gpd/cap); medium-strength is based on an approximate wastewater flowrate of 460 L/cap·d (120 gpd/cap); and high-strength is based on an approximate wastewater flowrate of 240 L/cap·d (60 gpd/cap).
[c] Values should be increased by amount of constituent present in domestic water supply.

NUTRIENT SOURCES

Nitrogen and phosphorus are essential nutrients to the growth of living organisms. Although some micronutrients (including iron) are also necessary for growth, nitrogen and phosphorus are of vital importance to living organisms.

SOURCES OF NITROGEN. Nitrogen is a naturally occurring element that is essential for growth and reproduction in living organisms. It is the key component of proteins and nucleic acids, and, without them, no life can exist. Nitrogen is the most abundant compound in the atmosphere. The gaseous nitrogen (N_2) consists of two nitrogen atoms and compromises 79% of the air volume.

This large amount of nitrogen in the atmosphere, however, is not readily available to most organisms. Certain groups of organisms assimilate nitrogen gas and make it available to other organisms. This process is termed nitrogen fixation. Lightning contributes to nitrogen fixation. However, most of the nitrogen fixation is either of biological or industrial origin. In biological nitrogen fixation, atmospheric nitrogen is converted to ammonia by enzymes. The major group of nitrogen-fixing organisms (diazotrophs) live in close proximity to plant roots and obtain energy from the plants. Industrial fixation produces ammonium and nitrate from the air through various chemical processes.

The major sources of nitrogen are of plant, animal, and human origin (decaying plant material and animal and human wastes); industrial and agricultural origin; and atmospheric origin. Nitrogen compounds in human and animal waste are associated with protein and nucleic acids. Ammonia is formed as a result of protein and nucleic acid decomposition. Volatile organic nitrogen is released to atmosphere during plant decay. Industrial emissions and fuel combustion contributes gaseous nitrous oxides and nitric acid. Many forms of nitrogen are used for agricultural purposes as fertilizer. The common nitrogen compounds used in fertilizers are urea, ammonium phosphate, ammonium sulfate, and ammonium nitrate. Atmospheric deposition can also contribute to the nitrogen balance. The relative contribution of nitrogen to surface waters varies greatly depending on the demographics of the watershed. As an example, Table 2.2 summarizes the relative contribution of nitrogen sources to Chesapeake Bay and Long Island Sound (U.S. EPA, 1993).

The most common forms of nitrogen in wastewater are ammonia (NH_3), ammonium ion (NH_4^+), nitrogen gas (N_2), nitrite (NO_2^-), nitrate (NO_3^-), and organic nitrogen. Municipal wastewater primarily contains ammonium and organic nitrogen, whereas some industrial wastewater contains appreciable amounts of nitrate-

TABLE 2.2 Distribution of nitrogen sources in Chesapeake Bay and Long Island Sound (adapted from U.S. EPA, 1993).

Chesapeake Bay	%	Long Island Sound	%
Point sources	23	Wastewater treatment plants	44
Animal wastes	4	Industry	2
Atmospheric ammonium	14	Atmospheric	12
Atmospheric nitrate	25	Coastal runoff	6
Fertilizers	34	Combined sewer overflows	1
		Tributaries	35
Total	100	Total	100

nitrogen. In domestic wastewater, approximately 60% of the nitrogen is in ammonium form, and 40% of nitrogen is in organic form. Organic nitrogen consists of a complex mixture of amino (NH_2^-) compounds, including amino acids and proteins. Organic nitrogen is easily converted to ammonium via bacterial decomposition in a process referred to as ammonification. Hydrolysis of urea transforms organic nitrogen to ammonium. Organic nitrogen is determined using the Kjeldahl method, where the solution is boiled to drive off ammonia before digestion (see Chapter 12). If the boiling step is omitted, then the measured nitrogen contains both organic and ammonia-nitrogen and is referred to as total Kjeldahl nitrogen (TKN) (APHA et al., 1998). Table 2.3 shows the forms and definitions of the various nitrogen species (Metcalf and Eddy, 2003).

SOURCES OF PHOSPHORUS. Phosphorus is an integral component in the process of energy metabolism used by cells. Phosphorus is also a key component of the cellular membrane. It is an essential nutrient for plants and microorganisms. Phosphorus is found in lawn fertilizers, manure, detergents and household cleaning products, and in human and animal waste. Surface waters receive phosphorus from domestic and industrial discharges and natural runoff.

TABLE 2.3 Forms of nitrogen and their definitions (adapted from Metcalf and Eddy, 2003).

Compound	Abbreviation	Form	Definition
Ammonia-nitrogen	NH_3-N	Soluble*	NH_3-N
Ammonium-nitrogen	NH_4^+-N	Soluble	NH_4^+-N
Total ammonia nitrogen	TAN	Soluble*	NH_3-N + NH_4^+-N
Nitrite	NO_2^--N	Soluble	NO_2^--N
Nitrate	NO_3^--N	Soluble	NO_3^--N
Total inorganic nitrogen	TIN	Soluble*	NH_3-N + NH_4^+-N + NO_2^--N + NO_3^--N
Total Kjeldahl nitrogen	TKN	Particulate, soluble*	Organic N + NH_3-N + NH_4^+-N
Organic nitrogen	Organic N	Particulate, soluble*	TKN - NH_3-N + NH_4^+-N
Total nitrogen	TN	Particulate, soluble*	Organic N + NH_3-N + NH_4^+-N + NO_2^--N + NO_3^--N

* In neutral pH range, gas form of ammonia (NH_3-N) is very negligible.

The chemical forms of phosphorus found in aqueous solution are orthophosphate, polyphosphates (condensed phosphates), and organic phosphates (phospholipids and nucleotides). The orthophosphates may be in the form of phosphoric acid (H_3PO_4) dihydrogen phosphate ($H_2PO_4^-$), hydrogenophosphate (HPO_4^{2-}) and phosphate ion (PO_4^{3-}), The phosphate species and their relative abundance change as a function of solution pH. The orthophosphate concentration in wastewater refers to sum of all orthophosphate species. By convention, all the measured quantities are reported as phosphorus and not as phosphates. Therefore, the plant operator must be careful when analyzing and reporting the phosphorus values. Phosphorus concentration is calculated by dividing PO_4 values by approximately 3. For example, if a wastewater contains 10 mg/L phosphorus (P) in its influent, then the phosphate (PO_4) content is approximately 30 mg/L.

Phosphorus in wastewater can be categorized into the following two major groups, based on their physical characteristics:

(1) Soluble phosphorus, and
(2) Particulate phosphorus.

The major part of the soluble phosphorus is orthophosphate. Particulate phosphorus is either biodegradable or nonbiodegradable. The particulate definition relies on which size filter is used during filtering. One generally accepted method uses 1.0-micron filters, whereas another method uses 0.45-micron filters to separate soluble and particulate fractions. Table 2.4 summarizes the forms and typical concentrations of phosphorus in United States wastewater (Sedlak, 1991).

Orthophosphates are readily available for organisms without further breakdown. Polyphosphates can be converted to orthophosphates via hydrolysis reactions, which are generally slow. In conventional wastewater treatment, without biological phosphorus removal, approximately 5 to 10% of the phosphorus is removed during primary settling and secondary clarification. Approximately 20 to 25% of the phosphorus is taken up in the activated sludge process during bacterial growth. Therefore, the final effluent of a conventional wastewater plant can contain 3 to 4 mg/L phosphorus. The organic phosphates are generally present in lower concentrations in domestic wastewaters. Their removal by biological and chemical processes is very difficult.

TABLE 2.4 Forms and typical concentration of phosphates in U.S. wastewater.

Phosphate form	Typical concentration, mg/L as P
Orthophosphate (PO_4^{3-}, HPO_4^{2-}, $H_2PO_4^-$, and H_3PO_4)	3 to 4
Condensed (poly) phosphates (e.g., pyrophosphate, tripolyphosphate, and trimetaphosphate)	2 to 3
Organic phosphates (e.g., sugar phosphates, phospholipids, and nucleotides)	1

EFFECTS OF NUTRIENTS ON RECEIVING WATERS

The excessive accumulation of nutrients discharges to surface waters can pose serious ecological problems that affect the health of aquatic life and consequently that of humans and animals. There are several major effects associated with the discharge of nutrient-containing streams to receiving waters. These include (a) eutrophication, (b) ammonia toxicity (U.S. EPA, 1993), and (c) nitrate contamination of groundwater. These are discussed in the following section.

EUTROPHICATION. Eutrophication is the excessive growth of plant and algae in receiving waters. The major concern with regard to eutrophication is its effect on water quality and aquatic life. As plants and algae die and decay, the resulting excessive respiration reduces the dissolved oxygen concentration in the water column.

The primary conditions that stimulate plant or algal growth are the presence of macronutrients (nitrogen and phosphorus) and sufficient carbon dioxide and light energy (U.S. EPA, 1993). In the absence of any macronutrients, excessive growth does not occur. Therefore, nitrogen and phosphorus are the two key compounds for the control of eutrophication. One of the most common control methods is to determine the growth-limiting nutrient (either nitrogen or phosphorus) and implement controls to reduce their release to the environment from both point and nonpoint sources. Point-source controls are the subject of this manual. Non-point-source controls include primarily the implementation of nutrient management plans and best management practices for agriculture and in rural and urban development. In some cases, both nitrogen and phosphorus removal is desired to control algal growth. A phosphorus concentration (orthophosphate form) of 0.005 mg/L has been found to be a growth-limiting concentration (WEF and ASCE, 1998). Other control methods include stream shading, vegetation removal, and oxygenation of surface waters.

AMMONIA TOXICITY. The molecular or un-ionized form of ammonia nitrogen is toxic to fish and other aquatic life. The effect can be acute (fish mortality) or chronic (effect on reproduction or health). Molecular free ammonia (NH_3) and ionized ammonium ion (NH_4^+) are in equilibrium in aqueous solution, where their relative percentages are a function of pH and temperature. The ionic strength of the solution also has an effect on the ammonia species. As the ionic strength increases, the fraction of the un-ionized form decreases. In a number of studies by the U.S. Environmental Protection Agency (U.S. EPA), it was shown that an un-ionized or free ammonia (NH_3) concentration of 0.1 to 10 mg/L resulted in acute toxicity for salmonid and nonsalmonid fish species (U.S. EPA, 1993). The maximum one-hour average in-stream concentration of ammonia permissible in a three-year period is under 1.0 mg/L (U.S. EPA, 1993).

NITRATE IN GROUNDWATER. Treatment systems that discharge to groundwaters have the potential to contaminate the groundwater with nitrates. This can occur directly by the discharge of nitrates in the effluent or by the discharge of ammonia, which then is nitrified in the soil column as rainwater brings in dissolved oxygen. The public health concern associated with nitrates is the potential for a blood disorder called methemoglobinemia, which affects infants. The nitrates would preferentially bind to the hemoglobin, thus preventing its association with oxygen. The result is suffocation, which is also why the condition is referred to a "blue baby" syndrome.

WASTEWATER CHARACTERISTICS

CARBONACEOUS MATERIALS. The organic carbon content in wastewater is commonly measured in terms of the BOD, which is a measure of the amount of oxygen consumed during the biochemical oxidation of the organic matter. Actually, there are several concurrent processes that occur. As the organic matter is oxidized, the products of this oxidation are used to create new cell mass and to maintain cells. Finally, when all of the waste organic matter is used up, the cells consume their own cell tissue to obtain energy through a process of endogenous respiration. The oxygen required to take these reactions to completion is referred to as the ultimate BOD (UBOD). However, nitrification can also occur in a BOD test. In other words, the oxidation of both the carbon and, if the plant is nitrifying, the nitrogen in the form of ammonia, contribute to the BOD value, as shown in Figure 2.1. Thus, effluent discharge permits are sometimes written in terms of the carbonaceous BOD (CBOD) which is determined by completing the BOD test with a chemical added that inhibits nitrification. The CBOD is generally approximately 80% or more of the total BOD value.

Another common measurement of organic content is chemical oxygen demand (COD), which is the amount of oxygen consumed during a laboratory procedure that chemically oxidizes the organic matter in the wastewater. The COD is generally much greater than the BOD because some of the carbon in a typical municipal wastewater is in a form that is not available for biological uptake. Typically, the ratio of COD to BOD is in the range 2.0 to 2.2. A higher ratio may be indicative of the presence of industrial wastes that can contain significant concentrations of refractory or

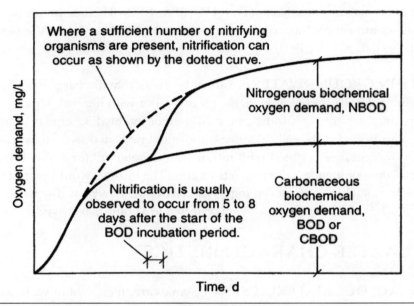

FIGURE 2.1 Forms of BOD (Metcalf and Eddy, 2003).

nonbiodegradable wastes. Higher values may also indicate that some stabilization or biological uptake of the carbon is occurring in the sewer system. This may be the case in a collection system with steep slopes and which is thus aerobic or in collection systems in warmer climates.

A low COD/BOD value (high percentage of carbon that is biodegradable) can occur in collection systems with long detention times, especially in warmer temperatures, as a result of anaerobic fermentation of the wastewater. Fermentation will solubilize more of the organic carbon, making it more readily biodegradable. A low value can also be indicative of industrial contributions to the wastewater where an industry is discharging highly soluble and biodegradable waste.

The COD (and BOD) may be divided into fractions that are biodegradable and nonbiodegradable, as illustrated in Figure 2.2. The biodegradable fraction can be further subdivided into that which is readily biodegradable and that which is slowly biodegradable. The readily biodegradable COD fraction is comprised of the smaller

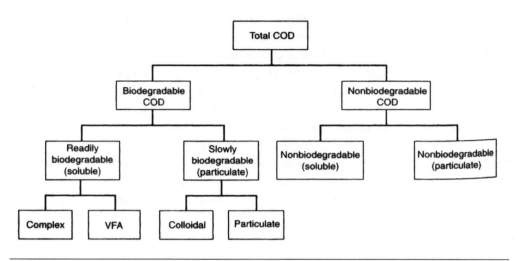

Figure 2.2 Fractionation of COD (VFA 4 = volatile fatty acid) (Metcalf and Eddy, 2003).

molecules, such as volatile fatty acids (VFAs) and other forms of dissolved or soluble COD, that are quickly assimilated by biomass. The slowly biodegradable COD is comprised of larger, more complex forms of carbon that must be broken down before they can be used by the cells. The readily biodegradable portion is assumed to be soluble, while the slowly biodegradable is considered particulate. Note that, unlike BOD, some forms of COD are not biodegradable. The nonbiodegradable soluble COD will pass through the treatment plant and appear in the effluent. The particulate form of the nonbiodegradable COD will be incorporated to the sludge.

The concentration of BOD in the raw wastewater will vary depending on the nature of the sewer shed and the collection system. Newer collection systems will tend to have higher concentrations because they would be typically subject to less infiltration and inflow, which dilutes the BOD concentration. Areas where garbage grinders are in use can have higher BOD values because of the addition of solid waste to the wastewater. The potential effect of industrial waste has already been discussed. Typical concentrations of BOD are shown in Table 2.1.

Another way to consider the organic content of a wastewater is based on the per capita generation. Typically, each person generates from 0.08 to 0.09 kg/d (0.18 to 0.19 lb/d) of BOD. If the number of people connected to the collection system, or population equivalents (PE), is known, then the kilograms or pounds per day of BOD per capita may be calculated, based on the influent wastewater BOD sampling. This number may be compared to the typical range as an indication of whether or not there is anything unusual about the wastewater.

NITROGEN. All of the reactions involving nitrogen in a wastewater treatment plant occur naturally in the environment. These transformations between the different forms of nitrogen are illustrated by the nitrogen cycle shown in Figure 2.3.

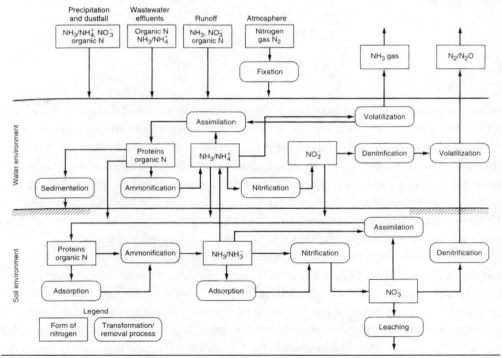

FIGURE 2.3 Nitrogen transformations (Metcalf and Eddy, 2003).

Nitrogen exists in wastewater in a variety of forms, from the most reduced form, which is ammonia, to the most oxidized form, which is nitrate. Nitrate is the product of the nitrification process in which ammonia is oxidized to nitrate. Ammonia, which is soluble, exists in equilibrium as both molecular ammonia (NH_3) and as ammonia in the form of the ammonium ion (NH_4^+). The relative concentration of each depends on the pH and temperature, with higher pH values and temperatures favoring the formation of molecular ammonia. It is the molecular form of ammonia that is toxic.

Nitrogen in raw wastewater is typically comprised of ammonia and organic nitrogen. Generally, there is little or no oxidized nitrogen present (nitrite or nitrate). The presence of oxidized nitrogen would be indicative of an industrial contribution, such as, for example, by a textile industry or a munitions manufacturing company. The combination of ammonia, which is an inorganic form of nitrogen, and the organic nitrogen is the TKN, which refers to the laboratory procedure used to measure it. The TKN value in raw wastewater is typically in the range 25 to 45 mg/L. The ammonia and organic nitrogen content of the TKN is generally 60 and 40%, respectively. Organic nitrogen derives from complex molecules, such as amino acids, proteins, nucleotides, and urea. Typical ranges for each of these constituents are shown in Table 2.1.

Total nitrogen (TN) consists of the sum of the ammonia and organic nitrogen (TKN) plus the oxidized forms of nitrogen (nitrite and nitrate). As stated previously, because typical domestic wastewater contains no nitrite or nitrate, the TKN value is generally indicative of the TN value of the raw wastewater. The forms of nitrogen, however, which are included in the TN, will change as the wastewater flows through the treatment plant. The forms of nitrogen are illustrated in Figure 2.4. The organic nitrogen will be hydrolyzed biologically in the aerobic portion of the treatment process to ammonia (ammonification).

Some of the organic nitrogen, however, is refractory and will remain as organic nitrogen. The particulate form may be captured and removed, but the soluble portion will pass through to the effluent. In a plant that nitrifies, the ammonia will be oxidized to nitrate. In a plant that denitrifies, the nitrate will be reduced to nitrogen gas and be removed from the process. These processes are discussed in greater detail in Chapter 3.

As with the various forms of carbon, the nitrogen forms can be divided into particulate and soluble forms and further subdivided into biodegradable and non-biodegradable (refractory) forms of each. A portion of the influent organic nitrogen will be soluble and refractory, meaning it will not be captured and removed by any

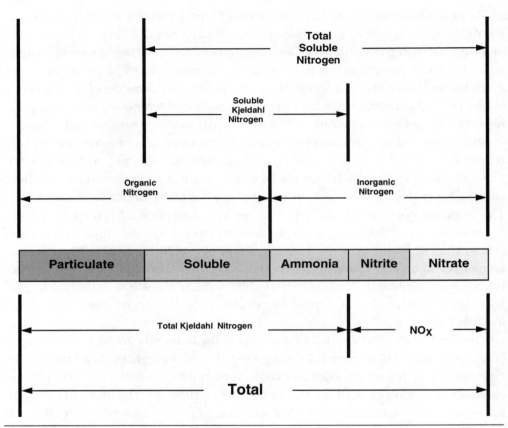

FIGURE 2.4 Forms of nitrogen.

of the settling or filtration processes, and it will not be biologically degraded in the process. Biological processes in the wastewater treatment plant can contribute to the nonbiodegradable particulate organic nitrogen as nitrogen is incorporated to cell mass. Portions of the cell, when they decay and are broken down, contribute to the nonbiodegradable organic nitrogen. The effect on a plant with a TN permit can be significant, especially if there is a requirement for low levels of TN (<6 mg/L), because refractory soluble organic nitrogen (SON) will pass through to the effluent. Certain industries, notably textile plants or wastewater from dye operations, can contain several mg/L SON that is refractory and will contribute directly to the effluent TN.

The ratio of BOD to TKN is of significance, particularly if the plant is required to denitrify. The typical range of BOD/TKN in the raw wastewater is 4:1 to 5:1 for domestic wastes. Lower ratios may indicate contributions from industrial waste, and higher values indicate that the waste may be nutrient-deficient, possibly also a result of dilution of the domestic waste by an industrial discharger. Nutrient deficiency will inhibit biological growth necessary for waste treatment and possibly encourage the growth of filaments (see Chapter 8). Low BOD/TKN ratios affect nitrogen removal because denitrification requires an organic source of carbon to proceed. Based on the stoichiometric relationship, 2.86 g COD are required per gram of nitrate denitrified. However, depending on the form of carbon and the operating mean cell residence time (MCRT), the actual amount (mass) will be approximately 4 mg of carbon, expressed as BOD, to denitrify 1 mg of nitrate. However, the actual amount depends on the specific type of carbon in the BOD and the plant operating conditions. Thus, efficient denitrification requires approximately four or more times as much BOD as TKN in the influent to the biological process, recognizing that some particulate BOD and TKN will be removed in the primary clarifiers.

A high BOD/TKN ratio can also negatively affect nitrogen removal by reducing the efficiency of the nitrification process. An excessively high BOD/TKN ratio results in a greater relative growth of heterotrophic biomass, which sequesters nitrogen that could otherwise be nitrified and denitrified. Also, the ratio of BOD/TKN influences the fraction of the biomass that is comprised of nitrifiers. The fraction of nitrifiers in a typical mixed liquor is typically low (less than 20% of the active biomass), but higher BOD/TKN values will further decrease the nitrifier ratio. Another way in which the BOD/TKN ratio can affect nitrification is related to the fact that both the nitrifiers (autotrophic) and the heterotrophic compete for resources. As the BOD/TKN ratio increases, the autotrophs are effectively forced deeper into the floc, where the mass-transport of substrate into the floc becomes more limited.

To simplify the relationship between the different nitrogen compounds, it is common practice to place all of these compounds on the same, "as-nitrogen" basis. This approach allows for all nitrogen species to be placed on an equal basis relative to the number of nitrogen atoms present in the particular compound, such that 1 g NH_4^+-N (ammonia as nitrogen) is converted to 1 g NO_2-N (nitrite as nitrogen), which is then converted to 1 g NO_3-N (nitrate as nitrogen), neglecting assimilation into new biomass formed. Total Kjeldahl nitrogen measurements are generally presented in an "as-nitrogen" basis. The conversion of any nitrogen species to an "as-nitrogen" basis is accomplished by multiplying the mass or concentration of that species by the molecular weight of nitrogen (14 g/mol) and dividing by the molecular weight of that compound. For example, ammonium with a molecular weight of 18 g/mol is converted to ammonium "as-nitrogen" by eq 2.1.

$$= \frac{18 \text{ g NH}_4^+ \times 14 \text{ g nitrogen/mol}}{18 \text{ g ammonium/mol}} = 14 \text{ g NH}_4^+\text{-N} \qquad (2.1)$$

PHOSPHORUS. The TP concentration in raw wastewater typically ranges from 4 to 8 mg/L, depending on a number of factors, including the contribution by industrial dischargers and the nature of the drinking water supply. In general, the phosphorus levels decreased after the detergent ban was imposed, but there are still some cleaning agents that contain phosphorus. Some drinking water suppliers use a form of phosphorus as a corrosion inhibitor and thus will contribute phosphorus to the raw wastewater.

The TP concentration is comprised of both inorganic and organic forms. The inorganic forms, which are soluble, include orthophosphate and polyphosphates. The orthophosphate form (PO_4^{3-}) is the most simple form of phosphorus and accounts for 70 to 90% of the TP. It is the form that is available for biological metabolism without further breakdown. It is also the form that is precipitated by metal salts in a chemical phosphorus removal system. Polyphosphates consist of more complex forms of inorganic orthophosphates that are generally synthetic in nature. The polyphosphates are broken down to orthophosphates during the treatment process.

Organically bound phosphorus can be in both a soluble and particulate form. Organically bound phosphorus includes a wide variety of more complex forms of phosphorus that derive from proteins, amino acids, and nucleic acids that, to some extent, are degraded and are present as waste products. Organic phosphorus is also contributed by a variety of industrial and commercial sources.

The organically bound phosphorus can be further subdivided into biodegradable and nonbiodegradable fractions. The soluble, nonbiodegradable fraction will pass through the treatment plant and be discharged in the effluent at a concentration equal to its concentration in the influent. The particulate, nonbiodegradable form, if not settled, will be removed with the sludge. The complex organic phosphorus compounds that are biodegradable are hydrolyzed within a wastewater treatment process to orthophosphate.

The ratio of BOD to TP is significant, particularly to a plant that has incorporated some form of enhanced biological phosphorus removal (EBPR). Ratios less than 20 can indicate potential problems achieving effective EBPR because the organisms responsible for the uptake of phosphorus require and adequate amounts of carbon; more specifically, they require carbon in the form of VFAs. See Chapter 5 for an in-depth discussion.

Phosphorus is an essential nutrient for biological growth and, if not present in sufficient quantities, can inhibit growth and reduce the efficiency of a biological treatment process. This can be an issue in treatment processes that incorporate enhanced primary settling, because the additional coagulation and settling involved by the addition of chemical salts to the primary clarifiers to improve solids removal will also reduce the amount of TP available for the downstream biological processes. Phosphorus inhibition may also continue to be an issue as treatment plants update for low levels of effluent nitrogen and phosphorus, particularly with plants that use a denitrification filter downstream of final clarifier or filter for phosphorus removal. Although there have been cases where plants have successfully operated in this manner, there have also been instances where inhibition by phosphorus deficiency was suspected as the cause for reduced performance.

SOLIDS. Solids in wastewater can be divided into suspended and dissolved fractions and further subdivided into volatile and nonvolatile (fixed) fractions, as shown in Figure 2.5. The total solids (TS) in a raw wastewater consists of all solids that remain after a sample has been evaporated and dried. However, the coarse solids, such as rags and debris, are first removed before analysis. The total suspended solids (TSS) is a somewhat arbitrary characterization, but generally refers to the portion of the TS retained in a glass fiber filter. The total dissolved solids (TDS) are those that pass through a nominal 2.0-m pore-size filter, as described by *Standard Methods* (APHA et al., 1998). Smaller pore sizes have also been used to characterize TDS, such as the Whatman glass fiber filter, with a nominal pore size of 1.58 m. The TDS is comprised of both colloidal and dissolved solids.

The raw wastewater suspended solids concentration can range from 100 to 350 mg/L. A reasonable estimate of the expected amount of TSS present in the influent can be made based on the per capita contribution, which is typically approximately 0.08 kg (0.18 lb) per capita. Thus, if the actual population served by the treatment plant is known, then the total kilograms (or pounds) per day of solids and the concentration of solids can be estimated.

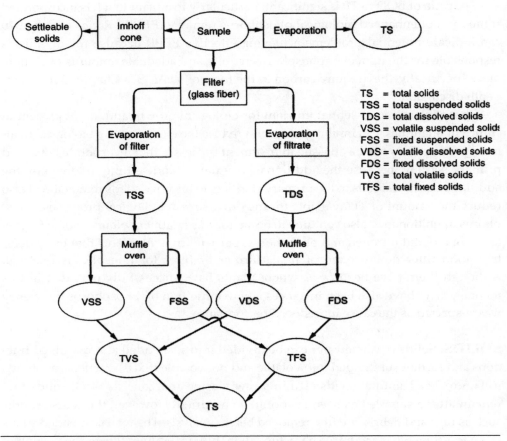

FIGURE 2.5 Solids fractionation (Metcalf and Eddy, 2003).

The influent solids are comprised of an inert (fixed) or nonvolatile fraction and a volatile fraction. The volatile fraction is primarily biodegradable, although not completely, because there are also inert volatile solids. The biodegradable fraction of the volatile suspended solids (VSS) contributes to the BOD, nitrogen, and phosphorus load, and is typically 70 to 80% of the TSS, with a higher per percentage indicative of greater organic content.

The higher end of the range is typical for domestic wastes, whereas contributions of inert solids by combined sewer systems can lower the ratio. Other sources of waste that can lower the ratio include the discharge of water treatment plant sludge into the wastewater collection system, certain industrial wastes, or the discharge of significant amounts of partially stabilized waste, such as septage.

The solids can be physically differentiated by particle size and classified, from smallest to largest, as dissolved, colloidal, and suspended solids. Generally, particles smaller than 10^{-3} μm are considered dissolved, and particles greater than 1 μm are considered suspended. Particles between approximately 10^{-3} and 1 μm are considered colloidal. Both colloidal and suspended solids can be removed by coagulation and settling processes, but dissolved solids containing fractions of nonbiodegradable carbon, nitrogen, and phosphorus will pass through the treatment plant and be discharged with the effluent.

TEMPERATURE. The temperature of the raw wastewater varies, of course, seasonally and is important because of the significant effect that temperature has on all biological processes. The optimum temperature for bioactivity is 25 to 35° C, although organisms are capable of adapting to operation outside this range. The temperature variation is fairly predictable, depending on the geographic location of the plant. Minimum monthly temperatures can vary greatly, but can be as low as 3° C or even less. Areas where a significant snow melt occurs in the spring can experience such low influent temperatures.

Cold temperatures particularly affect nitrification because the organisms responsible for nitrification have slow growth rates. As their growth rates are further reduced at cold temperatures, the solids retention time must be increased to maintain the nitrifiers in the system. Because of this, there are wastewater treatment plants that can completely nitrify in the summer, but that loose nitrification in the winter.

Temperature can have other indirect affects on the wastewater characteristics and thus the treatment plant performance. As mentioned previously, EBPR requires an adequate supply of VFAs that are fermentation products of more complex forms of carbon. During periods of warm weather, in sewers that are relatively flat (little dissolved oxygen entrainment) and with long detention times, anaerobic conditions promoting the formation of VFAs are favored. However, during cold temperature periods, fermentation will decrease, and there may not be an adequate supply of VFAs for EBPR.

Cold temperatures can also increase the dissolved oxygen carried by the wastewater, which can negatively affect the performance of the denitrification process if dissolved oxygen is brought into the anoxic zone. Considerably warm temperatures decrease the concentration of dissolved oxygen in the wastewater at the same time that the rate of biological activity increases. This can result in very low levels of dissolved oxygen.

PH. The pH is a measure of the hydrogen-ion concentration, and is important because there is a narrow range that is suitable for most biological activity. That range is 6 to 9, and most wastewaters fall within this range, unless there is some unusual industrial contribution.

ALKALINITY. There are a number of compounds that can contribute to the alkalinity in a wastewater. Primarily alkalinity consists of hydroxides (OH), carbonates, and bicarbonates of various inorganics, such as calcium. In fact, alkalinity is typically reported in terms of milligrams per liter of calcium carbonate ($CaCO_3$) equivalents. The amount of alkalinity present in the raw wastewater depends largely on the nature of the water supply in the area and on the characteristics of the groundwater, which may infiltrate the collection system. Typically surface waters contain greater alkalinity than do groundwater.

Alkalinity buffers the pH in a wastewater treatment process and is the source of inorganic carbon that is required by the nitrifying organisms as they oxidize ammonia to nitrite and nitrate. If there is not sufficient alkalinity available, then nitrification can be inhibited. However, even with enough alkalinity to allow nitrification to proceed, the amount of alkalinity remaining may not be sufficient to adequately buffer the system, resulting in a drop in the pH. Depressed pH levels, aside from potentially causing a permit violation, can inhibit nitrification.

Carbon dioxide (CO_2) is a normal byproduct of aerobic metabolism; thus, as BOD is oxidized in a biological process, CO_2 is released into the mixed liquor. This forms carbonic acid, which is normally buffered by the alkalinity, but again, without adequate alkalinity, the pH may decrease. It has been observed that, in plants using fine-bubble aeration systems, a higher level of alkalinity should be maintained in the mixed liquor as a buffer, compared to a course-bubble aeration system. The fine-bubble system, because of the greater pressure in the smaller bubble, can maintain equilibrium with a higher concentration of CO_2 in the mixed liquor. Thus, the pH can be maintained at a lower level. In addition, with the lesser degree of turbulence provided by the fine-bubble system compared to a coarse-bubble system, less CO_2 is stripped from the mixed liquor, thereby allowing more carbonic acid to be formed. Thus, some plants will incorporate a zone of coarse-bubble aeration to increase turbulence and strip CO_2 from the wastewater. High-purity oxygen (HPO) plants can have a problem with low pH and thus inhibited nitrification. The headspace above the mixed liquor in a HPO plant is confined to retain the oxygen and increase its partial pressure. However, CO_2 from biological respiration also accumulates in the head space, thus lowering the pH in the mixed liquor.

VARIATION IN FLOWS AND LOADS

The mass load that is received by a wastewater treatment plant is a product of both the flowrate and pollutant concentration. The mass loading will vary considerably over the course of a typical day and can be severe during peak flow periods. The variation can be more pronounced in smaller collection systems, where there is less storage capacity to dampen the effects. A typical diurnal flow and loading profile is shown in Figure 2.6.

The concentration does not necessarily increase with flowrate and, in fact, will often decrease as a result of dilution by the infiltration and inflow that is the cause of the high-peak flows. Therefore, it is important to determine the maximum month and peak day loadings based on daily data of flow and concentration and not on separate calculations of the maximum 30-day flow and maximum 30-day BOD concentration. This is because the flow and concentration do not necessarily peak at the same time. Typical peaking factors for raw wastewater are shown in Table 2.5. For example, if

TABLE 2.5 Ratio of load to the average annual load (Young et al., 1977).

Parameter	Maximum month	Peak day
BOD	1.30	1.59
TSS	1.37	2.28
NH_3-N	1.25	1.51
PO_4^-	1.27	1.57

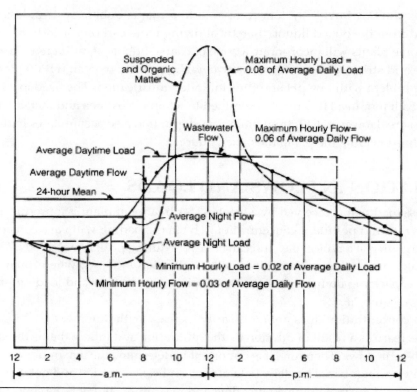

FIGURE 2.6 Hourly variation in flow and strength of municipal wastewater (from Fair, G. M., et al., *Elements of Water Supply & Wastewater Disposal.* 2nd ed., Copyright © 1971. Reprinted by permission of John Wiley & Sons, Inc., New York).

the average annual BOD load is 726 kg/d (1600 lb/d), then the maximum month BOD load would be approximately 944 kg/d (2080 lb/d) (1.30 × 726 [1.30 × 1600]). However, there can be considerable variability in these ratios, depending on the characteristics of the collection system.

Peak flows can affect treatment processes that are hydraulically limited, such as grit removal, primary clarification, and final settling. Peak flows can cause additional solids to be carried downstream to a filtration unit, which then increases the frequency of backwash required. This increases the internal return flow at a time when the plant may already be stressed hydraulically. Peak loads can affect the biological treatment systems that are process-limited. A peak BOD load, for example, may exceed the aeration capacity of an activated sludge process to the point where the dissolved oxygen level is depleted and treatment is incomplete.

Depending on the hydraulic and process capacity of the individual treatment units, flow equalization may be required. Flow equalization can be designed to accomplish both hydraulic and load equalization.

EFFECT OF RECYCLE FLOWS

REVIEW OF RECYCLE FLOWS. Internal plant recycles, such as from solids dewatering processes, supernatant from sludge digestion processes, thickening systems, and filter backwashes can have a significant effect on the treatment plant process and can introduce tremendous variability in the treatment plant flows and loads.

Anaerobic digestion, through fermentation processes, will release high concentrations of ammonia. High levels of phosphorus may also be released, particularly in plants that operate a EBPR process because the waste activated sludge that is processed will contain high levels of stored polyphosphate. Some of this phosphorus may be released under anaerobic conditions.

Return flows have the potential to upset the carbon to nitrogen ratio (as measured by the BOD:TKN ratio) or the carbon to phosphorus ratio (as measured by the BOD:TP ratio) that are typically required for effective biological nutrient removal. This is particularly true for plants that serve as a regional treatment facility for sludge streams or which receive large volumes of septage. Depending on the type of receiving facilities and pretreatment facilities provided, such facilities can introduce very high loads of nitrogen into the treatment process. A peak load of ammonia introduced back to the head of a treatment process may exceed the nitrification capacity of the biological process, thereby resulting in a spike of ammonia in the effluent. The treatment plant would generally be designed to accommodate its internal recycle flows and loads, but it is not unusual for plants to be upgraded with new solids handling processes, which, as a result, changes the nature of the recycles. For example, a plant that switches from aerobic digestion to anaerobic digestion will have very different recycle characteristics.

Plants that are removing phosphorus biologically need to be careful in their handling of waste sludge, particularly the waste activated sludge. If the sludge is held in an anaerobic condition, such as in a holding tank, gravity thickener, or anaerobic digestion, then return flows from these processes can have high concentrations of phosphorus. These heavy phosphorus loads can overwhelm the ability of the process to remove biologically, especially if there is not enough readily degradable carbon available.

Backwash flows are often introduced to the head of a treatment plant, which, if not properly managed, can affect the performance of the biological process. The backwash will contain a high concentration of inert solids. If these solids are not removed in a grit tank or primary settling tank, then they will pass through to the activated sludge process, where they will effectively decrease the volatile content of the mixed liquor suspended solids (MLSS). A lower mixed liquor volatile suspended solids concentration decreases the performance of the process because there is less active biomass, at any given MLSS concentration, to do the work of nitrification and denitrification.

Plants that incinerate their sludge may be introducing cyanide from the scrubber blow-down water to their process through the recycle flows. Depending on the concentration, the cyanide will be toxic or at least inhibitory to the nitrifying organisms.

MANAGEMENT OF RETURN FLOWS. Development of a mass balance of the wastewater treatment process is key to understanding the magnitude of the flows and loads contained in the recycle flows. Even small flows, with very high concentrations of a particular wastewater constituent, can have a significant effect on downstream treatment processes. Additional sampling of internal process and recycle flows may be required for a period of time to develop a mass balance, but the effort is critical to developing a strategy to effectively manage your recycle flows.

Management of plant recycle typically involves controlling the rate or schedule for returning them to the treatment process and the location at which they are returned. Equalization of the flows and loads may be all that is required. Plants with excessive recycle loads, such as may be introduced by regional sludge treatment facilities, may need to consider sidestream treatment processes to reduce the load of carbon or nitrogen being introduced. Plants which receive large volumes of septage (greater than 10% of the average daily flow) may consider discharging the septage into a gravity thickener or directly into the waste sludge stream rather than the main treatment process.

EFFECT OF EFFLUENT PERMIT REQUIREMENTS

The effluent discharge permit requirements certainly have a direct effect on the type of wastewater treatment facility that is designed and built, but the permit requirements also affect the operating requirements. The basis on which the discharge requirements were developed determines, to some extent, the amount of flexibility that the operator has in achieving the permit requirements.

In some cases, more stringent permit limits may be achieved through relatively simple operating changes. For example, a lower level of phosphorus may be achieved by increasing the chemical dosage for precipitation. Some limited degree of denitrification may be obtained by incorporating an anoxic zone to the beginning of an activated sludge reactor and by using the RAS as a nitrate recycle. However, generally, as permit levels become more stringent, treatment capacity is reduced in the existing facilities or additional treatment volume or unit processes must be added to retain the design capacity.

There are basically two types of permits or concepts for developing a discharge permit. They are (1) a technology-based permit, and (2) a water-quality-based permit. The variations of each and the implications they have for plant operations are discussed below.

TECHNOLOGY-BASED PERMITS. With this type of permit, the effluent concentration required is based on a specific level of treatment that is technically achievable. The level may or may not be related to the assimilative capacity of the receiving stream or on any water quality standards. For example, the basis of technology may be "best available" or "limit of technology." The time basis of the permit has implications regarding how the plant is operated, as discussed below.

Monthly Average. An effluent permit based on a 30-day average allows little flexibility to operate at any level other than safely below the permit level. Thus, the unit processes must be operated at optimum efficiency at all times. If a few samples indicate a high level of TN, for example, then the plant must operate at a higher level of performance for the rest of the month to maintain compliance with the monthly average. Typically, there are allowances for higher weekly or even daily results that provide some flexibility in response to peak daily or weekly influent loads.

Annual Average. An effluent permit that is based on an annual average over the period of a year allows some flexibility to operate at or near the permit level. If there is an excursion above the annual average, then there may still be time to operate at a higher level of treatment to maintain the required annual average. Of course, if the annual average approaches the limit of technology, then there again remains no margin for error and the plant must continuously operate at maximum efficiency. If, however, the annual average is 8 mg/L TN, for example, then the plant could perhaps achieve 6 mg/L in the summer and allow higher numbers in the winter, when the colder temperatures make biological nitrogen removal more difficult.

Seasonal Permit. A seasonal permit that enforces a different permit level in the summer versus the winter allows the plant to be operated in a "less aggressive" manner in the "off-season". For example, a plant with a seasonal nitrification permit may be able to operate in the winter at a lower MLSS level than otherwise would be required for a year-round permit. A plant with a seasonal TN permit of 3 mg/L during the summer months and no limit during the winter, for example, could turn the methanol feed off during the winter.

Seasonal nitrification, however, is associated with several operating concerns, because a plant can not simply "turn on" or "turn off" nitrification. The transition between the two states can result in the production of nitrites that are not further oxidized to nitrates by the biomass. This temporary condition of excess nitrites is evidenced by a large jump in the chlorine demand as the chlorine oxidizes the nitrites.

Transitioning from the non-nitrifying season to nitrification requires planning in advance of the permit date required for ammonia removal. The nitrifying organisms grow slowly, requiring that the MCRT be increased to maintain a population of nitrifiers in the biomass. The MCRT is increased incrementally over a period of time by controlling wasting and increasing the MLSS levels. Depending on the temperature, it can easily take a month for a plant to establish complete nitrification.

WATER-QUALITY-BASED PERMIT. Water-quality-based permits are based on the assimilative capacity of the receiving stream and are generally based on an annual average load allocation from the plant. As the flows increase, the level of treatment required to meet the permit increases. At lower flows, below design capacity, the allowable effluent concentration is greater. In this situation, the treatment process can be operated in a manner that demands less performance from the system than is required at design capacity. For example, depending on the specific permit requirements, perhaps the MLSS can be reduced or the nitrate recycle flowrate could be reduced. It is recognized that this somewhat oversimplifies the day-to-day operation of a wastewater treatment plant and that one does not simply "dial" in a number that one wishes to achieve in the effluent. However, an annual load allocation allows some flexibility in the operation of the plant seasonally if the treatment plant is capable of operating below the load allocation on an average monthly basis or is not yet at design capacity. In some states, water-quality-based limits are enforced in terms of a concentration limit, which therefore governs how the treatment plant must be operated.

REFERENCES

American Public Health Association; American Water Works Association; Water Environment Federation (1998) *Standard Methods for the Examination of Water and Wastewater*, 20th ed.; American Public Health Association: Washington, D.C.

Metcalf and Eddy (2003) *Wastewater Engineering: Treatment and Reuse*. G. Tchobanoglous, F. L. Burton, H. D. Stensel (Eds.); McGraw-Hill: New York.

Sedlak, R. (1991) *Phosphorus and Nitrogen Removal from Municipal Wastewater*, 2nd ed.; Lewis Publishers: Boca Raton, Florida.

U.S. Environmental Protection Agency (1993) *Nitrogen Control Manual*, EPA-625/R-93-010; Office of Research and Development, U.S. Environmental Protection Agency: Washington, D.C.

Water Environment Federation (1998) *Design of Municipal Wastewater Treatment Plants*, 5th ed., Manual of Practice No. 8; Water Environment Federation: Alexandria, Virginia.

Young, J. C.; Thompson, L. O.; Curtis, D. R. (1977) Control Strategy for Biological Nitrification Systems. *J. Water Pollut. Control Fed.*, **51**, 1824.

Chapter 3

Nitrification and Denitrification

Introduction	35
Wastewater Characteristics	35
Assimilation	35
Hydrolysis and Ammonification	35
Nitrifier Growth Rate	37
Nitrification	37
Process Fundamentals	37
Stoichiometry	37
Nitrification Kinetics	38
Biomass Growth and Ammonium Use	38
Wastewater Temperature	41
Dissolved Oxygen Concentration	42
pH and Alkalinity	42
Inhibition	43
Flow and Load Variations	43
Suspended-Growth Systems	43
General	43
Determining the Target $SRT_{aerobic}$	50
Example 3.1—Single Sludge Suspended-Growth Nitrification	51
Single Sludge Systems	52
Separate Sludge Systems	53
Attached Growth Systems	55
General	55
Trickling Filter	57
Rotating Biological Contactors	62
Biological Aerated Filter	65
Coupled Systems	67
Denitrification	68
Process Fundamentals	68
Stoichiometry	70
Denitrification Kinetics—Biomass Growth and Nitrate Use	71
Example 3.2—Single Sludge Suspended-Growth Postdenitrification	73
Carbon Augmentation	74
Separate-Stage Denitrification	75
Suspended-Growth Systems	75

Attached-Growth Systems	76
Denitrification Filter	76
Moving Bed Biofilm Reactor	77
Combined Nitrification and Denitrification Systems	78
Basic Considerations	78
Suspended-Growth Systems	78
Wuhrmann Process	79
Modified Ludzack-Ettinger Process	79
Bardenpho Process (Four-Stage)	82
Sequencing Batch Reactors	84
Cyclically Aerated Activated Sludge	86
Oxidation Ditch Processes	86
Countercurrent Aeration	87
Hybrid Systems	87
Introduction	87
Integrated Fixed-Film Activated Sludge	87
Descriptions Of Integrated Fixed-Film Activated Sludge Processes	88
Rope-Type Media	88
Sponge-Type Media	90
Plastic Media	90
Rotating Biological Contactors	92
Operational Issues	92
Rope-Type Media	92
Media Location	92
Growth	92
Kinetic	92
Worms	93
Media Breakage	93
Adequate Dissolved Oxygen Level	94
Mixing	94
Access To Diffusers	94
Odor	94
Sponge Media	94
Screen Clogging	94
Sinking Sponges	95
Loss Of Sponges	95
Taking Tank Out of Service	96
Loss of Solids	96
Air Distribution System	96
Plastic Media	96
Startup Procedures	96
Growth	97
Worms	97
Media Breakage	97
Media Mixing	97
Accumulation of Growth	97
Screen Clogging	97
Foaming	98
Taking Tank Out of Service	98
Membrane Bioreactor	98
Secondary Clarification	100
Suspended Growth	100
Denitrification	101
Flocculation Problem 1	101
Flocculation Problem 2	101
High Sludge Blanket	101
Hydraulic Problem	102
Attached Growth	102
References	102

INTRODUCTION

Microorganisms are resilient and adaptable and will proliferate wherever they have the opportunity. The environmental conditions that exist, intentionally or not, in the process flow train will govern the biology that predominates in that part of the system. The goal of the operator of any biological treatment system is to manipulate the environment of that system to achieve the desired result. Consideration must be given to each portion of every process tank to avoid the creation of environmental conditions that are unfavorable to process goals.

WASTEWATER CHARACTERISTICS

A comprehensive discussion of wastewater characteristics is included in Chapter 2. This section describes a few wastewater characteristics that influence nitrification, denitrification, and total nitrogen removal.

ASSIMILATION. Nitrogen serves as an essential nutrient for all living organisms, including the heterotrophic bacteria that remove organic pollutants from wastewater. Therefore, as the ratio of organic pollutant to nitrogen increases, a greater percentage of nitrogen is removed via bacterial growth and reproduction (i.e., assimilation into new cell mass). The quantity of nitrogen removed per unit of biochemical oxygen demand (BOD) or chemical oxygen demand (COD) removed depends on a number of variables associated with the process configuration and mode of operation. The nitrogen removed from the process via assimilation into new biomass can be estimated based on the quantity of new volatile suspended solids (VSS) formed as listed in Table 3.1 (Metcalf and Eddy, 2003). It is cautioned that much of the nitrogen removed via assimilation may be recycled back into the process through sludge handling sidestreams and require retreatment. A more detailed discussion of side stream and recycle flows and loads associated with sludge processing systems is included in Chapter 10.

HYDROLYSIS AND AMMONIFICATION. The microorganisms that perform nitrification and denitrification are only able to act on the inorganic forms of nitrogen (ammonia, nitrite, and nitrate). Therefore, any portion of the influent nitrogen remaining in the organic form-either particulate or soluble-has the potential of passing through the process untreated. Generally, particulate organic nitrogen is incorporated to the solids that are removed by clarification or filtration. The effluent

TABLE 3.1 Elemental composition of bacterial cells.

Parameter	Percentage
Carbon	50%
Oxygen	22%
Nitrogen	12%
Hydrogen	9%
Phosphorus	2%
Sulfur	1%
Potassium	1%
Sodium	1%
Other trace compounds	2%

particulate organic nitrogen will be a function of the effluent total suspended solids (TSS), similar to the relationship listed in Table 3.1. For example, if the effluent contains 10 mg/L TSS that is 80% volatile, approximately 1 mg/L nitrogen will be discharged with the effluent TSS.

The release of organic nitrogen from a particulate to a soluble state occurs during the hydrolysis of particulate organic matter (i.e., particulate COD). Hydrolysis is the conversion of particulate organic material into forms that are small enough to be taken up and metabolized by bacteria. Soluble organic nitrogen is then converted to ammonia via a process known as ammonification. Both hydrolysis and ammonification are performed by a number of facultative microorganisms that can exist within the sewer system or in any part of the treatment process. Hydrolysis and ammonification rarely limit nitrification rates in suspended-growth systems; however, fixed-growth systems may experience effects as a result of the relatively short duration that suspended solids are retained in these systems. Effluent soluble organic nitrogen measured at most nitrogen removal facilities is less than 1 mg/L, but has been measured as high as 3 to 5 mg/L, in some circumstances (WERF, 2003).

NITRIFIER GROWTH RATE. For suspended-growth systems, the rate at which nitrifiers reproduce determines the primary design and operating criteria of aerobic solids retention time ($SRT_{aerobic}$). It is becoming more common that nitrifier growth rate is being considered by design engineers as a key process parameter specific to each facility (WERF, 2003). The net nitrifier growth rate is a function of the nitrifier growth rate (u_n) minus the nitrifier decay rate (b_n) (i.e., the amount of new nitrifiers grown minus the amount that have died). Many previous literature sources generally considered the nitrifier decay rate to be negligible (U.S. EPA, 1993a).

NITRIFICATION

PROCESS FUNDAMENTALS. Nitrification is the two-step biological conversion of ammonia to nitrite and then to nitrate under aerobic conditions. The conversions to nitrite and nitrate involve two specific groups of autotrophic bacteria: Nitrosomonas and Nitrobacter. Autotrophic bacteria, specifically chemoautotrophic bacteria, differ from the heterotrophic bacteria that consume organic material (BOD) in that chemoautotrophic bacteria use carbon dioxide as their carbon source and specific inorganic chemicals as a source of energy for growth. In the case of Nitrosomonas and Nitrobacter, the inorganic chemicals used are ammonia and nitrite, respectively.

An alternative to the use of complete nitrification is the emerging use of sidestream treatments processes, such as the Sharon process, that intentionally prevent the oxidation of nitrite. Discussion of sidestream treatment processes is included in Chapter 10.

STOICHIOMETRY. The stoichiometry of the biochemical reactions associated with nitrification defines the proportion of reactants and products involved with this process. Understanding stoichiometry is important because it defines the basic inputs and outputs for each of the steps in the process and can determine which of these inputs will limit the reaction.

The stoichiometric equation that defines the molar ratios for the oxidation of ammonium (NH_4^+) to nitrite (NO_2^-) by *Nitrosomonas* is the following:

$$NH_4^+ + 1.5\, O_2 \Rightarrow NO_2^- + 2\, H^+ + H_2O \qquad (3.1)$$

Similarly, the stoichiometric equation that describes the oxidation of nitrite to nitrate (NO_3^-) by *Nitrobacter* is the following:

$$NO_2^- + 0.5\, O_2 \Rightarrow NO_3^- \tag{3.2}$$

These reactions also generate biomass associated with the growth of *Nitrosomonas* and *Nitrobacter* (i.e., nitrifiers). Unlike heterotrophs, *Nitrosomonas* and *Nitrobacter* obtain the carbon for cell growth from an inorganic source-carbon dioxide. The total yield of nitrifiers for the combination of both steps in the process is substantially lower than for heterotrophs, generally ranging from 0.06 to 0.20 g VSS/g NH_4^+-N oxidized (U.S. EPA, 1993a). The stoichiometric equations presented herein do not reflect that a portion of the nitrogen is incorporated to the VSS associated with the growth of nitrifiers. However, because the quantity of nitrogen removed via assimilation into nitrifier biomass is typically less than 2% of the ammonia nitrified, it generally can be ignored with regard to the total amount of ammonium removed. The overall stoichiometric expression for the two-step nitrification process combines eqs 3.1 and 3.2, as follows:

$$NH_4^+ + 2\, O_2 \Rightarrow NO_3^- + 2\, H^+ + H_2O \tag{3.3}$$

Converting the mole ratios in the stoichiometry above to a mass basis results in the following equation:

$$18g\, NH_4^+ + 64g\, O_2 \Rightarrow 62g\, NO_3^- + 2g\, H^+ + 18g\, H_2O \tag{3.4}$$

Converting the nitrogen species in eq 3.4 to an "as-nitrogen" basis, as discussed previously in the Nitrogen section of Chapter 2, results in the following:

$$14g\, NH_4^+\text{-N} + 64g\, O_2 \Rightarrow 14g\, NO_3^-\text{-N} + 2g\, H^+ + 18g\, H_2O \tag{3.5}$$

This approach simplifies the tracking of nitrogen compounds through the nitrification and denitrification processes, such that one part of ammonia is converted to one part nitrate. The two grams of hydrogen ions formed in eq 3.5 consume 100 grams of alkalinity (as calcium carbonate [$CaCO_3$]). Table 3.2 presents the oxygen and alkalinity relationships that are typical of the nitrification process.

NITRIFICATION KINETICS. *Biomass Growth and Ammonium Use.*
Growth of *Nitrosomonas* and *Nitrobacter* is the result of the oxidation of ammonia and nitrite, respectively. It is generally considered that the growth of either of these species of bacteria is limited by the concentration of the respective substrate (food source) as defined by the Monod equation (U.S. EPA, 1993a) illustrated as follows:

TABLE 3.2 Key nitrification relationships.

Parameter	Coefficient	Units
Oxygen use	4.57	g O_2/g NH_4-N
Alkalinity consumption	7.14	g alkalinity (as $CaCO_3$)/g NH_4-N

$$u_n = u_{n\text{-max}} S_n/(K_n + S_n) \qquad (3.6)$$

Where

u_n = specific growth rate of microorganisms (g nitrifiers/g nitrifiers in system·d);
$u_{n\text{-max}}$ = maximum specific growth rate of microorganisms (g nitrifiers/g nitrifiers in system · d);
K_n = half-saturation coefficient for ammonium-nitrogen (mg/L); and
S_n = growth-limiting substrate (NH_4^+-N) concentration (mg/L).

In the case of nitrification, ammonia-nitrogen is the substrate (S_n), and the half-saturation coefficient (K_n) is defined as the concentration when the nitrifier growth rate is 50% of the maximum nitrifier growth rate ($u_{n\text{-max}}$), typically estimated at 1 mg/L NH_4^+-N. The maximum specific growth rate is the rate at which nitrifiers will reproduce when no limiting conditions exist (i.e, with an excess of dissolved oxygen or ammonia). Figure 3.1 illustrates the influence of ammonium-nitrogen concentration on nitrifier growth rate.

Nitrite typically does not accumulate in large concentrations in biological treatment systems under stable operation because the maximum growth rate of *Nitrobacter* is significantly higher than that of *Nitrosomonas*. As a result, the growth rate of *Nitrosomonas* generally controls the overall rate of nitrification. However, plants that transition into and out of nitrification may experience a condition where the growth rate of *Nitrobacter* is not able to keep up with that of *Nitrosomonas*, and effluent nitrite concentrations increase. This condition is known as "nitrite lock" and results in a significant increase in effluent chlorine demand (5.1 mg chlorine/mg nitrite). In situations when effluent permits contain both seasonal nitrification and seasonal disinfection requirements (using chlorine), it is advisable to establish complete nitrification before the start of chlorination.

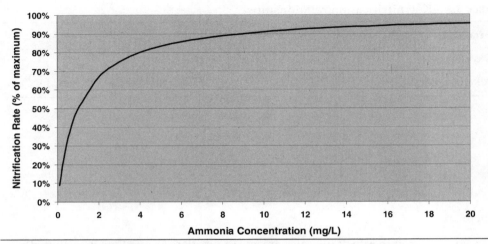

FIGURE 3.1 Influence of ammonia concentration on nitrification.

Bacterial growth rate-defined as the time required to double their population-of Nitrosomonas is 10 to 20 times slower that that for heterotrophic bacteria. As a result, the solids retention time (SRT) of a suspended-growth system operated to nitrify must be much longer than that typically necessary to maintain a stable heterotrophic bacterial population. The SRT in a biological system is typically defined as follows:

$$\text{SRT} = \frac{\text{(Mass of biological solids in the reactor}}{\text{Mass of biological solids leaving the system/d)}} \tag{3.7}$$

Because determination of the mass of active biological solids can be difficult, the overall solids maintained in and wasted from the reactor can be used to determine the SRT. At steady-state conditions, the solids leaving the system will equal the solids produced within the system. Therefore, if the unit solids production rate of the biological treatment system (i.e., the net yield [Y_n]), is known, the required mass of reactor solids can be determined. Typically, Y_n values are reported as the mass of solids produced per each unit mass of COD or five-day BOD (BOD$_5$) removed.

The growth rate of the nitrifying microorganisms in the system is related to the SRT, as follows:

$$\text{SRT} = 1/(u_n - b_n) = 1/u'_n \tag{3.8}$$

Where

u_n = specific growth rate of nitrifiers (g VSS added per day/g VSS in system);
b_n = endogenous decay coefficient for nitrifiers (g VSS destroyed per day/ g VSS in system); and
u'_n = net specific growth rate of nitrifiers (g VSS added per day/g VSS in system).

Previous references have generally considered the endogenous decay of nitrifiers (b_n) to be negligible, which results in u_n and u'_n values being equal (U.S. EPA, 1993a). More recent research (WERF, 2003) has identified that b_n may be more substantial than previously considered and may explain the wide variation in the value of u_n reported in literature (Grady and Lim, 1980).

A number of environmental factors significantly affect the nitrifier growth rates, which, in turn, influence the minimum required SRT required to allow for accumulation and maintenance of a sufficient population of nitrifying organisms. The effects of the major environmental factors that influence nitrification are presented in the following sections.

Wastewater Temperature. The growth rates of Nitrosomonas and Nitrobacter are particularly sensitive to the liquid temperature in which they live. Nitrification has been shown to occur in wastewater temperatures from 4 to 45°C (39 to 113°F), with an optimum growth rate occurring in the temperature range 35 to 42°C (95 to 108°F) (U.S. EPA, 1993a). However, most wastewater treatment plants operate with a liquid temperature between 10 and 25°C (50 and 77°F). It is generally recognized that the nitrification rate doubles for every 8 to 10°C rise in temperature. A number of relationships between maximum nitrifier growth rate (un-max) and wastewater temperature have been developed. The most commonly accepted expression for this relationship for wastewater over a temperature range from 5 to 30°C (41 to 86°F) is the following (U.S. EPA, 1993a):

$$u_{n\text{-max}} = 0.47\, e^{0.098\,(T-15)} \tag{3.9}$$

Where

$u_{n\text{-max}}$ = maximum specific growth rate of microorganisms (g nitrifiers/g nitrifiers in system · d);
T = wastewater temperature (°C); and
°C = (°F − 32)/1.8.

This relationship between maximum nitrifier growth rate and wastewater temperature is illustrated graphically in Figure 3.2.

Dissolved Oxygen Concentration. Nitrifiers are obligate aerobes, meaning they are only able to function under aerobic conditions. Consequently, the dissolved oxygen concentration in the bulk liquid can have a significant effect on the nitrifier growth rate. The value at which dissolved oxygen concentration reduces the rate of nitrification varies on a site-specific basis, depending on temperature, organic loading rate, SRT, and diffusional limitations. It is generally accepted that nitrification is not limited at dissolved oxygen concentrations greater than 2.0 mg/L. However, this may not be true for systems using high biomass concentrations, fixed-film systems, or in cyclical systems or transition zones between unaerated and aerated reactors, where dissolved oxygen concentrations less than 2.0 mg/L may affect nitrification.

pH and Alkalinity. Equations 3.4 and 3.5 describe the effect that nitrification has on the alkalinity levels within the treatment reactors. These equations illustrate that, in situations where low-alkalinity wastewater is being treated, the nitrification process can have a substantial effect on alkalinity levels and ultimately pH. Reactor pH levels have been shown to have a significant effect on the rate of nitrification (U.S. EPA,

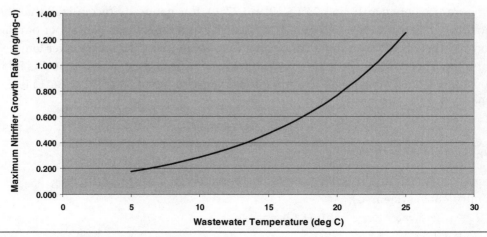

FIGURE 3.2 Influence of temperature on nitrification.

1993a) An important factor to consider with regard to the effects of pH on the nitrification process is the degree of acclimation that the particular system has achieved. Wide swings in pH have been demonstrated to be detrimental to nitrification performance, although acclimation generally allows satisfactorily performance with consistent pH control within the range 6.5 to 8.0 standard pH units. It is generally recommended that sufficient alkalinity be present through the reactors by maintaining a minimum effluent alkalinity of at least 50 and preferably 100 mg/L.

Inhibition. Nitrifiers are particularly susceptible to inhibition from a variety of organic and inorganic substances. Nitrifiers are particularly susceptible to wide fluctuations in the concentration of inhibitory substances, but may exhibit only minor effects if these substances are in low concentrations and consistently applied to the system. Table 3.3 presents a list of known organic compounds that have been identified as inhibitory to nitrification (WEF, 1998).

Nitrifier performance can also be significantly affected by heavy metals, including nickel (0.25 mg/L), chromium (0.25 mg/L), and copper (0.10 mg/L) (Metcalf and Eddy, 2003). Un-ionized or free ammonia can be inhibitory to *Nitrosomonas* and *Nitrobacter*, depending on temperature and pH conditions in the reactor. At a pH of 7.0 and temperature of 20°C, inhibition is expected to begin at an ammonia plus ammonium concentration of 1000 and 20 mg/L for *Nitrosomonas* and *Nitrobacter*, respectively (U.S. EPA, 1993a).

Flow and Load Variations. Provided that environmental conditions do not limit the growth of nitrifying organisms, the quantity or mass of Nitrosomonas and Nitrobacter that grow in the system will be a function of the applied ammonia load. As such, variations in flow and nitrogen load to the system that result in either a significantly reduced hydraulic retention time (HRT) or increased pollutant load may result in an increase in effluent ammonia. Short HRT systems, such as many fixed-film processes, are more likely to experience this reduction in process efficiency, even during normal diurnal variations. Longer HRT systems, such as extended aeration, sequencing batch reactor (SBR), and oxidation ditch activated sludge systems, are less likely to experience increased effluent ammonia levels because of variations in flow and ammonia load.

SUSPENDED-GROWTH SYSTEMS. *General.* Suspended-growth biological treatment systems operate in a fashion that allows control of the amount of biomass in the process and, therefore, the net growth rate of the biomass. The operating range

TABLE 3.3 Organic compounds reported as inhibitory to nitrification (Blum and Speece, 1991; Christensen and Harremoës, 1977; Hockenbury and Grady, 1977; Painter, 1970; Payne, 1973; Richardson, 1985; Sharma and Ahlert, 1977) *(continued)*.

Name	mg/L
Acetone	2000
Allyl alcohol	19.5
Allyl chloride	180
Allyl isothiocyanate	1.9
Allyl thiourea	1.2
AM (2-amino-4-chloro-6-methylprimidine)	50
Amino acids	1–1000
Aminoethanol	12.2
Aminoguanidine	74.0
2-Aminophenol	0.27
4-Aminophinol	0.07
Aminopropiophenone	43
Aminotriazole	70.0
Ammonium	1000
Aniline	7.7
1-Arginine	1.7
Benzene	13.00
Benzidine dihydrochloride	50.0
Benzocaine	100
Benzothiazzole disulphide	38.0
Benzylamine	100
Benzyldimethyldodecylammonium chloride	2.0
Benzylthiuronium chloride	40.0
2.2' Bipyridine	10.0
Bisphenol A	100
Bromodichloropropane	84.0
2-Bromophenol	0.35
4-Bromophenol	0.83
n-Butanol	8200
Cadmium	14.3
Carbamate	2
Carbon disulphide	35.0
Chlorine	1
Chlorobenzene	0.71, 500
Chloroform	18.0
2-Chloronaphthol	14.3
2-Chlorophenol	2.70
3-Chlorophenol	0.20

(continued)

TABLE 3.3 Organic compounds reported as inhibitory to nitrification (Blum and Speece, 1991; Christensen and Harremoës, 1977; Hockenbury and Grady, 1977; Painter, 1970; Payne, 1973; Richardson, 1985; Sharma and Ahlert, 1977) *(continued)*.

Name	mg/L
4-Chlorophenol	0.73
5-Chloro 1-pentyne	0.59
2-Chloro-6-trichloromethyl-pyridine	11.0
Chromium (III)	10
Copper	230
m-Cresol	01.–100
o-Cresol	11.4
p-Cresol	12.8
Cyanide	16.5
Cyclohexylamine	0.500
Di-allyl ether	100
1,2-Dibromoethane	50.0
Dibromethane	60
1,2-Dichlorobenzene	100
1,1-Dichloroethane	0.91
2,4-Dichloroethane	0.79
1,5-Dichloropentane	13.00
2,3-Dichlorophenol	0.42
2,3-Dichlorophenol	0.61
2,6-Dichlorophenol	8.10
3,5-Dichlorophenol	3.00
1,3-Dichloropropene	0.67
1,3-Dichloropropene	0.48
Dicyandiamide	250
Dicyclohexylcarbodiimide	10.0
Diethyl dithiothiosemicarbazide	0.1
Diguanide	50.0
Dimethylgloxime	140
Dimethylhydrazine	19.2
Dimethyl *p*-nitrosoaniline	19.0
Dimethyl *p*-nitrosoaniline	30
2,4-Dinitrophenol	37.0
Diphenylthiocarbazone	7.5
Dithio-oxamide	1
Dodecylamine	<1
Erythromycin	50.0
Ethanol	2400
Ethanolamine	100
Ethyl acetate	18
Ethylenediamine	100

(continued)

TABLE 3.3 Organic compounds reported as inhibitory to nitrification (Blum and Speece, 1991; Christensen and Harremoës, 1977; Hockenbury and Grady, 1977; Painter, 1970; Payne, 1973; Richardson, 1985; Sharma and Ahlert, 1977) *(continued)*.

Name	mg/L
Ethyl urethane	1000
Ethyl xanthate	10
Flavonoids	0.01
Guanidine	4.7
Hexamethylene diamine	85
Histidine	5
Hydrazine	58.0
Hydrazine sulphate	200
Hydrogen sulfide	50
8-Hydroxyquiniline mercaptobenzothiazole	1
Lauryl benzenesulphonate	118
Lead	0.500
l-Lysine	4.0
Mercaptobenzothiazole	3
Methanol	160
Methionine	9.0
n-Methylaniline	<1
Methylhydrazine	12.3
Methyl isothiocyanate	0.800
Methyl mercaptan	300
Methyl pyridines	100
2-Methylpyridine	100
4-Methylpyridine	100
Methylthiourea	0.455
Methyl thiuronium sulphate	1
Methylamine hydrochloride	100
Methylene blue	30
Monethanolamine	>200
N-serve	10
Napthylethylenediarmine dihydrochloride	23
Nickel	5.0
Ninhydrin	10.0
p-Nitroaniline	10.0
p-Nitrobenzaldehyde	50.0
Nitrobenzene	50.0
4-Nitrophenol	2.60
2-Nitrophenol	11.00
2-Nitrophenol	50.0
Nitrourea	1.0
Panthothenic acid	50
Pentachloroethane	7.90

(continued)

TABLE 3.3 Organic compounds reported as inhibitory to nitrification (Blum and Speece, 1991; Christensen and Harremoës, 1977; Hockenbury and Grady, 1977; Painter, 1970; Payne, 1973; Richardson, 1985; Sharma and Ahlert, 1977) *(continued)*.

Name	mg/L
Perchloroethylene phenol	5.6
Phenolics (substituted)	100
Phenolic acids	0.01
p-Phenylazoaniline	100
Potassium chromate	800
Potassium chlorate	2500
Potassium dichromate	6.0
Potassium thiocyanate	300
n-Propanol	20.0
Purines	50
Pyridine	10.0
Primidines	50
Pyruvate	400
Resurcinol	7.80
Skatole	7.0
Sodium azide	23.0
Sodium azide	20
Sodium arsenite	2000
Sodium chloride	35 000
Sodium cyanate	100
Sodium cyanide	1
Sodium dimethyl dithiocarbamate	13.6
Sodium chloride	35 000
Sodium cyanate	100
Sodium cyanide	1
Sodium dimethyl dithiocarbamate	13.6
Sodium methyl dithiocarbamate	0.90
Sodium pluoride	1218
Sodium methyldithiocarbamate	1
ST (sulfathiazole)	50
Strychnine	100
Sulphides	5.0
Tannin	0.01
Tetrabromobisphenol	100
1,2,3,4-Tetrachlorobenzene	20.00
1,2,4,5-Tetrachlorobenzene	9.80
1,1,1,2-Tetrachloroethane	8.70
1,1,2,2-Tetrachloroethene	1.40
1,2,3,5,6-Tetrachlorophenol	1.30
Tetramethylammonium chloride	2200
Tetramethyl thiuram disulfide	5

(continued)

TABLE 3.3 Organic compounds reported as inhibitory to nitrification (Blum and Speece, 1991; Christensen and Harremoës, 1977; Hockenbury and Grady, 1977; Painter, 1970; Payne, 1973; Richardson, 1985; Sharma and Ahlert, 1977) *(continued)*.

Name	mg/L
Thiamine	0.530
Thioacetamide	500
Thiocyanates	0.180
Thiosemicarbazide (Aminothiourea)	0.760
Thiourea	1
Thiourea (substituted)	3.6
1-Threonine	5
Threonine	50.0
2,4,6-Tribromophenol	7.70
2,4,6-Tribromophenol	50
2,4,6-Tribromophenol	2.5
2,2,2-Trichloroethanol	2.00
1,1,2-Trichloroethane	1.90
Trichloroethylene	0.81
Trichlorophenol	100
2,3,5-Trichlorophenol	3.90
2,3,6-Trichlorophenol	0.42
2,4,6-Trichlorophenol	7.90
Triethylamine	100
Trimethylamine	118
2,4,6-Trimethylphenol	30.0
1-Valine	1.8
Vitamins riboflavin, A-lipolic acid, B-pyridoxine HCL	50
Zinc	11.0

available is limited by the physical capacities of the system, most notably, aeration tank volume, aeration capacity, clarifier surface area, and return activated sludge pump capacity. The ability to control the overall biomass growth rate within the system provides the opportunity to manipulate the system to achieve the concurrent growth of heterotrophic (carbonaceous BOD [CBOD]-consuming) and autotrophic (nitrifying) bacteria.

Suspended-growth nitrification can be achieved in various reactor configurations; however, these configurations must be designed and operated to meet the following two overriding criteria:

(1) The biomass inventory must be retained in the system for a sufficient period to allow a stable population of nitrifiers to develop and be maintained in the process, and
(2) The HRT of the system must be such that the biomass provided is capable of reacting on the quantity of pollutant entering the system to the extent necessary to maintain permit compliance.

The $SRT_{aerobic}$ is the average period of time that any particle is retained in the aerated portion of a suspended-growth process reactor. The $SRT_{aerobic}$, also referred to as aerobic mean cell residence time or aerobic sludge age, is most commonly defined as the mass of solids in the aerated portion of the reactor tanks divided by the mass of solids removed (wasted) per day, as shown in eq 3.10.

$$SRT_{aerobic} = \text{Mass of MLSS in aerobic portion of reactor} / \text{mass of MLSS wasted per day} \quad (3.10)$$

This equation has been simplified based on two assumptions:

(1) Only the mixed liquor suspended solids (MLSS) inventory in the aerated portion of the system is considered to contribute to nitrification because nitrifier growth rate is assumed to be negligible at low or zero dissolved oxygen conditions.
(2) The mass of effluent TSS is ignored because it is significantly lower than the mass of MLSS in the waste sludge.

These two assumptions are generally true for most full-scale applications. However, it is noted that the mass of effluent TSS may be significant in situations when the effluent contains high concentrations of effluent TSS, such as during a secondary clarifier washout event or in activated sludge systems where the quantity of waste sludge generated is be very small as a result of the low organic loading, such as separate sludge nitrification systems. Separate sludge activated sludge systems are discussed in the Separate Sludge Systems section of this chapter.

The overall approach presented herein for operating a suspended-growth nitrification system is to determine the Target $SRT_{aerobic}$ using the equations presented and adjusting sludge wasting rates to maintain the actual $SRT_{aerobic}$ at or above the target

$SRT_{aerobic}$ within the system. It is noted that, because nitrification is sensitive to wastewater temperature, dissolved oxygen concentration, pH, and ammonia concentrations, each of these factors may have a significant affect on the target $SRT_{aerobic}$ determination.

DETERMINING THE TARGET $SRT_{aerobic}$. To determine the target $SRT_{aerobic}$, use the following steps:

(1) Step 1. Calculate maximum nitrifier growth rate ($u_{n\text{-max}}$) based on wastewater temperature (eq 3.9).

(2) Step 2. Calculate specific nitrifier growth rate (u_n) based on the reactor ammonia concentration (eq 3.6). Assuming pH values are stable and approximately neutral and dissolved oxygen concentrations of greater than 2 mg/L are maintained, proceed to Step 3.

(3) Step 3. Calculate the minimum aerobic SRT ($SRT_{aerobic\text{-min}}$) that equates to the SRT when nitrifiers are reproducing as fast as they are being wasted out of the system (eq 3.8).

(4) Step 4. Determine the process design factor (PDF) to account for the variability in influent wastewater characteristics and operation and a safety factor. The PDF is the product of a peaking factor (PF) and a safety factor (SF) and generally ranges between 1.5 and 3.0. The PF can be determined to account for the variability in process influent wastewater characteristics, internal plant recycles, plant operation (return activated sludge [RAS], waste activated sludge, and dissolved oxygen control), and effluent permit requirements and the physical process configuration. Highly variable influent wastewater characteristics; intermittent internal plant recycles; stringent permit limits; intermittent, manually controlled operation; and smaller process tanks warrant greater PF values. Consistent influent and internal plant recycles; less frequent (i.e., moving annual average) permit requirements; well-automated processes; and plants with larger, more forgiving process tanks or equalization allow lower PF values. The SF is generally a means of reflecting the level of uncertainty in design or process performance. Use of a SF may be warranted in situations with newer, less proven technologies and may be omitted if all variations to the process influent and operation are well-defined (U.S. EPA, 1993a).

(5) Step 5. Calculate the target $SRT_{aerobic}$ by multiplying the minimum $SRT_{aerobic}$ times the PDF.

Note that use of daily values for wastewater temperature and $SRT_{aerobic}$ should be avoided. Use of moving average values is recommended to dampen intermittent operations and sampling variability. Generally, a seven-day moving average is satisfactory to minimize the effect of variability without creating a significant "lag" effect on calculated values. Figure 3.3 illustrates daily versus seven-day moving average $SRT_{aerobic}$ for an operating facility. The graph in Figure 3.4 shows the calculated minimum and target $SRT_{aerobic}$ values versus temperature with an assumed PDF of 2.5.

EXAMPLE 3.1—SINGLE SLUDGE SUSPENDED-GROWTH NITRIFICATION.
The following is an example for single sludge suspended-growth nitrification.

Given:

Secondary influent (SI) conditions:
   Average daily flow = 4000 m³/d (1.06 mgd)
   SI BOD_5 = 150 mg/L
   SI total Kjeldahl nitrogen (TKN) = 35 mg/L
   SI ammonia-nitrogen (NH_3-N) = 21 mg/L
   SI wastewater temperature = 15°C
   Maximum monthly/average flow ratio = 1.3
   Diurnal peak/average flow ratio = 1.2
Secondary effluent (SE) requirements:
   SE NH_3-N = 1 mg/L

Note that SI flow, BOD, and TKN loading should be based on the most stringent permit criteria (monthly, weekly, etc) anticipated for the time period.

(1) Step 1. Calculate nitrifier maximum growth rate:
$u_{n\text{-max}} = 0.47 \, e^{0.098(15-15)} = 0.47/\text{days}$
(2) Step 2. Calculate specific growth rate.
$u'_n = 0.47 \, (1 \text{ mg/L}/[1 \text{ mg/L} + 1 \text{ mg/L}]) = 0.235/\text{days}$
(3) Step 3. Calculate minimum aerobic SRT:
Minimum $SRT_{aerobic} = 1/u'_n = 1/0.235/d = 4.3$ days
Note that Minimum $SRT_{aerobic}$ is the SRT when nitrifiers are growing just as fast as they are wasted from the system and is an inherently unstable operating condition.
(4) Step 4. Determine the PDF:
PDF = PF × SF

PF = Maximum month flow factor × diurnal flow factor
PF = 1.3 × 1.2 = 1.56
The SF is intended to account for variability in secondary influent wastewater characteristics and process operation (dissolved oxygen control, sludge wasting, etc.) and process performance uncertainty.
SF = 1.5
PDF = 1.56 × 1.5 = 2.34

(5) Step 5. Calculate the Target SRT_a:
Target $SRT_{aerobic}$ = Minimum $SRT_{aerobic}$ × PDF
Target $SRT_{aerobic}$ = 4.3 days × 2.34 = 10.1 days

Single Sludge Systems. Single sludge suspended-growth nitrification systems are defined as a process configuration including only one set of clarifiers per train of reactors for oxidation of both carbonaceous (CBOD) and nitrogenous (TKN and ammonia) pollutants. Because of this configuration, these systems generally operate

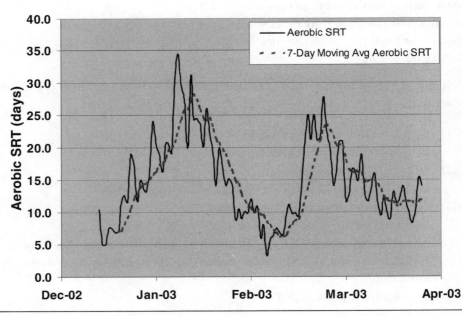

FIGURE 3.3 Daily versus seven-day moving average SRT.

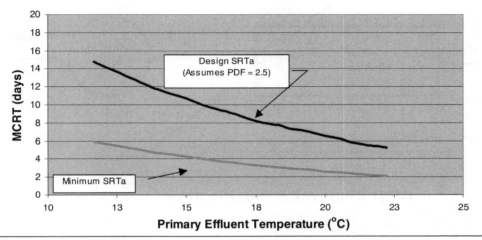

FIGURE 3.4 Nitrification aerobic SRT requirements.

at higher CBOD to TKN (CBOD:TKN) ratios than separate sludge systems. Because of the common practice of incorporating denitrification and biological phosphorus removal to nitrifying systems, single sludge biological nutrient removal (BNR) systems have become the predominate approach when nitrification is required. Figure 3.5 is a schematic of a single sludge nitrification system.

Separate Sludge Systems. Separate sludge suspended-growth systems, as shown in Figure 3.6, are defined as a process configuration that includes a separate set of reactors and clarifiers for each step of the process, typically for CBOD removal and nitrification. Separate sludge nitrification is characterized by low CBOD:TKN ratios (typically less than 5:1 and often less than 2:1). This influent loading to the nitrification system typically results in particularly low waste activated sludge generation rates. In some situations, a portion of the primary effluent is bypassed to the separate sludge nitrification system to provide sufficient loading to generate more biomass than is lost to the effluent TSS.

Separate sludge nitrification systems have been implemented for a variety of reasons, including the following:

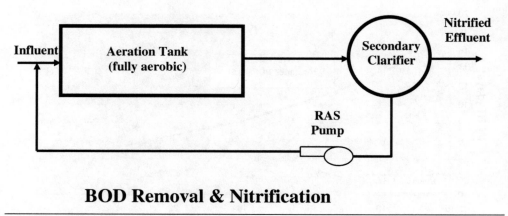

FIGURE 3.5 Single sludge nitrification.

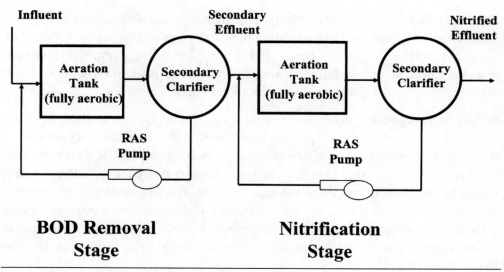

FIGURE 3.6 Separate sludge nitrification.

- To provide tertiary treatment following an upstream fixed-growth (trickling filter) process;
- To protect the nitrification system from an influent wastewater that contains materials toxic to nitrifiers; and
- To allow optimized operation of the separate carbonaceous and nitrifying activated sludge systems

In general, separate sludge suspended-growth nitrifying systems have become less common in recent years, primarily as a result of the increased cost associated with the construction of separate clarifiers for each stage and the tendency to integrate BNR systems together. Separate stage systems have experienced a number of operational challenges, primarily related to very low biomass growth rates and "weak" floc structure that is susceptable to floc shearing.

ATTACHED GROWTH SYSTEMS. *General.* Biological wastewater treatment processes with the biomass attached to some type of inert media are termed fixed-film, attached-growth, or fixed-growth reactors (Figure 3.7). True fixed-growth systems are differentiated from coupled treatment systems that incorporate a separate fixed-growth reactor followed by an activated sludge process without intermediate clarification and hybrid systems that include suspended- and fixed-growth processes within the same reactor.

Fixed-growth biological treatment processes are generally very easy to operate and are very resilient to shock loads, but also have somewhat less flexibility than suspended-growth treatment systems. Unlike suspended-growth nitrification systems, the onset and accumulation of autotrophic bacteria that accomplish nitrification is limited by the growth of heterotrophic bacteria that predominate until the substrate (food source) for those microorganisms is mostly exhausted. Nitrosomonas and Nitrobacter generally will not become a significant portion of the biomass growth on the media surface until soluble BOD_5 is less than 15 mg/L or CBOD is less than 20 mg/L. Therefore, CBOD removal and nitrification are generally considered sequential processes in fixed-growth systems.

In other words, if nitrification is to occur in a fixed-growth reactor, competition from the heterotrophs for oxygen and space on the media must be reduced. This can be accomplished via upstream CBOD removal processes or within the fixed-growth system by providing sufficient surface area to first reduce the CBOD and then the ammonia load. In general terms,

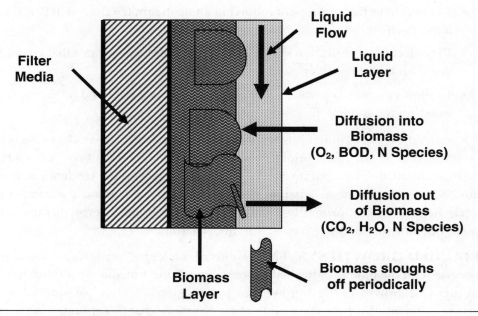

FIGURE 3.7 Biofilm schematic.

- An organic loading rate of 7.3 to 9.8 g/m²·d (1.5 to 2.0 lb CBOD/d/1000 sq ft) (media surface area) will reduce the CBOD to less than 20 mg/L so that nitrification can commence; and
- An ammonia-nitrogen loading rate of 1.0 to 2.0 g/m²·d (0.2 to 0.4 lb TKN/d/1000 sq ft) (media surface area) will reduce the ammonia-nitrogen to less than 2 mg/L.

These values are simply approximations for a multitude of fixed-growth reactors (U.S. EPA, 1993a). The actual performance of any particular system will depend on the type and quantity of media in use, wastewater characteristics, and a multitude of factors, including temperature, dissolved oxygen concentration, and pH.

Another considerable difference between suspended- and fixed-growth systems is the influence that diffusion has on performance. Soluble matter (soluble BOD, ammonia, and oxygen) must diffuse into bacteria to be used. In suspended-growth systems, biomass is suspended within liquid containing substrate (BOD) and an electron acceptor (oxygen or nitrate) on all sides. Diffusion of these materials and other necessary nutrients can enter the full 360 degrees around the floc particle. Fixed growth, by its very nature, has much less biomass surface area in contact with the passing liquid or gas to allow diffusion into the biomass, making the diffusion of substrate and electron acceptor into the bacteria much more likely to influence the pollutant removal rate.

Trickling Filter. Trickling filters are aerobic fixed-growth reactors consisting of open media that supports biomass growth. Settled primary effluent or fine screened wastewater is distributed at the top of the filter and allowed to "trickle" down through the media. Air flows through the media either concurrently or countercurrently with the liquid driven by either convection or by low-pressure fans. There should be no standing liquid within the trickling filter reactor that would hinder the flow of air throughout the entire height. The media supports a mixed aerobic biomass that removes organic material and, if conditions allow, ammonia by sorption and assimilation into the biomass. Excess biomass periodically "sloughs" off the trickling filter media and is typically removed via secondary clarifiers. Figure 3.8 shows a schematic of a trickling filter.

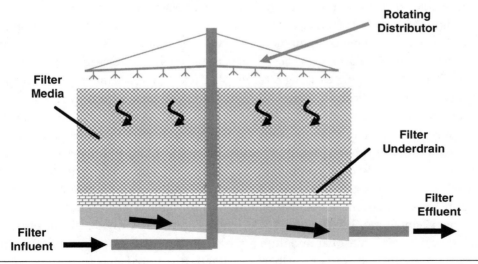

FIGURE 3.8 Trickling filter schematic.

Historically, trickling filters have been classified based on hydraulic and organic loading rates, as listed in Table 3.4. Trickling filter media typically consists of rock or one of a number of types of plastic media. Rock media trickling filters are generally limited to a depth of less than 3 m (10 ft) because of weight, but plastic media trickling filter can be more than 7.6 m (25 ft) deep (U.S. EPA, 1993a). Plastic media offers much greater available surface area per unit volume and less weight than rock trickling filters allowing taller and more efficient filters. Table 3.5 presents a comparison of the physical properties of various types of trickling filter media.

Because the biomass growing on trickling filters is not submerged, the media must be continuously wetted. Therefore, recirculation is commonly used to maintain consistent flow across the media and provide for continuous operation of hydraulically driven distribution mechanisms. Recirculation is also reported to improve performance resulting from reseeding of media with biomass, dampening variations of influent wastewater concentration, and improving flow distribution across the filter. In the past, many variations of recirculation arrangements have been used; however, many of these configurations are no longer being designed or used (WEF and ASCE, 1998). Clarification of filter recycle before recirculation has been shown to improve trickling filter performance as a result of TSS removal. However, improved trickling filter performance can easily be negated if the influent plus recirculation hydraulic loads negatively affect clarifier operation. Therefore, recirculation through clarifiers should only be practiced if clarifiers have been sized to handle the recycle flows or only during periods when influent plus recycle flows do not exceed clarifier capacity or negatively affect performance.

Figures 3.9a and 3.9b illustrate four more commonly used recirculation configurations in use today. Other configurations similar to configurations 3.9b through 3.9d may also incorporate flow through a clarifier depending on available clarifier capacity (WEF and ASCE, 1998). During trickling filter retrofit projects, consideration should be given to replacing hydraulically driven flow distribution with mechanically driven, variable speed flow distribution, and the addition of forced ventilation. The addition of covers to trickling filters can reduce temperature effects and improve cold weather performance.

Ammonium-nitrogen, dissolved oxygen concentration, and temperature factors can act individually or in combination to reduce the observed nitrification rate in trickling filters. It is generally believed that nitrification rates are reduced as a result of substrate limitations starting at ammonium-nitrogen concentrations somewhere between 2 and 4 mg/L (WEF and ASCE, 1998). Figure 3.10 provides the approximate upper and lower limits of ammonium removal rates measured at five nitrifying trickling filter facilities (Okey and Albertson, 1989). It is noted that the values used to

TABLE 3.4 Historical classification of trickling filters.

Design characteristics	Low or standard rate[a]	Intermediate rate[a]	High rate[a]	Super rate[a]	Roughing
Media	Stone	Stone	Stone	Plastic	Stone/plastic
Hydraulic loading					
mgd/ac[b]	1 to 4	4 to 10	10 to 40	15 to 90	60 to 180
gpd/sq ft[c]	25 to 90	90 to 230	230 to 900	350 to 2100	1400 to 4200
Organic loading					
lb BOD$_5$/ac-ft[d]	200 to 600	700 to 1400	1000 to 1500	—	—
lb BOD$_5$/d/1000 cu ft[e,f]	5 to 15	15 to 30	30 to 150	<=300	>100
Recirculation	Minimum	Usually	Always	Usually	Not normally required
Filter flies	Many	Varies	Few	Few	Few
Sloughing	Intermittent	Intermittent	Continuous[g]	Continuous[g]	Continuous
Depth, ft[h]	6 to 8	6 to 8	3 to 8	<=40	3 to 20
BOD removal, %[i]	80 to 85	50 to 70	40 to 80	65 to 85	40 to 85
Effluent quality	Well nitrified	Some nitrification	No nitrification	Limited nitrification	No nitrification

[a]Obsolete terminology.
[b]mgd/ac × 9353 = m^3/ha·d.
[c]gpd/sq ft × 0.0407 = m/d.
[d]lb/d/ac-ft × 0.36 = kg/1000 m^3·d.
[e]Excluding recirculation.
[f]lb/d/1000 cu ft × 1.602 = kg/100 m^3·d.
[g]May be intermittent up to a total hydraulic rate of between 0.7 and 1.0 gpd/sq ft.
[h]ft × 0.3048 = m.
[i]Including subsequent setting.

TABLE 3.5 Comparative physical properties of trickling filter media.

Media type	Nominal size, in. x in.[a]	Unit weight, lb/cu ft[b]	Specific surface area, sq ft/cu ft[c]	Void space, %	Application[d]
Plastic (bundle)	24 x 24 x 48	2 to 5	27 to 32	>95	C, CN, N
	24 x 24 x 48	4 to 6	42 to 45	>94	N
Rock	1 to 3	90	19	50	CN, N
Rock	2 to 4	100	14	60	C, CN, N
Plastic (random)	Varies	2 to 4	25 to 35	>95	C, CN, N
	Varies	3 to 5	42 to 50	>94	N
Wood	48 x 48 x 1.875	10.3	14		C, CN

[a] in. x 25.4 = mm.
[b] lb/cu ft x 16.02 = kg/m^3.
[c] sq ft/cu ft x 3.281 = m^2/m^3.
[d] C = CBOD5R; CN = CBOD5R and NODR; N = tertiary NODR.

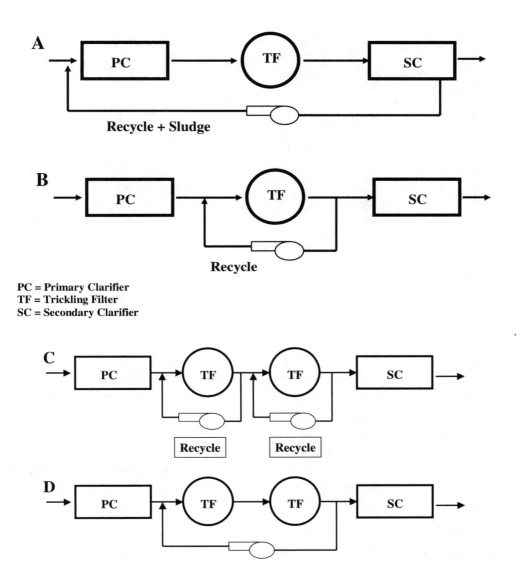

FIGURE 3.9 Trickling filter recirculation layouts (WEF, 1998).

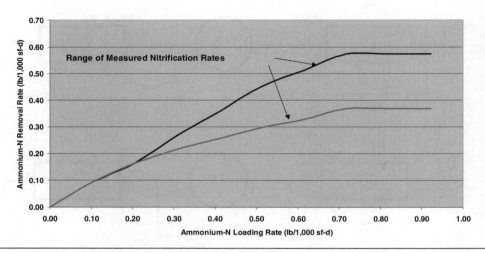

FIGURE 3.10 Trickling filter nitrification (adapted from Okey and Albertson, 1989).

develop Figure 3.10 have not been corrected for temperature or dissolved oxygen concentration. It has been hypothesized that ammonium or dissolved oxygen limitations can mask the effect of temperature on the -rowth nitrification process (Okey and Albertson, 1989). As the nitrification produces hydrogen ions, the pH of the trickling filter effluent can be depressed. This may result in a reduction in nitrification rates in the portions of the filter affected. The effects of lower pH on process performance will be limited in situations were recirculation rates are low. pH levels can be kept in check by alkalinity addition. A minimum of between 50 and 100 mg/L effluent alkalinity (as $CaCO_3$) is recommended.

Rotating Biological Contactors. Rotating biological contactors (RBC) consist of a series of circular plastic disks mounted on a horizontal shaft. The most common configuration consists of the shaft mounted above a tank and the disks approximately 40% submerged in the wastewater. In some situations, RBCs are almost fully submerged, and oxygen transfer is accomplished fully by diffused aeration. The shaft is rotated at 1 to 2 rpm, alternately exposing the plastic disks to wastewater and air (WEF and ASCE, 1998). Bacteria and microorganisms attach themselves to the disks and form a biomass covering the media surface. The microorganisms respond to the environmental conditions present, and the types and numbers of organisms present will vary from stage to stage. A stage may consist of a portion of one shaft, one complete shaft, or even multiple shafts. Excess biomass periodically "sloughs" off the RBC media and is typically removed via secondary clarifiers. Figure 3.11 shows a schematic of a RBC system.

Unlike trickling filters, RBCs cannot readily be used for heavy organic loading or "roughing" applications. Early in their history, RBCs experienced a number of structural shaft failures associated with excessive and unequal biomass buildup on the disks. Shaft designs were modified to alleviate this problem, and RBCs are well-suited to provide a high level of secondary treatment, nitrification, and, in fully submerged applications, denitrification. The RBCs are typically provided in standard 8.2-m- (27-ft-) long shafts with disks that are 3.7 m (12 ft) in diameter. Media densities range from 9290 m^2 (100 000 sq ft) per shaft to 16 722 m^2 (180 000 sq ft) per shaft. The RBCs used for CBOD removal are typically configured with a lower media density of 9290 m^2 (100 000 sq ft) per shaft to avoid media clogging, and use of higher density media is typically reserved for CBOD polishing or nitrification.

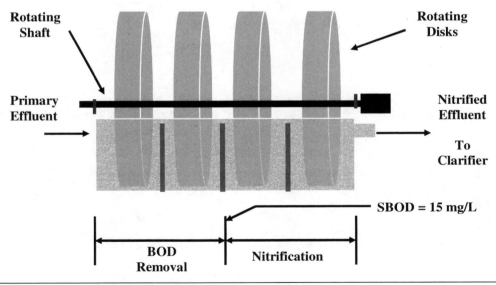

FIGURE 3.11   Rotating biological contactor schematic.

As with other attached-growth systems, nitrification using RBCs is subject to the wastewater characteristics entering the reactor. Nitrification will not commence until CBOD is reduced sufficiently to allow the nitrifiers to compete successfully on the media surface against heterotrophic bacteria. Removal of CBOD in a RBC is typically limited by a maximum first-stage BOD_5 loading of 24 to 29 $g/m^2 \cdot d$ (5 to 6 lb/d/1000 sq ft), a soluble BOD loading of 12 $g/m^2 \cdot d$ (2.5 lb/d/1000 sq ft) on any RBC stage, and an overall BOD_5 loading of approximately 10 $g/m^2 \cdot d$ (2 lb/d/1000 sq ft) to reduce effluent concentrations sufficiently to commence nitrification. The area requirements used for nitrifier growth cannot be met until the organic (CBOD) loading has been reduced. Therefore, in determining the overall media requirements for a facility, the amount of media needed for ammonium removal must be added to the amount needed for CBOD removal.

Once the CBOD is reduced sufficiently to start nitrification, the following procedure can be used to determine the area requirements and number of RBC shafts needed to reduce the effluent ammonia concentration:

(1) Determine the ammonium-nitrogen removal rate for the effluent ammonia concentration required from Figure 3.12;
(2) If the wastewater temperature is less than 13°C (55°F), calculate the temperature correction factor that needs to be applied to the RBC surface area required from Figure 3.13; and

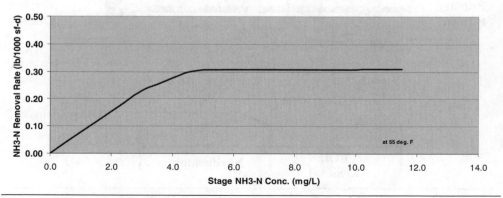

FIGURE 3.12 Rotating biological contactor nitrification rates.

(3) Apply the temperature correction factor to the ammonium removal rate to determine the area required for nitrification.

In the event that the effluent concentration required is less than 5 mg/L, the area requirement should be determined in two steps: (1) from the influent concentration down to 5 mg/L, and then (2) from 5 mg/L to the effluent concentration required (U.S. EPA, 1993a).

The RBC systems are generally configured with multiple stages, particularly when low effluent BOD_5 or ammonia limits are required. Table 3.6 presents two examples of RBC manufacturer's staging recommendations.

Biological Aerated Filter. Biological aerated filters (BAF) are a fixed-growth biological treatment process that combines aerobic biological treatment with filtration, eliminating the need for separate solids removal. The BAFs can be configured either as upflow or downflow units, with either a fixed or floating bed of media. The BAFs provide a physical configuration for biomass to either attach to or trapped between the filter media. Air is sparged into the bottom of the filter to provide an aerobic environment, allowing the biomass to oxidize CBOD or ammonia as it passes through the filter. As the biomass within the filter builds up, liquid flow is restricted, creating an increase in headloss through the filter (WEF, 2000). The filter is periodically backwashed to remove excess solids; however, the backwash is intentionally not so vig-

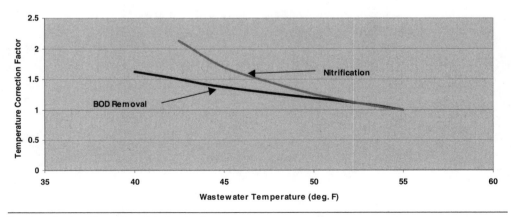

FIGURE 3.13 Rotating biological contactor temperature correction.

TABLE 3.6 Rotating biological contactor manufacturer recommended staging (U.S. EPA, 1993b).

	Carbon oxidation		Nitrification	
	Effluent BOD$_5$	Number stages	Effluent ammonia-N	Number stages
Envirex	>25 mg/L	1	5 mg/L	1
	15 to 25 mg/L	1 to 2	<5 mg/L	Based on first-order kinetics
	10 to 15 mg/L	2 to 3		
	<10 mg/L	3 to 4		
Lyco	<40 % removal	1	<40 % removal	1
	35 to 65% removal	2	35 to 65% removal	3
	60 to 85% removal	3	60 to 85% removal	3
	80 to 90% removal	4	80 to 90% removal	4

Nitrification and Denitrification

orous as to remove all the biological solids present; most are retained to provide for continued treatment of the wastewater. Figure 3.14 presents an upflow fixed-bed BAF configuration.

COUPLED SYSTEMS. A coupled process can be defined as a process configuration that combines two different treatment processes. Most combined processes consist of a fixed-growth reactor followed by a suspended-growth reactor. The most commonly used coupled processes currently in use are the trickling filter/solids contact and roughing filter/activated sludge (RF/AS) processes. Figure 3.15 shows one configuration of a coupled process, the RF/AS process.

The combination of the two processes is often a means of cost-effectively upgrading an existing facility, but may also be used for new designs. Using dual processes offers the opportunity to use the best attributes and limit the weaknesses of each reactor. As most coupled process configurations include a trickling or roughing filter as the initial process, they are not necessarily an optimum configuration for total nitrogen removal. However, upgrading a facility using a coupled process often is a cost-effective alternative to implementing nitrification.

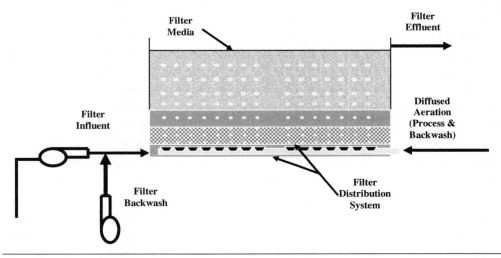

FIGURE 3.14 Biological aerated filter.

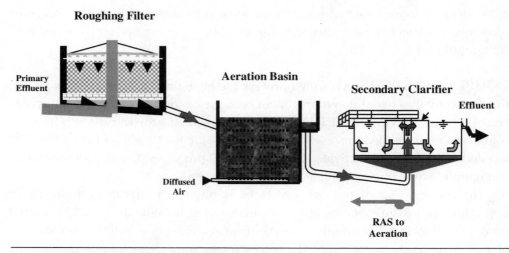

FIGURE 3.15 Roughing filter/activated sludge process.

DENITRIFICATION

PROCESS FUNDAMENTALS. Denitrification involves the biological reduction of nitrate and/or nitrite to nitrogen gas in the absence of dissolved oxygen. A biological environment with little or no dissolved oxygen but with nitrate or nitrite present is referred to as anoxic. In this process, bacteria use the oxygen contained in the nitrate or nitrite molecules to metabolize organic carbon. Unlike nitrification, denitrification is performed by a wide range of heterotrophic bacterial species, many of which are commonly found in typical biological treatment processes-even those not designed to remove nitrogen. These organisms are common because they are facultative and can use oxygen, nitrate, or nitrite as their terminal electronic acceptor (oxygen source). These organisms will use oxygen preferentially when it is available, but will change to use nitrate or nitrite whenever dissolved oxygen is in low supply or absent. Oxygen is the preferred terminal electron acceptor because the microorganisms are able to metabolize substrate more efficiently using oxygen compared to nitrate and nitrite, resulting in a greater amount of energy available to the bacteria and biomass produced per unit of carbonaceous material treated. Table 3.7 lists the major processes commonly included in biological nitrogen removal treatment with the corresponding carbon sources, electron donors, electron acceptors, and products.

TABLE 3.7 Electron donors and acceptors for carbon oxidation, nitrification, and denitrification (adapted from Metcalf and Eddy, 2003).

Process	Type of bacteria	Carbon source	Electron donor	Electron acceptor	Products
Aerobic oxidation	Aerobic heterotrophic	Organic compounds (CBOD)	Organic compounds (CBOD)	Oxygen (O_2)	Carbon dioxide, water (CO_2, H_2O)
Nitrification	Aerobic autrotrophic	Carbon dioxide (CO_2)	Ammonium, nitrite (NH_4^+, NO_2^-)	Oxygen (O_2)	Nitrite, nitrate (NO_2^-, NO_3^-)
Denitrification	Facultative heterotrophic	Organic compounds (CBOD)	Organic compounds (CBOD)	Nitrate (NO_3^-)	Nitrogen, carbon dioxide, water (N_2, CO_2, H_2O)

STOICHIOMETRY. Heterotrophic bacteria can accomplish denitrification using a wide variety of substrates (food sources). The following compounds are commonly used as the carbon sources for denitrification of wastewater:

- Organics present in domestic or industrial wastewater (CBOD);
- Methanol;
- Ethanol;
- Acetic acid; and
- Food processing organic waste materials (sugars).

The stoichiometric equation for denitrification using methanol is as follows:

$$6\ NO_3 + 5\ CH_3OH + H_2CO_3 \Rightarrow 3\ N_2 + 8\ H_2O + 6\ HCO_3^- \tag{3.11}$$

Converting this to a mass basis reduces this equation (in grams) to the following:

$$372g\ NO_3 + 160g\ CH_3OH + 62g\ H_2CO_3 \Rightarrow \\ 84g\ N_2 + 144g\ H_2O + 366g\ HCO_3^- \tag{3.12}$$

Reducing the nitrogen species to "as-nitrogen" results in the following:

$$84g\ NO_3\text{-}N + 160g\ CH_3OH + 62g\ H_2CO_3 \Rightarrow \\ 84g\ N_2 + 144g\ H_2O + 366g\ HCO_3^- \tag{3.13}$$

One gram of methanol has the theoretical oxygen equivalent of 1.5 g oxygen. Therefore, the above mass-based "as-nitrogen" equation provides the key relationships related to denitrification included in Table 3.8.

In addition to providing the energy for the reduction of nitrate, the methanol or CBOD consumed also is used to create new biomass. As a result, more methanol or CBOD is required than is presented above to reduce each unit of nitrate. The amount of new biomass generated and the portion used for denitrification are specific to each compound. The experiments conducted by McCarty et al. (1969) provide the basis of the stoichiometric equations for methanol consumption using nitrate, nitrite, and oxygen. These equations are as follows (U.S. EPA, 1975):

$$NO_3^- + 1.08\ CH_3OH + 0.24\ H_2CO_3 \Rightarrow \\ 0.04\ C_5H_7NO_2 + 0.48\ N_2 + 1.23\ H_2O + HCO_3^- \tag{3.14}$$

$$NO_2^- + 0.67\ CH_3OH + 0.53\ H_2CO_3 \Rightarrow \\ 0.056\ C_5H_7NO_2 + 0.47\ N_2 + 1.68\ H_2O + HCO_3^- \tag{3.15}$$

TABLE 3.8 Key nitrification relationships.

Parameter	Coefficient	Units
Methanol consumed	1.91	g methanol/g NO_3-N
Oxygen demand consumed	2.86	g COD/g NO_3-N
Alkalinity generated	3.57	g alkalinity (as $CaCO_3$)/g NO_3-N

$$O_2 + 0.93\ CH_3OH + 0.056\ NO_3^- \Rightarrow$$
$$0.056\ C_5H_7NO_2 + 1.04\ N_2 + 0.59\ H_2CO_3 + 0.056\ HCO_3^- \quad (3.16)$$

These three equations can be converted into a methanol dose based on nitrate, nitrite, and dissolved oxygen (DO) concentrations, as follows:

$$\text{Methanol dose (mg/L)} = 2.47\ NO_3^--N + 1.53\ NO_2^--N + 0.87\ DO \quad (3.17)$$

Provided that dissolved oxygen discharged or recycled to a denitrification reactor is minimal, the "rule of thumb" that 3 mg/L BOD_5 is removed for every 1 mg/L NO_3^--N denitrified is a reasonable approximation.

DENITRIFICATION KINETICS—BIOMASS GROWTH AND NITRATE USE.
The rate of denitrification is related to the growth of heterotrophs that use nitrate as their terminal electron acceptor (oxygen source) and are influenced by many factors related to influent wastewater characteristics, process configuration, process operation, and if supplemental substrate addition is used. Denitrification is affected by nitrate, carbon source, and dissolved oxygen concentrations in accordance with separate Monod-type expressions. However, because the half-velocity constants for nitrate and carbon source (CBOD or methanol) are very low, it is unlikely that either of these parameters will significantly affect the rate of denitrification.

Dissolved oxygen concentration has a more significant effect on denitrification rate, as defined by the Monod-type equation, as follows:

$$q_D = q_{D\text{-max}} [K_o/(K_o - S_o)] \tag{3.18}$$

Where

q_D = nitrate removal rate (g NO_3^--N/g VSS·d);
q_D-max = maximum nitrate removal rate (g NO_3^--N/g VSS·d);
K_o = half-saturation constant for dissolved oxygen (mg/L); and
S_o = dissolved oxygen concentration (mg/L).

Typical K_o concentrations are reported to range from 0.10 to 0.25 mg/L, indicating the sensitivity of the process with regard to oxygen concentration.

Several researchers (Barnard, 1974; Ekama and Marais, 1984) have identified that predenitrification using the CBOD in the influent wastewater proceeds at three separate and distinct rates, depending on the organic carbon substrate source available. The first phase of denitrification was associated with rapid denitrification using readily degradable organics. The second phase involved the use of particulate and more complex substrates, and the third denitrification rate was associated with endogenous respiration. Simplified empirical methods have also been developed to estimate the denitrification rate in predenitrification single sludge reactors. In this discussion, predenitrification is defined as denitrification using carbon sources available in raw or settled wastewater before nitrification, such as in a Modified Ludzack-Ettinger configuration, as described later in this section. Burdick et al. (1982) estimated that the specific denitrification rate (SDNR) in a predenitrification system was related to food/mixed liquor volatile suspended solids ratio (F/MLVSS), as follows:

$$SDNR_1 = 0.03 (F/M_1) + 0.029 \tag{3.19}$$

Where

$SDNR_1$ = specific denitrification rate in predenitrification zone (g NO_3-N rem/g MLVSS·d); and
F/M_1 = food/mass ratio in predenitrification zone (g BOD_5/g MLVSS·d).

Absent a supplemental carbon source, denitrification processes in a secondary or post anoxic zone are fueled by endogenous respiration and proceed at a rate that is much lower than in the first (predenitrification) anoxic zone. Two equations have been developed for use in describing the rate of denitrification in postdenitrification reactors (U.S. EPA, 1993a).

$$SDNR_2 = 0.12 \, SRT^{-0.706} \tag{3.20}$$

$$\text{SDNR}_2 = (0.175\, A_n)/(Y_{max}\, \text{SRT}) \qquad (3.21)$$

Where

SDNR_2 = specific denitrification rate in post-anoxic zone (g NO_3-N rem/g MLVSS·d);
SRT = solids retention time or sludge age (1/d);
A_n = net amount of oxygen required across activated sludge system (g O_2/g BOD_5 rem); and
Y_{max} = net TSS produced across activated sludge system (g TSS/g BOD_5 rem).

Typically, eq 3.20 is more commonly used because some of the parameters included in eq 3.21 are more difficult to obtain. Denitrification is affected by temperature changes in a fashion similar to other heterotrophs. Various Arrhenius θ values have been measured describing the effect of temperature on denitrification during a number of site-specific studies, yielding a range from 1.03 through 1.20 (U.S. EPA, 1993a). Equation 3.22 can be used to estimate the influence of temperature on denitrification.

$$\text{SDNR}_T = \text{SDNR}_{20}\, \theta^{T-20} \qquad (3.22)$$

Where

SDNR_T = specific denitrification rate at temperature T,
SDNR_{20} = specific denitrification rate at 20°C, and
θ = Arrehnius value.

EXAMPLE 3.2—SINGLE SLUDGE SUSPENDED-GROWTH POSTDENITRIFICATION.
The following is an example for single sludge suspended-growth postdenitrification.
Given:

Postdenitrification zone influent (PDI) conditions:
  Average daily flow = 4000 m³/d (1.06 mgd)
  Postdenitrification zone tank volume = 750 m³ (0.20 mil. gal)
  PDI NO_3-N = 4 mg/L
  SI wastewater temperature = 15°C
  Maximum monthly/average flow ratio = 1.3
Secondary effluent requirements:
  SE NO_3-N = 1 mg/L

Note that SI flow, BOD, and nitrogen loading should be based on the most stringent permit criteria (monthly, weekly, etc.) anticipated for the time period.

(1) Step 1. Calculate average mass of nitrate to denitrify:
NO_3-N (kg/d) = 4000 m^3/d × (4 − 1) g/m^3 = 12 kg/d (26.5 lb/d)

(2) Step 2. Calculate SDNR$_2$ at operating temperature using eq 3.20 and 3.22:
Assume overall system SRT based on acceptable range needed for upstream nitrification plus any other reactor tanks. A 15-day overall reactor SRT was assumed for the initial calculation.
SDNR$_2$ (20°C) = 0.12 (15 d)$^{-0.706}$ = 0.018 g NO_3-N removed/g MLSS·d
Correct for operating temperature using eq 3.22:
SDNR$_2$ (15°C) = 0.018 g NO_3-N removed/g MLSS·d (1.10)$^{15-20}$ = 0.011 g NO_3-N removed/g MLSS·d.

(3) Step 3. Calculate required MLVSS concentration in postdenitrification reactor:
MLVSS mass = (12 kg NO_3-N /d)/(0.011 kg NO_3-N removed/g MLSS·d) = 1091 kg MLVSS required (2405 lb MLVSS)
Average month MLVSS concentration = 1091 kg/750 m^3 = 1.455 kg/m^3 = 1455 mg/L
Multiply average month condition MLVSS by maximum month/average Ratio to determine maximum month MLVSS concentration required. If calculated value is less than MLVSS required to nitrify during maximum month condition, use nitrification MLVSS concentration.

The response of denitrifying bacteria and typical heterotrophic organisms to variations in pH, alkalinity, and the addition of inhibitory compounds will be similar, with one difference. A denitrifying environment will tend to increase the buffering capacity of the system. As denitrification proceeds, alkalinity is generated, improving the buffering capacity of the system.

CARBON AUGMENTATION. Nearly all separate-stage denitrification reactors and even some denitrification reactors within single sludge systems use some form of supplemental carbon or substrate augmentation. Most separate-stage denitrification processes follow CBOD and nitrification processes. Therefore, there is a very limited quantity of biodegradable organic carbon available to serve as the substrate for denitrification. The form of substrate, and, in particular, the ease or difficulty of biodegradation of that substance, is a key parameter related to the rate and efficiency of the denitrification process.

A number of substances have historically been used to augment the rate of the denitrification process, including the following (U.S. EPA, 1993a):

- Methanol,
- Ethanol,
- Acetate,
- Molasses,
- Soft drink wastes, and
- Brewery wastes.

The preferred material for carbon augmentation should be readily biodegradable, free of nutrients (especially nitrogen), stable at storage conditions, and relatively inexpensive. Methanol universally satisfies that criteria, although, in certain site-specific cases, other materials have been satisfactory alternatives.

Different process configurations are more or less sensitive to the accurate control of methanol dosing. Generally, systems with lower HRTs, such as fixed-film systems, are most susceptible to methanol overdosing and resulting methanol breakthrough. Methanol breakthrough to the effluent can result in CBOD permit violations, while underdosing can result in increased effluent nitrate values. Typically, the required methanol dose is approximately 2.5 and 3 times the nitrate-nitrogen removed. Carbon augmentation can often overcome dissolved oxygen levels discharged to the denitrification reactor. However, excessive dissolved oxygen levels transferred to the denitrification reactor will reduce efficiency and capacity of the process and result in increased operating cost associated with greater chemical addition requirements and sludge production. A separate discussion of operation of chemical feed systems is included in Chapter 8.

SEPARATE-STAGE DENITRIFICATION. *Suspended-Growth Systems.* Suspended-growth denitrification processes that accomplish CBOD removal, nitrification, and denitrification with the same biomass (i.e., single sludge) are classified as integrated nitrogen removal processes and are presented in detail in the Combined Nitrification and Denitrification Systems section. These systems expose the same biomass to different conditions either on the basis of tank volume or time and share the same set of secondary clarifiers. Systems that have a combined CBOD and nitrification step followed by a separate denitrification systems are classified as two-sludge systems. When all three processes (CBOD removal, nitrification, and denitrification) are configured as separate-stage systems with their own sets of clarifiers, the system is classified as a three-sludge system.

Stand-alone suspended-growth denitrification systems are classified as separate-sludge systems because the biomass is segregated from the remaining processes by providing a solids separation step between the preceding process and the denitrification reactor. In addition, the suspended-growth denitrification system has its own set of clarifiers and return activated sludge pump system. Continuous flow activated sludge denitrification systems are typically operated with a 2- to 3-hour detention time and include a small aerated zone after the denitrification zone to release any contained nitrogen gas bubbles and oxidize any excess methanol (Jenkins and Hermanowicz, 1989).

Attached-Growth Systems. DENITRIFICATION FILTER. The downflow deep-bed denitrification filter is a combination attached-growth biological process and effluent filter. As a result, the process design is based on two separate criteria; filtration and biological denitrification. The process consists of a block underdrain system that supports a coarse gravel support layer and a deeper sand layer. The combination of gravel and sand layer can range from 2.4 to 3.7 m (8 to 12 ft) in depth, depending on process requirements. Because of this configuration, no separate secondary clarifier or solids separation process is necessary.

A supplemental carbon source, such as methanol, is added to nitrified effluent entering this system. Heterotrophic bacteria accumulate and reproduce within the sand media to convert the nitrate to nitrogen gas using the methanol as the substrate. As nitrified effluent passes through the filter, solids are captured in the filter, and nitrogen gas is generated via denitrification, both filling the voids between the sand particles and increasing headloss. On a periodic basis, the backwash pumps will perform a "bump" or nitrogen release cycle to release the nitrogen gas that accumulates within the filter. Either on a timed basis or when filter headloss increases to a predetermined setpoint, the filter goes through a backwash process, where a combination of air and water are used to scour the solids from the filter and restore hydraulic capacity without removing all the biological solids necessary to maintain treatment efficiency. Backwash water is taken from a clearwell, which holds a predetermined volume of previously filtered effluent. Dirty backwash spills into a trough in the upper region of the filter and flows to a mudwell and then is pumped back to the plant headworks. Figure 3.16 shows a schematic of a downflow denitrification filter.

MOVING BED BIOFILM REACTOR. A moving bed biofilm reactor (MBBR) uses small, plastic carrier elements to support biomass growth in either an aerobic or anoxic environment. The carrier elements are retained in the reactor via the use of screens or sieves. Aerobic MBBRs are typically aerated using coarse-bubble diffused aeration, and anoxic units are mixed with slow speed submersible mixers. A "true" MBBR does not require backwashing, use any return sludge flow or use a suspended-growth component to achieve treatment goals. A modified version of the MBBR process is a hybrid type system, combining activated sludge treatment with plastic carrier elements in an integrated fixed-film activated sludge system (IFAS). A description of the hybrid MBBR process is included in the Hybrid Systems section.

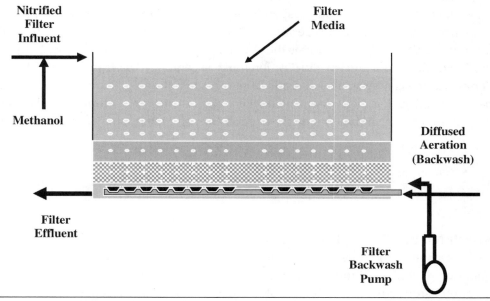

FIGURE 3.16 Deep bed denitrification filter.

The biofilm carrier elements are constructed of polyethylene and have a specific gravity slightly less than 1.0. The elements are commonly shaped like small cylinders with cross-fins inside. The plastic carrier elements provide a very high protected media surface area per unit volume. Carrier elements can be retrofitted into existing wastewater treatment plant basins and can be configured with separate anoxic and aerobic zones to achieve a high level of total nitrogen removal.

COMBINED NITRIFICATION AND DENITRIFICATION SYSTEMS

BASIC CONSIDERATIONS. Combined nitrification and denitrification treatment achieves CBOD removal with nitrification and denitrification in a single sludge configuration using linked reactors in series and a common set of secondary clarifiers. This consolidation of processes offers the potential of reduced capital and operating costs and, at the same time, a greater operational challenge related to the proper balancing that is sometimes required between the different process environments.

Combined nitrification and denitrification processes can be classified in various ways, including fixed- versus suspended-growth, flow regime, staging of process sequences, and method of aeration. One common thread that is consistent throughout all of these processes is the alternating environments that sequentially are used to nitrify and denitrify. Nitrification must be completed, at least partially, before denitrification can be achieved. Aside from outside influences that negatively affect performance, such as high dissolved oxygen recycles and toxic compounds, there are three parameters that can limit denitrification and, ultimately, total nitrogen removal.

- Nitrate;
- Substrate (CBOD); and
- Denitrification capacity (biomass and/or detention time).

SUSPENDED-GROWTH SYSTEMS. As part of achieving the objectives of the various stages of treatment, suspended-growth systems must properly balance the quantity of biomass in the reactors versus the capacity of the secondary clarifiers. It is counterproductive if a high level of nitrogen removal can be achieved within the reactors if periodic solids washouts are occurring in the secondary clarifiers. Although the subject of sludge bulking and foaming is presented elsewhere in this manual, it cannot be emphasized enough that the environmental conditions created in the aeration/anoxic reactors can have a significant effect on secondary clarifier performance and capacity.

Wuhrmann Process. The Wuhrmann process configuration, as shown in Figure 3.17 is a single sludge nitrification system with the addition of an unaerated anoxic reactor between the aerobic nitrifying reactor and the secondary clarifiers. This configuration places the denitrification reactor after the nitrification step; however, the lack of available carbonaceous substrate at this point in the process significantly limits the denitrification rate of this configuration. The Wuhrmann process, when applied in the past, also suffered from high effluent turbidity and was determined to be unsuitable for full-scale implementation. Two enhancements to the Wuhrmann Process that would allow for successful nitrogen removal would be the addition of a supplemental carbon source to the anoxic zone and incorporating a small aerobic zone between the anoxic reactor and the secondary clarifiers to strip any nitrogen gas and oxidize any remaining organic matter.

Modified Ludzack-Ettinger Process. The Ludzack-Ettinger process was developed to take advantage of the carbonaceous substrate available in the influent wastewater by placing the anoxic zone upstream of the nitrification reactor, as shown in Figure 3.18 (Ludzack and Ettinger, 1962). In this configuration, nitrates included in the RAS flow are mixed with influent wastewater and reduced to nitrogen gas in a

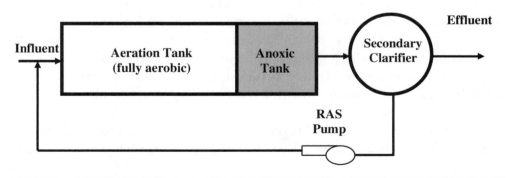

FIGURE 3.17 Wuhrmann process.

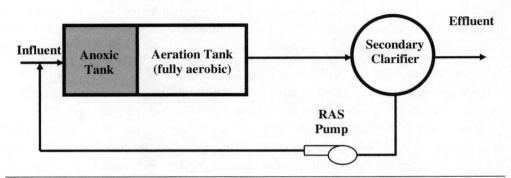

FIGURE 3.18 Ludzack-Ettinger process.

predenitrification reactor upstream of the main aeration basin. However, the total nitrogen removal efficiency of the Ludzack-Ettinger process is limited by the quantity of nitrate recycled back to the anoxic zone in the RAS flow. In response to that limitation, Barnard developed an improvement to the Ludzack-Ettinger process, identified as the modified Ludzack-Ettinger (MLE) process, which adds the recirculation of mixed liquor recycle (MLR) from the end of the aeration tank to the beginning of the anoxic tank, as shown in Figure 3.19 (Barnard, 1974). Figure 3.20 graphically depicts the amount of denitrification that can be achieved in a predenitrification reactor of an MLE process based on MLR flow (as a percentage of influent flow). It is noted that Figure 3.20 merely determines the denitrification that can be achieved if other factors, such as substrate availability or kinetic limitations, do not limit performance. Table 3.9 lists the key operating parameters of the MLE process.

Denitrification rates may be reduced at dissolved oxygen concentrations in the anoxic reactor as low as 0.2 mg/L (Randall et al., 1992). Therefore, dissolved oxygen inputs from sources such as MLR and backmixing from the aerobic reactor need to be minimized. The maximum effective MLR will be dependent on the dissolved oxygen included in the MLR flow and the oxygen demand (BOD) of the influent wastewater. Weaker strength wastewater and higher MLR dissolved oxygen concentrations are likely to limit the maximum beneficial MLR rate to less than 200%; higher strength wastewater and low MLR dissolved oxygen concentrations may allow beneficial use

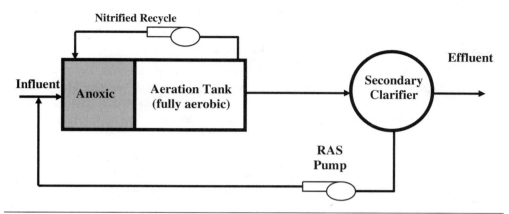

FIGURE 3.19 Modified Ludzack-Ettinger process.

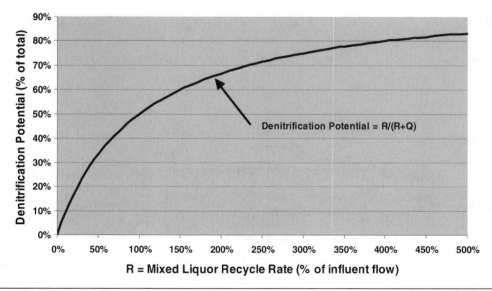

FIGURE 3.20 Nitrified recycle rate.

TABLE 3.9 Monitoring requirements for MLE process (modified from U.S. EPA, 1993a).

Reactor	Parameter	Rationale
Anoxic	Dissolved oxygen	Will reduce denitrification rate.
		Inadequate load limits denitrification.
	Nitrate-N	High nitrate recycled to aerobic zone may cause filamentous bulking.
Aerobic	Mixed liquor recycle	Control nitrate load.
		High DO may inhibit upstream denitrification.
	Dissolved oxygen	Low DO may inhibit nitrification.
	Alkalinity, pH	Nitrification consumes alkalinity.

of MLR rates as high as 400%. Often, MLR rates greater than 400% of influent flow are not beneficial to improving total nitrogen removal because of insufficient substrate availability, dissolved oxygen recirculation, and substrate dilution that reduces denitrification kinetics.

Bardenpho Process (Four-Stage). The four-stage Bardenpho process, shown schematically in Figure 3.21, incorporates the principles used for the MLE and Wuhrmann processes to create two anoxic zones to achieve a high level of total

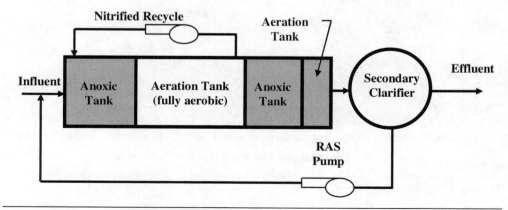

FIGURE 3.21 Four-stage Bardenpho process.

nitrogen removal. The first two stages of the Bardenpho process function similarly to an MLE process, although the primary anoxic zone in many Bardenpho facilities is often sized large enough to consistently accommodate at least 400% MLR rate without nitrate "bleed through." This primary anoxic zone removes the majority of nitrate generated in the process, and the secondary anoxic zone, located outside the MLE "loop," provides denitrification for that portion of flow that is not recycled to the primary anoxic zone. The fourth reactor zone in the Bardenpho process is an aerobic or reaeration reactor and serves to strip any nitrogen gas formed in the second anoxic zone, increase the dissolved oxygen concentration before secondary clarification, improve activated sludge flocculation, and reduce effluent turbidity. The Bardenpho process can generally achieve greater than 80% and often better than 90% total nitrogen (TN) removal (U.S. EPA, 1993a).

The Bardenpho process has been designed with many reactor configurations, including plug flow, complete mix, and oxidation ditch reactors. In the United States, the systems often use an oxidation ditch as the MLE portion of the system with separate complete mix reactors for the secondary anoxic and secondary aerobic (reaeration) zones. Table 3.10 lists the key monitoring requirements at different stages of the four-stage Bardenpho process (U.S. EPA, 1993a).

TABLE 3.10 Monitoring requirements for four-stage Bardenpho process (modified from U.S. EPA, 1993a).

Reactor	Parameter	Rationale
First anoxic	Dissolved oxygen	Will reduce denitrification rate.
		Inadequate load limits denitrification.
	Nitrate-N	High nitrate recycled to aerobic zone may cause filamentous bulking.
Aerobic	Mixed liquor recycle	Controls nitrate load.
	Dissolved oxygen	High DO may inhibit upstream denitrification. Low DO may inhibit nitrification.
Second anoxic	Nitrate	High nitrification in aerobic zone may overwhelm endogenous denitrification capacity, resulting in NO_3 in effluent.
	Dissolved oxygen	High DO will inhibit endogenous denitrification.

Sequencing Batch Reactors. Sequencing batch reactors are a variable volume suspended-growth treatment technology that uses time sequences to perform the various treatment operations that continuous treatment processes conduct in different tanks. The first activated sludge processes were actually SBRs, but the lack of automation forced a shift toward continuous treatment systems that required less operator intervention. With the advances in automated equipment and computer technologies over the last 20 years, SBRs have reemerged as a viable treatment technology.

The SBRs proceed through a series of phases for every cycle and will typically complete 4 to 6 cycles per day per SBR tank for domestic wastewater treatment. Between 50 and 75% of the liquid volume of the SBR containing the settled biomass is retained at the end of every cycle. The minimum liquid level determines the volume of sludge inventory that can be retained in the SBR reactor and the maximum volume of influent that can be accommodated for any individual cycle. The SBRs function using the following four basic phases:

(1) Fill. Wastewater is added to the SBR, raising the liquid level from the minimum level to a depth that corresponds to the amount of influent received during the fill phase time period. The aeration and mixing treatment steps generally commence during the fill phase.
(2) React. Biological processes are performed, including various aeration and mixing regimes, depending on process goals.
(3) Settle. Aeration and mixing are terminated, and biomass is allowed to settle.
(4) Decant. Clarified effluent is removed from the basin, aeration and mixing are off, and biomass is wasted as necessary.

The greatest challenges related to SBRs are related to the significant headloss through the system, difficulty in removing floating material from the SBR tanks, and intermittent decant that generally warrants equalization before downstream processes such as filtration and disinfection. Figure 3.22 illustrates the basic phases of an SBR cycle. The SBR control systems permit system operation to be configured to mimic almost any other suspended-growth reactor configuration. The SBRs can be configured and operated to act as a single sludge nitrification system, a multiphased cyclic aeration process, an MLE process, or a Bardenpho system. With sufficient volume to provide for the necessary reaction time, SBRs can achieve greater than 90% total nitrogen removal. Figure 3.23 presents a potential operating strategy for a high level of nitrogen removal.

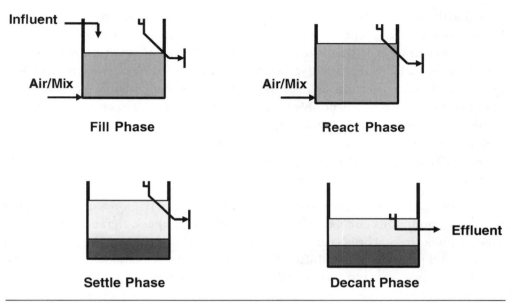

FIGURE 3.22 Sequencing batch reactor process cycle.

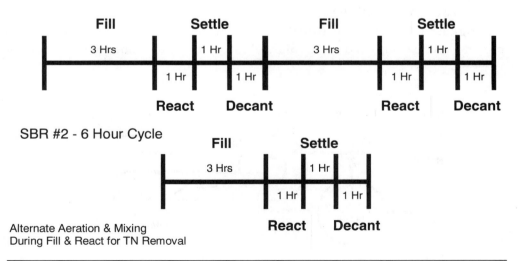

Alternate Aeration & Mixing
During Fill & React for TN Removal

FIGURE 3.23 Sequencing batch reactor cycle timeline—2 basin system.

Cyclically Aerated Activated Sludge. Many activated sludge reactor basins are easily converted from continuous aeration to cyclical aeration by cycling mechanical aerators on and off with timers or by alternating diffused aeration zones by electrically or pneumatically actuated valves. The key factor that will likely determine if cyclic aeration can function successfully is if the existing system has the ability to maintain a sufficient SRT to be able to consistently nitrify with intermittent aeration. If the aerobic SRT (total SRT times percentage of time aerating) is more than necessary to maintain nitrification, the cyclical aeration process can reduce effluent total nitrogen discharges. The SBRs and oxidation ditches can also function as cyclically aerated activated sludge processes.

Oxidation Ditch Processes. Oxidation ditch processes use looped channels that provide a continuous circulation of wastewater and biomass. Aeration can be provided with horizontal brush aerators, vertical shaft mechanical aerators that impart a horizontal liquid velocity, or diffused aeration with submersible mixers. Typically, aeration is concentrated to a few locations within the continuous loop to create variations in dissolved oxygen concentration along the length of the loop, ranging from a high dissolved oxygen just downstream of the aerator to low or no dissolved oxygen just upstream of the aerator. One oxidation loop configuration, the Carrousel system, is commonly configured as a MLE reactor or as the first two stages of a Bardenpho configuration. Figure 3.24 presents a typical oxidation configuration with an external secondary clarifier (WEF, 1998).

Another variation of the oxidation ditch process is the phased isolation ditch processes-Bio-Denitro and Bio-Denipho. Bio-Denitro is the configuration used for

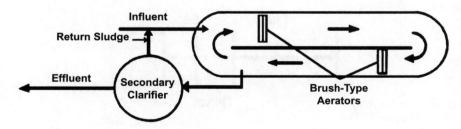

FIGURE 3.24 Oxidation ditch reactor.

biological nitrogen removal and Bio-Denipho for combined biological nitrogen and phosphorus removal. The Bio-Denitro process includes two oxidation ditches and secondary clarification. The two ditches are alternately fed wastewater and aerated, creating a time sequencing of aerobic and anoxic conditions to nitrify and denitrify. The process has a number of similarities to the SBR process, although the Bio-Denitro process discharges effluent continuously (U.S. EPA, 1993a; WEF, 1998).

Countercurrent Aeration. The countercurrent aeration process uses a moving-bridge diffused aeration system that rotates in the opposite direction as the liquid flow. Because the rotating assembly typically provides adequate mixing energy to maintain the MLSS in suspension, the system can operate without the need for the aeration system to mix the tank contents, allowing operation at low dissolved oxygen levels. Provided that the system is designed and operated with a sufficient SRTaerobic to maintain nitrification, this lower dissolved oxygen can enhance the denitrification potential of the system.

HYBRID SYSTEMS

INTRODUCTION. Hybrid treatment systems can be defined as any process that combines several technologies to create a new process, thereby integrating the advantages of each technology. The two types of hybrid processes that will be discussed in this section are IFAS and membrane bioreactors (MBR). The IFAS combines a fixed-film media with a suspended-growth activated sludge process. An MBR combines an activated sludge process with membrane filtration.

Each type of hybrid system can be incorporated to the various types of process configurations generally used for nitrification and denitrification and in treatment plants designed for biological phosphorus removal. There are many versions and manufacturers of these processes. Each process will be reviewed in this section, and general operational and maintenance issues will be discussed.

INTEGRATED FIXED-FILM ACTIVATED SLUDGE. Integrated fixed-film activated sludge (IFAS) processes include any wastewater treatment system that incorporates some type of fixed-film media within a suspended-growth activated sludge process. The media systems used vary greatly and include some form of rope or looped strand media, sponge cuboids, plastic wheels, or packing material of various types. The media can be free floating in the mixed liquor; or, in some manner, fixed in the aeration basin on cages or in frames; or, in the case of RBCs, on rotating plates. Literature describing research and development and pilot-scale testing of IFAS systems provides examples of all types of media systems. Commercial applications of IFAS systems have gravitated toward the use of a few specific media systems, which will be discussed further in this section.

The basic principal behind the IFAS process is to expand treatment capacity or upgrade the level of treatment by supplementing the biomass in a suspended-growth activated sludge process by growing additional biomass on fixed-film media contained within the mixed liquor. The additional biomass allows a higher effective rate of treatment within the existing process tanks, thus making tank volume available to incorporate denitrification or biological phosphorus removal within the same tank volume. The advantages of using this type of process to expand or upgrade a wastewater treatment plant are briefly summarized below.

- Additional biomass for treatment without increasing solids loading on final clarifiers;
- Higher rate treatment processes possible, thus allowing greater treatment in a smaller space;
- Improved settling characteristics;
- Reduced sludge production;
- Simultaneous nitrification and denitrification;
- Similar operation to conventional activated sludge;
- Minimal additional operating cost; and
- Improved resistance to toxic shock and washout.

These benefits have been reported in the literature and observed at actual full-scale installations.

Descriptions of Integrated Fixed-Film Activated Sludge Processes. ROPE-TYPE MEDIA.
Rope-type media, also referred to as looped-cord or strand media, takes the form of a woven rope with protruding loops that provided a surface for growth of biomass. The media is manufactured of a polyvinyl or polyethylene material. There are several variations of this type of media by different manufacturers (see Figure 3.25). Manufacturers of the media systems include Ringlace Products (Portland, Oregon), Brentwood Industries (Reading, Pennsylvania), Biomatrix Technolo-

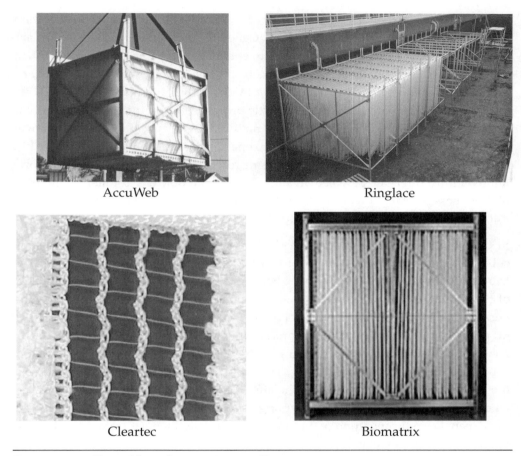

FIGURE 3.25 Examples of rope-type media.

gies (Lincoln, Rhode Island), and EIMCO Cleartec (Salt Lake City, Utah). The Ringlace and Biomatrix products are strung vertically in racks that contain many strands of media. The Brentwood industries product (AccuWeb) is woven into a meshlike pattern that is also hung in a frame. The EIMCO Cleartec media is also produced in sheets, but with a somewhat different design concept than the other rope-type media. The Cleartec system is optimized for nitrification by selecting for a thinner, more uniform biomass.

This media system has been used for carbonaceous BOD removal, nitrification, and, to a limited extent, for denitrification. Generally, the media is installed in the aerobic zone of an activated sludge process to enhance BOD removal and for nitrification. The media can be used in any type of process configuration, including, for example, conventional activated sludge for BOD removal and nitrification, MLE, the anaerobic/oxic and anaerobic/anoxic/oxic process, the Virginia Initiative process, and the Bardenpho process. The media is not used to enhance phosphorus removal, although it can be installed in the aerobic zone of processes that do incorporate biological phosphorus removal. The organisms responsible for biological phosphorus removal must alternate between an aerobic and anaerobic environment. The fixed-film biomass is retained within one zone (aerobic) and thus will not select for biological phosphorus-removing organisms.

SPONGE-TYPE MEDIA. Sponge media is a free-floating media comprised of small cuboids with a specific gravity close to that of water and, with good mixing, are distributed throughout the mixed liquor. Screens are required at the downstream ends of the activated sludge basin to retain the sponge cuboids within the media zone. Because the media will migrate with the flow toward the downstream screen, an air lift pump is required to continuously return media to the upstream end of the media zone. The amount of biomass growing on the sponge media will vary and must be controlled to prevent it from sinking. Therefore, an impingement plate is located at the discharge of the air lift pump. The effect of the media on the plate controls the amount of biomass retained by the sponges. An air knife is provided at the base of the screen. Periodic operation of this air knife will scour biomass from the screen and reduce the possibility of clogging.

Several manufacturers have developed sponge-media systems, including the LinPor by Lotepro (State College, Pennsylvania) and the Captor process (Figure 3.26).

PLASTIC MEDIA. There have been several applications of a fixed-plastic media, such as trickling filter media or caged packing material, in an activated sludge process. The problem generally experienced with this type of media is the potential for clogging within the media as a result of rags and other coarse solids. Although there are some common and successful applications of fixed-plastic media in industrial applications, the trend with municipal applications has been to use some form of free-floating plastic media, also referred to as biomass carriers (Figure 3.27). The plastic media is manufactured from high-density polyethylene (HDPE). There are several manufacturers of this type of media. Although they each have their own spe-

FIGURE 3.26   Examples of sponge-type media.

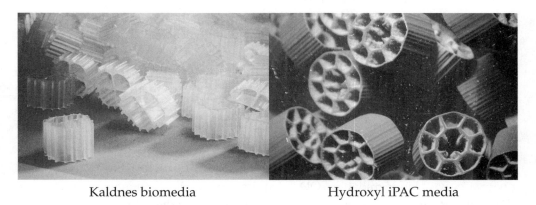

FIGURE 3.27   Examples of plastic media.

cific dimensions, they each loosely resemble a "wagon wheel." The biomass grows on the surface but is abraded from the outside surface of the media, leaving the active biomass on the inside of the wheel. The primary manufacturers of free-floating plastic media system include Kaldnes (Providence, Rhode Island) and Hydroxyl Systems (Victoria, British Columbia, Canada).

With a specific density slightly less than that of water, the media is distributed throughout the mixed liquor with the aid of aeration in the aerobic sections or with submersible-type mixers (slow speed or banana-blade type) in anoxic sections. These media types work with both coarse- and fine-bubble aeration. The HDPE will not, under normal operation, degrade or require regular replacement, especially if media is manufactured with the inclusion of UV inhibitors.

This type of media requires a retaining screen typically 1 mm larger than the media's smallest significant dimension. Screen types can be of the flat-panel type or flanged-cylindrical type. The screen design should incorporate sufficient screen area to minimize headloss. An air knife may be required on some installations to continuously scour the screen.

Upfront fine screening (1 or 2 mm smaller than the retaining screen) is recommended to keep inorganic debris from accumulating in the basin.

ROTATING BIOLOGICAL CONTACTORS. Several RBC plants have been upgraded by the addition of a return sludge flow and diffused aeration, which essentially converts the pure fixed-film RBC process to a hybrid of activated sludge and fixed film. Some plants have used a submerged biological contactor (SBC), which is essentially a RBC with a greater level of submergence. The RBCs typically operate with the media approximately 40% submerged, whereas SBCs operate at 70 to 90% submergence. The greater degree of submergence reduces the load imposed by the media and biomass on the shaft, which allows greater surface area to be installed on an SBC. The SBC can be driven by submerged aeration, which provides process air. If driven mechanically, the SBC can operate anoxically for denitrification.

Operational Issues. *ROPE-TYPE MEDIA.* **Media Location**. GROWTH. An adequate supply of biodegradable COD is required to establish and maintain growth on the media. If the media is located too far downstream in the activated sludge basin, there may not be enough substrate to establish adequate growth for effective treatment. If the media is located too far upstream, the type of growth may be excessively heavy, which promotes anaerobic conditions, and the growth would be dominated by heterotrophic-type organisms.

KINETIC RATE. As with any fixed-film system, the kinetics is dependent on the substrate concentration. If the goal of the media system is to enhance nitrification, then it is important to locate the media in a region of the activated sludge basin where it will be exposed to a concentration of ammonia that maximizes the nitrification rate. This can vary, depending on the media and the wastewater characteristics, but previous research has indicated that the optimum range is approximately 2 to 8 mg/L. If the media is exposed to higher concentrations of ammonia, there is probably also a high concentration of COD, which would tend to encourage the growth of heterotrophic organisms over the autotrophs that are responsible for nitrification.

Because kinetics are also temperature-dependent, the optimum location may also vary seasonally. However, experience has not shown that relocation of the media seasonally is necessary, once the media has been located approximately in the area of the tank where effective treatment is achieved. Adjustment in the mean cell residence time can be made within limits to influence the amount of carbon and ammonia that is directed at the media zone of the process.

Worms. Several rope-type media IFAS systems have experienced a bloom of a type of redworm population that feeds on the biomass on the media, thereby reducing or eliminating treatment. Limited operating experience under these conditions provides some insight to the cause for their bloom and means of treatment; however, the exact conditions that promote their growth are not well-documented or understood. The worms are obligate aerobes. Thus, high levels of dissolved oxygen favor their growth. They may be encouraged to grow when operating conditions change from a higher load and low dissolved oxygen condition to a lower load and higher dissolved oxygen condition. The high-load condition could establish a fairly thick growth of biomass which, once the load was reduced and the dissolved oxygen increased, would provide an ideal environment for the worms.

Control of the worms consists of creating an anoxic condition in the area of the worm bloom by turning the air off for several hours. In addition, the RAS should be chlorinated. This will kill the worms, but the treatment must be repeated to coincide with the egg cycle (approximately every two weeks).

A certain, limited population of redworms will normally exist in a well-functioning IFAS process. They will be visible as small colonies the size of a dime or quarter located randomly on the media strands.

Media Breakage. The breakage of media strands is not generally a problem during normal operation. However, excessively heavy growth or over-aeration can contribute to conditions that would stress the media strands. Media breakage is more likely to occur when the media rack is removed from a tank by overhead crane for inspection or for maintenance of the aeration diffusers located below the media. Prevention requires the use of heavier duty media frames that will not flex to the point that would cause the media to break.

Adequate Dissolved Oxygen Level. Activated sludge processes typically operate with a dissolved oxygen level of 2.0 mg/L or slighty greater. Fixed-film processes benefit from higher levels of dissolved oxygen because all substrates must penetrate into the biofilm, and higher levels of substrate in the liquid provide greater driving force into the biofilm; the greater the penetration of the substrate into the biofilm, the more biomass that is effectively doing the work. However, IFAS systems operate well at the same level of dissolved oxygen, or perhaps slightly higher levels, as pure suspended-growth systems. It is more important that adequate mixing be provided to drive the mixed liquor, which contains the dissolved oxygen, into and throughout the media section.

Mixing. The media racks submerged within a suspended-growth reactor present an increased resistance to normal flow patterns, and the mixed liquor will tend to flow around the racks rather than penetrate completely through the media. The aeration pattern, whether using coarse- or fine-bubble diffusers, should be arranged to establish a crossflow pattern that will circulate the mixed liquor into the media. Deflection baffles can be added to the edge of the aeration basin to force mixed liquor flowing along the sides to be diverted into the media. Also, if the media is located just downstream of a baffle wall, openings in the baffle wall can be arranged to distribute flow across the complete cross-section of the tank.

Access to Diffusers. Access to the grid of aeration diffusers located below a media section is required periodically for maintenance. Access can be gained by either providing a means of relocating the media racks within the tank, such as by sliding them on rails, or by lifting the racks with an overhead crane. Because access to the diffusers is not required very frequently, it may be more cost-effective and simpler to remove the media with a crane. In this case, lifting cables attached to the frames and tethered to the sides of the basin must be provided. Note that the media can deteriorate if subjected to UV radiation, so the media must be protected from exposure to sunlight.

Odor. Odors are not an issue unless a tank is dewatered and a media section with biomass is exposed to the air. Under these conditions, the odors can be quite severe. It is suggested that, before taking the tank offline, waste flow to the tank be stopped and aeration continued for a period of time to "burn off" the active biomass. This will not eliminate any potential source of odors, but it may reduce the potential.

SPONGE MEDIA. **Screen Clogging.** The sponge media will naturally tend to migrate downstream toward the screen. A possible exception to this is where the aeration basin has been divided into media zones that resemble small, complete-mix basins and sufficient mixing energy has been provided to ensure that the sponges are completely mixed throughout the zone, regardless of the flow through the basin.

The operator should observe the in-basin screens on a daily basis. If there is an indication that the liquid level is building up on the upflow side of the screens, the operator should check the air knives to make sure they are operating and injecting sufficient air and the air lift recirculation pumps to ensure they are operating satisfactorily. Also, the forward flow through the basin should be checked to make sure that it is not too high. Flows may be excessive because of high wet weather flows, poor distribution of flows through multiple basins, or excessive nitrate recycle rates.

If these are within the normal range, the water level in the tank should be lowered 0.9 to 1.5 m (3 to 5 ft) for a few hours to check the screens for biofilm buildup. This may occur if the air knife was not operating properly, the roll pattern within the tank was too gentle, or if the sponges were starved of organic carbon for an excessive length of time. The latter, in combination with low mixing, leads to a buildup of long-tailed stalked ciliates, which appear as a "skin" on the sponges. When this skin sloughs off, the material can temporarily plug the screen. The screen will then have to be cleaned with a hose on the downstream surface. The intensity of the roll pattern and the air flow through the air knife would then need to be increased in the future.

Consideration should be given to a bypass as a safety feature in the event of a clogged screen. The bypass could be a side overflow weir into an adjacent tank or a pipe in the basin that allows flow from an upstream zone to a downstream zone, thus bypassing the media zone.

Sinking Sponges. Excessive growth on the sponge media can cause the sponges to sink to the bottom of the tank. This is typically managed by the use of a submersible pump that operates periodically on a timer. The pump is located in the media zone, and, as sponges pass through the volute of the pump, turbulence removes some of the biomass. The operation of the pump should be adjusted to adequately control growth, but not to strip too much biomass from the media. Typically, the cleaning pump is run in one basin at a time, which increases the MLSS of the system by 100 to 300 mg/L. A schedule for operating the cleaning pump should be developed, based on the capacity of the pump and the size of the basin.

Loss of Sponges. The sponge media will continuously abrade against the walls of the aeration basin. Over a period of time, the cuboids become rounded, and surface area is lost. Typically, it is necessary to replace a few percent of the media, at least for the first few years.

Taking Tank Out of Service. If the fill volume of a sponge-media type IFAS system is relatively low, which would be in the range of 20 to 25%, the sponges may be left in the basin as it is dewatered. After dewatering, the sponges could be pushed to a side to gain access to the diffusers. At a 40% fill volume, it is recommended that the media be pumped from the cell to another tank that does not contain media or distribute them to the basins remaining in operation. Distribution of the sponges and return of the media to the dewatered tank is managed by accounting for the sponges on a mass balance basis. If the approximate concentration of the sponge media is known, the number of sponges can be accounted for by metering the volume pumped.

The media should not be exposed to sunlight for extended periods of time, because UV radiation will cause the media to disintegrate.

Loss of Solids. Should there be a washout of MLSS in a storm event, the operator can take advantage of the biofilm in the sponge system to replenish the MLSS. By squeezing biomass out of the media through operation of a submersible pump, the operator may be able to recover the process more quickly.

Air Distribution System. Air is required not only for the process, but also to operate the air knife and air lift pump. If these are fed from a common manifold, a situation could develop where the system becomes unstable hydraulically. If the air supply to the air knife and air lift pumps is taken from a common manifold that supplies air to the diffusers in all the reactors in the system, then even a small increase in head will cause the air to redistribute in the manifold to equalize pressure. Thus, at a time when the systems needs more air, the system will deliver less. It is recommended, therefore, that consideration be given to an independent blower system, preferably a positive displacement blower, to supply air to the air lift pumps and air knives.

PLASTIC MEDIA. **Startup Procedures.** During initial installation, the media has a tendency to float on the water surface until being thoroughly "wetted" out, although greater airflow into the aeration basin will promote media mixing. Airflow rates may be reduced once the biofilm is established. Depending on the wastewater temperature, the media will show signs of performance in two to four weeks from startup.

Foaming can occur during the initial weeks of startup. During this time, an antifoam chemical may be used to mitigate foaming issues, or, if equipped, the plant may use its foam abatement system. Excessive foaming generally ceases once the microbiology is established.

Growth. Media located in the upstream section of the system will typically have a thicker biofilm, with a greater portion of the biofilm attributed to heterotrophs. Media in downstream sections will have a thinner biofilm, with a higher proportion of autotrophs.

Because the process uses free-floating media, the biofilm will adjust itself to the conditions around it. During periods of extended high loads, the biofilm will get thicker, while, during periods of extended low loads, the biofilm will get thinner. Typically, the whole aeration basin will contain the free-floating media. Therefore, there is no need to relocate media to other sections of the basin in response to changing load patterns.

Worms. Worm growth is not a problem with free-floating media because of the constant motion and separation of the media. The organisms are not easily transported between carriers because of their separation in space, and they cannot tolerate the turbulence of the biomass carriers elements.

Media Breakage. The breakage of media is typically not a problem unless the media type is operated in open tanks and does not contain UV inhibitors. In this case, breakage may occur. However, the media pieces will pass through the retaining screen and float to the surface in the downstream clarifier.

Media Mixing. The aeration pattern, whether using coarse- or fine-bubble diffusers, should be arranged to establish single or multiple roll patterns to ensure that the media is fully mixed in the basin and that no dead zones are apparent.

Accumulation of Growth. The operator should monitor the accumulation of growth inside the hollow plastic media. The determination of biomass densities (grams of biomass per square meter of protected surface area) should periodically be performed. If the media appears to be plugged, the organic loading rate may be excessive. To rectify this, the organic loading may have to be reduced, or additional media may have to be installed. Also, the roll pattern could be increased or media with a larger diameter installed in the section of the tank that is subjected to the highest organic load. The hollow cylinder type of design allows for a media fill volume of up to 70%.

SCREEN CLOGGING. Retention screens are required on the aeration basin effluent port to prevent media loss or migration from the aeration basin. Screens should be equipped with air knives to encourage media scouring and to avoid excessive headloss resulting from accumulations on the screen surface. Preventative maintenance schedules should include screen inspections and cleaning.

FOAMING. As indicated previously, foaming may occur during system startup. However, foaming may occur as a result of operational issues, especially when excessive airflow rates are used. Dissolved oxygen concentrations should be monitored, and airflow rates can be decreased if they are greater than 3.0 mg/L. Because aeration also serves to keep the media mixed, operators must remain mindful that substantial reductions in airflow may affect media movement.

Suspended-growth parameters, such as MLSS concentration, F/M, and low influent flowrates, can also contribute to foaming issues. These parameters should be closely monitored to ensure an optimized treatment scheme.

TAKING TANK OUT OF SERVICE. If the fill volume of the plastic-media-type IFAS system is relatively low (20 to 25%), then the media may be left in the basin as it is dewatered. After dewatering, the media could be pushed to the side to gain access to the diffusers. At fill volumes greater than 25%, it is recommended that the media be pumped from the media cell to another tank that does not contain media or distribute them to the basins remaining in operation.

Distribution of the media to other tanks during servicing and return of the media to the dewatered tank is managed by using a secchi disk. The disk can be sunk into the basin during aeration. Aeration is then stopped, and the media depth is determined by lifting the secchi disk. Levels are determined when the disk meets resistance from the media. Following servicing, media depths within the basins can be returned to initial levels, based on the level readings.

MEMBRANE BIOREACTOR. Membrane bioreactors use membrane-type filtration units, instead of clarifiers, that are placed either directly in the activated sludge basin or are located outside of the basin. Because a secondary clarifier is not required for liquid solids separation, and thus the clarifier solids loading is not an issue, this process is able to operate at very high MLSS levels (8 to 18%), although typically it is operated closer to 10 000 mg/L. There are a number of manufacturers of this type of process, including ZENON Environmental, Inc. (Oakville, Ontario, Canada), USF/Memcor (Sydney, Australia), Enviroquip (Austin, Texas), Kubota (Osaka, Japan), and Mitsubishi (Orange County, California). The membranes vary from hollow tube fibers to flat panels. The MBR systems are designed either as outside/in systems that are vacuum-driven or as inside/out systems that are pressure-driven. The vacuum systems pull the treated effluent through the membrane, whereas the pressure-driven systems pump the mixed liquor through the membrane.

The most significant operational issue with MBRs is the potential for membrane biofouling. The higher the MLSS level and the higher the flux rate (which is the flowrate through the membrane per unit of surface area), the greater the potential for fouling. Also, as the potential for fouling increases, the frequency of cleanings required increases. The need to clean a membrane is determined by the pressure differential across the membrane.

The initial buildup of solids on the membrane surface improves performance. However, as the solids continue to accumulate and the flowrate is increased to compensate for increased pressure loss across the membrane, the operation can become unstable. The result of attempting to increase the flowrate further compresses the solids on the surface, thus again increasing the pressure drop.

There are several operational changes that can be implemented in an effort to minimize biofouling. Fouling may be reduced by increasing the turbulence induced by the coarse-bubble aeration system beneath the membranes, which reduces the thickness of the biofilm.

There is evidence that the primary cause of biofouling is a result of the extracellular polymeric substances (EPS) that form on the surface of the bacteria that comprise the biofilm. The EPS tends to clog the smaller pores in the cake layer surrounding the membrane. Reducing the amount of EPS reduces filtration resistance. The amount of EPS is influenced by the SRT, with the degree of EPS formation increasing at very low and high SRTs.

Improvements can also be realized by adding flocculant, which creates larger particles in the cake layer, thereby providing larger pores in the biofilm and reducing the filtration resistance.

There are several cleaning cycles required. Typically, there is a frequent but short duration backwash using permeate or filtered wastewater to flush solids from the membrane surface. Periodically, a more extensive backwash is performed with chemical cleaning agents. The more extensive cleaning operation, where the membrane is removed from service and washed in a chemical bath, is much less frequent and typically required only once every 6 to 12 months, depending on the waste and operating conditions. Most of the backwash and cleaning operations can be automated.

Membrane integrity can be monitored with online turbidity meters. If a membrane were to be ruptured, the turbidity will suddenly increase.

With MBR systems configured in a BNR process, care should be taken in handling the RAS flows. This is because the MLSS is typically highly aerated, and the RAS can return high levels of dissolved oxygen to the anoxic zone.

Foaming is frequently experienced in the bioreactor; however, because clarification is performed by a membrane, the foam or high sludge volume index (SVI) values are of little consequence.

SECONDARY CLARIFICATION

Based on empirical measurements by McCarty (1970), effluent TSS from biological treatment units contain approximately 12% nitrogen and 2.2% phosphorus. This translates into 1.2 mg/L TN for every 10 mg/L effluent TSS. With the exception of some of the biological filter processes and membrane bioreactors, the biological nitrogen removal processes presented herein depend on the efficient capture of biological solids by secondary clarifiers. It is critical that secondary clarifiers are operated to achieve a consistently high level of treatment to both protect the water environment and avoid the high cost of downstream tertiary treatment.

Although many of the basic principles of suspended- and fixed-growth secondary clarification are the same, the units have some key differences related to the upstream biological processes.

SUSPENDED GROWTH. Suspended-growth (activated sludge) reactors and clarifiers must be designed and operated as a coordinated system to provide a consistently high level of performance. With the incorporation of nutrient removal to many existing treatment facilities, many secondary clarifiers are being challenged well beyond their original design capacity. Even with the construction of new and expanded processes, many facilities are forced to operate with longer SRTs and resulting high MLSS concentrations. It is crucial that the operations staff responsible for process control decisions enhance their understanding of clarifier capacity and operating procedures.

The Water Environment Research Foundation (Alexandria, Virginia) Clarifier Research Technical Committee activated sludge secondary clarifier evaluation protocol defines a step-by-step procedure for evaluating secondary clarifier capacity and troubleshooting poor clarifier performance. Poor activated sludge secondary clarifier performance has been associated with one or more of the following conditions (WERF, 2001):

- Sludge blanket denitrification,
- Sludge particle flocculation problems,
- High sludge blankets caused by solids overload, and
- Poor clarifier hydraulics.

Of the potential problem conditions noted, operators have potential operational response measures to at least three of the conditions identified. The actual measures undertaken depend on the site-specific conditions that exist at the time of the problem. A few examples and possible solutions are as follows (WERF, 2001):

Denitrification.
Cause: Elevated nitrate-nitrogen concentration in reactor effluent.
Possible solutions: Lower SRT to eliminate nitrification (generally not an option).
Reduce clarifier sludge blanket.
Increase speed of collector mechanism.
Increase RAS flow.
Decrease number of clarifiers in service.
Increase dissolved oxygen concentration in clarifier influent.
Modify process to denitrify in upstream reactors.

Flocculation Problem 1.
Cause: Floc breakup resulting from a mechanical issue.
Possible solutions: Design or construct flocculation zone in clarifier.
Add chemical flocculant.

Flocculation Problem 2.
Cause: Poor floc formation resulting from a biology problem.
Possible solutions: Alter mode of operation, check for influent toxicant.
Add chemical flocculant.

High Sludge Blanket.
Cause: Solids loading to clarifier exceeds capacity.
Possible solutions: Perform state-point analysis.
Alter RAS flow.
Decrease solids inventory.

Alter mode of operation to reduce SVI.
Design or construct changes to improve SVI or increase solids loading capacity.

Hydraulic Problem.
Cause: Physical configuration of clarifier.
Possible solutions: Perform clarifier dye testing, hydrodynamic modeling.
Design or construct clarifier improvements.

ATTACHED GROWTH. Secondary clarifiers for attached growth systems have not received the same level of interest or research as those for activated sludge systems. Regardless, these units are no less critical to the performance of their associated upstream biological reactors. In fact, two of the conditions identified as challenges to suspended-growth system clarifier performance, flocculation, and hydraulic problems, have been identified as the primary cause of problems with attached growth system clarifier performance (WEF, 2000). Investigational techniques and potential solutions to attached-growth secondary clarifier performance problems are similar to the items identified for suspended-growth systems.

REFERENCES

Barnard, J. L. (1974) Cut N and P without Chemicals. *Water Wastes Eng.*, **11**, 41–44.

Blum, D. J.; Speece, R. E. (1991) A Database of Chemical Toxicity to Environmental Bacteria and Its Use in Interspecies Comparison and Correlations. *Water Environ. Res.*, **63**, 198.

Burdick, C. R.; Refling, D. R.; Stensel, H. D. (1982) Advanced Biological Treatment to Achieve Nutrient Removal. *J. Water Pollut. Control Fed.*, **54**, 1078–1086.

Christensen, M. H.; Harremoës, P. (1977) Biological Denitrification of Sewage: A Literature Review. *Prog. Water Technol.*, **8**, 509.

Ekama, G. A.; Marais, G. v. R. (1984) *The Influence of Wastewater Characteristics on Process Design.* Chapter Three-Theory, Design and Operation of Nutrient Removal Activated Sludge Processes. Water Research Commission: Pretoria, South Africa.

Grady, C. P. L.; Lim, H. C. (1980) *Biological Wastewater Treatment: Theory and Applications.* Marcel Dekker: New York.

Hockenbury, M. R.; Grady, C. P. L. (1977) Inhibition of Nitrification—Effects of Selected Organic Compounds. *J. Water Pollut. Control Fed.*, **49**, 768.

Jenkins, D.; Hermanowicz, S. W. (1989) Principles and Practice of Phosphorus and Nitrogen Removal from Municipal Wastewater. In *Principles of Chemical Phosphorus Removal*, Sedlak, R. I. (Ed.); Soap and Detergent Association: New York.

Ludzack, F. T.; Ettinger, M. B. (1962) Controlling Operation to Minimize Activated Sludge Effluent Nitrogen. *J. Water Pollut. Control Fed.*, **34**, 9.

McCarty, P. L. (1970) Phosphorus and Nitrogen Removal in Biological Systems. *Proceedings of the Wastewater Reclamation and Reuse Workshop*, Lake Tahoe, California, 226.

McCarty, P. L.; Beck, L.; St. Amant, P. (1969) Biological Denitrification of Wastewaters by Addition of Organic Materials. *Proceedings of the 24th Purdue Industrial Waste Conference*; Purdue University: Lafayette, Indiana.

Metcalf and Eddy (2003) *Wastewater Engineering: Treatment and Reuse*, 4th ed.; G. Tchobanoglous, F. L. Burton, H. D. Stensel (Eds.); McGraw-Hill: New York.

Okey, R. W.; Albertson, O. E. (1989) Diffusions Role in Regulating Rate and Masking Temperature Effects in Fixed Film Nitrification. *J. Water Pollut. Control Fed.*, **61**, 510.

Painter, H. A. (1970) A Review of Literature on Inorganic Nitrogen Metabolism in Microorganisms. *Water Res. (G.B.)*, **4**, 393.

Payne, W. J. (1973) Reduction of Nitrogenous Oxides by Microorganisms. *Bacteriol. Rev.*, **37**, 409.

Randall, C. W.; Barnard, J. L.; Stensel, H. D. (1992) *Design and Retrofit of Wastewater Treatment Plants for Biological Nutrient Removal*. Technomic Publishing: Lancaster, Pennsylvania.

Richardson, M. (1985) *Nitrification Inhibition in the Treatment of Sewage*. The Royal Society of Chemistry. Burlington House: Thames Water, Reading, Pennsylvania.

Sharma, B.; Ahlert, R. C. (1977) Nitrification and Nitrogen Removal. *Water Res. (G.B.)*, **11**, 897.

U.S. Environmental Protection Agency (1975) *Process Design Manual for Nitrogen Control*, EPA-625/1-75-007; National Environmental Research Center, U.S. Environmental Protection Agency: Washington, D.C.

U.S. Environmental Protection Agency (1993a) *Nitrogen Control Manual*; EPA-625/R-93-010; Office of Research and Development; U.S. Environmental Protection Agency: Washington, D.C.

U.S. Environmental Protection Agency (1993b) *Process Design Manual for Nitrogen Control*; EPA-625/R-93-010; U.S. Environmental Protection Agency: Washington, D.C.

Water Environment Federation (2000) *Aerobic Fixed-Growth Reactors*. Special Publication; Water Environment Federation: Alexandria Virginia.

Water Environment Federation (1998) *Biological and Chemical Systems for Nutrient Removal*. Special Publication; Water Environment Federation: Alexandria Virginia.

Water Environment Federation (1998) *Design of Municipal Wastewater Treatment Plants*, 5th ed.; Manual of Practice No. 8; Water Environment Federation: Alexandria, Virginia.

Water Environment Research Foundation (2001) *WERF/Clarifier Research Technical Committee (CRTC) Protocols for Evaluating Secondary Clarifier Performance*. Water Environment Research Foundation: Alexandria, Virginia.

Water Environment Research Foundation (2003) *Methods for Wastewater Characterization in Activated Sludge Modeling*. Water Environment Research Foundation: Alexandria, Virginia.

Chapter 4

Enhanced Biological Phosphorus Removal

Introduction and Basic Theory of Enhanced Biological Phosphorus Removal	106	Anaerobic/Oxic and Anaerobic/Anoxic/Oxic Configurations	130
Basic Enhanced Biological Phosphorus Removal Theory	107	Modified Bardenpho Configuration	132
Basic Enhanced Biological Phosphorus Removal Design Principles	108	University of Cape Town, Modified University of Cape Town, and Virginia Initiative Process Configurations	133
Operational Parameters	113	Johannesburg Configuration	135
Influent Composition and Chemical-Oxygen-Demand-to-Phosphorus Ratio	113	Oxidation Ditches, BioDenipho, and VT2 Configurations	136
Solids Retention Time and Hydraulic Retention Time	117	Hybrid Systems	138
Temperature	122	*PhoStrip Configuration*	138
Recycle Flows	127	*Biological Chemical Phosphorus and Nitrogen Removal Configuration*	139
Internal Recycles	127	Process Control Methodologies	140
Plant Recycles	129	Influent Carbon Augmentation	140
Types of Enhanced Biological Phosphorus Removal Systems	129	*Volatile Fatty Acid Addition*	140
Suspended-Growth Systems	130	*Pre-Fermentation*	140

Solids Separation and Sludge Processing	143	Unique New Designs	149
Chemical Polishing and Effluent Filtration	144	McAlpine Creek Wastewater Management Facility of Charlotte, North Carolina	150
Case Studies	146	Traverse City Regional Wastewater Treatment Plant, Traverse City, Michigan	150
The Lethbridge Wastewater Treatment Plant, Alberta, Canada	146	References	153
Durham, Tigard, Oregon, Clean Water Services	148		

INTRODUCTION AND BASIC THEORY OF ENHANCED BIOLOGICAL PHOSPHORUS REMOVAL

Enhanced biological phosphorus removal (EBPR) relies on the selection and proliferation of a microbial population capable of storing orthophosphate in excess of their biological growth requirements. This group of organisms, referred to as the phosphate-accumulating organisms (PAOs) rely on operational conditions that impose a selective advantage for them, while putting the other groups in a temporary disadvantage with respect to access to food (i.e., substrate). Once this is achieved by implementation of special design and operational conditions, PAOs gain the selective advantage to grow and function, and the result is the excessive accumulation of orthophosphate in mixed liquor. With proper mixed liquor wasting, the phosphate removal can then be achieved.

Enhanced biological phosphorus removal research has clearly shown that the PAOs are a subset of the microbial population that the wastewater engineers and plant operators have been familiar with-the heterotrophs. Heterotrophic organisms, which make up the greater majority of the activated sludge biomass found at most secondary treatment plants, rely on organic carbon sources for growth and require oxygen for energy generation. Unlike the autotrophic organisms (e.g., the nitrifiers), which use carbon dioxide (CO_2) in its soluble form $HCO_3^=$ (i.e., bicarbonate alkalinity) as their carbon source for growth and have a strict need for oxygen, heterotrophs can grow under the absence of oxygen when other electron acceptors, such as nitrate (NO_3^-), are present. Hence, PAOs can grow under both anoxic and aerobic conditions, as long as the one and only requirement for their proliferation is met: presence of an anaerobic zone at the head of the secondary treatment units, where no electron acceptor is present, and the secondary influent is introduced.

In this chapter, the EBPR theory will be presented with the engineering aspects for reliable design and operation of biological treatment systems for phosphorus removal. Because the development of the EBPR systems has relied on laboratory and full-scale research studies, such findings will be presented. However, examples of full-scale EBPR plant design and operational information will also be given to guide the reader in selecting the more suitable configuration and design for their case at hand.

BASIC ENHANCED BIOLOGICAL PHOSPHORUS REMOVAL THEORY

Biological phosphorus removal is a biological process in which alternation of anaerobic and aerobic stages favors biophosphorus (bio-P) or PAOs, which are the heterotrophic organisms that are responsible for biological phosphorus removal. In the anaerobic stage, bio-P organism do not grow, but convert readily available organics material (i.e., acetate and propionate) to energy-rich carbon polymers called poly--hydroxyalkanoates (PHAs). Biophosphorus organisms use energy during acetate uptake and its conversion to PHA. This energy is generated through breakdown of polyphosphate (poly-P) molecules, which results in an increase in phosphate concentration in the anaerobic stage (i.e., phosphorus release). Polyphosphate is made up of many phosphate molecules combined together. Magnesium and potassium ions are concurrently released to the anaerobic medium with phosphate. In addition, for bio-P organisms to produce PHA, a substantial amount of reducing power is required. The reducing power is generated from the breakdown of glycogen, another form of internal carbon storage (Erdal et al., 2004; Filipe et al., 2001; Mino et al., 1987). In the aerobic zone, PAOs can oxidize previously stored PHAs to obtain energy. The energy and the carbons are used for growth and maintenance requirements. Under aerobic conditions, energy reserves are restored through phosphate uptake and polymerization. The effluent from the EBPR reactors is now low in phosphorus, and all the phosphorus stored in the biomass can now be wasted through regular solids wastage. This results in a net phosphorus removal from the system and the wastewater. In addition, some of the energy and carbon is used to restore the glycogen stores for the reactions to continue when mixed liquor is recirculated to the head of the anaerobic zones. The events that take place in the anaerobic and aerobic stages are summarized in Table 4.1.

BASIC ENHANCED BIOLOGICAL PHOSPHORUS REMOVAL DESIGN PRINCIPLES

Successful EBPR system operation depends on the presence of the following:

(1) Sufficient readily available organic carbon and phosphorus in the secondary influent;
(2) A correctly sized anaerobic "selector" zone preceding the latter zones, where sufficient electron acceptors, such as oxygen or nitrate, will be maintained for growth; and
(3) Sufficient amount of cations, such as magnesium and potassium, to facilitate release and uptake of phosphate.

These three requirements will ensure that PAOs are selected in the anaerobic zones by their ability to store the readily available substrate, such as the volatile fatty acids (VFAs) in the form of PHA granules. To accomplish this, PAOs need energy,

TABLE 4.1 Major events observed in anaerobic and aerobic zones of a EBPR plant.

Process or compound	Anaerobic zone	Aerobic zone
Readily available substrate (i.e., acetate)	Used	Used*
Phosphate	Released	Taken up
Magnesium and potassium	Released	Taken up
PHA	Stored	Oxidized
Glycogen	Used	Restored

*Note: Readily available substrate is taken up before the aerobic stage in a properly designed EBPR plant.

which is readily obtained from the breakdown of the high energy phosphate bonds of the poly-P granules. Once all the readily available substrate is exhausted in the anaerobic zone, the aerobic reactions; which include uptake of released phosphate, consumption of PHAs, and new cell growth; can proceed. For a balanced growth environment, biomass wastage (Q_w) from the system must take place in accordance with the design solids retention time (SRT).

A main indicator of the presence of a well-established PAO population is the typical release and uptake patterns of the EBPR. Figure 4.1 illustrates these patterns, which are observed in individual sections of the generic EBPR system shown in the figure. The significance of each constituent and the deviations from these typical patterns will be presented in the following sections.

As can be deduced from Figure 4.1, although the apparent release and uptake response that is observed in successfully designed and operated EBPR systems is quite simple, the microbiological reactions and growth balances that take place in different zones of the systems depend on complicated interactions.

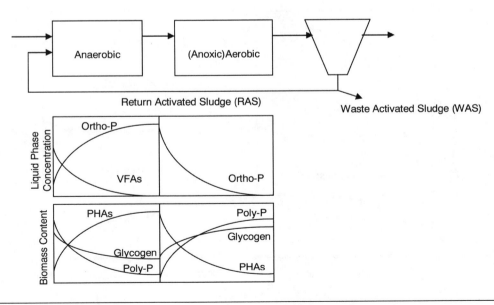

FIGURE 4.1 Typical concentration patterns observed in a generic EBPR system.

The simplest reaction (simple with respect to the involved biochemistry) is the poly-P breakdown in the anaerobic zone and reassembly in the (anoxic) aerobic zones. The poly-P breakdown involves enzymatic hydrolysis of the high energy poly-P bonds with the help of appropriate polymeric breakdown enzymes (such as poly-P kinase [ppK]), and transfer to the outside of PAO cells. This transfer across the cell membrane is where the cations, such as magnesium (Mg^{+2}) and potassium (K^{+1}), become crucial. Because each phosphate molecule (PO_4^{-3}) contains three negative charges, it is not possible for the charged molecules to pass through the cell membrane. However, when the phosphate molecules bond with the positively charged magnesium and potassium, they become neutralized and then can be transferred across the cell membrane. The requirements for each cation are such that commonly encountered concentrations in domestic wastewater are generally sufficient for EBPR requirements. For wastewaters with significant industrial contribution, the design engineer or the plant operator must be cautious to ensure adequate cation feed to

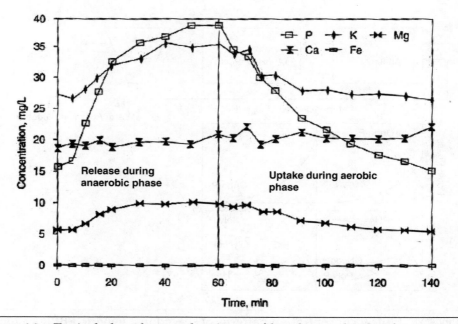

FIGURE 4.2 Typical phosphate and cation profiles observed in batch experiments conducted on EBPR biomass (WEF, 1998).

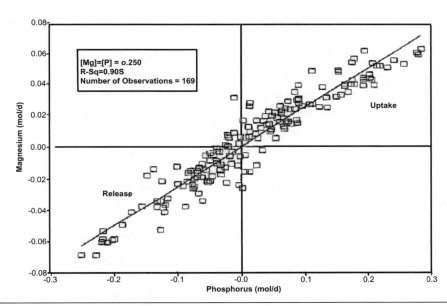

FIGURE 4.3 Typical release and uptake of magnesium and phosphate observed in an EBPR system (WEF, 1998).

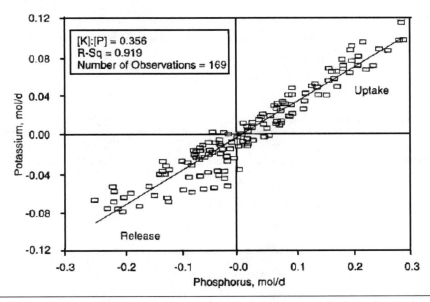

FIGURE 4.4 Typical release and uptake of potassium and phosphate observed in an EBPR system (WEF, 1998).

EBPR systems. Figures 4.2 through 4.4 illustrate the involvement of the two cations in EBPR systems.

The ability of the PAOs to accomplish anaerobic VFA uptake and to store PHA polymers is the main mechanism through which they gain selective advantage in EBPR systems. As the municipal wastewater travels through the collection system, the complex organic material undergoes fermentation reactions that break down the larger organic molecules into smaller ones. The smallest molecules the organic material can be broken into are the VFAs. The VFAs commonly found in fermented wastewater are listed in Table 4.2.

The PHAs formed from the VFAs present in the secondary influent are stored in granules that can be used as a carbon source for growth and maintenance energy generation later, when electron acceptors, such as oxygen or nitrate, are present. The most common types of PHAs are poly-β-hydroxybutyrate (PHB) and poly-β-hydroxyvalerate (PHV). When the PAOs are storing VFAs as PHA granules, they use the energy generated from poly-P breakdown, and thus they are also storing energy besides carbon for growth. When PHAs are broken down, acetate and propionate are the most commonly released short-chain fatty acids, and they can conveniently be used in growth and energy-generating metabolic reactions. Because excess energy is put into the storage of PHAs, excess energy is released when short-chain fatty acids are released from PHAs. This excess energy is stored for later use in high-energy phosphate bonds of poly-P, which is shown to increase in EBPR biomass as PHA content decreases, as shown in Figure 4.1.

TABLE 4.2 Volatile fatty acids typically found in fermented wastewater.

Volatile fatty acid	Chemical formula	Phosphorus uptake/ VFA COD consumed
Acetic acid	$CH_3\text{-}COOH$	0.37
Propionic acid	$CH_3\text{-}CH_2\text{-}COOH$	0.10
Butyric acid	$CH_3\text{-}CH_2\text{-}CH_2\text{-}COOH$	0.12
Isobutyric acid	$CH_3\text{-}CH_2\text{-}COOH\text{-}CH_3$	0.14
Valeric acid	$CH_3\text{-}CH_2\text{-}CH_2\text{-}CH_2\text{-}COOH$	0.15
Isovaleric acid	$CH_3\text{-}CH_2\text{-}COOH\text{-}CH_2\text{-}CH_3$	0.24

The other internal storage product introduced in Figure 4.1, glycogen is one of the most common intracellular carbon storage products among all the living organisms. Under conditions of environmental stress, such as feast or famine, where there is either an excess or lack of substrate, organisms choose to store glycogen. It is merely a polymer of glucose molecules branched together, and it can be used for a number of different purposes when needed, such as cellular osmotic pressure adjustment, temperature-resistant chemicals production, or consumption as substrate in starvation times. In the case of EBPR microbiology, glycogen involvement has been observed since the early 1980s and has been investigated, to some extent, to understand whether it is a mere stress reaction of the EBPR biomass or it actually is a requisite part of the EBPR biochemistry. Recent research showed that, although there are nonPAOs that can also store glycogen without phosphate storage, glycogen metabolism is actually a required part of EBPR (U. G. Erdal, 2002; Z. K. Erdal, 2002; Erdal et al., 2003a).

Bringing all these reactions together are the delicately intertwined biochemical reactions that have been investigated by the researchers. In this chapter, before the design and operational issues are presented, these biochemical interactions will be presented shortly to provide a basis for the readers' understanding of the reasons of potential problems that can be encountered and solutions that have been developed for EBPR systems.

OPERATIONAL PARAMETERS

The forthcoming discussion of the EBPR biochemistry has shown that, for the EBPR systems to function properly, microbial reactions must proceed in favor of the PAO population. In this section, a number of the operational and design parameters that effect PAO enrichment of EBPR sludge will be presented.

INFLUENT COMPOSITION AND CHEMICAL-OXYGEN-DEMAND-TO-PHOSPHORUS RATIO. The importance of the VFAs in EBPR metabolism was presented in the Basic Enhanced Biological Phosphorus Removal Design Principles section. Municipal wastewater fermented in the collection systems is generally a good source of VFAs for EBPR operation. Advanced primary treatment practiced at some treatment plants must be looked at carefully, because at plants where EBPR is going to be implemented, adequate quantities of organic material must be supplied to support PAO functions. In some cases where sufficient carbon substrate is not available, carbon augmentation of secondary influent is practiced. Carbon augmentation can be either by bringing in industrial wastewater that contains VFAs or by on-site sludge fermentation to generate VFAs to be fed into the EBPR system, as is presented later in this chapter and in Chapter 10. Plant recycles also have significant bearing in EBPR systems operation, because they can also contain some VFAs that can be sufficient to ameliorate the phosphorus functions.

The influent-chemical-oxygen-demand- (COD) (or biochemical oxygen demand [BOD]) to-total-phosphorus ratio (influent COD:P) is crucial for proper design of the phosphorus removal systems. Whether a system is limited by COD (or BOD) or phosphorus determines the extent to which PAOs can function and excess phosphorus can be taken up from the solution. Earlier research has shown that the influent COD:P ratio correlated very well with the EBPR biomass total phosphorus (TP) content and phosphorus-removal functions (Kisoglu et al., 2000; Liu et al., 1997; Punrattanasin, 1999; Schuler and Jenkins, 2003). Figures 4.5 through 4.8 illustrate the effect of different feed COD:P ratios, as shown in these pilot-scale studies.

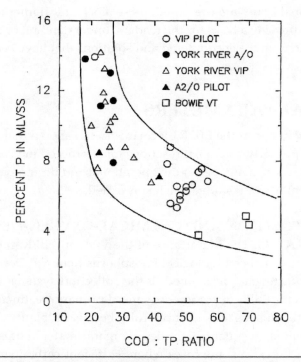

FIGURE 4.5 Observed COD:TP ratio effect on mixed liquor PAO enrichment and phosphorus storage at different EBPR plants (WEF, 1998).

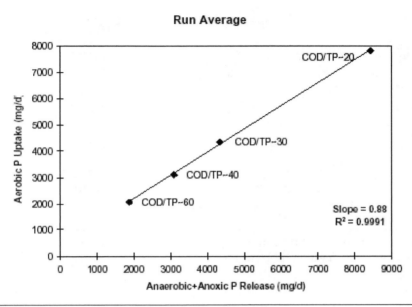

FIGURE 4.6 Effect of influent COD:TP ratio on phosphorus release and uptake (Punrattanasin, 1999).

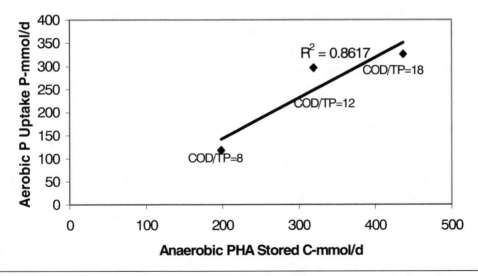

FIGURE 4.7 Effect of influent COD:TP ratio on PHA storage and phosphorus uptake (Kisoglu et al., 2000).

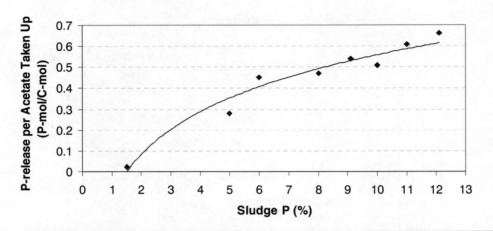

FIGURE 4.8 Data collected by Liu et al. (1997) using pilot-scale EBPR plants fed with VFA at various feed COD:P ratios.

Because EBPR systems rely on enrichment of PAOs, to be able to select the phosphorus accumulators, the system must be fed with a minimum amount of phosphorus that can be taken up in correlation with the amount of substrate COD available in the feed. Through laboratory- and full-scale studies, it was shown that a minimum of 20 to 1 five-day biochemical oxygen demand (BOD_5) to TP ratio is required to enrich a population of PAOs that can achieve secondary effluent soluble phosphorus concentrations of 1 mg/L. Because total COD (TCOD) is a better approximation of the actual ultimate BOD, use of secondary influent COD measurements to estimate the feed organic content is also widely practiced. In those cases, it was seen that a secondary influent TCOD:TP ratio of 45:1 can conservatively ensure secondary effluent soluble phosphorus concentrations of 1 mg/L. However, as indicated earlier, any operational condition that results in reduction of the TCOD before the secondary treatment system must be closely monitored to allow appropriate TCOD:TP ratio in the EBPR systems feed.

Seasonal variations in BOD:COD ratios and wastewater VFA content must be closely investigated as part of the wastewater characterization that must precede EBPR design.

SOLIDS RETENTION TIME AND HYDRAULIC RETENTION TIME.

Solids retention time and hydraulic retention time (HRT) are two very important operational parameters in a biological system, because they define separately the time a solid particle and a water drop spend in the biological reactors. Depending on individual biochemical reactions, the solids retention time (i.e., sludge age) in the system dictates the rate at which the biological system operates. The HRT, on the other hand, determines the contact time between the solution phase that contains the substrates and nutrients and the biomass, which is made up of microorganisms scavenging food material for growth.

In an EBPR system, because the emphasis is on the anaerobic storage of the readily available substrate and subsequent consumption of the stored carbon, anaerobic and aerobic SRT and HRT gain importance to allow just enough time for the EBPR reactions to take place. The substrate storage metabolism of the PAOs give them advantage over other heterotrophs, and, although the performance of the other heterotrophic reactions rely more on the system sludge age, in EBPR systems, performance cannot be defined solely based on SRT and HRT. The importance of the feed COD:P ratio shapes the microbial composition of the EBPR sludge and the effluent phosphorus levels that can be attained.

Figure 4.9 illustrates the combined effect of the observed yield, as dictated by the system SRT and the feed COD:P on the phosphorus content of the biomass. As the observed yield increases for a constant COD:P value, biomass phosphorus content decreases. In other words, at lower SRT values, less phosphorus is stored in the biomass. Conversely, as more SRT is allowed, the sludge will be more robustly enriched by PAOs, as reflected in higher biomass phosphorus values. If, at the same SRT and observed yield value, the feed COD:P ratio is lowered (i.e., increased phosphate or decreased COD in the feed), the biomass phosphorus content will be observed to increase. As feed phosphate increases, more of it can be stored in the biomass, given that the feed COD is sufficient, in result improving the biomass PAO enrichment. Again, because of the strong effect of the feed characteristics and COD:P ratio on the biomass enrichment and microbial composition, results from two different plants cannot be reliably compared in terms of EBPR performance.

It was shown, through years of field and laboratory experience, that EBPR systems can operate at SRT values greater than three days. At SRT values between 3 and 4 days, the effluent quality becomes poor, and chemical polishing can be needed. At SRT values greater than 4 days and at temperatures greater than 15°C, nitrification will get into the picture, and process configurations that include anoxic zones for

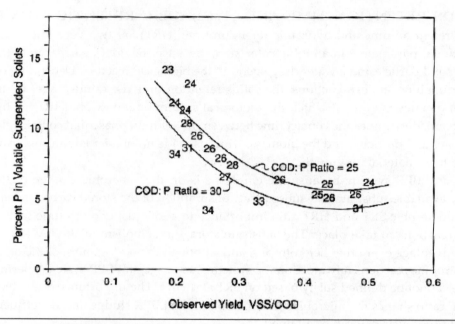

FIGURE 4.9   Effect of EBPR biomass observed yield and SRT on mixed liquor volatile suspended solids phosphorus content (WEF, 1998).

denitrification of nitrate in system recycles must be used. As the SRT is increased to a level where endogenous reactions become significant (i.e., increased biomass decay), secondary release of phosphorus may lead to decreased performance at given feed VFA and COD values, and phosphorus removal at these higher SRT values can be increased if the feed COD can be increased. Thus, seasonal variations in feed COD and phosphorus can be handled by varying the operational conditions, and the operators can determine the best operational strategy for their influent quality and wastewater treatment plant setup.

To explain the SRT and EBPR performance, a number of researchers conducted experiments under various operational conditions. McClintock et al. (1991) showed that, at a temperature of 10°C and an SRT of 5 days, EBPR function of a given activated sludge system would "washout" before other heterotrophic functions do. The washout SRT is a design parameter that defines a critical SRT point, below which no growth of biomass occurs (Grady et al., 1999). Mamais and Jenkins (1992) showed that there is a washout SRT for all temperatures over the range 10 to 30°C. It clearly indicates that, if the SRT-temperature combination is below a critical value, EBPR ceases before other heterotrophic functions.

The mechanisms leading to washout or cessation of EBPR activity before other heterotrophic functions halt was investigated by Erdal et al. (2003b and 2004), and, after examining the underlying biochemical methods, they showed that the main effect of system SRT in EBPR systems is on the PHA and glycogen polymerization reactions, descriptions of which were given earlier. The growth of ordinary heterotrophs growing on soluble substrate, independent of glycogen metabolism, were not affected in the same manner, and their functions lasted longer, down to shorter SRTs; whereas the growth of EBPR bacteria strictly depends on concurrent operation of internal storage consumption of PHA, poly-P, and glycogen.

Hence, in setting the system and individual zone SRT and HRT values, the design engineer must consider the VFA uptake, PHA polymerization and use, and glycogen breakdown and polymerization rates in corresponding zones of the system. At higher system SRT values, because the percentage of PAOs in the biomass increases, complete uptake of VFAs and their conversion to PHAs can be achieved under a given system setup. Consideration of all these intricate and highly interrelated reactions can be achieved through use of widely accepted EBPR modeling tools, such as International Water Association (London) ASM nos. 2 and 3 (IWA, 2002). The details of these and other models developed for better understanding and design of the BNR systems will be presented in Chapter 7 of this manual.

Phosphorus release and uptake rates in anaerobic and aerobic zones, respectively, must also be considered in selecting the overall system and individual zone HRT values, which can be adjusted by appropriate volume and flow adjustments. Figure 4.10 illustrates the effect of anaerobic HRT on system performance. It is clear that there is no benefit in maintaining anaerobic residence time beyond the completion of poly-P hydrolysis and subsequent phosphorus release. Figure 4.11 (Z. K. Erdal, 2002), illustrates the changes in internal storage products in time, as observed in batch tests conducted on EBPR biomass. The importance of the selection of adequate anaerobic contact time can be seen from the PHA data. The PHA polymerization continues even after the bulk solution VFAs are exhausted within the first hour of the anaerobic period. If sufficient time is not allowed, an adequate amount of PHA will not be available to support the desired phosphorus uptake in aerobic zones. The effect of excessive aeration was found to reduce EBPR efficiency as a result of

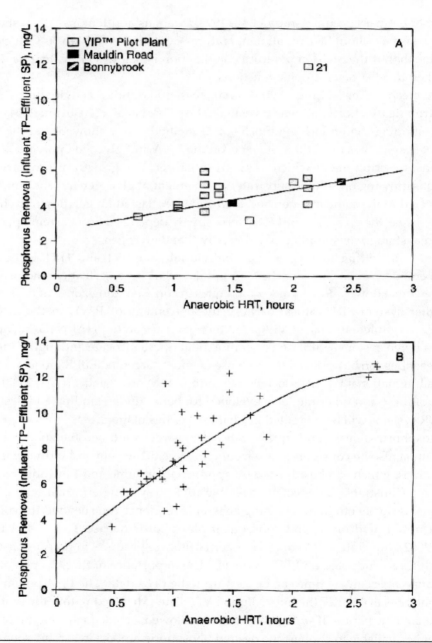

FIGURE 4.10 Effect of anaerobic HRT on system phosphorus removal performance (WEF, 1998).

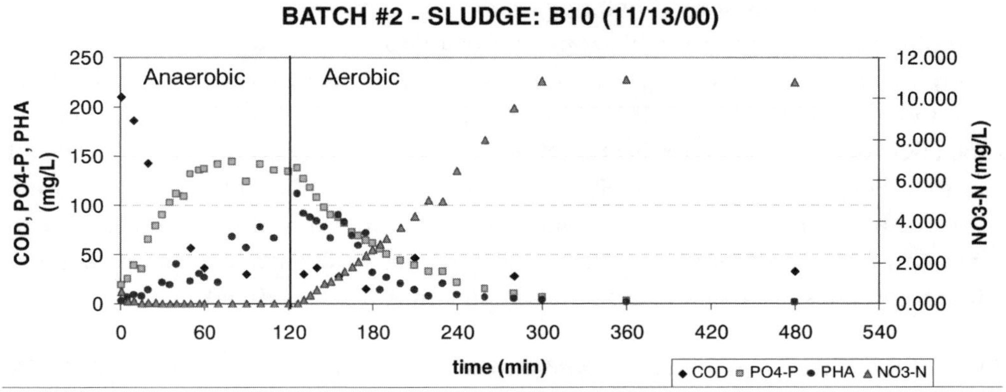

FIGURE 4.11 Data collected during batch tests performed on an enriched EBPR sludge (Z. K. Erdal, 2002).

depleting glycogen reserves through the aerobic stage, which, in turn, limits the PHA storage in the anaerobic zone, thereby reducing EBPR efficiency.

In BNR systems designed for both nitrogen and phosphorus removal, it is recommended that individual zones be designed to have variable sizing with provisions provided for mixing and aeration zones be expanded or reduced to achieve desired effluent phosphorus concentrations, depending on the influent characteristics and operational conditions.

At-full scale plants, VFA uptake is a relatively rapid reaction, requiring an anaerobic zone SRT of as low as 0.3 to 0.5 days. For the majority of the cases, this corresponds to a nominal anaerobic zone HRT of 0.75 hour or less. However, depending on the concentration of the mixed liquor biomass concentration, the required HRT will be different for different systems. For example, the HRT of the anaerobic zone of a University of Cape Town (UCT) system should be approximately twice as much as the HRT of an anaerobic/oxic (A/O) system (for system configurations, see later sections in this chapter). This is because of the fact that mixed liquor suspended solids (MLSS), rather than return activated sludge (RAS), which has a higher MLSS concentration, is recycled to the anaerobic reactor. For the same degree of VFA uptake in the two systems, the same mass of MLSS should be present in the anaerobic zone of the two systems; therefore, both systems should have approximately the same anaerobic SRT, but they will require different anaerobic HRTs.

The fermentation of readily biodegradable organic matter is a slower process, generally requiring an anaerobic zone SRT of 1.5 to 2 days. This corresponds to an anaerobic zone HRT of 1 to 2 hours or more, and 2 hours or more in the case of the UCT and modified UCT systems. This provides guidance on the required anaerobic zone HRT for a particular application. If the influent wastewater contains significant concentrations of VFAs, a relatively short anaerobic zone SRT and HRT can be used. If, on the other hand, significant fermentation is required in the anaerobic zone to generate VFAs, then a longer anaerobic zone SRT and HRT, or other carbon augmentation techniques presented later in this chapter, must be considered.

TEMPERATURE. The effects of temperature on the efficiency and kinetics of EBPR systems have been investigated for the past two decades, but the studies yielded contradictory results. Early researchers (Barnard et al., 1985; Daigger et al., 1987; Ekama et al., 1984; Kang et al., 1985; Sell, 1981; Siebrietz, 1983) reported that EBPR efficiency was unchanged at lower temperatures than at higher temperatures, over the range 5 to 24°C. Beatons et al. (1999); Brdjanovic et al. (1997); Choi et al. (1998); Jones and Stephenson (1996); and Marklund and Morling (1994) showed that cold temperatures adversely affect EBPR performance.

Contradictory to previous findings, Helmer and Kuntz (1997) and U. G. Erdal (2002) reported that, despite the slowing reaction rates, EBPR performance can be significantly greater at 5°C compared to 20°C. This shows that better system performance can be achieved as a result of reduced competition for substrate in the anaerobic zones and increased population of PAOs. The phosphorus content of EBPR biomass achieved by U. G. Erdal (2002) was up to 50% of volatile suspended solids (VSS) in the end of aerobic zone. Figure 4.12 illustrates the importance of acclimation and the resulting improved cold temperature operation of an EBPR system. Figure 4.13 shows an electron microscopy picture of the biomass enriched at these temperatures.

Very good EBPR performance can be achieved, as long as SRT values of 16 and 12 days are provided for 5 and 10°C, respectively. The system performance was not affected between 16 and 24 days and 12 and 17 days SRT for 5 and 10°C, respectively. High SRT operations increased the endogenous glycogen use, thereby consuming the available reducing power used for PHA formation in anaerobic stages.

Glycogen metabolism was found to be the most rate-limiting step in EBPR biochemistry at temperatures below 15°C. The pilot EBPR systems removed phosphorus until the complete shutdown of glycogen use and replenishment was observed. Despite the presence of available energy sources (poly-P and PHA), the shutdown of

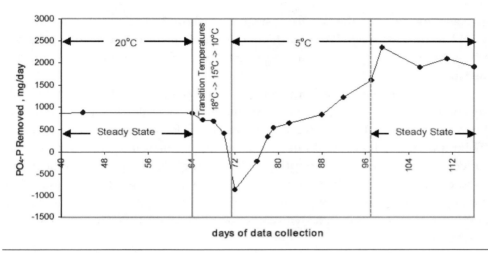

FIGURE 4.12 Effect of acclimation on cold-temperature performance of enriched EBPR populations (Erdal et al., 2002).

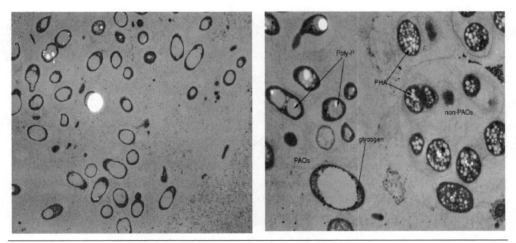

FIGURE 4.13 Comparison of two aerobic sludges obtained from 5 (left) and 20°C (right). In the case of the 5°C sludge, nearly all cells have the ability to store poly-P (empty circles). However, at 20°C, one distinct group stored poly-P (PAOs), whereas the other group did not (non-PAOs). Total magnifications are 7500 and 15 000X at 5 and 20°C, respectively.

the glycogen metabolism was the major reason for washout to occur. The shutdown of glycogen use through the anaerobic stage in washout SRTs prevented acetate use and PHA formation. While PAOs wash out of the system, ordinary heterotrophs can continue to grow, using the acetate passing through the anaerobic stages unconsumed and into aerobic stages.

Figure 4.14 shows the washout SRT values obtained in the U. G. Erdal (2002) study and the linear line developed by Mamais and Jenkins (1992). The reason for the differences are related to the differences in feed COD:P conditions. The former had COD-limiting operational conditions, whereas the Mamais and Jenkins study was conducted under phosphorus limitation (Erdal et al., accepted for publication). Therefore, it is clear that influent conditions do not only influence the system performance, but also influence the washout point of EBPR systems; and COD:P, SRT, and temperature are the parameters that define the EBPR biomass makeup and system performance.

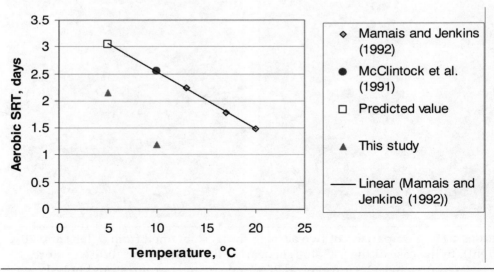

FIGURE 4.14 Effect of feed quality and nutrient limitation on EBPR aerobic washout SRT (Erdal et al., 2003b).

Figure 4.15 illustrates, contradictory to general belief, how well an EBPR system can operate under 5°C conditions, once adequate acclimation and operational conditions are provided.

In contrast to the improved EBPR performance under cold temperature conditions, at warmer temperatures, EBPR performance tends to slow down or diminish completely. However, similar to the case for cold temperature effects, the researchers report contradicting results. In the Mamais and Jenkins (1992) study mentioned earlier, both long- (13.5 to 20°C) and short-term (10 to 30°C) effects of temperature in continuous flow, bench-scale, activated sludge systems were investigated. The optimum temperature for aerobic phosphorus removal was reported to be between 28 and 33°C. Jones and Stephenson (1996) suggested that the optimum temperature was 30°C for anaerobic release and aerobic uptake of phosphate. Enhanced biological phosphorus removal was also observed at two extreme temperatures—5 and 40°C—

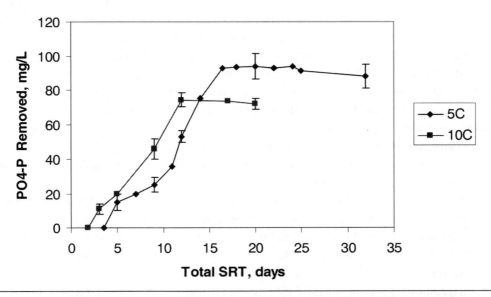

FIGURE 4.15 Effect of system SRT on phosphorus removal at cold temperatures (Erdal et al., 2003b).

but the efficiency of EBPR reduced significantly. Brdjanovic et al. (1997), in a laboratory-scale sequencing batch reactor, determined the short-term effects of temperature on EBPR performance and kinetics at 5, 10, 20, and 30°C. The optimum temperature for anaerobic phosphorus release and acetate uptake was found to be 20°C. However, a continuous increase was obtained for temperature values up to 30°C for aerobic phosphorus uptake. The stoichiometry of EBPR was found to be insensitive to temperature changes.

Panswad et al. (2003) achieved 100% phosphorus removal at all temperatures, except at 35°C. Phosphorus removal was inhibited also at 40°C (Panswad, 2000). The unchanged system performance at low temperatures can be attributed to relatively low influent COD concentrations (300 versus 400 mg/L) entering the anaerobic zones and to the selection of a longer anaerobic contact time (approximately 5 hours versus 1.5 to generally used 2 hours). Although the rates of all reactions decrease, providing enough anaerobic and aerobic contact times permit the reactions to go to completion. Unlike at the low temperatures, the EBPR performance of the systems was reduced at high temperatures in this study. The importance of the anaerobic and aerobic contact times on EBPR performance was discussed previously. Longer anaerobic contact times cause a decrease in phosphorus content and PHA storage by the EBPR biomass (Wang and Park, 1998). Similar observations were made in the Panswad (2000) and Panswad et al. (2003) studies. The biomass phosphorus content at 35 and 40°C decreased by more than 60% compared to what was observed at 5 and 15°C. Significant decreases in PHB content was also reported at these two high temperatures, which are almost at the boundary temperatures for the shift between mesophilic to thermophilic organisms. In boundary regions, the dominant organisms are generally considered to be unstable and cannot function effectively. Although no microecological study has been performed to prove it, it may be that population shifts from mesophilic to thermophilic favors the nonpolyphosphorus-accumulating organisms over polyphosphorus-accumulating bacteria.

Because the reason for degrading EBPR performance observed at warmer temperatures is related to the increased competition for substrates in nonaerated zones of the biological phosphorus removal systems (i.e., increased competition from non-PAOs that can accomplish anaerobic PHA storage and increased denitrification in anoxic zones), this further emphasizes the importance of the feed COD:P ratio that must be maintained to support PAO growth and the anaerobic contact time for uptake of VFAs by the PAOs. The cold temperature gives selective advantage to

PAOs to outcompete their mesophilic competitors. At high temperatures, the same bacteria prefer to use and accumulate glycogen to a greater extent. Although many researchers believe that it is a population shift from PAOs to glycogen-accumulating organisms (GAOs), no evidence exists to support this or to reject that the majority of the organisms go through a metabolic adaptation. However, in either case, increased glycogen-dependent metabolism (GAO proliferation or glycogen dependency) may lead to complete EBPR failure, especially in a full-scale system. The breakthrough of VFAs to anoxic and aerobic zones will indicate such a shift and will diminish phosphorus removal.

RECYCLE FLOWS. *Internal Recycles.* As explained up to this point, EBPR systems strongly rely on sufficient PHA storage using the feed VFAs under anaerobic conditions and subsequently on PHA breakdown and orthophosphate polymerization from the excess energy generated from PHA catabolism. Recycle flows internal to the EBPR system are mainly the RAS, recycles from anoxic zones to anaerobic, and recycles from the aerobic zones to anoxic zones (see next section for different EBPR configurations and associated flow recycles). In systems where internal recycles are used, sometimes at rates greater than the influent flow rates, close attention must be paid to the amount of oxygen and nitrate that is returned to the nonaerated zones.

The recycle of nitrate to the anaerobic zone will interfere with fermentation reactions, in the same way as would oxygen. Given the option of either using nitrate to oxidize readily biodegradable organic matter or fermenting the readily biodegradable organic matter, oxidation using nitrate will dominate as a result of the greater amount of energy microorganisms can obtain through oxidative reactions. This is another reason that the recycle of nitrate into the anaerobic zone can interfere with effective and efficient biological phosphorus removal.

The best way to monitor the oxygen and nitrates returned to the head of the EBPR systems is the "mass balance" method. The mass of phosphorus, oxygen, and nitrate can be calculated using eq 4.1. Once the mass going in and coming out are calculated for each reactor, the change in the reactor can be calculated from the difference between the two. Figure 4.16 illustrates the results of mass balance calculations performed for a UCT type EBPR system that consisted of two anaerobic (AN), two anoxic (AX), and three aerobic (AE) reactors.

$$\text{Mass difference} = \text{Mass in} - \text{Mass out} = [C_{in} \times Q_{in}] - [C_{out} \times Q_{out}] \qquad (4.1)$$

Where

M = mass of the specific constituent (lb/d or kg/d),
C = concentration of the specific constituent (mg/L), and
Q = flowrate (mgd or ML/d).

Figure 4.16 also illustrates the effect of the mixed liquor recycle from the end of the aeration zone (Ae3) to the first anoxic zone (Ax1). The recycle contained nitrate and oxygen, which contributed to the electron acceptor budget in the anoxic zone. Once the recycle was taken out of service, anoxic phosphate uptake decreased significantly, leading to the increase of aerobic uptake. By using the mass balance calculations, one can avoid the confusion that can be caused by using concentration values as a monitoring tool. Because the recycle flows also have a dilution effect on the plant influent at different points of the treatment system, mass loading must be used instead of concentrations.

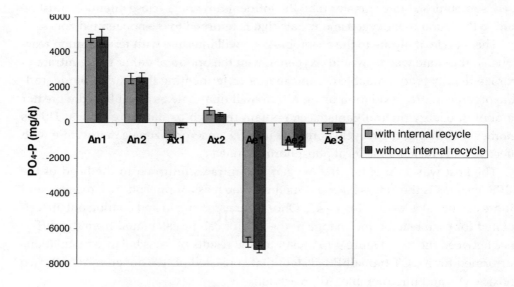

FIGURE 4.16 Phosphorus mass balance performed for phosphorus in a pilot-scale, VIP-type EBPR system (Z. K. Erdal, 2002). Positive values indicated phosphorus release, whereas negative values indicate phosphorus uptake.

Plant Recycles. Plant recycles from other treatment units, such as sludge-handling facilities supernatant, filtrate, centrate, and backwash flows, can contain high quantities of nitrogen and phosphorus, both particulate and soluble. A similar mass balance calculation can be performed to determine the effect of the recycle loads on EBPR systems.

Similar mass balance calculations are recommended to monitor nitrogen return to the EBPR units, especially in cases where seasonal nitrification is not avoidable or nitrification is part of the treatment scheme. Organic carbon (BOD or COD) return from sludge thickening operations, especially where sludge retention in the thickener tanks is long enough to allow fermentation reactions, must also be monitored. In some cases, where digester supernatant is recycled back to the biological units, inhibition of EBPR was observed. However, it is recommended that digester supernating be avoided and digested sludge be sent directly to the biosolids handling facilities. There are several potential mechanisms that may lead to part of the phosphorus that is released during digestion to be removed from the liquid phase. The formation of struvite ($MgNH_4PO_4$), brushite ($CaHPO_4 \cdot 2H_2O$), and vivianite [$Fe_2(PO_4)_3 \cdot 8H_2O$] are possible in an anaerobic digester and can lead to a significant degree of apparent reduction in the extent of phosphorus solubilization in the anaerobic digester. These mechanisms can be used to precipitate and further remove phosphorus from digester effluent sludges, if pH adjustments and magnesium addition are achieved in a controlled manner. For example, at the York River wastewater treatment plant in Virginia, only 27% of the phosphorus that enters the anaerobic digester leaves as soluble phosphorus, while the rest is precipitated by a combination of the three mechanism presented above (Randall et al., 1992). Because formation of these precipitates can also be problematic if not performed in a controlled manner (i.e., precipitation in pipes and valves downstream of the digesters leading to significant size reduction and scaling), such alternatives must be approached with caution.

TYPES OF ENHANCED BIOLOGICAL PHOSPHORUS REMOVAL SYSTEMS

Biological nutrient removal, namely nitrogen and phosphorus removal, can be achieved in separate or single sludge systems; in suspended-growth systems, where special media that allows fixed-film growth is introduced; or in fixed-film reactors that are operated in parallel, in series, or in absence of suspended-growth reactors. Although phosphorus removal can be achieved independently from nitrogen

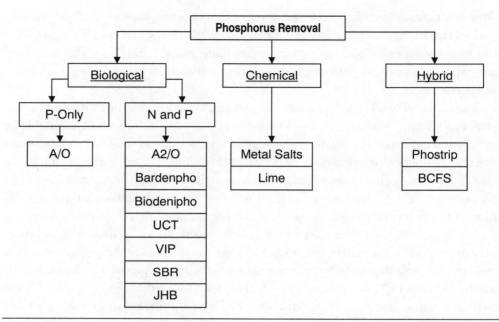

FIGURE 4.17 Summary of processes developed for phosphorus removal.

removal, industry experience and extensive research that has been conducted in the United States and Europe have shown that systems design for simultaneous removal of these nutrients are more economic and reliable.

Figure 4.17 summarizes different types of well-understood and widely accepted phosphorus removal methods. There are other methods and configurations that are developed as variants from any one of the listed methods, and they will be briefly described as needed in the next two sections. Although all methods are listed in this figure, for the purposes of this manual, only the biological processes will be further explored.

SUSPENDED-GROWTH SYSTEMS. *Anaerobic/Oxic and Anaerobic/Anoxic/Oxic Configurations.* The anaerobic/oxic (A/O) process was developed originally as the Phoredox system in South Africa in 1974 for biological phosphorus removal (Barnard, 1974). Later, it was patented in the United States under the name "A/O" (Spector, 1979). It is very similar to a conventional activated sludge setup, with the addition of an anaerobic zone where the secondary influent is introduced. The RAS from the bottom of the secondary clarifiers is also returned to the anaerobic zone for mixed liquor recycle. This configuration allows for the selection of PAOs in the anaerobic zone and is used where nitrogen removal is not required. However, in nitrifying systems, RAS will contain nitrates that will result in consumption of a portion of the influent COD for denitrification purposes, because the nitrification taking place in the aerobic zone will result in considerable nitrate generation. Hence, the first zone of the A/O process actually operates as an anoxic/anaerobic zone, and the distribution between the two oxidation states depends on the amount of nitrate and oxygen returned to the head of the system in the RAS stream.

A variant of A/O configuration is the anaerobic/anoxic/oxic (A^2/O) process, also developed in South Africa in 1974 and later patented in the United States, where an anoxic zone is introduced between the anaerobic and the aerobic zones for simultaneous nitrogen and phosphorus removal. For denitrification to proceed, a nitrate-rich mixed liquor recycle (NO_3-R) from the end of the aerobic zone is included in this configuration. In other words, an A^2/O system is a modified Ludzack-Ettinger configuration with an anaerobic zone as the initial zone. Because, in an A^2/O system, the anoxic

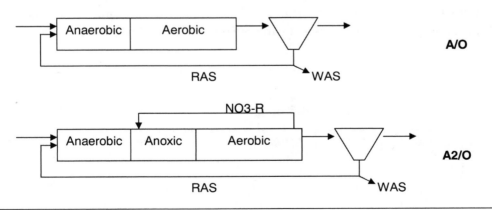

FIGURE 4.18 A/O and A^2/O process configurations.

zone allows nitrogen removal, nitrate returned to the anaerobic zone in RAS will not be as high as it would be in an A/O system operated under similar conditions.

The anaerobic and anoxic zones can be introduced to existing plants with relative ease if retrofitting is needed. In that case, the existing aeration system will also need to be moved and modified to achieve nonaerated zones in accordance with the desired configuration. Figure 4.18 presents an illustration of both configurations.

The presence of an anoxic zone has a number of consequences on nutrient removal. As explained earlier, PAOs are heterotrophic bacteria with the ability to use either oxygen or nitrate, depending on the redox potential in a particular basin. Therefore, once the mixed liquor passes into the anoxic zone following the anaerobic zone, the stored PHAs and any remaining soluble COD will be consumed, the latter one being consumed by PAOs and nonPAOs for denitrification. Thus, some phosphorus uptake will be observed in accordance with the amount of nitrate present and anoxic residence time allowed in system design.

To avoid secondary phosphate release (i.e., release in excess of what can be taken up in the aerobic zone), SRTs in each zone of the treatment must be adequate. Also, to avoid release in the secondary clarifiers, sludge loading and withdrawal rates must be selected appropriately to not allow EBPR sludge to accumulate and undergo endogenous degradation in the clarifier bottom. This will increase the secondary effluent phosphorus concentration, and also the phosphorus in RAS returned to the anaerobic zone will affect the balance established between the COD and phosphorus removal.

Modified Bardenpho Configuration. The modified Bardenpho configuration (Figure 4.19) was also developed in South Africa in 1978 as a modification of the A^2/O and Bardenpho processes for the purpose of achieving effluent phosphorus

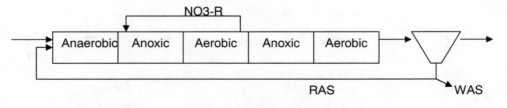

FIGURE 4.19 Modified Bardenpho process configuration.

and nitrogen concentrations of less then 1 and 3 mg/L, respectively. The Bardenpho process was originally developed for nitrogen removal, and, with the addition of an anaerobic zone at the head of the process configuration, successful phosphorus removal was ensured.

Similar to A²/O, the modified Bardenpho has an internal mixed liquor nitrate recirculation stream (NO$_3$-R), providing continued feed of nitrate into the anoxic zone. Similar phosphorus removal response is achieved in the modified Bardenpho process; however, significantly lower effluent nitrogen concentrations can be achieved. Again, depending on the COD:P ratio of the influent, the PHAs stored in the anaerobic zone will determine how much phosphorus will be taken up in the later zones under the presence of the electron acceptors nitrate and oxygen.

University of Cape Town, Modified University of Cape Town, and Virginia Initiative Process Configurations. The original UCT process was developed in South Africa as a modification of the A²/O process, with the main purpose of minimization of the adverse effects of the nitrate return to the anaerobic zone (Figure 4.20). In this way, PAOs were given the advantage of full access to readily available COD for PHA storage. In the case where RAS is recycled to the anaerobic zone, nitrates present in the RAS stream (generally operated at 80 to 100% of the influent

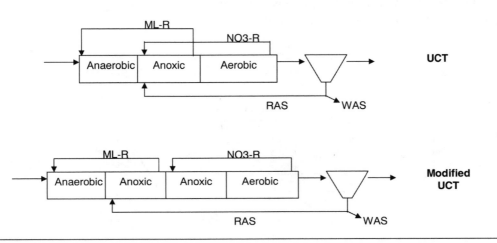

FIGURE 4.20 The UCT and modified UCT process configurations.

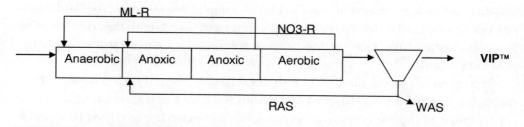

FIGURE 4.21 The VIP process configuration.

flow) would cause the denitrification reactions to consume readily available COD that would otherwise be stored as PHA by PAOs, increasing the energy potential of the organisms to take up more phosphorus in the anoxic and aerobic zones.

The modified UCT process configuration came about following the efforts for further minimizing the nitrate recycle back to the anaerobic zone. By adding a second anoxic zone, where the internal NO_3-R is returned, and using the first anoxic zone as the source reactor for the mixed liquor recycle to the anaerobic zone ensures that, in well-designed and well-operated systems, no nitrates are returned to the anaerobic reactors. These configurations best compartmentalize the secondary treatment units with respect to their intended purposes, leaving a smaller margin for performance decline resulting from operational problems.

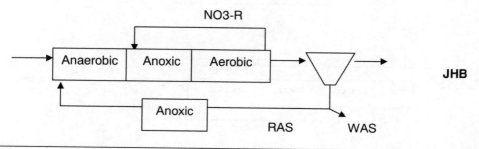

FIGURE 4.22 The JHB process configuration.

A variation of the UCT configuration is the Virginia Initiative Process (VIP) process, which was developed at the Hampton Roads Sanitation District in Virginia (Daigger et al., 1987). The main difference between the two processes is the location of the mixed liquor recycle and the way the two are operated (Figure 4.21). The VIP process maximizes the nitrogen removal, removes nitrate completely in the anoxic zone, and thus, at the same time, allows additional anaerobic time for PHA storage. This, in turn, results in greater phosphorus uptake potential in the aerobic zone that follows. The system is operated as a high-rate system, resulting in significant reduction in reactor volume.

Johannesburg Configuration. The Johannesburg (JHB) process is a modification of the modified UCT process (Figure 4.22). The anoxic zone, where RAS return takes place in the modified UCT, is on the RAS line and operates as a dedicated denitrification tank for minimizing the nitrate recycle to the anaerobic zone. Because there is no dilution effect from the plant influent on the concentrated RAS flow, this reactor is much smaller than the anoxic zones of the UCT processes. However, the denitrification capacity is limited as a result of the lack of available COD. In certain cases, a sludge recycle from the end of the anaerobic zone to the anoxic reactor on the RAS line is used to recycle back leftover COD to improve denitrification.

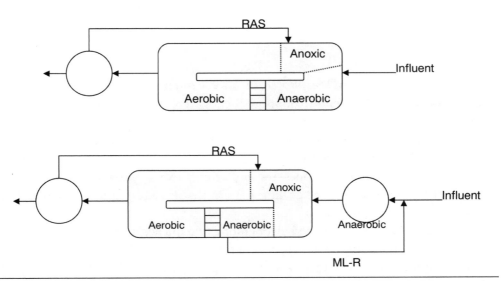

FIGURE 4.23 Oxidation ditch designs for nitrogen and phosphorus removal.

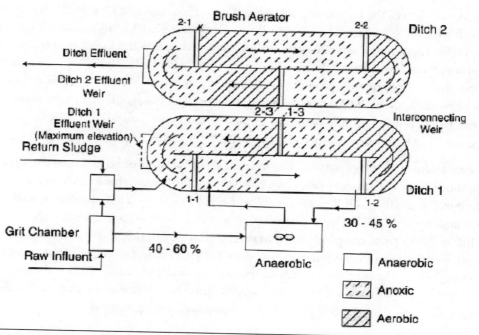

FIGURE 4.24 VT2 process configuration.

Oxidation Ditches, BioDenipho, and VT2 Configurations. Aside from the reactor-type activated systems, oxidation ditches were also studied to modify their configurations into phosphorus removal systems. Oxidation ditches are constructed in a racetrack setup, where the secondary influent and mixed liquor are circulated around a center barrier. Studies have shown that parts of the ditches can be operated as anaerobic and/or anoxic zones through control of oxygen transfer, thereby allowing the anaerobic/aerobic cycling pattern that is required for enrichment of PAOs in the sludge (Figure 4.23).

The VT2 design was developed by operating two ditches in series with dedicated zones (Figure 4.24). This system is capable of producing total nitrogen and total phosphorus concentrations less then 4 and 0.7 mg/L, respectively, all year, without effluent filtration.

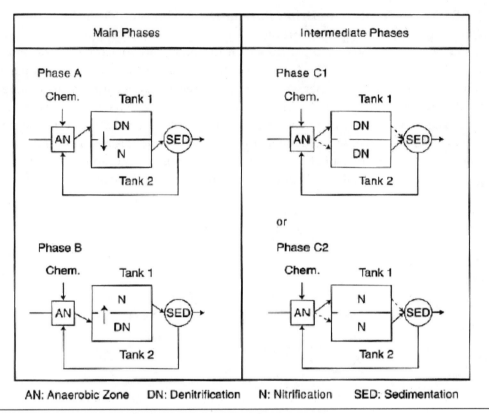

FIGURE 4.25 BioDenipho process configuration and operation cycles.

The BioDenipho process configuration developed in Denmark also makes use of ditch pairs, operated on similar EBPR basics as the other configurations introduced herein. Different from the VT2 system, the ditches are operated in alternating mode (Figures 4.25 and 4.26).

The BioDenitro system, developed in the 1970s, made use of cycling the operation of the aerobic and anoxic phases. BioDenipho, a modification of this system, includes a dedicated anaerobic reactor, where the system influent is introduced and added to the head of the units, and the following two reactors are cycled between aerobic and anoxic operation.

HYBRID SYSTEMS. *PhoStrip Configuration.* This is a sidestream biological-chemical phosphorus removal process, where a phosphate stripper tank is used instead of the anaerobic zones included in the other biological-only phosphorus removal processes (Figure 4.27). The phosphate stripper reactor is designed to operate as an anaerobic sludge thickener where phosphate release takes place. Once the supernatant is removed, the sludge is returned to the RAS line and back into the aeration basins. The supernatant is treated with lime in a separate reactor, phosphorus is chemically precipitated out of the solution in a settler, and the liquid is returned to the head of the system. Unless phosphate recovery is desired, this configuration does not have significant benefits over the biological-only treatment configurations.

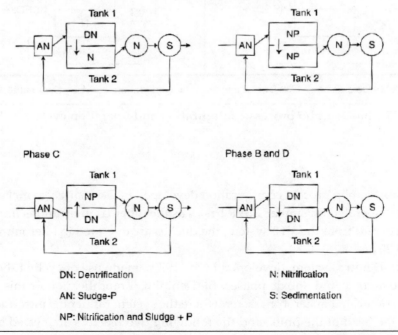

FIGURE 4.26 BioDenipho trio process configuration and operation cycles.

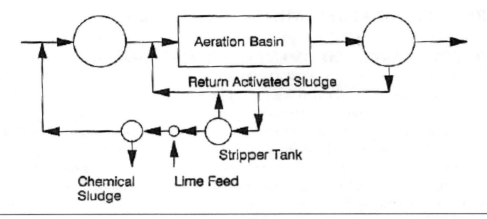

FIGURE 4.27 PhoStrip process configuration.

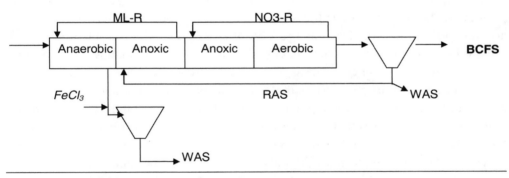

FIGURE 4.28 BCFS process configuration.

Biological Chemical Phosphorus and Nitrogen Removal Configuration. One modification of the modified UCT process developed in Netherlands is the biological chemical phosphorus and nitrogen removal (BCFS) process (Figure 4.28). It allows additional phosphorus removal from the system through chemical precipitation in a dedicated tank. Because the chemical addition is accomplished in the sludge line and a separate tank, the chemical sludge does not have to be returned to the main system, which is a benefit over the configurations that include a chemical polishing step in the secondary clarifiers. By this process, inert material return accumulation in the activated sludge basins is prevented.

PROCESS CONTROL METHODOLOGIES

INFLUENT CARBON AUGMENTATION. At wastewater treatment plants where plant influent is weak with BOD values approximately 200 mg/L and less, and year round temperature variation is great with winter mixed liquor temperatures of 17°C or below, fermentation reactions in the anaerobic zones of the EBPR systems will be significantly slow. This will result in reduced plant performance resulting from insufficient anaerobic PHA storage to support the subsequent aerobic poly-P storage. Only for cases where wastewater BOD values and the winter temperatures are higher, the potential for sufficient fermentation in the anaerobic zones can indicate merit in enlarging the anaerobic zones of the EBPR system. Otherwise, an external carbon augmentation step can be well-justified.

Volatile Fatty Acid Addition. In cases where influent does not contain sufficient VFAs to support PAO enrichment in EBPR systems, external VFA addition can be performed. Because acetate is the most efficient VFA (see Table 4.2), in most cases, acetate addition is considered as the VFA of choice. In industrial facilities where acetate-laden waste streams are generated, utilities and/or wastewater treatment plant operators can arrange for transport of waste streams to the EBPR plant.

A mixture of short-chain VFAs that can be taken up and metabolized into PHA polymers can also be used for augmentation of influent COD. It must be remembered that formic acid (COOH) that contains only one carbon is not a suitable substrate for the PAOs. It cannot be combined with other formic acid molecules to make up PHA polymers. Polymerizing formic acid and other VFAs is not thermodynamically favorable; hence, formic acid should not be considered as an external source. Acetate, being at the center of all the metabolic reactions, is very easily used, and other VFAs with carbon numbers of up to five (valeric and isovaleric acids) can be easily manipulated to be converted to PHAs polymers.

Pre-Fermentation. Primary sludge fermentation is a cost-effective method of augmenting the supply of readily biodegradable carbon (principally VFAs acetic and propionic acids) in plants designed for biological phosphorus and nitrogen removal. Fermentation stabilizes the bio-P process and significantly increases dentrification rates in BNR processes. Fermenters are particularly cost-effective in large plants in temperate climates, in plants that treat low-organic-strength wastewater, and in plants required to meet stringent effluent total phosphorus and nitrogen limits. However, not all BNR plants need fermentation. Many plants in warm, dry climates operate well without fermentation. In smaller plants, it may be more economic to achieve the effluent limits through the addition of iron or aluminum salts or commercially available acetic acid.

Acid fermentation of complex organic material present in wastewaters also occurs in the collection system before the wastewater reaches the treatment plant. The degree of fermentation is affected by several factors, primarily the HRT and wastewater temperature. In general, long flat sewers favor VFA generation, whereas short steep sewers decrease the retention time and create conditions for reaeration of the wastewater. Force mains encourage acid fermentation because little or no reaeration takes place. High infiltration during storms dilutes the wastewater, reduces the retention time, and increases reaeration, all of which reduce in-pipe fermentation. Like all anaerobic processes, the reaction rates for acid fermentation increases with increasing wastewater temperatures. This is clearly seen by comparing the seasonal changes in the raw wastewater BOD:COD ratio. A high BOD:COD ratio is generally associated with a high VFA concentration, and the ratio is highest when wastewater temperatures are high and flowrates are low.

On-site fermenters are not always required in many BNR processes, as a result of the characteristics of the sewer system and of the wastewater itself. For example, many of the earlier plants in the United States that were reported to remove phosphorus consistently had slow flowing sewers with a large number of pump stations, forced mains, and inverted siphons in the collection systems. These units all act as natural fermenters by increasing the concentration of VFA in the wastewater entering the treatment process.

The four fermenter configurations widely used at EBPR wastewater treatment plants form the basis of all sidestream primary sludge fermenters. They are the following:

(1) Activated primary sedimentation tanks,
(2) Complete mix fermenter,
(3) Single-stage static fermenter, and
(4) Two-stage complete mix/thickener fermenter.

Details of sludge fermentation using these configurations and incorporation of pre-fermentation to EBPR plants will be further described in Chapter 10. One must remember that each fermenter design is unique and is generally tailored to meet the specific requirements of the plant. As new fermenters are brought into service, plant staff should optimize the operation of the units to meet the BNR requirements of the process and minimize operating problems such as odors, corrosion, and blockages. As the results of these full-scale optimization studies become available, designers and operators will be able to incorporate this new information to the design and operation of future units.

The key parameter for monitoring the performance of primary sludge fermenters is the VFA concentration in the fermenter supernatant. This is best measured by gas chromatography or high-performance liquid chromatography, as these methods provide accurate information about the concentration of individual VFAs present. The distillation method is a reasonable method for measuring the total VFA concentration, but tends to be inaccurate at concentrations below 100 mg/L. The concentration of soluble COD in the fermenter supernatant provides a reasonable indication of the VFA concentration. The redox potential in the sludge blanket can indicate the level of anaerobic activity in the fermenter and whether optimal conditions for acid fermentation or methane and sulphide formation are being maintained. The pH of the sludge blanket can indicate good VFA production, but is somewhat influenced by the natural alkalinity of the wastewater.

The two principal control parameters for the operation of primary sludge fermenters are the fermenter SRT and HRT. The fermenter SRT is controlled by adjusting the solids inventory and the sludge wastage rate. By increasing the fermenter SRT, the growth of slower growing fermentative organisms is favored, and more complex molecules and higher acids are produced. Conversely, decreasing the SRT favors the growth of faster growing organisms, resulting in simpler biochemical pathways and the production of acetic acid and, to a lesser extent, propionic acid. The ratio of VFA produced per VSS added to a fermenter has a fairly broad range, from 0.05 to 0.3 g VFA/g VSS added. The use of fermenters can significantly change the characteristics of the influent wastewater and make them more amenable for the establishment of a stable EBPR operation.

The fermenter HRT is controlled by adjusting the primary sludge and elutriation water (added to wash and separate the released soluble VFAs from the particulate matter as an overflow stream) pumping rates. Increasing the HRT increases the available time for the conversion of solubilized substrates to VFAs. The HRT should be increased if there is insufficient hydrolysis of the particulates. However, too long an HRT results in the production of complex molecules and higher acids.

Enhanced Biological Phosphorus Removal 143

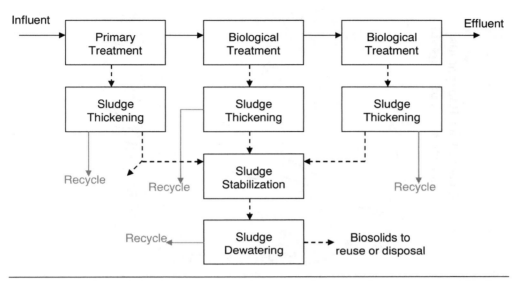

FIGURE 4.29 Recycle streams generated at typical wastewater treatment plant solids handling facilities.

SOLIDS SEPARATION AND SLUDGE PROCESSING. As mentioned earlier, the recycle stream at a wastewater treatment plant has significant bearing on biological systems operation. Figure 4.29 illustrates the recycle streams that are generated at typical wastewater treatment plant solids handling facilities. The quality of these streams varies based on the technology used in the solids processing operations. For example, sludge thickening using belt filter dewatering generally generates two times more recycle flow (filtrate) compared to centrifuge dewatering because of the amount of wash water used in the dewatering operation. Total recycle streams can even amount to 20 to 30% of the plant influent, in some instances.

The importance of the EBPR feed COD:P ratio and the amount of nitrate in the internal recycle streams were discussed earlier. The wastewater treatment plant recycle streams generated from the solids processing units typically contain high ammonia and phosphorus concentrations, especially if the recycle stream was collected following sludge digestion. These concentrations can be as high as 900 to 1100

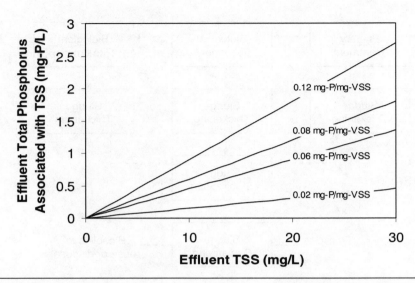

FIGURE 4.30 Contribution of the effluent TSS to the total phosphorus in the effluent for different phosphorus contents in the MLSS (assuming that the VSS/TSS is 75%).

mg/L of ammonia and 100 to 150 mg/L of phosphorus, depending on the struvite and phosphorus precipitates formation in the digestion systems and associated appurtenances, and thus 50 to 60% of the released phosphorus can be retained in dewatered biosolids and/or taken out in the form of struvite precipitates.

CHEMICAL POLISHING AND EFFLUENT FILTRATION. Figure 4.30 shows the effect of effluent total suspended solids (TSS) concentration on the effluent total phosphorus concentrations for mixed liquor samples with different phosphorus contents (CH2M Hill, 2002).

Although the soluble phosphate concentration may be below 0.1 mg/L, if the TSS in the effluent is 10 mg/L and the phosphorus content of the mixed liquor is 0.06 mg/mg VSS, the effluent total phosphorus concentration is expected to be approximately 0.5 to 0.6 mg/L. Hence, the phosphorus associated with the TSS is one of the main factors that affect the total phosphorus concentrations that can be attained at wastewater treatment plants that include EBPR systems, and chemical polishing is recommended, especially at plants where stringent effluent total phosphorus concentrations are imposed.

Chemical addition is generally accomplished in the secondary clarifiers and/or effluent filters, depending on the EBPR effluent phosphorus concentrations that can be achieved at a particular wastewater treatment plant and the discharge requirements. Alum, ferric, or lime are the three chemicals most widely used for this purpose. Chapters 2, 7, and mainly 9 presents the details of chemical feed requirements, and reader is referred to those chapters for more detailed discussion of chemical phosphorus precipitation.

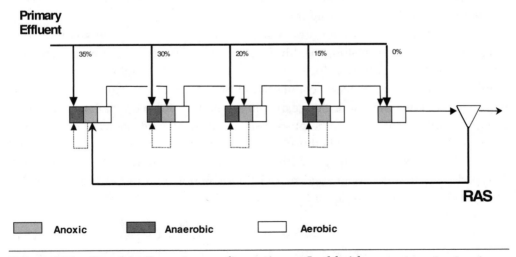

FIGURE 4.31 Step bio-P reactor configuration at Lethbridge wastewater treatment plant, Alberta, Canada (Johnson et al., 2003).

CASE STUDIES

THE LETHBRIDGE WASTEWATER TREATMENT PLANT, ALBERTA, CANADA. A novel reactor configuration was used at the Lethbridge wastewater treatment plant that allowed the plant to be upgraded from a conventional activated sludge system to a fully nitrifying EBPR system, without needing to construct additional aeration tanks and secondary clarifiers. When plants are required to nitrify, the SRT must be high enough to allow nitrifiers to grow in the system. An increase in the SRT leads to an increase in the MLSS concentration, given that the aeration tank volume is maintained constant. As a consequence of the increase of MLSS concentration, the solids loading rate to the secondary clarifiers is also increased and may limit the amount of flow that can be treated in the plant. It is not uncommon that, when a plant is upgraded for nitrification, the secondary treatment capacity has to be derated (note that secondary clarifier capacity is only one of the possible limiting factors; aeration capacity may also be limiting when upgrading a plant for nitrification).

The Lethbridge plant was required to nitrify and produce an effluent with a total phosphorus concentration less than 1.0 mg/L on a monthly basis, and these effluent requirements were to be achieved without the construction of additional aeration tanks and secondary clarifiers. These restrictions led to the development of a new bioreactor configuration, which is called "step bio-P" (Crawford et al., 1999). A schematic of the bioreactor configuration is presented in Figure 4.31.

At Lethbridge, the influent flow is stepped in four stages, as shown on the diagram. Because the return sludge is diluted by the primary effluent flow in stages, the average MLSS concentration in the total bioreactor is 18% higher than the MLSS would be at full dilution using conventional BNR technology. This means that the bioreactor contains 18% more sludge inventory and has 18% more biological capacity at a given MLSS concentration because of step feeding. It is this increase in biological capacity that has allowed the Lethbridge wastewater treatment plant to be converted to BNR in the existing tankage, without derating the flow capacity of the plant. Pumped recirculation of nitrified mixed liquor is not required at the Lethbridge plant, because the nitrified liquor flows out of each stage directly into the anoxic zone of the subsequent stage.

The step bio-P system consists essentially of a series of UCT systems, into each of which a certain fraction of the primary effluent is directed. In the last stage, an anaerobic zone was not included, because it was found through simulation that it would

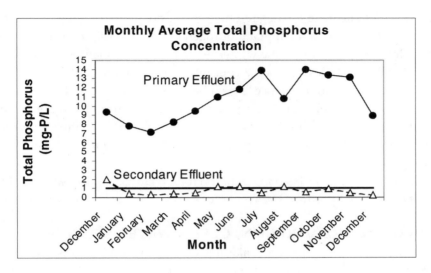

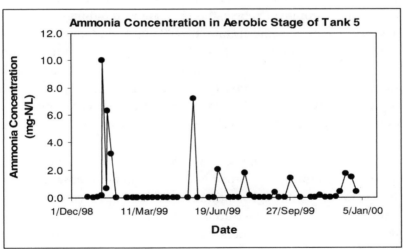

FIGURE 4.32 Monthly average total phosphorus concentration in the plant effluent and ammonia concentration in the aerobic stage of the last tank of the Lethbridge wastewater treatment plant (Johnson et al., 2003).

not be required to achieve the desired degree of phosphorus removal, and, for this particular wastewater treatment plant, nitrogen removal is a significant challenge, as a result of low temperatures in the winter. Also, primary effluent is not added to the final pass to allow for a low ammonia concentration in the final effluent.

The plant is operational since December 1998, and Figure 4.32 presents the monthly average total phosphorus and ammonia concentrations in the last pass of the system recorded in the initial year of operation. The plant is not using chemical polishing for phosphorus removal, and the secondary effluent is not filtered before discharge.

Although the plant has been in operation for a relatively short period of time, the quality of the effluent produced has met the requirements that were set for this plant.

DURHAM, TIGARD, OREGON, CLEAN WATER SERVICES. To protect water quality in the slow-moving Tualatin River, treatment plants operated by Clean Water Services (Oregon) must meet stringent nutrient limits. During the summer permit season (May through October), the 151 400-m^3/d (40-mgd) Durham facility must achieve an effluent phosphorus limit of 0.07 mg/L, and, during late summer, it must meet an ammonia-nitrogen limit as low as 0.4 mg/L. The nutrient limits were initially established in 1989. The activated sludge process was configured with a series of anaerobic, anoxic, and aerobic cells and two internal recycle streams to provide EBPR, BOD removal, nitrification, and denitrification. The denitrification step was not needed for permit compliance, but was used to recover alkalinity and reduce the cost of lime addition.

During the initial years of operation, Clean Water Services relied heavily on chemical precipitation for phosphorus removal. Initial alum dosages in 1993 averaged 170 mg/L; however, through continuous process improvements and optimization, the alum consumption was reduced to 60 mg/L by 1997. To address the VFA issue, three primary sludge thickeners were converted to the unified fermentation and thickening (UFAT) process developed and patented by Clean Water Services. The process uses naturally occurring bacteria in wastewater to provide the fuel to biologically remove the phosphorus from wastewater. Return of the supernatant from the UFAT tanks to the activated sludge process sufficiently augmented VFA levels to allow consistent biological uptake of phosphorus.

Several steps were taken to reduce the introduction of oxygen to the anaerobic zone, including elimination of primary-effluent, flow-split structures that entrained air; trimming the mixer blades in the anaerobic zones to minimize air entrainment resulting from vortex formation; and investigation of nitrate levels in the return activated sludge (which turned out to be negligible). Also, the newest basin was designed as a plug-flow reactor (PFR). The PFR provided more stable performance, increased capacity, and was better able to attenuate peak ammonia loadings. The other basins will be converted to plug flow in the future. Finally, the relative sizes of the anaerobic, anoxic, and aerobic basins were optimized to eliminate sludge bulking issues and maximize treatment capacity.

The improved EBPR system now achieves orthophosphorus concentrations of 0.2 to 0.3 mg/L in the secondary effluent, and the full plant achieves total phosphorus concentrations of 0.04 to 0.05 mg/L in the final effluent. Final effluent ammonia-nitrogen levels are typically below 0.05 mg/L, and nitrate-nitrogen concentrations range from 5 to 7 mg/L. Alum is used in the secondary system only when EBPR in that train is producing effluent greater than 0.5 mg/L soluble phosphorus. The dosage has dropped to as low as 20 mg/L for effluent polishing. Lime usage has declined slightly. Overall sludge production has been reduced substantially, and the elimination of alum in the primary sludge has achieved a more stable solids handling system.

UNIQUE NEW DESIGNS. In this section, two new EBPR design approaches that combine a number of existing nutrient removal technologies will be presented. These wastewater treatment plants are currently under construction; however, their design and planned operational strategies carry potential value in their uniqueness, and the

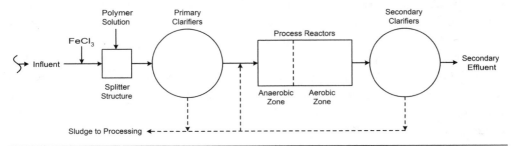

FIGURE 4.33   McAlpine Creek Wastewater Management Facility of Charlotte, North Carolina, liquid process train (Goodwin et al., 2003).

design engineers and plant operators would find them interesting as new hybrid systems.

McAlpine Creek Wastewater Management Facility of Charlotte, North Carolina. A chemically enhanced primary treatment (CEPT) system in combination with EBPR is currently being built as the most cost-effective means to implement phosphorus removal at the McAlpine Creek Wastewater Management Facility of Charlotte, North Carolina, at a design capacity of 242 240 m^3/d (64 mgd). The target effluent limit equates to 1 mg/L of total phosphorus at the National Pollutant Discharge Elimination System permitted effluent flow of 242 240 m^3/d (64 mgd).

Although the primary means of phosphorus removal will be an A/O configuration activated sludge system, CEPT will be available to supplement the process by removing load from the aeration basins (Figure 4.33). By using this process configuration, Charlotte Mecklenburg Utilities was able to avoid adding additional process basins and can instead locate new chemical feed buildings in more convenient locations.

The unique characteristics of this treatment configuration are the online analyzers and probes that were added to help plant staff monitor and control the new chemical feed systems and plant performance. For both the north and south plants, orthophosphate and TSS are monitored upstream of the primary clarifiers, with the orthophosphate signal used to flow pace the ferric chloride feed pumps and the TSS signal used for information. The polymer solution feed rate is tied to the ferric chloride feed rate, but operations staff have the ability to remotely adjust the polymer feed rate independent of the metal salt. Orthophosphate, TSS, and alkalinity will be measured in the primary effluent, with the signals used for information. Plant staff will be able to determine the TSS removal across the primary clarifiers and can remotely adjust the ferric dosage for optimal removal. The downstream orthophosphate signal will let staff know if there is a danger of removing too much phosphorus. The alkalinity analyzer signal, when combined with the signals from new pH probes in the aeration basins, will inform the plant staff if supplemental alkalinity is needed. Orthophosphate and ammonia analyzers are also provided downstream of the secondary clarifiers before the filters, and additional ferric chloride feed points provided upstream of the secondary clarifiers and effluent filters will be used when effluent polishing is needed.

Traverse City Regional Wastewater Treatment Plant, Traverse City, Michigan. Annual growth rates for Grand Traverse County will exert increasing

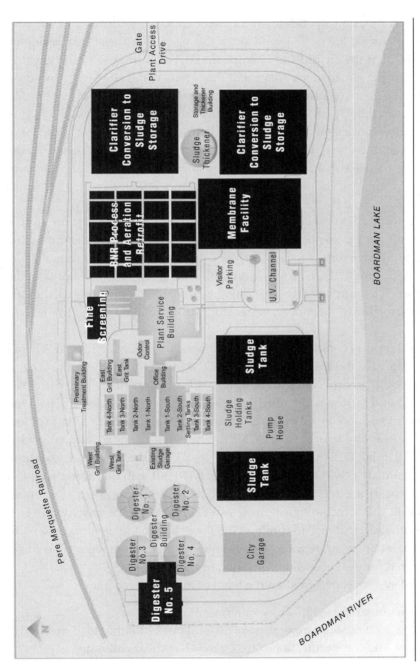

FIGURE 4.34 Process additions and improvements designed for the Traverse City Wastewater Treatment Plant (Crawford and Lewis, 2003).

demand on the area's wastewater treatment capacity. Capacity to treat a 40% increase in BOD load is expected to be required over the next 25 years. Area residents expressed a desire that the city and townships make maximum use of the existing Traverse City Regional Wastewater Treatment Plant (RWWTP) site before building new collection systems or constructing a new treatment plant. At the current site, no opportunity to expand the plant footprint exists. As shown in Figure 4.34, the facility is bordered by railroad tracks to the north, Boardman Lake to the south, Boardman River to the west, and the recently developed Hull Park and regional library to the east. Additionally, increase in effluent loadings to surface waters was another concern; in fact, the community voluntarily decided to reduce effluent loadings to levels much lower than suggested by the Department of Environmental Quality.

The current permit also establishes nutrient limits of 11 mg/L for ammonia-nitrogen and an effluent phosphorus limit of 1 mg/L. To provide a desirable discharge to the natural environment, the Traverse City has set voluntary target objectives of 4 mg/L TSS, 4 mg/L BOD, 1 mg/L NH_3-N, and 0.5 mg/L phosphorus in the effluent. The space constraints and these stringent discharge requirements led to the selection of a combination of membrane bioreactor (MBR) technology and EBPR techniques.

A unique BNR process configuration was developed by CH2M HILL and is being used for the first time for the Traverse City RWWTP design. The biological process is a variation of the UCT process configuration, with adjustments made because of the highly oxygenated and nitrate-rich nature of the mixed liquor recirculation. As in the UCT process, the anoxic and anaerobic zones and recycles are configured to promote the formation of PAOs and to accomplish EBPR. The existing anoxic and anaerobic zones, totaling 27.9 mil. L (7.37 mil. gal) and therefore 23% of the total bioreactor tank volume will be retained, as will the existing 28 mil. L (18 mil. gal) of aerated zones. A new tank will be constructed to accommodate the membrane equipment, containing 24.6 mil. L (6.5 mil. gal). The biological process will continue to be provided in two parallel trains, with the capability to combine the flows before the membrane separation stage or to maintain separate trains throughout the membrane separation stage, as might be required for testing and optimization of the process.

The biological system is designed on the basis of following criteria:

(1) Solids retention time = 12.7 days, minimum;
(2) Minimum wastewater temperature = 13°C (at peak monthly loading); and
(3) Maximum MLSS concentration = 10 000 mg/L at the membranes.

In recognition of the occasional variation in VFA content of the wastewater, the existing provisions for metal salt addition are retained in the new design. Project construction was completed in 2004. .

REFERENCES

Barnard, J. L. (1974) Cup P and N without Chemicals. *Water Wastes Eng.*, **11** (7) 33.

Barnard, J. L.; Stevens, G. M.; Leslie, P. J. (1985) Design Strategies for Nutrient Removal Plant. *Water Sci. Technol.*, **17** (11/12), 233–242.

Beatons, D.; Vanrolleghem, P. A.; vanLoosdtrecht, M. C. M.; Hosten, L. H. (1999) Temperature Effects in Bio-P Removal. *Water Sci. Technol.*, **39** (1), 215–225.

Brdjanovic, D.; van Loodsdrecht, M. C. M.; Hooijmans, C. M.; Alaerts, G. J.; Heijnen, J. J. (1997) Temperature Effects on Physiology of Biological Phosphorus Removal. *J. Environ. Eng.*, **123** (2), 144–153.

CH2M Hill, Denver, Colorado (2002) A Review of Phosphorus Removal Technologies in North America. White Paper.

Choi, E.; Rhu, D.; Yun, Z.; Lee, E (1998) Temperature Effects on Biological Nutrient Removal System with Municipal Wastewater. *Water Sci. Technol.*, **37** (9), 219–226.

Crawford, G.; Lewis, R. (2003) Traverse City Membrane Bioreactor Facility: The Largest in North America and a Sustainable Solution for the Future. *Proceedings of the 76th Annual Water Environment Federation Technical Exhibition and Conference* [CD-ROM], Los Angeles, California, Oct 11–15; Water Environment Federation: Alexandria, Virginia.

Crawford, G. V.; Elliott, M. D.; Black, S. A.; Daigger, G. T.; Stafford, D. (1999) Step Feed Biological Nutrient Removal: Design and Operating Experience. *Proceedings of the 72nd Annual Water Environment Federation Technical Exposition and Conference* [CD-ROM], New Orleans, Louisiana, Oct 2–6; Water Environment Federation: Alexandria, Virginia.

Daigger, G. T.; Randall, C. W.; Waltrip, G. D.; Romm, E. D. (1987) Factors Affecting Biological Phosphorus Removal for the VIP Process, a High-Rate University of Cape Town Type Process. In *Biological Phosphate Removal from Wastewaters*, R. Ramadori (Ed.); Pergamon Press: Oxford, England.

Ekama, G.; Marais, G.; Siebritz, I. (1984) Biological Excess Phosphorus Removal. In *Theory, Design and Operation of Nutrient Removal Activated Sludge Processes*, Water Research Commission: Pretoria, South Africa.

Erdal, U. G.; Erdal, Z. K.; Randall, C. W. (accepted for publication) The Mechanisms of Enhanced Biological Phosphorus Removal Washout and Temperature Relationships. *Water Environ. Res.*

Erdal, U. G. (2002) *The Effects of Temperature on EBPR Performance and Microbial Community Structure.* Ph.D. Dissertation; Virginia Polytechnic Institute and State University: Blacksburg, Virginia.

Erdal, U. G.; Erdal, Z. K.; Randall, C. W. (2002) Effect of Temperature on EBPR System Performance and Bacterial Community. *Proceedings of the 75th Annual Water Environment Federation Technical Exhibition and Conference* [CD-ROM], Chicago, Illinois, Sep 28–Oct 2; Water Environment Federation: Alexandria, Virginia.

Erdal, U. G.; Erdal, Z. K.; Randall, C. W. (2003a) The Competition between PAOs (Phosphorus Accumulating Organisms) and GAOs (Glycogen Accumulating Organisms) in EBPR (Enhanced Biological Phosphorus Removal) Systems at Different Temperatures and the Effects on System Performance. *Water Sci. Technol.*, **47** (11), 1–8.

Erdal, U. G.; Erdal, Z. K.; Randall, C. W. (2003b) The Mechanisms of EBPR Washout and Temperature Relationships. *Proceedings of the 76th Annual Water Environment Federation Technical Exhibition and Conference* [CD-ROM], Los Angeles, California, Oct 11–15; Water Environment Federation: Alexandria, Virginia.

Erdal, Z. K. (2002) *An Investigation of Biochemistry of Biological Phosphorus Removal Systems.* Ph.D. Dissertation; Virginia Polytechnic Institute and State University: Blacksburg, Virginia.

Erdal, Z. K.; Erdal, U. G.; Randall, C. W. (2004) Biochemistry of Enhanced Biological Phosphorus Removal and Anaerobic COD Stabilization. *Proceedings of International Water Association, Biennial Conference*, Marrakech, Morocco, September; International Water Association: London.

Filipe, C. D. M.; Daigger, G. T.; Grady, C. P. L. Jr. (2001) Effects of pH on the Rates of Aerobic Metabolism of Phosphate-Accumulating and Glycogen-Accumulating Organisms. *Water Environ. Res.*, **73**, 213–222.

Goodwin, S. J.; Neeley, K.; Eller, K.; Gullet, B.; Wagoner, D.; Daigger, G. (2003) Cost-Effective Phosphorus Removal at Charlotte's McAlpine Creek Wastewater Management Facility. *Proceedings of the 76th Annual Water Environment Federation Technical Exhibition and Conference* [CD-ROM], Los Angeles, California, Oct 11–15; Water Environment Federation: Alexandria, Virginia.

Grady, C. P. L. Jr.; Daigger, G. T.; Lim, H. C. (1999) *Biological Wastewater Treatment*, 2nd ed.; Marcel Dekker, Inc.: New York.

Helmer, C.; Kunst, S. (1997) Low Temperature Effects on Phosphorus Release and Uptake by Microorganisms in EBPR Plants. *Water Sci. Technol.*, **37** (4–5), 531–539.

International Water Association Task Group on Mathematical Modeling for Design and Operation of Biological Wastewater Treatment (2002). *Activated Sludge Models ASM1, ASM2, ASM2D and ASM3*. Scientific and Technical Report No. 9; IWA Publishing: London.

Johnson, B. R.; Daigger, G. T.; Crawford, G.; Wable, M. V.; Goodwin, S. (2003) Full-Scale Step-Feed Nutrient Removal Systems: A Comparison between Theory and Reality. *Proceedings of the 76th Annual Water Environment Federation Technical Exhibition and Conference* [CD-ROM], Los Angeles, California, Oct 11–15; Water Environment Federation: Alexandria, Virginia.

Jones, M.; Stephenson, T. (1996) The Effect of Temperature on Enhanced Biological Phosphorus Removal. *Environ. Technol.*, **17**, 965–976.

Kang, S. J.; Hong, S. N.; Tracy, K. D. (1985) Applied Biological Phosphorus Technology for Municipal Wastewater by the A/O Process. *Proceedings of the International Conference on Management Strategies for Phosphorus in the Environment*; Selper, Ltd.: London.

Kisoglu, Z.; Erdal, U. G.; Randall, C. W. (2000) The Effect of COD/TP Ratio on Intracellular Storage Materials, System Performance and Kinetic Parameters in a BNR System. *Proceedings of the 73rd Annual Water Environment Federation Technical Exposition and Conference* [CD-ROM], Anaheim, California, Oct 14–18; Water Environment Federation: Alexandria, Virginia.

Liu, W.; Nakamura, K.; Matsuo, T.; Mino, T. (1997) Internal Energy-Based Competition between Polyphosphate- and Glycogen-Accumulating Bacteria in Biological Phosphorus Removal Reactors-Effect of P/C Feeding Ratio. *Water Res.,* **31** (6), 1430–1438.

Mamais, D.; Jenkins, D. (1992) The Effects of MCRT and Temperature on Enhanced Biological Phosphorus Removal. *Water Sci. Technol.,* **26** (5–6), 955–965.

Marklund, S.; Morling, S. (1994) Biological Phosphorus Removal at Temperatures from 3 to 10°C-A Full-Scale Study of a Sequencing Batch Reactor Unit. *Can. J. Civ. Eng.,* **21**, 81–88.

McClintock, S. A.; Randall, C. W.; Pattarkine, V. M. (1991) Effects of Temperature and Mean Cell Residence Time on Enhanced Biological Phosphorus Removal. *Proceedings of the 1991 Specialty Conference on Environmental Engineering,* Krenkel, P. A., Ed.; Reno, Nevada, Jul 10–12; American Society of Civil Engineers: Reston, Virginia; 319–324.

Mino, T.; Arun, V.; Tsuzuki, Y.; Matsuo, T. (1987) Effect of Phosphorus Accumulation on Acetate Metabolism in the Biological Phosphorus Removal Process. *Proceedings of the IAWPRC Specialized Conference,* Rome, Italy, Sep 28–30; International Association on Water Quality: London; 27–38.

Panswad, T., Department of Environmental Engineering, Chulalongkom University, Bangkok, Thailand (2000) Personal communication.

Panswad, T.; Doungchai, A.; Anotai, J. (2003) Temperature Effect on Microbial Community of Enhanced Biological Phosphorus Removal Systems. *Water Res.,* **37** (2), 409–415.

Punrattanasin, W. (1999) *Investigation of the Effects of COD/TP Ratio on the Performance of a Biological Nutrient Removal System.* M.S. Thesis; Virginia Polytechnic Institute and State University: Blacksburg, Virginia.

Randall, C. W.; Barnard J. L.; Stensel, D. H. (1992) *Design and Retrofit of Wastewater Treatment Plants for Biological Nutrient Removal.* Technomic Publishing Co.: Lancaster, Pennsylvania.

Schuler, A. J.; Jenkins, D. (2003) Enhanced Biological Phosphorus Removal from Wastewater by Biomass with Varying Phosphorus Contents, Part I: Experimental Methods and Results. *Water Environ. Res.,* **75** 485–498.

Sell, R. L. (1981) Low Temperature Biological Phosphorus Removal. Paper presented at the 54th Annual Water Pollution Control Federation Technical Exposition and Conference, Detroit, Michigan, Oct 4–9.

Siebrietz, I. P. (1983) *Biological Excess Phosphorus Removal in the Activated Sludge Process.* Ph.D. Dissertation; University of Cape Town: South Africa.

Spector, M. L. (1979) U.S. Patent 4 162 153, July 24.

Wang, J.; Park, J. K. (1998) Effect of Wastewater Composition on Microbial Populations in Biological Phosphorus Removal Processes. *Water Sci. Technol.,* **38** (1) 159–166.

Water Environment Federation (1998) *Biological and Chemical Systems for Nutrient Removal.* Special Publication; Water Environment Federation: Alexandria, Virginia.

Chapter 5

Combined Nutrient Removal Systems

Combined Nitrogen and Phosphorus Removal Processes — 162	General Remarks about the Various Process Configurations — 174
Flow Sheets for Combined Nutrient Removal — 162	Interaction of Nitrates and Phosphorus in Biological Nutrient Removal Plants — 178
The Five-Stage Bardenpho Process — 165	Process Control Methodologies — 182
Phoredox (A^2/O) Process — 166	Effect of Oxygen — 182
The University of Cape Town and Virginia Initiative Processes — 166	Temperature Effects — 183
	pH Effects — 184
Modified University of Cape Town Process — 168	Sufficient Dissolved Oxygen in the Aeration Zone — 185
Johannesburg and Modified Johannesburg Processes — 168	Chemical Oxygen Demand to Total Kjeldahl Nitrogen Ratio — 186
Westbank Process — 169	Selection of Aeration Device — 187
The Orange Water and Sewer Authority Process — 171	Clarifier Selection — 187
Phosphorus Removal Combined with Channel-Type Systems — 171	Effect of Chemical Phosphorus Removal on Biological Nutrient Removal Systems — 189
	Primary Clarifiers — 189
Cyclical Nitrogen and Phosphorus Removal Systems — 173	Secondary Clarifiers — 190
	Tertiary Filters — 190

Process Selection for Combined Nitrogen and Phosphorus Removal	192	*Problem*	203
		Correction	203
		Problem	203
Effluent Requirements	192	*Correction*	203
Phosphate Removal but No Nitrification	192	*Problem*	203
		Correction	204
Phosphate Removal with Nitrification but No Denitrification	193	*Problem*	204
		Correction	204
Phosphate Removal with Nitrification Only in Summer	194	Plant Designed for Nitrification and Denitrification	204
		Problem	204
High-Percentage Nitrogen and Phosphate Removal	195	*Correction*	204
		Problem	204
Benefits from Converting to Biological-Nutrient-Removal-Type Operation	196	*Correction*	205
		Problem	205
		Correction	205
Reliable Operation	196	Plant Designed for Phosphorus Removal Only	206
Restoring Alkalinity	198	*Problem*	206
Improving the Alpha Factor	199	*Correction*	206
Improving Sludge Settleability	199	*Problem*	206
		Correction	206
Troubleshooting Biological Nutrient Removal Plants	202	*Problem*	207
		Correction	207
Plant not Designed for Nitrification but Nitrification in Summer Causes Problems	202	Plant Designed for Ammonia and Phosphorus Removal	207
		Problem	207
Problem	203	*Correction*	207
Correction	203	*Problem*	207
Problem	203	*Correction*	207
Correction	203	*Problem*	207
Problem	203	*Correction*	208
Correction	203	*Problem*	208
Plant Designed for Nitrification but No Denitrification	203	*Correction*	208
		Problem	208

Correction	208
Problem	208
Correction	208
Problem	208
Correction	208
Retrofitting Plants for Nutrient Removal	208
Return Activated Sludge and Internal Recycle Rates	212
Minimizing the Adverse Effect of Storm Flows	213
Foam Control	214
Waste Sludge and Return Stream Management	215
Summary	216
Case Studies	216
City of Bowie Wastewater Treatment Plant (Bowie, Maryland)	216
Facility Design	216
Effluent Limits	216
Wastewater Characteristics	217
Performance	217
Potsdam Wastewater Treatment Plant (Germany)	217
Facility Design	217
Performance	217
Goldsboro Water Reclamation Facility, North Carolina	217
Facility Design	217
Effluent Limits	218
Wastewater Characteristics	218
Performance	218
South Cary Water Reclamation Facility, North Carolina	218
Facility Design	218
Effluent Limits	218
Wastewater Characteristics	219
Performance	219
North Cary Water Reclamation Facility, North Carolina	219
Facility Design	219
Effluent Limits	219
Wastewater Characteristics	220
Performance	220
Wilson Hominy Creek Wastewater Management Facility, North Carolina	220
Facility Design	220
Effluent Limits	220
Wastewater Characteristics	220
Performance	221
Greenville Utilities Commission Wastewater Treatment Plant, North Carolina	221
Facility Design	221
Effluent Limits	221
Wastewater Characteristics	221
Performance	222
Virginia Initiative Plant, Norfolk	222
Facility Design	222
Effluent Limits	222
Wastewater Characteristics	222
Performance	222
References	223

COMBINED NITROGEN AND PHOSPHORUS REMOVAL PROCESSES

In the previous chapter, the mechanism of phosphorus removal was explained, and the possible interference of nitrates in the mechanism for biological phosphorus removal (BPR) was discussed. This chapter will focus on the interaction between nitrates and phosphorus removal in combined systems.

While the removal of phosphorus in a combined nitrogen and phosphorus removal plant can be achieved both chemically and biologically, the biological alternative has a number of significant advantages, such as considerably lower operating costs, less sludge production, and little or no added chemicals in the sludge. Total nitrogen (TN) removal in a combined nitrogen and phosphorus removal wastewater plant is most commonly achieved in a two-zone system through nitrification (under aerobic conditions) and denitrification (under anoxic conditions).

Because of the possible interference of nitrates on the anaerobic conditions required for the BPR process, at least three zones or periods in intermittent systems (anaerobic, anoxic, and aerobic) are required to provide the different environmental conditions for combined nitrogen and phosphorus removal. Additionally, the possible interferences between nitrogen and phosphorus removal processes often mean that additional zones or periods are required, for example, for the removal of nitrate in the recycled activated sludge (Keller et al., 2001). In this chapter, combined nitrogen and phosphorus removal systems are discussed, with the operational conditions, process control strategies, and troubleshooting, followed by some successful case studies of combined nitrogen and phosphorus removal.

FLOW SHEETS FOR COMBINED NUTRIENT REMOVAL

Biological processes for removal of nitrogen and phosphorus from wastewater may be incorporated to the typical activated sludge secondary treatment process with relative ease. All combined nitrogen and phosphorus removal flow sheets have certain stages in common, namely the following:

- An anaerobic zone free of dissolved oxygen and nitrate,
- An anoxic zone for denitrification of nitrates formed during nitrification, and
- An aerobic nitrification zone for the conversion of ammonia to nitrates.

Sometimes the nitrification zone is designed to achieve a degree of simultaneous nitrification and denitrification (SND) (see the Phosphorus Removal Combined with Channel-Type Systems section).

There are several variations of combined nitrogen and phosphorus removal processes configurations based on influent characteristics, effluent limits, and desired operating conditions. All of these processes include the same basic anaerobic/anoxic/aerobic components to achieve enhanced biological phosphorus uptake and nitrification/denitrification. The biological transformations and the functions achieved in each zone are given in Table 5.1. Filtration of the final effluent through sand or other media is required for the removal of particulate matter when low effluent nitrogen and phosphorus concentrations are required. The function of sand filtration could be combined with attached growth denitrification for further reduction of soluble nitrates in the effluent. In a later discussion, the interaction of nitrates and phosphates in these filters will be discussed.

The choice of a particular process type from the various process schemes and their modifications for the combined removal of nitrogen and phosphorus is mostly dependent on the influent wastewater characteristics, nutrient removal requirements, site, and cost constraints.

The dominant TN and total phosphorus (TP) removal technologies being implemented by wastewater treatment plants producing low nitrogen and phosphorus effluents are based on nitrification/denitrification and enhanced biological phosphorus removal (EBPR) processes, including the Bardenpho process and the University of Cape Town (UCT) process followed by or combined with physicochemical methods. In both cases, enhancement of the nitrogen and phosphorus removal by chemical addition (i.e., methanol for denitrification and acetate; or fermentate or precipitants for phosphorus removal) is practiced to achieve low TN and TP effluents on a sustainable basis. Solids separation from effluents by filtration following secondary sedimentation in the activated sludge process significantly aids in the production of low nitrogen- and phosphorus-containing effluents by reducing particulate nitrogen and phosphorus fractions in the effluent. To further lower the phosphorus concentration to very low levels, postchemical precipitation followed by filtration and/or membrane filtration of the tertiary effluents are possible methods. The speciation and reactivity of the remaining phosphorus in the secondary effluent would be critical to achieve <100 g/L.

The most common process configurations used to achieve both nitrogen and phosphorus removal are shown in the flow sheets referenced in the following sections.

TABLE 5.1 Summary of BNR process zones.

Zone	Biological transformation	Functions	Zone required for
Anaerobic	Uptake and storage of VFAs by PAOs*	Selection of PAOs	Phosphorus removal
	Fermentation of readily biodegradable organic matter by heterotrophic bacteria		
Anoxic	Denitrification	Conversion of nitrate to nitrogen gas	Nitrogen removal
	Alkalinity production	Selection of denitrifying bacteria	
Aerobic	Nitrification	Conversion of ammonia to nitrite and nitrate	Nitrogen removal
	Metabolism of stored and exogenous substrate by PAOs	Nitrogen removal through gas stripping	Phosphorus removal
	Phosphorus uptake	Formation of polyphosphate	
	Alkalinity consumption		

*VFAs = volatile fatty acids and PAOs = phosphate-accumulating organisms.

THE FIVE-STAGE BARDENPHO PROCESS. Barnard (1976) proposed a number of flow sheets for combined nitrogen and phosphorus removal. They all consisted of adding an anaerobic zone ahead of either a high rate non-nitrifying aeration basin (later called anaerobic/oxic [A/O]) or ahead of an modified Ludzack-Ettinger (MLE) process (later called anaerobic/anoxic/oxic [A^2/O]) or ahead of the four-stage Bardenpho process, which was then referred to as the modified or five-stage Bardenpho process. The aim was to ensure that the nitrates in the return activated sludge (RAS) could be reduced to avoid interfering with the BPR process. The five-stage Bardenpho process is illustrated in Figure 5.1. All recycle rates are defined in terms of the average influent flowrate, Q. The four-stage Bardenpho process described in Chapter 3 can reduce effluent nitrates to less than 2 mg/L. There will thus be little nitrates in the RAS to interfere with the mechanism for phosphorus removal.

In the five-stage Bardenpho process, influent municipal wastewater combines with return sludge from the secondary clarifier of a wastewater treatment plant in an anaerobic zone. In this zone, phosphorus is released, and substrate is stored by the phosphorus-accumulating bacteria, in the absence of nitrate. The flow then enters an anoxic zone, where it combines with a recycle stream containing nitrate from nitrified mixed liquor recycled from the end of the aerobic zone. In this unaerated (anoxic) zone, nitrate is reduced to nitrogen gas. The flow then enters an aerobic zone, where ammonia nitrogen is oxidized to nitrate. Mixed liquor is recycled at a rate of between 2 and 5 times the average influent flowrate, Q. The remainder of the

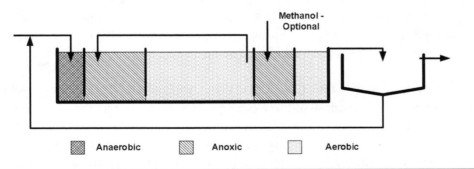

FIGURE 5.1 Five-stage (modified) Bardenpho process.

mixed liquor enters a second anoxic zone for further nitrate reduction, then a reaeration zone, and then passes to the final clarifiers. Phosphorus is removed from the process in accordance with the theory described in the previous chapters, along with the sludge wasted either from the clarifier underflow or the last aerobic zone (waste activated sludge). The final aerobic zone serves to take up phosphorus and to prevent the occurrence of anaerobic conditions and the associated release of phosphorus from the sludge to the final effluent.

The intent of the original five-stage Bardenpho process was first to reduce nitrogen to very low levels, such as in Florida, where the requirement is for effluent TN concentrations of less than 3 mg/L. At this low level of the effluent nitrate concentration, coupled with some unavoidable denitrification in the sludge blanket, there is no danger of recycling nitrates to the anaerobic zone through the RAS.

PHOREDOX (A^2/O) PROCESS. This process consists of the MLE process for nitrogen removal, with an anaerobic zone in front for phosphorus removal (Figure 5.2). The RAS is recycled to the anaerobic zone. The anaerobic zone is followed by an anoxic zone for nitrogen removal through denitrification. Even where only ammonia removal is required, the anoxic zone is often included to reduce the nitrate loading on the anaerobic zone through the RAS flow streams. Otherwise, the nitrate concentration in the RAS could reduce the efficiency and effectiveness of phosphorus removal.

Mixed liquor is recycled from the aerobic zone to the anoxic zone at a rate of $2Q$ to $5Q$. Nitrogen removals of 40 to 70% have been achieved with this process. Because nitrate concentrations between 5 and 10 mg/L may still be present in the RAS, the phosphorus removal capability of the Phoredox process is reduced unless there is a surplus of volatile fatty acid (VFA) that can denitrify the nitrates with a sufficient amount left to activate the phosphorus removal mechanism. This process has largely been replaced with others that allows for denitrification of nitrates in the RAS.

THE UNIVERSITY OF CAPE TOWN AND VIRGINIA INITIATIVE PROCESSES. The purpose of these configurations is to protect the anaerobic zone against nitrates, even when the effluent nitrates are high. The UCT process consists of anaerobic, anoxic, and aerobic zones, with the RAS directed to the anoxic zone for achieving denitrification (Ekama and Marais, 1984) (Figure 5.3). The inlet wastewater flows directly to the anaerobic zone, which provides a source of organic matter to the anaerobic zone. Nitrified mixed liquor from the aerobic zone is pumped to the anoxic

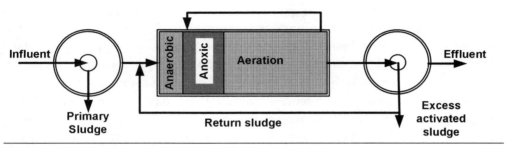

FIGURE 5.2 Three-stage Phoredox ($A^2/0$) process.

zone at rates of $2Q$ to $4Q$ to increase nitrogen removal through denitrification. Denitrified effluent from the end of the anoxic zone is recycled to the anaerobic zone at a rate of $1Q$ to provide the microorganisms needed there for phosphorus removal to occur. The Virginia Initiative Process (VIP) process is a high-rate version of the UCT process (Daigger et al., 1988). This process is designed as a high-rate process, and all zones consist of at least two cells in series. The idea behind these processes is that, when there is not sufficient carbon in the influent for nitrogen and phosphorus removal, preference could be given to phosphorus removal. For example, if the effluent nitrate is high, as are the nitrates in the RAS, the nitrates in the RAS could be denitrified in the anoxic zone before it is recycled back to the anaerobic zone. If the anoxic zone cannot remove all the nitrates, the mixed liquor recycle can be reduced to ensure that the nitrates at the end of the anoxic zone will be low, thus ensuring that

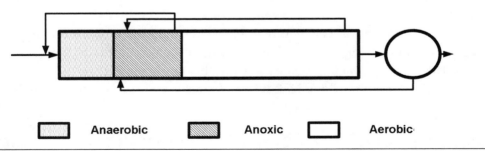

FIGURE 5.3 The UCT and VIP process configurations.

nitrates will not be recycled to the anaerobic zone. The negative side of the process is that it is difficult to control the nitrates in the effluent of the anoxic zone, and, when the zone runs out of nitrates too soon, secondary release of phosphorus can take place. This led to the development of the modified UCT process. The philosophy behind the UCT process is that, when there is not sufficient carbon for both nitrogen and phosphorus removal, preference should be given to phosphorus removal. The mixed liquor recycle from the aeration to the anoxic zone could be reduced to ensure that all nitrates are removed before passing the mixed liquor to the anaerobic zone. This may then result in higher effluent nitrates. If, in a situation like this, there is a need for further reduction in nitrates, a post-attached growth denitrification system, such as denitrifying sand filters or moving bed biofilm reactor, could be used with methanol for denitrification.

MODIFIED UNIVERSITY OF CAPE TOWN PROCESS. In this process, the anoxic zone is divided into two cells (Figure 5.4). The first anoxic zone receives the effluent of the anaerobic zone and the RAS. The denitrified mixed liquor is then recycled to the anaerobic zone. The first anoxic zone is therefore required to reduce only the nitrates in the RAS and can be better controlled. The second anoxic cell receives the mixed liquor from the aerobic zone, where bulk denitrification occurs. Accurate control of the mixed liquor recycle is not a requirement. Primary influent mixed with denitrified anoxic recycle from the first anoxic cell is routed into the anaerobic zone, where multistaged compartments are used to minimize nitrate recycle to the anaerobic zone. Flow from the anaerobic zone then enters the first anoxic cell; the flow from the first anoxic cell then enters the second anoxic cell. From there, it flows to the aerobic zone, where nitrification occurs. The nitrified aerobic mixed liquor is recycled back to the second anoxic cell for denitrification. It is not important that the mixed liquor recycle to the second anoxic cell be accurately controlled because it will not affect the return of nitrates to the anaerobic zone.

The modification was very effective, and a large number of modified UCT plants are in operation, giving very good performance of removal of phosphorus to low levels. The only disadvantage of the process is that the mixed liquor concentration in the anaerobic tank is only one-half of that of other processes, requiring double the basin volume for the same anaerobic mass fraction.

JOHANNESBURG AND MODIFIED JOHANNESBURG PROCESSES. The City of Johannesburg, South Africa, solved their problem of nitrates in RAS by

Combined Nutrient Removal Systems

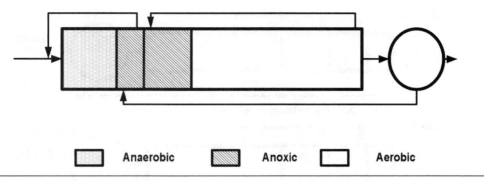

FIGURE 5.4 Modified UCT process.

providing an endogenous pre-denitrification zone, as illustrated in Figure 5.5 (Pitman, 1992). The mixed liquor suspended solids (MLSS) concentration in the pre-anoxic zone is typically approximately twice that of the MLSS concentration in the following zones, which results in endogenous denitrification taking place at a much higher rate, reducing the nitrates to very low levels (Johannesburg [JHB] process). This was later improved by adding a recycle stream from the end of the anaerobic zone to the preanoxic zone to use readily biodegradable compounds not taken up by the phosphate-accumulating organisms (PAOs) for denitrification (modified JHB [MJHB] process). This recycle stream may be as low as 10% of the influent flow, ensuring a cheap way of providing more carbon for denitrification before the anaerobic zone. Because of the high concentration of solids in the predenitrification zone, the volume of this zone may be as little as 25% of the first anoxic zone of the modified UCT process.

WESTBANK PROCESS. The Westbank process was developed at the same time as the JHB process; however, because of the low influent VFA in British Columbia, primary sludge was fermented to produce a stream of VFA that was discharged directly to the anaerobic zone (Figure 5.6). The RAS was also returned to a preanoxic zone, with some of the influent (5 to 10%) added to the RAS to assist in denitrification. From $0.6Q$ to $1.2Q$ of the remainder of the influent was passed to the anaerobic zone with a stream of fermentate. The Q is the average dry weather flow. Flow in

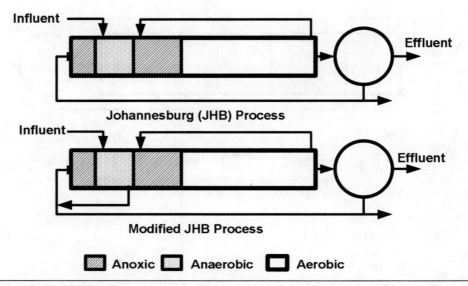

FIGURE 5.5 Johannesburg and modified Johannesburg processes.

excess of 1.2Q was passed directly to the main anoxic zone to maintain a minimum retention time in the anaerobic zone for VFA uptake and phosphorus release. Controlled denitrification of the RAS is possible in a small preanoxic zone.

The advantage of the process is that endogenous denitrification is used through the high concentration of solids in the preanoxic zone, assisted by some influent carbon from approximately 5 to 10% of the feed. The MLSS in the preanoxic zone and in the anaerobic zone will be higher than in the reminder of the tank, depending on the percentage of feed discharged to these sections. For example, if the production of VFA in the fermenter is sufficient, the percentage of the feed going to the anaerobic zone could be reduced to increase the retention time in the anaerobic zone. There is thus a great degree of flexibility. Effluent nitrates may range from 3 to 8 mg/L, but this will not have an effect on the ability of the plant to remove phosphorus. The Westbank plant in Canada is presently operated with no primary effluent to the anaerobic zones.

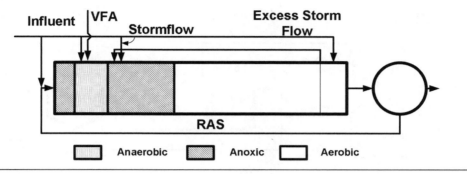

FIGURE 5.6 Westbank process.

THE ORANGE WATER AND SEWER AUTHORITY PROCESS. This process consists of anaerobic, anoxic, and aerobic zones, as in most other nitrogen and phosphorus removal processes (Kalb et al., 1990) (Figure 5.7). However, the nutrition (anaerobic) zone is provided in a sidestream reactor. The primary settled wastewater flows to trickling filters and then to the aerobic zone, where biochemical oxygen demand (BOD) is reduced and ammonia is converted to nitrate or nitrite; then to the anoxic zone for denitrification; and finally to another aerobic zone for stripping of nitrogen gas generated by denitrification. Primary sludge is fermented to increase production of VFAs. Return activated sludge and fermenter supernatant are combined in the nutrition (anaerobic) zone to facilitate phosphorus release. The fermented sludge is passed to anaerobic digesters. Phosphorus uptake occurs in the aerobic zone, and concentrations in the aerobic effluent are typically less than 1.0 mg/L (Kalb and Roeder, 1992).

PHOSPHORUS REMOVAL COMBINED WITH CHANNEL-TYPE SYSTEMS. Looped channel systems such as Pasveer, Carousel, and Orbal systems are described in the Design of Municipal Wastewater Treatment Plants (WEF, 1998b). A feature of channel systems is that they use point-source aerators, such as surface aerators and brush aerators that push the mixed liquor around the looped channel, passing it through aerated and unaerated zones resulting in various degrees of nitrogen removal by simultaneous nitrification and denitrification (SND). Theoreti-

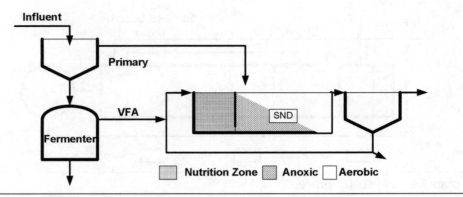

FIGURE 5.7 Orange Water and Sewer Authority process.

cally, all that would be needed for phosphorus removal is to add an anaerobic zone in front and perhaps a reaeration zone after the channel, as was suggested by Barnard (1976). Channel systems are excellent for nitrogen removal when the solids retention time (SRT) is in excess of 15 to 20 days. At such SRTs, they provide more flexibility and longer periods of anoxic conditions between aeration points. At shorter SRTs (8 to 12 days), it is more difficult to balance the SND in the process to achieve complete ammonia removal as well as a high rate of nitrate reduction. The addition of an anoxic zone ahead of the channel with mixed liquor recycle from the channels to the anoxic zone, have resulted in more reliable operation and control (deBarbadillo et al., 2003). These arrangements allow for optimizing ammonia removal in the aerated channel and achieve a degree of SND in addition to the removal of nitrates in the anoxic zone. For phosphorus removal, an anaerobic zone is added ahead of the anoxic zones to get a three-stage Phoredox process (A^2O) with the channel as the aeration basin.

Phosphorus removal to well below 1 mg/L also requires that the SRT be kept as short as possible. At longer SRTs (in excess of 15 days), there is more endogenous breakdown of the sludge; thus, less sludge production and the phosphorus content of the sludge must be higher for the same degree of removal. Also, some endogenous release may take place when all the carbon has been removed and the nitrates are less than 2 mg/L.

The advantages of using a channel system as an aeration basin in biological nutrient removal (BNR) plants is well-demonstrated by the success of the five-stage Bardenpho plants that have a Carrousel channel system as an aeration basin. The TN is removed to below 3 mg/L, while phosphorus can be removed to low values. Some case studies will be discussed later. When using surface or other point-source aerators in the aerobic zone of any plant instead of fine-bubble aeration, a higher degree of nitrate reduction takes place as a result of SND. This may reduce the nitrate in the RAS to sufficiently low levels so that the three-stage Phoredox (A^2/O) process could be used without resorting to a preanoxic zone for reducing the nitrates in the RAS. The degree of SND is increased when using slow-speed surface aerators because of the higher pumping action of the larger aerators that are required. Numerous plants with surface aerators function efficiently in this mode of operation because of the ability to pump mixed liquor to the underaerated zone while aerating (deBarbadillo et al., 2003).

CYCLICAL NITROGEN AND PHOSPHORUS REMOVAL SYSTEMS.

Sequencing batch reactors (SBRs) are extremely flexible, relatively inexpensive, and very effective treatment systems for small- to medium-sized facilities. The ability to vary the duration of the aerated and unaerated phases of the SBR cycle provides the flexibility to obtain optimum nitrogen and phosphorus removal. Compared to channel systems, SBR has great flexibility and good performance when the SRT is over 20 days.

Sequencing batch reactors for phosphorus removal consist of a large batch reactor with a reseeding anaerobic zone. The typical sequence for phosphorus removal incorporates a number of stages, namely the following:

- An anoxic idle period, during which nitrates in the sludge is reduced;
- Feed to an anaerobic zone, with recycle of sludge to this zone from the anoxic stage;
- Aeration of the main reactor, while still feeding to the anaerobic zone;
- Aerobic reaction;
- Settling; and
- Decanting in the main tank.

Figure 5.8 shows two basins of a multiple basin configuration. Each process occurs at a different time rather than in a different tank compared to conventional, continuous-flow, activated sludge processes. After decanting, the sludge is again kept under anoxic conditions to reduce nitrates. Mixed liquor is pumped from under the decanters to the anaerobic zone, where it is contacted with the feed that contains VFA, and phosphorus is released. The mixed liquor then overflows into the batch reactor during a mixing and no-aeration (anoxic) period, when denitrification takes place. This is followed by the aeration stage, during which nitrification and phosphorus uptake takes place. Endogenous denitrification takes place during settling and decant. It is essential that the remainder of the nitrates be removed during reaction and settling phases to achieve better phosphate removal in the next cycle (Kazmi and Furumai, 2000).

GENERAL REMARKS ABOUT THE VARIOUS PROCESS CONFIGURATIONS. Typical design parameters for the commonly used combined BNR systems are presented in Table 5.2. In these BNR processes, nitrogen and phosphorus removal capability is a function of the percentage and content of mixed liquor recycle rate to

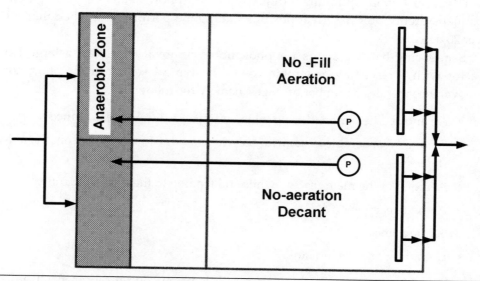

FIGURE 5.8 Example of SBR for nitrogen and phosphorus removal.

the anoxic zone and RAS recycle rate to the anaerobic zone. Because the RAS contains some nitrate, the rate should be as low as possible, which would bring less nitrate back to the anaerobic zone. Screw pumps for RAS should be avoided, if at all possible, because they serve as an excellent aeration device, introducing oxygen to the anaerobic zone.. Also, the control of the RAS rate is more cumbersome and must be strictly controlled for each final clarifier. Uncaring operators tend to increase the RAS rates to the flow that the pumps can handle. This led to RAS recycle rates of 4Q, with loss of phosphorus removal in some plants.

Lower BPR efficiency is observed for systems with longer SRTs. At long SRTs, the PAOs are in a more extended endogenous phase, which will deplete more of their intracellular storage products, resulting in less efficient acetate uptake and storage of intermediate products in the anaerobic zone and less uptake of phosphorus in the aerobic zone (Stephens and Stensel, 1998). The SRT in a combined nitrogen and phosphorus removal plant should be just sufficient to allow for reliable nitrification. The SRT should also be varied in winter and summer, in accordance with the minimum requirements for nitrification at different temperatures.

While there are many processes with many different names, nothing should withhold plant operators from integrating processes or borrowing from other processes where necessary to get the best results. Following are a few examples:

- Converting an existing channel system that has reliable nitrogen removal to a nitrogen and phosphorus removal process will require the addition of an anaerobic basin to which the RAS is returned. If the channel system is operated for low ammonia and there are some nitrates in the effluent, a preanoxic zone could be provided to get rid of the nitrates before entering the anaerobic zone.

- Collection systems in flat tropical areas like Florida generally have slow-flowing sewers, and there is a need for pumping long distances. Forced mains are excellent generators of VFA, and it is possible that there is sufficient VFA in the influent so that nitrates in the RAS may not be a problem in terms of phosphorus removal. For example, good phosphorus removal was obtained in the old aeration plant at Reedy Creek, Florida, by switching off aerators in the first one-half of the first pass out of four passes. While nitrates were recycled to the first pass in the RAS, there was sufficient VFA in the influent to remove the nitrates and still leave sufficient for phosphorus removal. Denitrification takes some time, and, during this period, the PAOs can take up suffi-

TABLE 5.2 Typical design parameters for commonly used biological nitrogen and phosphorus removal processes.[a]

Design parameter or process	SRT, d[b]	MLSS, g/L	HRT, h[c] Anaerobic zone	HRT, h[c] Anoxic zone	HRT, h[c] Aerobic zone	RAS, % of influent	Internal recycle, % of influent
A²/O	5 to 25	3 to 4	0.5 to 1.5	1.5 to 2.5	4 to 8	25 to 100	200 to 400
UCT	8 to 25	3 to 4	1 to 3	2 to 4	4 to 12	80 to 100	100 to 200 (anoxic) 100 to 200 (aerobic)
VIP	5 to 10	2 to 4	1 to 2	2 to 4	4 to 6	80 to 100	100 to 200 (anoxic) 100 to 300 (aerobic)
Bardenpho (five-stage)	10 to 20	3 to 4	0.5 to 1.5	1 to 3 (first stage) 2 to 4 (second stage)	4 to 12 (first stage) 0.5 to 1 (second stage)	50 to 100	200 to 400
SBR	20 to 40	3 to 4	1.5 to 3	1 to 3	2 to 4		

[a]Adapted from WEF, 1998b.

[b]Total SRT, including anaerobic and anoxic zones.

[c]The HRT refers to the retention time of the zones in terms of the average flow. Thus, an HRT of 1 hour refers to the volume that will be filled in 1 hour at average flow.

cient VFA to ensure the good operation of the phosphorus removal mechanism. The plant was operated in this way until it was taken out of service.

- The Clark County plant in Nevada was designed as an MLE for only nitrogen removal. The plant consisted of anoxic and aerobic zones, with mixed liquor recycle from the aerobic to the anoxic zone. There were phosphorus and ammonia standards, but not one for TN. By switching off the mixed liquor recycle pumps, the anoxic zones became anaerobic and, because there was sufficient VFA in the influent, very good phosphorus removal averaging 0.3 mg/L was obtained.

- In the Westbank plant in British Columbia, the Westbank configuration is in place. However, the production of VFA in the thickener is very good, and there is little VFA in the influent. The operators bypass most of the primary effluent to the anoxic zone and feed mostly fermentate to the anaerobic zone, with very good results. This is approaching an aspect of the methodology used by the Orange Water and Sewer Authority (OWASA) process.

- On the other hand, OWASA found that the nitrates in the RAS upsets phosphorus removal and is now using a preanoxic zone similar to that developed in Westbank. When feeding only fermentate to the anaerobic zone, the retention there may be too long, resulting in secondary release of phosphorus. This could be reduced by again increasing the flow of primary effluent to the anaerobic zone, optimizing each process according to the specific wastewater characteristics.

- One of the options for the Eagles Point plant in Cottage Grove, Minnesota, is to run a BNR plant with no anoxic zone and only a preanoxic zone. There was no need to remove nitrates, and the effluent nitrate concentration is approximately 15 mg/L. All of the nitrates in the RAS can be removed in the preanoxic zone, and it does not affect the phosphorus removal. There is a thickener/fermenter to supply VFA to the anaerobic zone.

- The Iron Bridge plant in Orlando, Florida, was loaded higher than the rated design capacity and had insufficient aeration as a result of limitation of the Carrousel aerators. The plant owners installed fine-bubble aerators in some of the passes of the Carrousel system that was part of the five-stage Bardenpho process. They found that they had excess anoxic capacity and that the existing first and second anoxic basins could deal with the nitrates that would remain after aeration of some of the unaerated channels. Great improvement in effluent quality resulted.

INTERACTION OF NITRATES AND PHOSPHORUS IN BIOLOGICAL NUTRIENT REMOVAL PLANTS

When there is a sufficient supply of VFA in the feed or from the fermenter, the biology for phosphorus removal is very forgiving. However, in many plants, the VFA supply is just sufficient or may have to be augmented. When secondary release of phosphorus takes place, more VFA is needed to take up the excess phosphorus released. Secondary release takes place in any zone where there are unaerated conditions without a supply of VFA or nitrates. Secondary release results from the breakdown of the PAOs under conditions where no growth is possible. In the anaerobic zone, when there is VFA, bacteria will release phosphorus to gain energy for subsequent phosphorus uptake in the aeration zone. When the VFA runs out, phosphorus is released without any VFA uptake, and the energy will not be available for taking up the phosphorus again. Generally, after VFA uptake and phosphorus release, the PAOs can take up more phosphorus than was released. They thus take up all of the phosphorus released plus that in the feed. If there is secondary release, then they have to take up this surplus as well. The capacity to take up excess phosphorus is limited by the available VFA. If the VFA is limiting, secondary release must be minimized. Secondary release can take place in anaerobic zones that are too large. The graphs in Figure 5.9 show the effect on phosphorus release in anaerobic zones when the VFA runs out. The top curve represents combined primary and secondary release. The primary release stops when the VFA are all absorbed, and then there is only secondary release. The zone should therefore be just large enough for the completion of the primary release to minimize the secondary release. Adjusting the RAS recycle rate or the amount of flow going to the anaerobic zone can be used for optimization.

Secondary release can also take place in the anoxic zones when the nitrates run out. Therefore, if the first or second anoxic zones are too large and run out of nitrates, secondary release will take place. As long as nitrates are present, PAOs, which can use nitrates as the electron acceptor, will take up any released phosphorus. Without nitrates and oxygen, there is no means of taking up phosphorus that is released by the other organisms, and too much secondary release takes place. This could be determined from measuring soluble phosphorus profiles through the anoxic zone.

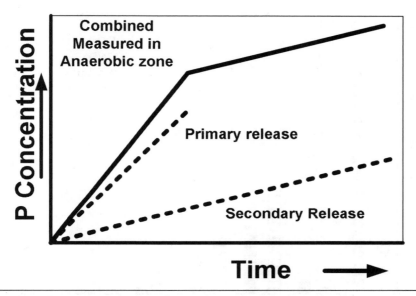

FIGURE 5.9 Primary and secondary release in the anaerobic zone.

Figure 5.10 is an example profile of a plant where, as a result of the low mixed liquor recycle rate, secondary release took place in the anoxic zone.

In Figure 5.10, the concentration of phosphorus in the anoxic zone is the same as that in the anaerobic zone of a three-stage Phoredox plant. The recycle of mixed liquor from the aeration to the anoxic zone should have diluted the phosphorus in the anoxic zone; however, because there is no reduction, there must have been secondary release. This could be overcome by increasing the mixed liquor recycle rate; the retention time in the anoxic zone will be reduced, and more nitrates and dissolved oxygen will be recycled to this zone. This will stop excessive secondary release and reduce the mass of released phosphorus that must be taken up in the aeration basin. In this particular case, the plant manager confirmed that only one out of three recycle pumps were operational.

Secondary release can take place in the second anoxic zone of a five-stage Bardenpho plant, as illustrated in Figure 5.11.

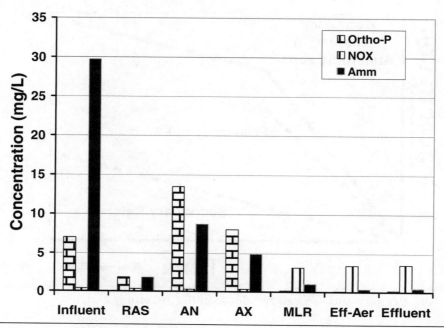

FIGURE 5.10 Profile of soluble phosphorus through Rooiwal, South Africa, plant.

In this case, the SND in the aeration basin, using draught tubes with surface aerators, was so efficient that there were no nitrates in the mixed liquor to the second anoxic zone. Phosphorus was released and could not be taken up again with aeration alone. If VFA were added to the second anoxic zone, phosphorus would be taken up in the last or reaeration zone. In this case, the solution was to aerate the second anoxic zone, and no further release took place.

Phosphorus may also be released when there is a low nitrate concentration in the mixed liquor to the final clarifiers. When a deep sludge blanket develops in the final clarifiers, nitrates will be denitrified. When the nitrates are low, phosphorus will be released. This release may not affect the effluent phosphorus directly, but may return a large portion of the released phosphorus back to the anaerobic zone, where there may not be enough VFA for uptake of this additional released phosphorus. At one plant, with a relaxed requirement for nitrogen removal but with a TP limit of less

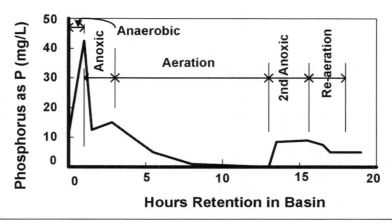

FIGURE 5.11 Secondary release of phosphorus in the second anoxic zone.

than 0.15 mg/L, the operator tried to reduce nitrates in the RAS by reducing the recycle rate. A sludge blanket formed in the clarifiers, and there was 10 mg/L of orthophosphorus in the RAS. Increasing the RAS rate reduced the secondary release of phosphorus, which resulted in the effluent orthophosphorus dropping from 1.5 to 0.08 mg/L. The main point is that, when the plant is operated for low effluent nitrates, in addition to low phosphorus, the sludge blanket should not be deep.

The counter argument for deep sludge blankets is that, when the nitrates in the effluent is high, there is an advantage in running a deeper sludge blanket at lower RAS rates because some denitrification will take place in the sludge blanket. However, carrying sludge blankets may lead to rising sludge in warmer climates, where the temperature rises above 22°C. At the same time, the reduced RAS rate will reduce the nitrates being returned to the anaerobic zone. When the preanoxic zone is adequate, this will not be an issue. The size of the preanoxic zone will depend much on the need for nitrate reduction. For example, when using a five-stage Bardenpho plant that is already producing a low effluent nitrate concentration, there is no need for a preanoxic zone. If there is a preanoxic zone, pass the full flow through it to prevent secondary release resulting from a lack of nitrates in the RAS. At the Eagle's Point plant in Minnesota, the plant is running in the two-stage anaerobic/nitrification mode while nitrifying. Nitrate concentration in the effluent runs approximately 15 to

17 mg/L. The RAS is passed through a preanoxic zone, where denitrification is complete before entering the anaerobic zone.

PROCESS CONTROL METHODOLOGIES

Factors such as temperature, pH, SRT, dissolved oxygen (DO) concentration, and inhibitors influence the extent of nitrification. The ratios of the total BOD (TBOD) or total chemical oxygen demand (TCOD) to total Kjeldahl nitrogen (TKN) are principal factors controlling the extent of denitrification.

The readily biodegradable COD (BOD) or rbCOD to phosphorus ratio in the secondary influent controls the reliability of phosphorus removal because most of the rbCOD will be converted to VFA in the anaerobic zone. The COD/TKN ratio will determine the process configuration for achieving good phosphorus removal, nitrification, and/or denitrification.

EFFECT OF OXYGEN. As discussed in the previous chapters, a DO concentration of more than 2 mg/L is required for optimal nitrifier growth. The growth rate of the nitrifying organisms is gradually reduced with lower DO. On the other hand, denitrification requires the absence of DO. In SND systems, nitrifiers will grow at lower DO values, but at a much slower rate, requiring a larger reactor. Denitrification will also take place in the low DO environment in the bulk liquid. The measured bulk liquid DO concentration does not represent the actual DO concentration within the activated sludge floc. Besides considering the effect of DO on nitrification, the DO concentration may also affect the denitrification rate. Wheatland et al. (1959) found that the denitrification rate at a 0.20 mg/L DO concentration was approximately one-half of the rate at a zero DO concentration. Wheatland et al. (1959) also noted that, as the DO concentration was raised to 2.0 mg/L, the denitrification rate was only 10% of the zero DO concentration rate. In most SND systems, there is not a uniform DO concentration through the tanks, but rather a gradient of dissolved oxygen.

In systems with formal anoxic zones with mixed liquor recycle from the nitrification zone, care must be taken to prevent the recycling of an excessive amount of oxygen in the dissolved form or bubbles being pumped or directed into the anoxic zone. The DO, at the point where mixed liquor recycling takes place, should be kept at a level of approximately 1 mg/L. In addition, a box could be constructed around the recycle pump intakes to prevent air bubbles from being swept into the pump intakes and pumped to the anoxic zone. If the box rises to approximately 0.9 m (3 ft) from the surface and is large enough, bubbles will not be transported down to the pump with the mixed liquor, and there will be some time for the DO to be used up before entering the anoxic zone.

The effect of dissolved oxygen on the uptake of phosphorus has not been studied as extensively. The PAOs are obligate aerobes and must have a positive DO concentration for phosphorus uptake. Anecdotal evidence from the operation of an Orbal plant indicated that a zone of DO in excess of 2 mg/L was necessary towards the end of the aeration basin to ensure good uptake of phosphorus.

A DO concentration of 0.2 mg/L and above has been reported to inhibit denitrification by activated sludge treating domestic wastewater (Dawson and Murphy, 1972). Nelson and Knowles (1978) reported that denitrification ceased in a highly dispersed growth at a DO concentration of 0.13 mg/L. The issue of DO concentration effects on denitrification in activated sludge systems is confounded by the fact that the measured bulk liquid DO concentration does not represent the actual DO concentration within the activated sludge floc. With point-source aeration and extended aeration, the energy input to the basin may be as low as 25 W/m^3 (125 hp/mil. gal), and large flocs form in the unaerated zones of the channels in the Carrousel system. This may be one of the reasons SND is much more effective at longer SRTs.

Dissolved oxygen in the feed to the plant or in the RAS should be minimized as a result of the detrimental effect of oxygen on the phosphorus removal mechanism in the anaerobic zones. Screw pumps for RAS or primary effluent to the anaerobic zone should be avoided. Backmixing from an aerated zone to an anaerobic zone should be avoided. Avoid channel flow metering in the RAS or feed line, in which there is a standing wave and air entrainment.

TEMPERATURE EFFECTS. The temperature effects on nitrification and denitrification are described in Chapter 4.

Biological phosphorus removal processes are relatively insensitive to temperature changes compared to other biological processes. Sell (1981) concluded that the responsible bacteria are Psychrophiles, which grow better at lower temperatures than the typical activated sludge BOD-removing bacteria, which have their maximum growth rate at 20°C. Laboratory-scale studies found that there was no difference in BPR efficiency at temperatures of 15 and 10°C. However, full- and pilot-scale studies have also shown that BPR can be affected by low temperature. Fermentation in the collection system will decrease with decrease in temperature. It can thus happen that, during cold weather, there may be insufficient VFA in the feed and that denitrification is not complete, interfering with phosphorus removal. A plant in Grimstad, Norway, is removing phosphorus consistently, even though the mixed liquor temperature is approximately 5°C for approximately 3 months of the year. However, VFA is supplied by a very efficient fermenter. There is some evidence that glycogen-accumulating organisms (GAOs) may successfully compete with PAOs at temperatures approaching 30°C. Phosphorus removal in the VIP plant at Hampton Roads, Virginia, becomes unstable during the summer when the temperature of the mixed liquor exceeds 30°C (WERF, 2005).

PH EFFECTS. The pH in BPR plants may vary from 6.6 to 7.4; however, indications are that there is a decline in efficiency of both nitrification and phosphorus removal when the pH drops to below approximately 6.9. There is evidence that GAOs may also compete well at pH values below 7, reducing the VFA available for PAOs. Higher pH values have not been adequately studied.

The effect of various operations on alkalinity in the mixed liquor is shown in Table 5.3. Hydrolysis of organic nitrogen will supply 3.6 g alkalinity/g nitrogen hydrolyzed. Nitrification consumes 7 mg/L of alkalinity for every 1 mg/L ammonia-nitrogen converted to nitrate-nitrogen. Approximately 3.5 mg/L is regained for every 1 mg/L nitrate-nitrogen reduced to nitrogen gas. When adding chemicals for phosphorus removal, more alkalinity is consumed. This may affect nitrification and require the addition of buffering chemicals to prevent the pH from dropping too far.

The most probable cause of lowering of alkalinity and lowering of pH values is when the plant is fully nitrifying without denitrification while treating a low-alkalinity wastewater. The pH can generally be adjusted back to the optimum range by including denitrification in the single sludge system. There are, however, instances where alkalinity must be added, not only for phosphorus uptake, but also because of the sensitivity of nitrifying organisms at low pH values. Low pH can be a problem when treating industrial wastewaters containing high concentrations of VFAs, such as acetic acid. Pully (1991) found that the anaerobic zones of both two- and three-stage BPR systems were only capable of adjusting the pH of a wastewater high in acetic acid from 4.5 to 5.5, even though COD removal was complete in the aerobic zone where the pH was 7.5. The RAS rate was equal to the influent flow. Consequently, BPR could not be established without pH adjustment of the wastewater, even though neutralization was not necessary for COD removal.

TABLE 5.3 Alkalinity consumed or produced by certain processes.*

Process	Alkalinity change, mg/L	Per mg/L of
Ammonification of organic N	+3.6	Organic N hydrolyzed
Nitrification	-7.1	Ammonia-N oxidized
Denitrification	+3.6	Nitrate-N reduced
Chlorination	-1.4	Chlorine added
Breakpoint chlorination	-1.4	Chlorine added
Dechlorination	-2.4	Sulfur dioxide added
Dechlorination	-1.4	Sodium bisulfite
Phosphorus removal	-5.6*	Aluminum added
Phosphorus removal	-2.7*	Iron added

*For reducing phosphorus to 1 mg/L. For lower effluent phosphorus values, proportionally more chemicals will be required.

SUFFICIENT DISSOLVED OXYGEN IN THE AERATION ZONE. The reactions in the aerobic zone determine the system phosphorus removal. There must be sufficient DO for the polyphosphorus (poly-P) bacteria to completely metabolize the stored organics to gain the energy for phosphorus uptake. The DO concentration should not limit the rate of oxygen transfer to the cells. There should also be enough DO or nitrates in the bioreactor effluent to prevent phosphorus release in the secondary clarifier, but the amount contained in the RAS should be near zero. Dissolved oxygen concentrations ranging from 1 mg/L at the influent end of the aerobic zone to 3.0 mg/L at the effluent end appear to be satisfactory for high-rate, plug-flow systems (Barnard and Nel, 1988). High SRT oxidation ditch systems have shown that performance is determined by the mass of oxygen transferred rather than the concentration maintained at any point in the tank. The DO concentration will vary from high to low, but the gradient will not be constant because of daily variations in load. It is very difficult to control the DO gradient with a probe in a fixed position. Instruments that determine the oxygen uptake rate, redox monitoring, or some other surrogate parameter, such as effluent turbidity or alkalinity, have been used instead of DO monitoring for aeration control purposes (Sen et al., 1990b). Phosphorus removal could not be achieved in an Orbal system in South Africa until the DO in a section of the last channel was raised to more than 1 mg/L.

A simple way of controlling the oxygen input to channel systems is to use online oxidation-reduction potential (ORP) meters controlled by a supervisory control and data acquisition (SCADA) system. The DO is increased and the ORP increases. The aeration is then reduced, and the decline in the ORP is monitored. When the mixed liquor runs out of nitrates, there is a discernable bend in the downward line as a result of the sharp drop in the redox potential, which is an indication that oxygen supply must be resumed.

There is also evidence that phosphorus removal in the aerobic zone does not begin until all of the VFAs have been removed from solution (Pattarkine, 1991). Therefore, if the VFAs are not completely assimilated in the anaerobic zone, the size of the aerobic zone should be enlarged to ensure completeness of the phosphorus removal reaction, in addition to soluble substrate removal. The breakthrough of readily available organics to the aerobic zone is also likely to stimulate filamentous growth, particularly if the aerobic zone is completely mixed rather than plug-flow. Organic breakthrough is unlikely to be significant when treating North American municipal wastewaters and can easily be avoided by including an anoxic zone ahead of the aerobic zone, if conditions are such that it would otherwise occur.

CHEMICAL OXYGEN DEMAND TO TOTAL KJELDAHL NITROGEN RATIO. Most wastes have a COD:TKN ratio on the order of 8 to 10:1 after primary sedimentation. When these ratios are low, all aeration of the influent should be avoided so that the carbon is not oxidized by the influent oxygen. These ratios may also be influenced by pretreatment, as indicated above. In addition, there are instances where the groundwater contains nitrates that infiltrate to the sewers or nitrates are added from an industrial source that upset the ratios. Phosphate uptake is less of a problem at high COD:TKN ratios, because most of the nitrates will be removed during treatment and do not interfere with the process for phosphorus removal. At the lower COD:TKN ratios, greater reliance on endogenous respiration is necessary to optimize use of the available carbon. This could take the form of a second anoxic zone or a preanoxic zone to reduce nitrates through endogenous respiration before entering the anaerobic zone. Where the organic carbon content of the influent is not sufficient and good nitrogen and phosphorus removal is required, primary sludge may be fermented and the supernatant directed to the anaerobic zone. Sludge could also be stored in the primary tanks for fermentation. Some VFA will diffuse into the liquid phase, but the sludge could also be centrifuged and the centrate returned to the anaerobic basin. A note of caution, however; polymers used for sludge conditioning can be toxic to the nitrifying bacteria, Nitrosomonas. Processes to deal with low COD:TKN ratios will be discussed later in this chapter.

SELECTION OF AERATION DEVICE. Point-source aerators such as vertical spindle surface aerators allow for a higher rate of SND in the aeration basin, where used in a channel system or in an open basin. This is reflected in current simulation models for BNR plants by preset configurations that have multiple aerobic and anoxic zones with high recycle rates representing the flow in the channel systems. In full-scale plants, nitrogen losses in the aeration basins of between 20 and 100% of the total have been observed. In a five-stage Bardenpho plant with fine-bubble aeration, treating mostly domestic wastewater, approximately 18% of the total influent nitrogen was lost in the aeration basin. In a similar plant with slow-speed surface aerators receiving a small fraction of industrial waste, up to 50% of the total influent nitrogen was lost in the aeration basin (Van Huyssteen et al., 1989). At the Orange County wastewater treatment plant in North Carolina (OWASA), it was possible to get 85% nitrogen removal in the plant without formal anoxic zones by cutting air to the jet aerators to achieve simultaneous nitrification and denitrification. While this latter instance was a deliberate cutting back of air, the experience with surface aerators in the Bardenpho processes mentioned previously happened in spite of aerators, fitted with draught tubes, running at maximum energy in 4.5-m-deep aeration tanks. The degree of denitrification possible can, in certain cases, drastically reduce the size of the anoxic zones or do away with the need for a second anoxic zone. When designing anoxic zones for plants with slow-speed surface aerators, there is a tendency to over-design the anoxic zones and create problems with the secondary release of phosphorus.

CLARIFIER SELECTION. Clarifier design is important because it has an influence on the RAS rate and effluent suspended solids. The more the RAS rate can be reduced while still reducing the sludge blanket in the clarifier to a minimum, the better for plant performance. At the same time, the clarifiers should be operated with some sludge blanket, but not so deep that release of phosphates to the liquid will result. When the sludge passes through the critical flux zone, little liquid is exchanged to the upper layers, and, if some phosphorus release takes place in this layer, it is not serious. This may be an option where there is no preanoxic zone available. Simultaneously, the remaining nitrate in the sludge may be reduced to zero before it can reach the anaerobic basin. In one such plant with scraped, conical clarifiers, the RAS rate was reduced to a point where none of the 3 mg/L of nitrate in the effluent appeared in the underflow. Excellent phosphorus removal was obtained, such that the final effluent contains less than 0.1 mg/L after filtration. In contrast, in another plant, with shallow clarifiers of less than 2.4 m (8 ft) deep and with suction lift mechanisms, a 150% sludge recycle to the anaerobic basin was required, which led to fresh and nitrate-rich sludge being recycled to the anaerobic zone. This also resulted in a reduction of the actual retention time in the anaerobic basin and poor phosphorus removal. In a sense, a preanoxic zone replaces the need for running with a sludge blanket.

The total suspended solids (TSS) concentration of the final effluent is important because it contains both organic nitrogen and phosphorus. Clarifiers should be designed with good flocculation wells and peripheral baffles. If the effluent TSS could be maintained below approximately 8 mg/L, it is possible to reduce the average effluent phosphorus to less than 0.5 mg/L and the TN to approximately 4 mg/L. In most plants, this may not be possible because of internal shortcomings or defective final clarifiers. When effluent concentrations lower than the above is required, some form of filtration or dissolved air flotation would be required for removal of suspended solids.

Control of scum is also important because a high concentration of scum on the final clarifiers generally results in higher effluent TSS. Scum could be prevented from reaching the final clarifiers by selective wastage of scum with the mixed liquor from the aeration basin. Alternatively, a full-radius scum skimmer is recommended. When scum is not positively removed from the surface of final clarifiers, it may result in accelerated growth of the scum-forming organisms. As an alternative to final effluent filtration, dissolved air flotation (DAF) of the total flow may be considered. In plants where DAF is used for solids removal, the scum baffles on the final clarifiers were omitted, resulting in less scum growth and accumulation.

Several factors must be considered in the selection of the appropriate biological nitrogen and phosphorus removal process. These are described in the following sections.

EFFECT OF CHEMICAL PHOSPHORUS REMOVAL ON BIOLOGICAL NUTRIENT REMOVAL SYSTEMS

When adding chemicals such as ferric chloride ($FeCl_3$) or alum to the nitrogen and phosphorus removal process for phosphorus removal down to approximately 1 mg/L, the chemical mass added may be stoichiometric to the phosphorus removed. This would require not much more than a 1:1 molar ratio of iron or aluminum to phosphorus. Translated to concentration, it would require approximately 1.8 g Fe/g phosphorus removed and approximately 1.1 g Al/g phosphorus removed. However, when adding chemicals for removals from approximately 5 mg/L down to 0.1 mg/L levels, side reactions will require very high dosages. Biological phosphorus removal combined with chemical polishing would require a fraction of the chemicals. It is thus possible that, while reducing the phosphorus to just less than 1 mg/L by chemicals may be competitive with biological removal, a combination of biological and chemical phosphorus removal may be cheaper when it is needed to get to the 0.1 mg/L level. There is ample evidence that, with a well-designed and operated BNR plant; it is possible to reduce the effluent soluble phosphorus to close to 0.1 mg/L. It does not take much more chemicals to polish this to much lower levels.

When relying solely on chemical phosphorus removal in a nitrification plant, the effect of the chemical addition on the alkalinity must be taken into account (see Table 5.3). Dose point location can be critical to successful system operation and chemical dosage minimization. The addition of chemicals to the different points in the BNR system and the way it affects the nitrogen and phosphorus removal are discussed below.

PRIMARY CLARIFIERS. In addition to precipitation of phosphorus compounds, metal salt addition upstream of the primary clarifiers enhances suspended solids and BOD removal in the primary clarifiers as a result of coagulation of suspended organic matter. Therefore, primary sludge will contain a greater amount of organic matter because it is captured with the inorganic flocs. Removal of organic material in the primary clarifier reduces the loading to the secondary treatment facilities, resulting in capital and operation and maintenance cost savings for secondary treatment. On the other hand, the removal of BOD and phosphorus from the dilute wastewaters could cause a low BOD:TKN, affecting nitrogen removal in the BNR process; cause phosphorus deficiency to the biological process, if too much phosphorus is removed by chemical precipitation; and cause loss of alkalinity needed in the biological process.

Chemical enhanced primary clarifiers will not remove the soluble COD, and some rbCOD could be generated by fermentation of some of the primary sludge for denitrification and phosphorus removal.

SECONDARY CLARIFIERS. Addition of metal salts upstream of the secondary clarifiers provides a high level of phosphorus removal when used in conjunction with BPR. At this point in the treatment process, phosphorus is typically in the orthophosphate form, which can be precipitated with the metal salt, or it is included with the biomass. The metal salt and phosphorus precipitate can be removed with the flocculent biomass in the secondary clarifier. Chemical addition could also enhance nitrogen removal because colloidal nitrogen is flocculated as a result of chemical addition and thereby reduces the total nitrogen in the effluent. The addition of chemicals to the secondary clarifiers increases the amount of inerts carried in the mixed liquor, as the RAS returns inorganic precipitates to the aeration basin. This is also a concern when chemicals, such as ferrous iron, are added directly to the aeration basin for phosphorus removal.

When supplementing biological phosphorus removal with chemical addition, the final clarifiers would be the ideal point of addition because chemicals would only precipitate the remaining orthophosphate in the liquid phase, resulting in low dosages to achieve high levels of phosphorus removal.

TERTIARY FILTERS. When low effluent phosphorus levels are required, effluent filtration will be necessary. If suspended solids are low from the secondary treatment system (less than 30 mg/L), secondary effluent with metal salt addition can be applied to the tertiary filter without additional clarification or polymer addition. The phosphorus precipitates will be removed in the tertiary filter. This also had advantageous effect on total nitrogen removal to flocculation of colloidal solids in the effluent before filtration, reducing the final effluent total nitrogen concentration.

The general curve in Figure 5.12 shows the advantage of using biological in conjunction with chemical for achieving low effluent phosphorus. Recent papers about the Durham plant of the Clean Water Services near Portland, Oregon (Baur et al., 2002), described a situation where they were adding chemicals in the primary tanks, the activated sludge unit, and in a final post chemical treatment plant, to reduce the effluent phosphorus to concentrations of less than 0.07 mg/L. They needed a chemical dosage of approximately 170 mg/L of alum. When they optimized the BNR plant for BPR, the chemical requirements dropped to only 25 mg/L. At the Pinery Water

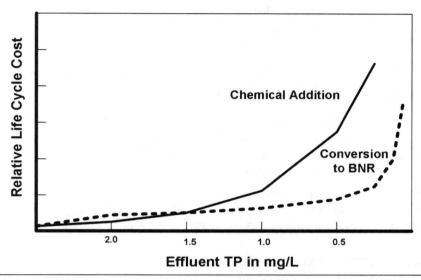

FIGURE 5.12 Relative life cycle cost for removal of phosphorus to low levels.

and Wastewater District near Denver, Colorado, the effluent standard requires that the TP be reduced to less than 0.03 mg/L. The district uses a five-stage Bardenpho plant to reduce the soluble phosphorus to less than 0.2 mg/L, then a chemical contact reactor with 70 mg/L of alum and sand filters to reduce the phosphorus to the required value.

Chemical addition may be a solution for phosphorus removal in the short term. However, in the long term, biological treatment combined with chemical addition will be more reliable and cost-effective to reduce the phosphorus to very low levels. When using anaerobic digestion, the major part of the phosphorus will be in the return streams. This will then require chemical addition to break the cycle. However, experience showed that chemical requirements for phosphorus removal in return streams are much less than that which would be required for in-line phosphorus removal (Goins et al., 2003). While BPR is a complex process, the operation is simple when the right conditions for BPR exist or is created. One of the main advantages of combined biological and chemical phosphorus removal is that the chemical polishing step will serve to coagulate and flocculate the effluent phosphorus-containing particles and react with the soluble phosphorus.

PROCESS SELECTION FOR COMBINED NITROGEN AND PHOSPHORUS REMOVAL

A nitrification/denitrification biological phosphorus removal (EBPR) [system involves three separate groups of microorganisms (poly-P-heterotrophs, non-poly-P heterotrophs, and nitrifying autotrophs) operating on a large number of chemical components in three distinct environmental regimes (aerobic zones; anoxic zones, where nitrate, but not oxygen, is present; and anaerobic zones, where nitrate and oxygen are excluded as far as possible). These features make for complex behavior, which has increased the level of difficulty in design, operation, and control (WEF, 1998a).

EFFLUENT REQUIREMENTS. The number of stages used in BNR plants will, to a large extent, be dictated by the effluent requirements. These requirements may entail the following:

(1) Phosphorus removal with no nitrification,
(2) Phosphorus removal with nitrification but no denitrification,
(3) Phosphorus removal with only partial denitrification,
(4) Phosphorus removal with nitrification only in summer, and
(5) Year-round nitrogen and phosphorus removal.

The ways in which all these requirements can be accommodated will be discussed in the following sections.

PHOSPHATE REMOVAL BUT NO NITRIFICATION. While this was covered in the previous chapter, the interference of unavoidable nitrification must be addressed. If no nitrification is required and the temperatures are not high, the simple two-stage high-rate Phoredox (A/O) process may be sufficient. However, with higher temperatures, some nitrate formation cannot be avoided even at low SRTs, and the RAS should be subjected to an anoxic stage to rid it of nitrates before mixing it with the influent wastewater, or a preanoxic zone should be installed to reduce the nitrates in the RAS. The flow diagram in Figure 5.13 may be used, or a variation of the UCT process (Figure 5.14), to remove sufficient nitrates to avoid upsetting the phosphorus removal process. The short SRT and high temperatures will ensure that the endogenous respiration rate is high, and the high concentration of solids in the anoxic stage will lead to a rapid removal of nitrates in the preanoxic zone of the first flow diagram. Only a very small preanoxic zone will be required.

PHOSPHATE REMOVAL WITH NITRIFICATION BUT NO DENITRIFICATION. In this case, it is possible to provide for a predenitrification zone followed by an anaerobic and a nitrification stage, as shown in Figure 5.13, but designed to ensure nitrification at all times. Where the nitrification stage of a BNR plant consists

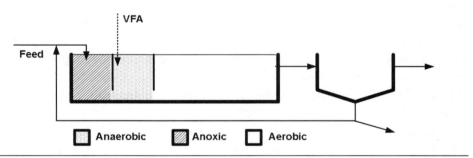

FIGURE 5.13 Flow diagram for phosphorus removal with occasional nitrification.

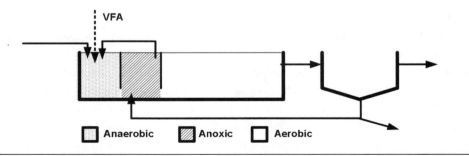

FIGURE 5.14 Alternative way to prevent nitrates to the anaerobic zone in high-rate plants with occasional nitrification.

of a channel system with simultaneous nitrification and denitrification, only a small anoxic zone may be required. Recent experience indicates better economy by providing for a small formal anoxic zone after the anaerobic zone with recycle from the aerobic zone to make use of the nitrates formed to save energy and restore alkalinity, even where nitrogen removal is not required.

The modified UCT process (Figure 5.4) could also be selected for this type of operation. Two mixed liquor recycle systems are required, as opposed to the option of using the JHB configuration (Figure 5.5), with nitrate reduction of the RAS by endogenous respiration and the addition of a small fraction of the incoming wastewater to the anoxic section.

PHOSPHATE REMOVAL WITH NITRIFICATION ONLY IN SUMMER.

When year-round nitrification is required, the plant should be designed to ensure nitrification at the coldest expected mixed liquor temperatures. When nitrification in summer only is required, the SRT can be reduced considerably, allowing for the loss of nitrification in the winter. This could lead to a great reduction in the overall cost of the plant, as will be shown later. The advantage is that the SRT is short, the active fraction of the mixed liquor is higher, and the resultant denitrification rate in summer is higher. This leads to shorter anoxic retention times and an overall saving in capital and operating costs. In effect, a layout, not dissimilar to that in Figure 5.13, may suffice. In winter, when nitrification is lost, the anoxic zone could be aerated to avoid secondary release of phosphates within the zone. The phosphate removal will not suffer because of loss of nitrification. In fact, it typically improves.

The only problem with this approach is the unstable operating conditions when nitrification is initiated during the spring and is lost during the onset of winter. The sludge tends to bulk during these transitions, for reasons explained later in this chapter. When operating in this mode, the operator should try to speed up the transition. This could be done by drastically reducing the SRT when the nitrification requirements allows it and by keeping the SRT low in the spring until the requirement for nitrification needs to be met, and then increasing the SRT to the maximum possible. Another reason for speeding up the transition is to prevent "nitrite lock" from incomplete nitrification. When there are more than three activated sludge units, the operator might consider running one with a reduced load to nitrify during the winter, and then recycle this nitrifier-rich mixed liquor back to the other modules when the ammonia standard kicks in.

HIGH-PERCENTAGE NITROGEN AND PHOSPHATE REMOVAL. For high levels of nitrogen and phosphate removal, the five-stage process is required, even though the removal of nitrates in the second anoxic zone is low. Alternatively, processes that do not remove nitrates completely, such as the UCT and Westbank processes, could be used with attached growth denitrification as the final stage. When a final effluent TN of less than 3 mg/L is required, the effluent nitrate must not exceed 1.5 mg/L to allow for some ammonia and nondegradable organic nitrogen that cannot be removed. Both ammonia removal and nitrate removal follow Monod kinetics, which makes it difficult to consistently meet low effluent standards.

Thus far, most instances of consistently good removals of both nitrogen and phosphorus have been performed by five-stage Bardenpho type plants having point-source aerators. The Kelowna wastewater treatment plant in British Columbia, Canada, will be used as an example. Aeration is supplied by a turbine aeration system, which allows turning down the DO without loss of mixing. This allows for a considerable amount of SND in the aeration basin. When the UCT model was applied to the plant during design, it predicted that the effluent nitrate concentration should be approximately 7 mg/L (Van Huyssteen et al., 1989). Yet, during actual performance, less than 1 mg/L ammonia and nitrate were maintained over months. This difference between theory and practice typically has been ignored; however, if not taken into account, one would have to come to a final conclusion that it is impossible to obtain a total nitrogen of less than 3 mg/L, in spite of the fact that this standard has been produced consistently by a number of plants, as can be seen from the plot in Figure 5.15.

Most of the plants that achieved such low TN values in the activated sludge plant used either surface or turbine aerators. Surface aerators were mostly used in Carrousel plants, forming the aeration portion of many five-stage Bardenpho plants in the United States. The effect of denitrification in the aeration basin will be demonstrated by using an example later in this chapter.

Table 5.4 summarizes the effects of the factors identified on nutrient removal process selection. The primary factor affecting process selection is the degree of nitrification and/or nitrogen removal desired for the process. If nitrification only or only a moderate degree of nitrogen removal (effluent total nitrogen of 6 to 12 mg/L) is desired, then either the Phoredox (A^2/O), UCT, or VIP process would typically be selected. Selection among these three options depends on projected wastewater characteristics ($TBOD_5$/TP ratio) and the degree of phosphorus removal capability

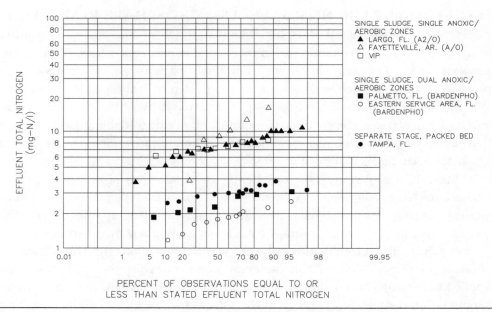

FIGURE 5.15 Removal of nitrogen in BNR plant in the United States.

required. The phosphorus removal capability of the Phoredox process is generally lower than that of the UCT or VIP process; however, when a preanoxic zone is added, it is equal or better. The Bardenpho process would be selected when extensive nitrogen removal (effluent values of approximately 3 mg/L) is desired or a tertiary denitrification attached growth system could be added to any of the other processes. While exceptions exist to the generalizations presented in Table 5.4, they represent a good starting point for preliminary biological process selection (Sedlak, 1991).

BENEFITS FROM CONVERTING TO A BIOLOGICAL-NUTRIENT-REMOVAL-TYPE OPERATION

RELIABLE OPERATION. There are a large number of high-rate plants, some specifically designed to avoid nitrification, first, because it was not required and,

TABLE 5.4 Nitrogen and phosphorus removal process selection.

Process	Nitrification	Nitrogen removal	Sensitivity to TBOD$_5$/TP ratio[a]
A/O	No	No[b]	Moderate
Phostrip	No	No[c]	Low
Phoredox (A^2/O)	Yes	6 to 8 mg/L[d]	High
JHB, Westbank	Yes	6 to 8 mg/L	Moderate
UCT/VIP	Yes	6 to 12 mg/L[d]	Moderate
Modified UCT	Yes	6 to 12 mg/L	Moderate
Bardenpho	Yes	3 mg/L	High

[a]All processes except Phostrip can benefit from using fermenters.
[b]Same degree of removal as achieved in conventional activated sludge facility.
[c]Used, in particular, if wastewater is fresh and low in readily biodegradable organic matter; can be used with any of the above processes.
[d]Approximate effluent concentration.

second, because it would use much more energy for nitrification. The warmer the mixed liquor temperature, the more likely that unwanted nitrification and resultant problems will be experienced. In one Brazilian plant, with mixed temperatures of up to 30°C, an attempt to run the plant at less than 1 day SRT still resulted in partial nitrification and denitrification in the final clarifiers. In Phoenix, Arizona, the 23rd Avenue plant was similarly operated at a very low SRT to avoid nitrification. When the plant was retrofitted with a fine-bubble system, operated at a higher SRT to allow for nitrification and provided with anoxic zones for denitrification, the overall energy consumption was less, in part, as a result of the fine-bubble system and, in part, as a result of the almost doubling of the alpha factor or the efficiency of oxygen transfer. Most of the soluble organic matter is removed in the anoxic or anaerobic zones, and there is less interference by such compounds with oxygen transfer.

Extended aeration plants are used in some form for smaller plants as a result of the ease of operation and the stable sludge produced. The longer SRT produces a reasonably stable sludge, which could be irrigated or dried on drying beds. In some cases, the sludge is further stabilized by storage in lagoons and the supernatant returned to the plant. The longer SRT makes nitrification inevitable, and the plant manager must learn to accept and operate the plant for nitrification and denitrification to gain the maximum benefit from a more stable system and lower energy cost.

RESTORING ALKALINITY. In areas where the water is soft (i.e., there is little buffer capacity), it is not possible to nitrify without also denitrifying to restore the alkalinity. Unless there is sufficient buffer capacity, the pH will drop, and nitrification will be lost. The concentration of the influent nitrogen and the residual buffer capacity will dictate the degree to which the nitrates must be reduced. For the Tai Po plant in Hong Kong and the Rotorua plant in New Zealand, the highest concentration of nitrates that could be tolerated in the aeration basin was 6 mg/L. When this value was exceeded, the alkalinity is reduced to the point where the pH drops suddenly, and ammonia appears in the effluent as a result of loss of nitrification. The automatic response to the appearance of ammonia is to raise the DO in the aeration basin. The increase in DO means that more oxygen is recycled to the anoxic basin, resulting in less nitrate reduction, which, in turn, results in a lowering of the pH and inhibition of nitrifiers and thus more ammonia. Lowering the DO level reduces oxygen transfer to the anoxic zones and encourages simultaneous nitrification and denitrification.

The same principles apply to aerobic digesters, which may operate at SRTs to 40 days. The microorganisms break up as a result of the lack of food and release nutrients, such as ammonia, which is then further converted to nitrates, with a further drop in the pH. An aerobic digester in Pewaukee near Milwaukee, Wisconsin, had a nitrate content of 550 mg/L, and caustic was added to neutralize the acids. This could have been achieved by simply switching off the air until all the nitrates were reduced by the endogenous respiration of the heterotrophic organisms. Such actions would have resulted in power savings and savings in caustic.

At the Shek Wu Hui plant in Hong Kong, there is a contribution from a leachate treatment plant. The leachate treatment plant consisted of a decanted aerated lagoon, which was operated in a batch mode. Aeration was done by means of floating surface aerators. After some period of aeration, the aerators would be switched off, and a period of settlement produced a clear supernatant, which was decanted to a balancing pond. All aerators were running in the aeration mode, and tons of alkali were added to the lagoon to neutralize the pH. The discharge consent was for 2000 mg/L of nitrate-nitrogen and less than 10 mg/L of ammonia-nitrogen in 5 ML/d (1.3 mgd).

It was deemed necessary to install a methanol plant at the main treatment plant to denitrify the nitrates from this source, because there would not be sufficient carbon in the domestic wastewater to denitrify this source of nitrates and that from the domestic waste. Upon examination of the actual nitrates in the discharge from the leachate treatment plant, it appeared that the effluent nitrates ranged from 40 to 1200 mg/L, while they did not exceed the ammonia requirement. It was clear that, on some days, under-aeration led to a high degree of SND. Nitrates were reduced to low levels by using the carbon in the feed. Our advice was to train the operators of the leachate plant to switch off the aerators closest to the feed point and watch the ammonia and nitrates. This led to savings in power and chemicals, and the consent limit was reduced to 200 mg/L, which allowed the designers to eliminate the methanol plant as a carbon source for nitrate reduction at the main treatment plant. The remainder of the nitrates was reduced in the pumped mains and in the primary tanks of the main treatment plant, keeping those areas "fresh" and reducing odors.

Chances are that when the SRT is sufficiently long and the temperature high, while the DO is above approximately 0.6 mg/L and ammonia appears in the effluent, the problem is low pH caused by nitrate formation. The curve in Figure 5.16 shows approximately the transition line between nitrification and no nitrification at various temperatures and SRT. When the plant is operated on the safe side of this line and ammonia appears in the effluent, the operator should look at the pH of the mixed liquor.

IMPROVING THE ALPHA FACTOR. The alpha factor is the energy required for dissolving oxygen in water divided by the energy required for dissolving oxygen in the process water. It is thus a measure of the interference of the wastewater constituents, mostly in solution, with oxygen transfer to the liquid. With the provision of an anaerobic zone and/or anoxic zones, much of the compounds that may cause the reduction in the alpha factor is either adsorbed in the anaerobic zone or reduced in the anoxic zone, using nitrates already in solution, making it easier to get the oxygen in solution in the aeration section following the anoxic section. Thus, the use of nitrates in the first contact zone has a beneficial effect beyond that of replacing oxygen as the electron acceptor. As mentioned before, the alpha factor at the 23rd Avenue plant in Phoenix was doubled by including anoxic selectors.

IMPROVING SLUDGE SETTLEABILITY. This is a more controversial advantage and, judging from some reports, may be a mixed boon. This topic will be dis-

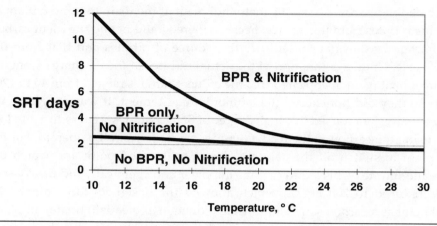

FIGURE 5.16 Relationship between SRT, BPR, and nitrification.

cussed more thoroughly in other sections of this manual. Generally, it is considered that the provision of an anaerobic contact zone will increase the density of the sludge. The reason is the selection for floc-forming bacteria, such as the PAO organisms. Also, the inorganic phosphorus in the cells weighs them down, and they tend to form clusters, as opposed to filaments. Anaerobic selectors have been installed successfully in plants that had bulking tendencies to improve the sludge volume index (SVI). Biological phosphorus removal is not a requisite for a selector to function, but occurs in most cases.

Bulking is common in BNR plants, especially those that rely on a high degree of SND. The most common filamentous organism causing bulking in BNR systems has been found to be Microthrix parvicella (Martins et al., 2004). In extended aeration plants, there is generally no primary sedimentation, and the heavier sludge helps the settling of the filamentous sludge, giving it a double advantage. The filaments filter the solids, while the primary sludge weigh it down, so that the SVIs do not get out of control and ensuring a high effluent quality. With primary sedimentation, there is a greater danger of excessive bulking. Microthrix parvicella is the most common bulking agent in channel systems, and they thrive where there are low DO gradients.

In the higher energy input plants, good SND was still observed where surface aerators were used and good SVIs were possible. The graph in Figure 5.17 shows the effect that switching off aerators to improve denitrification had on the SVI at the JHB Goudkoppies plant. There were 12 evenly spaced aerators. When three were turned off, the SVI still remained at approximately 100 mL/g; however, when two more were switched off, the SVI shot up to 400 mL/g. With so many switched off, the mixing was compromised, and zones of low DO resulted in filamentous growth. When the aerators were switched on again, the SVI returned to normal.

Bulking has also been experienced when a plant is on the edge of its nitrification capacity. In high-rate plants at high temperature, the plant may move slowly into the nitrification mode at the beginning of summer. This is generally associated with an unstable period of bulking, as can be seen from Figure 5.18. The same thing may happen when the plant is losing nitrification in the fall. After the loss of nitrification, the SVI will settle down and return to normal.

There are several ways of overcoming the problems with the intermediate situation. One method would be to keep the SRT as low as possible and avoid nitrification

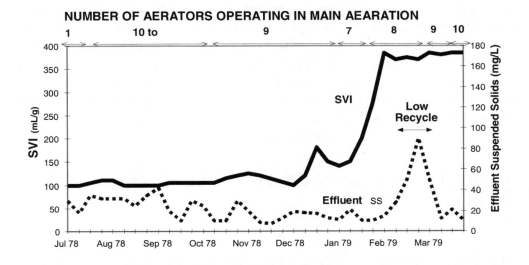

FIGURE 5.17 Effect of under-aeration on the SVI.

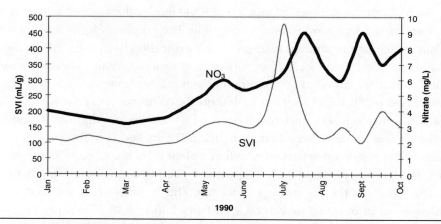

FIGURE 5.18 The SVI at start of nitrification in summer in Bonnybrook, Calgary, Alberta, Canada.

altogether. If this is not possible, the operator should hold the SRT down until the temperature reaches a value where nitrification could be sustained, then increase it as rapidly as possible by not wasting sludge for a week or two. The third possibility is to use one module in plants with multiple modules to keep nitrification going through winter by loading at a lower rate. Then, the operator should transfer nitrifying sludge over to the other units when nitrification could be sustained. In the fall, the operator should use the reverse of this strategy by wasting sludge at a high rate when ammonia starts to appear in the effluent.

TROUBLESHOOTING BIOLOGICAL NUTRIENT REMOVAL PLANTS

The control strategies will be discussed for various effluent quality objectives.

PLANT NOT DESIGNED FOR NITRIFICATION BUT NITRIFICATION IN SUMMER CAUSES PROBLEMS. First establish the nature of the problem.

Problem. The pH may drop as a result of low alkalinity, and the sludge deflocculates and contributes to suspended solids in the effluent

Correction. Either lower the SRT to try and wash out the nitrifying organisms or partition the first 20% of the plant and provide it with stirrers. Use a high rate of RAS recycle or install recycle pumps. For surface aeration plants, simply switch off some aerators, preferably close to the inlet. Alternatively, add alkalinity.

Problem. Too little dissolved oxygen in the plant when nitrification begins.

Correction. Switch on more blowers to raise the DO, or create anoxic zones, as above, to use more of the nitrates formed.

Problem. Nitrification cannot be maintained year-round, but serious bulking occurs when the plant moves into or out of nitrification.

Correction. If nitrification is unavoidable, the best policy is to operate the plant around it, trying to maintain nitrification year-round. This is not always possible. If it is not, be sure that the SRT is kept low as the liquid warms up, then do not waste sludge for a period when nitrification starts. This will ensure rapid startup of nitrification. Alternatively, run one module at a high SRT, even during the winter, to supply seed nitrifying organisms for the startup period. An aerobic digester will also have a store of nitrifiers for kick-starting nitrification. When the plant is loosing nitrification, do the reverse. Keep the SRT high up to a point, and then waste down to low levels to wash out the nitrifiers quickly.

PLANT DESIGNED FOR NITRIFICATION BUT NO DENITRIFICATION.

Problem. The pH drops to below 6.5, and deflocculation takes place.

Correction. Install anoxic zones as above, with or without mixed liquor recycle. Switch off surface aerators near inlet.

Problem. The oxygen supply is insufficient at the upstream end of the aeration basin during peak ammonia demands.

Correction. Install anoxic zones with mixed liquor recycle of at least twice the influent flow or step-feed the flow to a lower point to spread the oxygen demand. Anoxic zones may be created where the feed is entered.

Problem. No funds for installing anoxic zones.

Correction. Alternate aeration to allow some denitrification to take place in the aeration basin. When using surface aerators, switch off some during the off peak hours. Add alkalinity in the form of caustic or lime.

Problem. The SRT is long enough, pH >7 but less than 9, DO more than 1 mg/L, but loss of nitrification occurs.

Correction. If the ammonia steadily increases until there is no nitrification or increases over a period of time and then slowly decreases, suspect a toxic discharge. Upon first noticing the increase, take a sample of the activated sludge on consecutive days for possible analyses. Heavy metals in the feed will accumulate in the sludge and can be detected. Knowing the type of metal may lead one to the source. If the problem persists and easy detection is not possible, look at possible dischargers of heavy metals, such as nickel and chrome from chrome platers or paint shops or zinc from galvanizers. Otherwise, do comparative nitrification tests on different sewer branches to find out where the toxins originate from. Get help.

PLANT DESIGNED FOR NITRIFICATION AND DENITRIFICATION.

Problem. Denitrification insufficient, and nitrates in effluent are too high.

Correction. Determine the COD:TKN in the settled wastewater. If the ratio is higher than 8, lower the DO in the aeration basin, especially towards the point from where mixed liquor is recycled, to reduce the mass of oxygen recycled to the anoxic zone. The lower DO will also encourage a degree of simultaneous nitrification and denitrification. The air supply rate at the end of the aeration basin may have been set for mixing, resulting in high dissolved oxygen levels. Reduce the air supply rate to below the required mixing rate and monitor. Many operators are running the aeration rate substantially below the accepted value for mixing with no ill results.

Enlarge the anoxic zone, if possible, and increase the mixed liquor recycle.

If the COD:TKN is less than 7, add a carbon source, such as methanol, to the anoxic zone. Acid fermentation of sludge will also provide an additional carbon source. Much more carbon than nitrogen is removed in the primary tanks. When fermenting this sludge, the resultant supernatant is rich in carbon and low in nitrogen. This helps to redress the balance. This will also increase phosphorus removal.

Problem. The DO is already high, the temperature is high, and ammonia peaks keep appearing in the effluent.

Correction. Check for slugs of ammonia-rich streams, such as from the return streams. Check if the SRT is sufficient and, if not, increase by wasting less sludge. Check the pH. If the pH is too low, the nitrates in the mixed liquor may be too high, causing a drop in the pH value, and loss of nitrification is a result of the lower pH. Lower the DO in the aeration basin, which may have the effect of lowering the nitrates being recycled to the anoxic basin, which, in turn, will raise the pH and lower the ammonia in the effluent or add alkalinity.

Problem. The effluent shows a diurnal pattern of ammonia, consisting of a peak during certain hours of every day.

Correction. This is perfectly normal, but should be watched. Whereas carbonaceous material is adsorbed onto the heterotrophic bacteria, the nitrifying organisms cannot store any ammonia. There could be sharp peaks of total nitrogen in the diurnal pattern, as shown as an accentuated example from one plant in Figure 5.19.

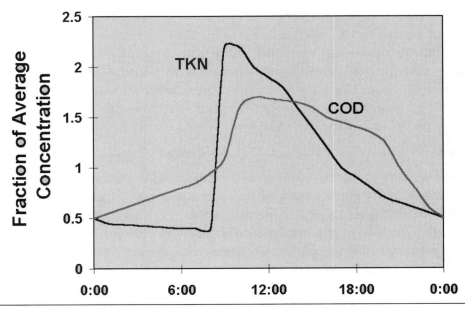

FIGURE 5.19 Typical diurnal pattern of nitrogen in influent.

Such sharp peaks will result in peaks of ammonia that could be observed in the effluent a few hours after the peak in the influent reaches the plant. A diurnal pattern of the effluent may show no ammonia in the effluent except at approximately 14:00 hours, when a sharp peak of up to 4 mg/L may appear in the effluent. This will still keep the average for the day low, but should be no cause for alarm. On the contrary, an operator may use this as an indicator of how close he is operating the plant to failure. Produce a number of typical profiles for your plant, then sample only at the peak every day and analyze the grab sample. If this peak keeps steady at 4 mg/L, there is no cause for alarm. When it steadily increases, it will give a timely warning before the composite sample shows anything.

PLANT DESIGNED FOR PHOSPHORUS REMOVAL ONLY.

Problem. The plant is a two-stage Phoredox (A/O) plant and works well when it is cold; however, when the temperature rises, the phosphorus removal is impaired.

Correction. Nitrates may form during the warm weather and be present in the RAS being recycled to the anaerobic zone. Check the release of phosphorus in the anaerobic zone. Phosphorus should be released to approximately four times the concentration in the feed. If the release is reduced in summer, this may be caused by nitrates being recycled. Reduce the SRT, and reduce the RAS rate. The lower RAS rate will lead to a higher density sludge, which will allow some denitrification in the final clarifiers and reduce the nitrates associated with the RAS stream. Alternatively, use a small preanoxic zone to achieve endogenous breakdown of the nitrates. Lower the DO in the aeration basin to discourage nitrification, but not so low to encourage release of phosphorus in the final clarifiers.

Problem. The removal of phosphorus is inconsistent and cannot be controlled.

Correction. If this cannot be contributed to nitrates that form in the plant, then check if there is sufficient carbon in the feed. As a first cut, the BOD:P should be > 20, the COD:P > 38, and the rbCOD:P > 15. Approximately 4 to 6 mg/L VFA is required for every milligram per liter of phosphorus to be removed. However, VFA analyses are difficult to perform, and the results from commercial laboratories can differ by 100%. Most of the rbCOD in the influent will be converted to VFA in the anaerobic zone of a BNR reactor, and it is thus a good surrogate from VFA contents. The flocculated and filtered COD test could be used in any laboratory to estimate the rbCOD. If there is not sufficient rbCOD, consider the installation of a fermenter to increase the production. As above, check the phosphorus concentration in the anaerobic zone. If the release is not high, it could be from lack of VFA or too much oxygen entering the anaerobic zone.

Problem. The phosphorus release in the anaerobic zone is above the minimal requirements, but the removal is dismal.

Correction. Check the phosphorus in the RAS. If the RAS rate is too low, secondary release may take place in the bottom of the final clarifiers. Increasing the RAS rate to reduce this release has helped one plant to achieve very good overall phosphorus removal. Also, check the release curve from a batch, and make sure that the secondary release is limited.

PLANT DESIGNED FOR AMMONIA AND PHOSPHORUS REMOVAL.

If there is no need for nitrogen removal, but the plant must remove ammonia and phosphorus, then denitrification is also required to reduce the nitrates to the anaerobic zone. Thus, whether nitrogen removal is required or not, the plant must be operated as one designed for the removal of some nitrogen and phosphorus removal. Nitrates in the RAS could be reduced in an internal anoxic or preanoxic zone.

Problem. Nitrate removal is not sufficient and is interfering with phosphorus removal.

Correction. Reduce nitrates in effluent and thus also in RAS. Lower the DO in the aeration basin, increase the recycle of mixed liquor to an anoxic zone, if there is one, and reduce the RAS rate to the optimal value. A plant may function well with a mixed liquor recycle more than double the optimal value. However, this will recycle much more nitrates to the anaerobic zone. Reducing the RAS rate will increase the sludge concentration in the final clarifiers and reduce the nitrate recycle. If the nitrate concentration in the RAS is still too high, see if it is possible to partition the upstream end of the anaerobic zone to reduce the nitrates from the RAS through endogenous predenitrification (JHB, Westbank). If the nitrate concentration is not reduced sufficiently by these means, add a small fraction of the flow or recycle approximately 10% of the flow from the end of the anaerobic basin to the preanoxic basin, as shown in Figure 5.20.

Problem. There is not sufficient rbCOD or VFA in the feed.

Correction. Install fermenter for primary sludge and feed supernatant to anaerobic basin or add a carbon source, such as acetate.

Problem. Phosphorus removal is lost with every rainstorm.

Correction. Too much flow through the anaerobic basin. Bypass storm flows around the anaerobic basin to the anoxic basin.

Problem. Good release of phosphorus in the anaerobic basin, but removal not good.

Correction. Look at secondary release of phosphorus. Take a profile of phosphorus removal through the entire reactor. The release in the anaerobic basin should have a profile as shown in Figure 5.11. If the second slope of the release goes on for too long, there may be too much secondary release in the anaerobic zone. Reduce retention by increasing the RAS of reduce the size of the zone. If there is an increase in the soluble phosphorus through the anoxic zone, there is a possibility of secondary release. Increase the mixed liquor recycle because the nitrates are running out in the anoxic zone or decrease the size of the anoxic zone. If there is a high concentration of phosphorus in the RAS, this means secondary release is taking place in the final clarifiers, and the RAS rate should be increased.

Problem. Phosphorus uptake is rapid in the aeration basin, but release takes place towards the end of the aeration basin.

Correction. The SRT is too long. Reduce the SRT by wasting more sludge. This may be a typical summer phenomenon. Raise the SRT again in winter.

Problem. Scum accumulates in the aeration basin or in the anoxic zones.

Correction. Close any bottom openings between the anoxic and aeration zone and restrict the overflow from the anoxic to the aeration zone to induce a small drop in liquid level to allow the scum to move to the aeration zone. Allow scum to pass through the aeration zones. Remove scum at the end of the aeration zone to a solids separation step. Do not recycle any scum back to the bioreactor. DO NOT TRAP OR RECYCLE SCUM.

Problem. The sludge settles well, but there is a problem with solids in the effluent. The solids contain phosphorus, and the phosphorus limits cannot be met.

Correction. Look at the clarifier operation. Determine if hydraulic improvements can result in better flocculation and prevent gravity currents from rising up to the effluent weirs by installing a peripheral baffle. Consult clarifier experts.

RETROFITTING PLANTS FOR NUTRIENT REMOVAL. There are numerous examples of benefits derived from retrofitting plants for nutrient removal.

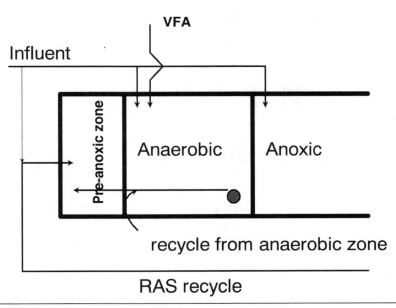

FIGURE 5.20 Improvement to deal with nitrates in RAS.

In some cases, such as Phoenix, Arizona, and Barueri in Sao Paulo, Brazil, where the temperature in summer exceeds 30°C, the plants were not designed for nitrification, but it could not be avoided. The solution was to partition approximately 25% of the influent end and to install mixed liquor recycle pumps to supply nitrates to an initial anoxic zone. This relieved the pressure on the oxygen demand in the influent and increased the alpha factor for better oxygen transfer.

At the Reedy Creek plant which served Disney World in Florida, the warm weather ensured nitrification at all times, but no provision was made for denitrification. It was not considered necessary because the remaining nitrogen was removed adequately in the wetland to which the effluent was discharged. However, approximately 50% phosphorus removal was observed in the aeration plant, but nothing through the wetland. Switching off some aerators in the first pass of the four-pass system resulted in the phosphorus removal, as shown in Figure 5.21.

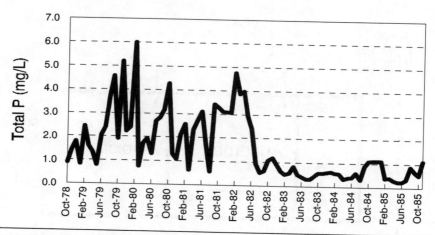

FIGURE 5.21 Phosphorus removal at Reedy Creek, Florida, by switching off aerators.

In New York, partial nitrogen removal is required, and, at the Tallman Island plant, this was achieved by introducing anoxic zones to a step-feed plant. This is shown in Figure 5.22. The step-feed mode allows the plant to carry higher solids inventory and thus a higher SRT. The higher SRT allows nitrification in this high-rate plant, but at the cost of ammonia breaking through, as a result of the short time for nitrification in the last pass of the plant. Thus, only partial removal is possible. In this case, they must achieve 50% removal, and that has been shown to be possible in this mode.

Sometimes a high degree of nitrogen removal is required in summer and partial nitrification in winter. In Edmonton (Alberta, Canada), nitrogen and phosphorus removal was required in summer, but the ammonia standards were lifted to 10 mg/L in winter. It was possible to use a semi-step-feed mode to achieve this goal in winter, while switching back to a more conventional mode in summer to achieve a higher degree of overall nutrient removal (Figure 5.23).

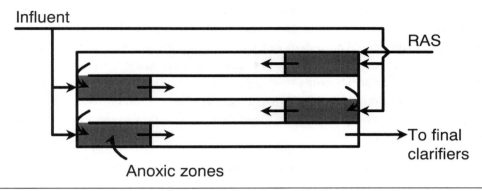

FIGURE 5.22 Conversion of high-rate plant to partial nitrogen removal at Tallman Island, New York.

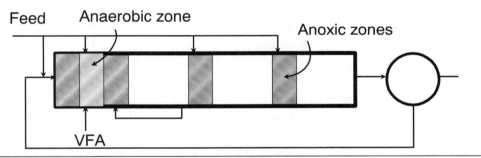

FIGURE 5.23 Step feed approach for partial nitrification in summer.

There are numerous ways in which plants could be retrofitted for nutrient removal, depending on the local conditions and effluent requirements. The use of computer simulations goes a long way towards analyzing a number of options to get as much out of the existing basins as possible. In the case of the Edmonton plant, it was possible to retrofit a high-rate plant under extreme cold weather conditions to remove nitrogen and phosphorus biologically, while derating the plant to 85% of its previous capacity. The plant consisted of a four-pass system, and the installation of partitions was relatively simple. One module of the plant was retrofitted and operated for a full calendar year to demonstrate the feasibility of this approach.

Biological nutrient (nitrogen and phosphorus) removal by nitrification and denitrification and biological excess phosphorus removal has provided significant benefits to reduce nitrogen and phosphorus levels in wastewater. However, the system will provide BNR at considerably reduced cost, if the following significant difficulties can be overcome:

(1) It is a tendency to develop bulking sludge as a result of filamentous organism proliferation.
(2) Specific control of the low food-to-microorganism ratio filaments, inter alia types 0092, 0041, 0675, and *Microthrix parvicella* (Jenkins et al., 1993) is of major importance; these filaments cause practically all of the bulking problems in BNR systems. In BNR systems, other factors, such as unaerated mass fraction, frequency of alternation between anoxic and aerobic conditions, and the DO, nitrate, and nitrite concentrations at the anoxic-aerobic transition, may be more important in causing bulking than the F/M or the SRT (Ekama and Wentzel, 1999).
(3) It is a requirement for a sludge age (>15 days, 8 days aerobic SRT) to ensure year-round nitrification.
(4) The limitation of the nitrogen and phosphorus removal placed on the system by the influent wastewater characteristics, in particular, the readily biodegradable COD (rbCOD) fraction and the TKN/COD and P/COD.
(5) The problems arise in the treatment of the phosphorus-rich waste sludge.

RETURN ACTIVATED SLUDGE AND INTERNAL RECYCLE RATES. In smaller wastewater treatment plants, low head centrifugal pumps may be used for mixed liquor recycle; however, in larger plants, slow-speed axial flow pumps are essential. With good design, the physical pumping head may be as little as 80 mm (3 in.), and all the energy will be consumed by friction losses. Slow-speed axial flow pumps reduce the energy for recycling mixed liquor to a few kilowatts. Also, induced turbulence and the introduction of DO are reduced. This is important because the recycle is generally directed to anoxic or anaerobic zones. Submerged mixers used as recycle pumps have also become popular, especially where the transfer is through the wall from one basin to the next.

The massive volume of mixed liquor to be recycled dictates the plant layout. If the plant layout could be arranged such that the end of the aeration zone and the start of the anoxic zone could be next to one another, no long large channels would be needed, and the energy consumption could be kept to a minimum. Although this arrangement makes it difficult to measure the recycle flow, accurate measurement is not that important. At the startup of the plant, when the designer must check the recycle rate, and as required thereafter, stream velocity meters could be used.

MINIMIZING THE ADVERSE EFFECT OF STORM FLOWS. Wastewater treatment plants serving combined sewer systems typically receive excess flows of relatively dilute wastewater during major storm or snowmelt events. The performance of BNR can be adversely affected in at least three ways by such events.

(1) The concentration of organic matter in the incoming wastewater during a storm event can be quite low, thus not providing sufficient soluble organic readily biodegradable carbon in the anaerobic zone to stimulate the BPR mechanism.
(2) The hydraulic retention time in the anaerobic zone is decreased by the high incoming flowrate of the wastewater.
(3) A large storm will flush the collection system of settled organic matter, thus eliminating a source of short-chain VFAs (SCVFAs) in the raw wastewater that is a stimulant to the PAOs.

There are three measures that can be taken to achieve stable phosphorus removal under such circumstances.

(1) The inclusion of step-feed capability in the bioreactor to bypass all but a fixed portion of the flow around the anaerobic zone, thereby preserving the anaerobic retention time for the proper functioning of the bio-P organisms.
(2) The provision of a preanoxic zone to remove all nitrates from the RAS, mostly through endogenous respiration, but assisted by a small dose of primary effluent, if needed.
(3) The addition of a supplemental source of SCVFAs directly to the anaerobic zone. Supplemental SCVFAs can be produced at a wastewater treatment plant by fermentation of the primary sludge (Oldham and Abraham, 1994; Rabinowitz, 1994). A suitable odor control system must be installed when using on-site fermentation. An alternative option is to keep a supply of VFA in store to assist the phosphorus-removal mechanism during storm flows.

FOAM CONTROL. The following items are proposed for consideration as foam and scum control measures in the design of any BNR system:

- The careful hydraulic design of over and under baffles; inverted chimneys to direct the flow to the bottom of the following basin; and peak flow overtopping weirs, as appropriate, in the partitions between the various zones of a bioreactor. There should be no backmixing from the aerobic to the anoxic zone, and foam must spill freely from anoxic to aerobic zones. Partitions within any zones should be just below the surface to avoid contact areas for scum to grow. Figure 5.24 shows an anoxic zone with two partitions to create a meandering flow. The partitions are only approximately 25 mm (1 in.) below the surface, allowing the main flow to go around, while keeping the surface clear.

- The implementation of continuous selective wasting for the preferential removal of foam and scum organisms from the aeration basin as part of the waste activated sludge (WAS) stream. No recycling of this foam to the plant should be tolerated because it will reseed the foaming organisms.

- The elimination of dead ends, dead corners, and other quiescent zones in channels or bioreactors where there is a potential for foam and scum to accumulate.

- The careful design of secondary sedimentation tank inlet wells and flocculation wells to allow for the passage of floating material.

- The installation of an effective scum removal system on secondary sedimentation tanks, preferably a full radius skimmer.

- The avoidance of opportunities for recycling foam and scum organisms to the mainstream treatment train from sidestream solids processing facilities.

- The installation of chlorine sprays, as necessary, at localized points of foam and scum collection and/or accumulation to kill these organisms and prevent them from causing problems in either mainstream or sidestream treatment processes. Alternatively, there are some polymers that could be sprayed on the foam. The foam loses its hydrophobic properties and mixes with the sludge and can be wasted with the waste sludge.

Bulking problems are more difficult, but indications are that, by ensuring a low nitrate concentration at the end of the anoxic zone and sufficient aeration in the first section of the aerobic zone, the stirred sludge volume index can be reduced to below 90 mL/g (Casey et al., 1993).

WASTE SLUDGE AND RETURN STREAM MANAGEMENT. It is important to minimize the amount of phosphorus that is released and returned from side-stream sludge processing steps to the mainstream treatment train. It is likewise important to control the TKN content of return streams. There are several articles in the recent technical literature on the topic of minimizing phosphorus release and return (Jardin and Pöpel, 1994 and 1996; Matsuo, 1996; Nyberg et al., 1994; Pitman et al., 1991; Pöpel and Jardin, 1993; Sen et al., 1991). The basic principle is to maintain aerobic conditions in a bio-P WAS before thickening and dewatering steps to minimize the potential for phosphorus release and return to the mainstream treatment train. In this regard, WAS thickening steps, such as dissolved air flotation and gravity belt thickening, are appropriate and proven technologies. Likewise, dewatering steps, such as centrifugation and filter belt pressing, are also appropriate technologies. The prolonged storage of bio-P WAS under unaerated conditions should be avoided. Storage of bio-P WAS in contact with primary sludge (which is rich in SCVFAs) should also be avoided. Successful dewatering of blended bio-P WAS with primary sludge using a centrifuge has been reported, as long as the bio-P WAS is

FIGURE 5.24 Surface of anoxic zones in Southwest plant, Jacksonville, Florida.

introduced to the primary sludge in a tee joint, as close as possible to the feed point of the centrifuge to minimize the time of contact before dewatering.

Unlike most heterotrophs, the autotrophic nitrifying organisms cannot store ammonia. When large fluctuations in TKN loadings on a nitrifying system pass through the bioreactor operating in a nitrifying mode, there will not be enough autotrophic organisms in the mixed liquor to handle the peak, and ammonia will appear in the effluent, especially during cold temperature operation. The intermittent fluctuating nature of the TKN load does not allow the nitrifying organisms to grow to a sufficient population to provide adequate treatment.

SUMMARY

A good understanding of the mechanism of BNR makes it possible to better understand and comprehend the possibilities of improving performance. Sometimes a simple change to the operating procedure can go a long way towards achieving goals in nutrient removal, which could also result in better reliability and operability, while reducing the operating and running cost of the plant.

CASE STUDIES

CITY OF BOWIE WASTEWATER TREATMENT PLANT (BOWIE, MARYLAND). *Facility Design.* The City of Bowie Wastewater Treatment Plant (Maryland) is an extended aeration activated sludge plant, using the oxidation ditch concepts as the main biological treatment component. The major plant processes include denitrification/nitrification activated sludge. The plant was modified for biological nitrogen and phosphorus removal to achieve a year-round monthly average of 1 mg/L TP and 8 mg/L of TN.

Effluent Limits. The National Pollutant Discharge Elimination System (NPDES) permit limits are as follows:

CBOD	<30 mg/L (monthly)
TSS	<30 mg/L (monthly)
TP	1.0 mg/L (yearly)
TN	8.0 mg/L (yearly)

DO 5.0 mg/L (monthly

Wastewater Characteristics.

Flow	8100 m^3/d (2.14 mgd)
CBOD	187 mg/L
TSS	217 mg/L
TP	5.5 mg/L
NH$_4$-N	20.2 mg/L

Performance.

CBOD	7.3 mg/L
TSS	6.7 mg/L
TP	0.24 mg/L
TN	8.7 mg/L

POTSDAM WASTEWATER TREATMENT PLANT (GERMANY). *Facility Design.* The Potsdam facility was designed to treat primary settled wastewater (90 000 population equivalent; 21 100 m^3/d) in a four basin configuration, with anaerobic digestion of the waste sludge mixture by using cyclic activated sludge technology (Demoulin et al., 2001). Each basin is integrally connected to allow operation on a 4-hour, dry weather cycle and a 3-hour, wet weather cycle. Each cycle includes time sequences for fill, aeration, settle (no inflow), and effluent withdrawal (decanting), with all sequences contributing to the reaction time for EBPR and cocurrent nitrification/denitrification. The selector in each basin is designed to allow maintenance or variation to oxic, anoxic, and anaerobic retention time components.

Performance. Total nitrogen removal is up to 92%. The average effluent total phosphorus concentration is 0.38 mg/L, without the addition of precipitants.

GOLDSBORO WATER RECLAMATION FACILITY, NORTH CAROLINA. *Facility Design.* The City of Goldsboro Water Reclamation Facility (North Carolina) was originally constructed as an extended aeration oxidation ditch and was retrofitted to an aeration basin with fine-bubble diffusers and biological nutrient removal capabilities in 1994. Process flexibility was incorporated to the plant design so that several nutrient removal alternatives could be operated. However, the plant, with a

40 900 m^3/d (10.8 mgd) flow, has been optimized as an A^2/O process (Sadler et al., 2002).

Effluent Limits. The NPDES permit limits are as follows:

BOD$_5$ (summer/winter)	4/8 mg/L (monthly)
TSS	30 mg/L (monthly)
TP	1.0 mg/L (quarterly)
TN	4.4 mg/L at design flow (monthly)

Wastewater Characteristics. The average influent characteristics from January 1999 to May 2002 are as follows:

Flow	3293 m^3/d (8.7 mgd)
BOD$_5$	159 mg/L
TSS	235 mg/L
NH$_4$-N	12.5 mg/L
TKN	24.7 mg/L
TP	3.1 mg/L

Performance.

TN	<4.0 mg/L
TP	<1.0 mg/L

SOUTH CARY WATER RECLAMATION FACILITY, NORTH CAROLINA.

Facility Design. The South Cary Water Reclamation Facility (North Carolina) completed construction of an expansion and upgrade to a BNR facility in 1998. Plant unit processes include deep bed filters, UV disinfection, and aerobic digestion of biosolids to treat 48 400 m^3/d (12.8 mgd). Plant staff have experimented with several BNR flow sheets for plant optimization. The current operational flow sheet is a four-stage process with a sidestream fermentation recycle (Sadler et al., 2002).

Effluent Limits. The NPDES permit limits are as follows:

BOD$_5$ (summer/winter)	5/10 mg/L (monthly)
TSS	30 mg/L (monthly)

TP	2.0 mg/L (quarterly)
TN	4.6 mg/L at design flow (monthly)

Wastewater Characteristics. The average influent characteristics from January 1999 to May 2002 are as follows:

Flow	18 925 m^3/d (5 mgd)
BOD$_5$	177 mg/L
TSS	244 mg/L
NH$_4$-N	22.9 mg/L
TKN	44.9 mg/L
TP	6.7 mg/L

Performance.

TN	<4.0 mg/L
TP	<1.0 mg/L

NORTH CARY WATER RECLAMATION FACILITY, NORTH CAROLINA. *Facility Design.*

The North Cary Water Reclamation Facility (North Carolina) was expanded and upgraded to a nutrient removal capability in 1997, with the Kruger (a Veolia Subsidiary) phased oxidation ditch aeration basin design. The ability to alter the timing of air for cyclic aeration in the ditches allows flexibility in optimization. This facility includes deep bed filters, UV disinfection, and aerobic digestion of biosolids. The North Cary Water Reclamation Facility also includes a reuse facility (Sadler et al., 2002).

Effluent Limits. The NPDES permit limits are as follows:

BOD$_5$ (summer/winter)	4.1/8.2 mg/L (monthly)
TSS	30 mg/L (monthly)
TP	2.0 mg/L (quarterly)
TN	3.9 mg/L at design flow (monthly)

Wastewater Characteristics. The average influent characteristics from January 1999 to May 2002 are as follows:

Flow	22 331 m³/d (5.9 mgd)
BOD_5	216 mg/L
TSS	316 mg/L
NH_4-N	37.8 mg/L
TKN	52.0 mg/L
TP	7.5 mg/L

Performance.

TN	<4.0 mg/L
TP	<1.0 mg/L

WILSON HOMINY CREEK WASTEWATER MANAGEMENT FACILITY, NORTH CAROLINA.

Facility Design. The City of Wilson Hominy Creek Wastewater Treatment Plant (North Carolina) has recently finished an expansion and upgrade to a four-stage process for nitrogen and BOD removal. In 1994, an existing trickling filter was retrofitted to accommodate a sidestream POH process for EBPR. This facility includes gravity belt thickening, anaerobic digestion of biosolids, and lime stabilization (Sadler et al., 2002).

Effluent Limits. The NPDES permit limits are as follows:

BOD_5 (summer/winter)	5/10 mg/L (monthly)
TSS	30 mg/L (monthly)
TP	2.0 mg/L (quarterly)
TN	3.7 mg/L at design flow (monthly)

Wastewater Characteristics. The average influent characteristics from January 1999 to May 2002 are as follows:

Flow	35 580 m³/d (9.4 mgd)
BOD_5	195 mg/L
TSS	266 mg/L
NH_4-N	13.9 mg/L
TKN	25.0 mg/L
TP	4.1 mg/L

Performance.

TN	<4.0 mg/L
TP	<1.0 mg/L

GREENVILLE UTILITIES COMMISSION WASTEWATER TREATMENT PLANT, NORTH CAROLINA. *Facility Design.*

The Greenville Utilities Commission (North Carolina) operates two activated sludge processes. One process consists of three Schreiber completely mixed activated sludge (CMAS) systems with limited nutrient removal; the second process consists of 16 sidestream BPR cells and two mainstream aeration basins. The facility was designed with the capability to operate under numerous flow sheets. However, plant staff have optimized their process for a four-stage process with sidestream BPR. Their sidestream BPR process receives a fraction of influent flow and 70% of their RAS flow. The plant is currently underloaded hydraulically and for soluble substrate. Only one BNR train is in service and receives a maximum influent flow for one train (120%). The remaining influent flow is split to the Schreiber CMAS process trains. Effluents from both processes are combined before disinfection. The Greenville Utilities Commission wastewater treatment plant also includes deep bed filters with methanol feed capability (not in use), UV disinfection, and aerobic digestion of biosolids (Sadler et al., 2002).

Effluent Limits. The NPDES permit limits are as follows:

BOD_5 (summer/winter)	8/15 mg/L (monthly)
TSS	30 mg/L (monthly)
TP	1.0 mg/L (quarterly)
TN	5.0 mg/L (monthly)

Wastewater Characteristics. The average influent characteristics from January 1999 to May 2002 are as follows:

Flow	35 200 m^3/d (9.3 mgd)
BOD_5	156 mg/L
TSS	196 mg/L
NH_4-N	15.4 mg/L
TKN	28 mg/L
TP	4.9 mg/L

Performance.

TN	<4.0 mg/L
TP	<1.0 mg/L

VIRGINIA INITIATIVE PLANT, NORFOLK. *Facility Design.* The Virginia Initiative Plant (VIP) operates the VIP process with a 151 400-m^3/d (40-mgd) design capacity. The plant process train includes preliminary treatment, primary treatment, secondary treatment, and effluent disinfection (Randall and Ubay-Cokgor, 2000).

Effluent Limits. The NPDES permit limits are as follows:

NO_3-N (summer/winter)	1/3 mg/L (monthly)
TP	2.0 mg/L (monthly)

Wastewater Characteristics. The average influent characteristics from January 1999 to May 2002 are as follows:

Flow	133 232 m^3/d (35.2 mgd)
BOD_5	142 mg/L
NH_3-N	18.3 mg/L
TN	24.6 mg/L
TP	5.13 mg/L

Performance.

TN	8.1 mg/L
TP	0.42 mg/L

REFERENCES

Barnard, J. L. (1976) A Review of Biological Phosphorous Removal in Activated Sludge Process. *Water SA*, **2** (3), 136–144.

Barnard, J. L.; Nel, B. (1988) Biological Nutrient Removal on Trickling Filters. Poster presentation, IAWQ Biennial Conference, Rio De Janeiro, Brazil; International Association on Water Quality: London.

Baur, R.; Bhattarai, R. P.; Benisch, M.; Neethling, J. B. (2002) Primary Sludge Fermentation-Results from Two Full-Scale Pilots at South Austin Regional (TX, USA) and Durham AWWTP (OR, USA). *Proceedings of the 75th Annual Water Environment Federation Technical Exhibition and Conference* [CD-ROM], Chicago, Illinois, Sep 28–Oct 2; Water Environment Federation: Alexandria, Virginia.

Casey, T. G.; Ekama, G. A.; Wentzel, M. C.; Marais, G. v. R. (1993) Causes and Control of Low F/M Filamentous Bulking in Nutrient Removal Activated Sludge Systems. In *Prevention and Control of Bulking Activated Sludge*, Jenkins, D.; Ramadori, R.; Cingolani, L. (Eds.); Luigi Bazzucchi Centre: Perugia, 77–97.

Daigger, G. T.; Waltrip, G. D.; Roman, E. D.; Morales, L. M. (1988) Enhanced Secondary Treatment Incorporating Biological Nutrient Removal. *J. Water Pollut. Control Fed.*, **60**, 1833.

Dawson, R. N.; Murphy, K. L. (1972) The Temperature Dependency of Biological Denitrification. *Water Res.*, **6**, 71.

deBarbadillo, C.; Bates, V.; Cauley, C.; Holcomb, S.; Ebron, W.; Rexrod, S.; Barnard, J. (2003) Get the Nitrogen Out. *Water Environ. Technol.*, **15** (12), 52–57.

Demoulin, G.; Rudiger, A.; Goronszy, M. C. (2001) Cyclic Activated Sludge Technology—Recent Operating Experience with a 90,000 p.e. Plant in Germany. *Water Sci. Technol.*, **43** (3), 331–337.

Ekama, G. A.; Marais, G. v. R. (1984) Biological Nitrogen Removal. In *Theory, Design and Operation of Nutrient Removal Activated Sludge Processes*, Water Research Commission: Pretoria, South Africa.

Ekama, G. A.; Wentzel, M. C. (1999) Difficulties and Developments in Biological Nutrient Removal Technology and Modelling. *Water Sci. Technol.*, **40** (1), 127–135.

Goins, P.; Parker, D.; deBarbadillo, C.; Wallis-Lage, C. (2003) Building a Better Nutrient Trap: A Case Study of the McDowell Creek WWTP. *Water Environ. Technol.*, **15** (7), 54.

Jardin, N.; Pöpel, H. J. (1994) Phosphate Release of Sludges from Enhanced Biological P-Removal during Digestion. *Water Sci. Technol.*, **30** (6), 281–292.

Jardin, N.; Pöpel, H. J. (1996) Behaviour of Waste Activated Sludge from Enhanced Biological Phosphorus Removal during Sludge Treatment. *Water Environ. Res.*, **68**, 965–973.

Jenkins, D.; Daigger, G. T.; Richard, M. G. (1993) *Manual on the Causes and Control of Activated Sludge Bulking and Foaming,* 2nd ed.; Lewis Publishers: Chelsea, Michigan.

Kalb, K. R.; Williamson, R.; Frazier, M. (1990) Nutrified Sludge: An Innovative Process for Removing Nutrients from the Wastewater. *Proceedings of the 63rd Annual Water Environment Federation Technical Exposition and Conference,* Washington, D.C., Oct 7–11; Water Environment Federation: Alexandria, Virginia.

Kalb, K.; Roeder, M. (1992) Nutrified Sludge in the OWASA System. Paper presented at Annual Conference of the North Carolina Water and Pollution Control Association, Charlotte, North Carolina.

Kazmi, A. A.; Furumai, H. (2000) Filed Investigations on Reactive Settling in an Intermittent Aeration Sequencing Batch Reactor Activated Sludge Process. *Water Sci. Technol.,* **41** (1), 127–135.

Keller, J.; Watts, S.; Battye-Smith, W.; Chong, R. (2001) Full Scale Demonstration of Biological Nutrient Removal in a Single Tank SBR Process. *Water Sci. Technol.,* **43** (3), 355–362.

Martins, M. P.; Pagilla, K. R.; Heijnen, J. J.; van Loosdrecht, M. C. M. (2004) Bulking Filamentous Sludge—A Critical Review. *Water Res.,* **38**, 793–817.

Matsuo, Y. (1996) Release of Phosphorus from Ash Produced by Incinerating Waste Activated Sludge from Enhanced Biological Phosphorus Removal. *Water Sci. Technol.,* **34** (1/2), 407–415.

Nelson, L. M.; Knowles, R. (1978) Effect of Oxygen and Nitrate on Nitrogen Fixation and Denitrification by *Azospirillum brasilense* Grown in Continuous Culture. *Can. J. Microbiol.,* **24**, 1395.

Nyberg, U.; Aspegren, H.; Andersen, B.; Elberg Jorgensen, P.; la Cour Jansen, J. (1994) Circulation of Phosphorus in a System with Biological P-Removal and Sludge Digestion. *Water Sci. Technol.,* **30** (6), 293–302.

Oldham, W. K.; Abraham, K. (1994) Overview of Full-Scale Fermenter Performance; from Seminar on Use of Fermentation to Enhance Biological Nutrient Removal. *Proceedings of the 67th Annual Water Environment Federation Technical Exposition and Conference,* Chicago, Illinois, Oct 15–19; Water Environment Federation: Alexandria, Virginia.

Patterkine, V. M. (1991) *The Role of Metals in Enhanced Biological Phosphorus Removal from Wastewater.* Ph.D. Dissertation; Virginia Polytechnic Institute and State University, Blacksburg, Virginia.

Pitman, A. R. (1992) Process Design of new BNR extensions to Northern Works, Johannesburg. Paper presented at Biological Nutrient Removal Conference, Leeds University, Leeds, United Kingdom.

Pitman, A. R.; Deacon, S. L.; Alexander, W. V. (1991) The Thickening and Treatment of Sewage Sludges to Minimize Phosphorus Release. *Water Res.*, **25** (10), 1285–1294.

Pöpel, H. J.; Jardin, N. (1993) Influence of Enhanced Biological Phosphorus Removal on Sludge Treatment. *Water Sci. Technol.*, **28** (1), 263–271.

Pully, T. (1991) Investigation of Anaerobic-Aerobic Activated Sludge Treatment of a High-Strength Cellulose Acetate Manufacturing Wastewater. Unpublished research, Department of Civil Engineering, Virginia Polytechnic Institute and State University: Blacksburg, Virginia.

Rabinowitz, B. (1994) Criteria for Effective Primary Sludge Fermenter Design; from Seminar on the Use of Fermentation to Enhance Biological Nutrient Removal. *Proceedings of the 67th Annual Water Environment Federation Technical Exposition and Conference,* Chicago, Illinois, Oct 15–19; Water Environment Federation: Alexandria, Virginia.

Sadler, M.; Stroud, F. R.; Maynard, J. (2002) Evaluation of Several Wastewater Reclamation Facilities Employing Biological Nutrient Removal in North Carolina. *Proceedings of the 75th Annual Water Environment Federation Technical Exhibition and Conference* [CD-ROM], Chicago, Illinois, Sep 28–Oct 2; Water Environment Federation: Alexandria, Virginia.

Sedlak, R. (1991) *Phosphorus and Nitrogen Removal from Municipal Wastewater,* 2nd ed.; Lewis Publishers: Boca Raton, Florida.

Sell, R. L. (1981) Low Temperature Biological Phosphorus Removal. Paper presented at the 54th Annual Water Environment Federation Technical Exposition and Conference, Detroit, Michigan, Oct 4–9; Water Environment Federation: Alexandria, Virginia.

Sen, D.; Randall, C.W.; Grizzard, T. J.; Knocke, W. R. (1991) Impact of Solids Handling Units on the Performance of Biological Phosphorus and Nitrogen

Removal Systems. *Proceedings of the 64th Annual Water Pollution Control Federation Technical Exposition and Conference,* Toronto, Ontario, Canada, Oct 7–10; Water Pollution Control Federation: Alexandria, Virginia..

Stephens, H. L.; Stensel, H. D. (1998) Effect of Operating Conditions on Biological Phosphorus Removal. *Water Environ. Res.,* **70**, 360–369.

U.S. Environmental Protection Agency (1993) *Nitrogen Control Manual,* EPA-625/R-93-010; Office of Research and Development, U.S. Environmental Protection Agency: Washington, D.C.

Van Huyssteen, J. A.; Barnard, J. L.; Hendriksz, J. (1989) The Olifantsfontein Nutrient Removal Plant. *Proceedings of the International Specialized Conference on Upgrading of Wastewater Treatment Plants,* Munich, Germany, Sep 3–7; International Association on Water Quality: London, 1–8.

Water Environment Federation (1998a) *Biological and Chemical Systems for Nutrient Removal.* Special Publication; Water Environment Federation, Alexandria, Virginia.

Water Environment Federation (1998b) *Design of Municipal Wastewater Treatment Plants,* 4th ed.; Manual of Practice No. 8; Water Environment Federation: Alexandria, Virginia.

Water Environment Research Foundation (2005) *Factors Influencing the Reliability of Enhanced Biological Phosphorus Removal;* Project 01-CTS-3; Water Environment Research Foundation: Alexandria, Virginia.

Wheatland, A. B.; Barnett M. J.; Bruce, A. M. (1959) Some Observations on Denitrification in Rivers and Estuaries. *Inst. Sewage Purification. J. Proc.,* 149.

Chapter 6

Models for Nutrient Removal

Introduction	227	Use of Simulators for Plant Operation	230
History and Development of Models for Biological Nutrient Removal	227	Ease of Use	230
		Developing Data for Model Input	230
Description of Models	228	Using Simulators to Troubleshoot	231
Mechanistic Models	228		
Simulators	228	Reference	232

INTRODUCTION

Operators are now using models and simulators to understand nutrient removal processes, evaluate new designs, and optimize and troubleshoot. Models and simulators are becoming easier to understand and user friendly. This chapter briefly describes what they are and how they can be used. More detailed information can be found in Chapter 9 of *Biological and Chemical Systems for Nutrient Removal* (WEF, 1998).

HISTORY AND DEVELOPMENT OF MODELS FOR BIOLOGICAL NUTRIENT REMOVAL

Activated sludge systems become more complex as their function is expanded from carbonaceous removal alone to include nitrification, denitrification, and biological phosphorus removal. Typically, a nutrient removal system involves multiple reactors, some aerated and some not, and the internal circulation of mixed liquor between reactors. The number of biological reactions and the number of compounds involved in the process also increase correspondingly. This is because a nitrification/denitrification, biological phosphorus removal process involves three separate groups of microorganisms (polyphosphorus heterotrophs, nonpolyphosphorus heterotrophs, and nitrifying autotrophs) operating on a large number of chemical compounds in three distinct environmental regimes (aerobic; anoxic, where nitrate is present but there is no dissolved oxygen; and anaerobic, where there is neither nitrate or dissolved oxygen). These features make for complex behavior, which has increased the level of difficulty in design, operation, and control.

Given this complexity, the performance of any proposed system design can be determined only by experimentation (i.e., in pilot plants) or by a mathematical model that simulates the behavior accurately. Comprehensive experimentation to evaluate the influence of a wide range of parameters is costly and time-consuming. Increasingly, the mathematical modeling of system behavior is being used as a tool to facilitate design and to evaluate operation.

DESCRIPTION OF MODELS

MECHANISTIC MODELS. Activated sludge systems typically are represented using a mechanistic model. The best know mechanistic model is the one developed by the International Water Association (IWA) (London), formerly known as the International Association on Water Quality. These mechanistic models incorporate mathematical expressions that represent the biological processes occurring within the system. The models quantify both the kinetic (reaction rate and concentration dependence) and the stoichiometry (effect on the masses of compounds involved). This model describes the biochemical reactions within the activated sludge system and the manner in which these act on the various model compounds.

SIMULATORS. Simulators are really computer programs used to solve the IWA model, which represents the response of a given activated sludge system to changes in various parameters. Typically, this is done in two steps. First, the reactor configuration and the flow scheme must be specified, including the reactor sizes, influent characteristics, recycle flowrates, wastage rate, and other specifics. After this information is fixed, it is possible to perform mass balances over each reactor (or reactor zone) for each model. These mass balances constitute the state equations that relate the dependent variables (compound concentrations) to the independent variables, such as reactor volume. The mass balances form a set of simultaneous equations, which, when solved, characterize the system behavior. The simultaneous solution provides values of the concentrations in different reactors and time. In this way, the change in concentrations throughout the system is related to the input and output and conversions processes occurring within the system.

Simulating the response of a nutrient removal system based on a compressive biological model is mathematically complex and typically achieved with a computer program. A simulation program is useful for a number of reasons.

(1) System analysis and optimization. If a system model provides accurate predications of response behavior, then these predictions can be compared to observed responses in analyzing the operation of existing systems. Any discrepancies can be useful in identifying problems in operation. An accurate model also can be used to optimize performance of existing systems. Various operating strategies can be proposed and tested rapidly, without having to resort to potentially difficult practical evaluation.

(2) System design. A simulation program does not design a system directly. However, a simulation program can be a useful tool for the design engineer to evaluate proposed system designs rapidly. In addition, a dynamic model can provide valuable design information that has often only been available through empirical estimates. For example, a parameter such as peak oxygen demand can be obtained directly from the simulation program run under time-varying input patterns. This means that peak aeration capacity can be quantified accurately.

Table 6.1 lists the names and websites for five of the current software products that can be used for simulating biological nutrient removal (BNR) processes. The listing may exclude other products about which the author is unaware. Providing this information does not endorse any of these products.

It is important to emphasize that models are not perfect. This is particularly true for nutrient removal systems, in which there are many unresolved technical aspects that are subject to ongoing research. Perhaps the biggest limitation is the expense and effort required for comprehensive influent wastewater characterization. Furthermore, the nitrifier growth rate is a parameter with high variability, so using the

TABLE 6.1 Examples of BNR computer software.

Product name	Manufacturer (location)	Web site
BioWin	EnviroSim Associates, Ltd. (Flamborough, Ontario, Canada)	www.envirosim.com
EFOR	DHI Software (Hørsholm, Denmark)	www.dhisoftware.com/efor/
GPS-X	Hydromantis, Inc. (Cambridge, Ontario, Canada)	www.hydromantis.com
SimWorks	Hydromantis, Inc. (Cambridge, Ontario, Canada)	www.hydromantis.com

default value may not give accurate model predictions. The user must understand the kinetics of the process and how the various coefficients relate to the prediction of BNR performance.

USE OF SIMULATORS FOR PLANT OPERATION

EASE OF USE. The ease of simulator use is dependent on program configurations. Some are relatively easy but require training to use them effectively. Model use also requires an understanding of the theory of nutrient removal, including kinetic and stoichiometric relationships. Some terms may be difficult for operators to understand, such as half-saturation constant and specific growth rates; however, even without fully comprehending their meanings, operators can use simulators to optimize plant performance. Developers of these software models typically include extensive training as part of the purchase agreement.

DEVELOPING DATA FOR MODEL INPUT. Developing data for use with these models may be expensive, time-consuming, and somewhat complicated. The plant laboratory or a contract laboratory must determine both soluble and particulate components of chemical oxygen demand (COD) and total Kjeldahl nitrogen (TKN). For COD, the soluble and particulate portions are further characterized to determine biodegradable and non-biodegradable fractions.

The total organic nitrogen fraction of TKN is further characterized as soluble and particulate biodegradable and soluble and particulate non-biodegradable organic nitrogen. Procedures for these tests are available in the literature and through each software provider. A review of these procedure helps evaluate whether to make these determinations in house or if outsourcing is more appropriate.

USING SIMULATORS TO TROUBLESHOOT. As mentioned earlier, engineers and operators use simulators to assist in the design of nutrient removal systems. Another important function of the simulator is to help troubleshoot the process. Following are some examples:

(1) Many times, influent wastewater characteristics change from those used during design, an industry leaves or a new industry comes to town, for example. An operator can input the new characteristics to the simulator to determine how those changed characteristics have affected predicted effluent quality. Through the simulator, the operator can decide whether to take tanks offline or add tanks or whether to change the rate of internal recycle, the rate of methanol addition, or other parameters.

(2) The simulator can also be used to understand the effect of taking tanks out of service for maintenance. An operator can evaluate what time of year is the best time for preventive maintenance by entering in various temperatures to determine which temperatures favors the available volume. The operator can also determine if methanol addition would attenuate the effects of the reduced volume.

(3) The operator can evaluate potential for an existing plant to meet stricter effluent discharge requirements by operating at a higher mixed liquor suspended solids concentration or longer solids retention time.

(4) The operator can evaluate the ability of the plant to accommodate higher flows or loads and still meet the effluent requirements in lieu of a new capital expenditure.

Simulators give tools to the operator understand what the design engineer is proposing, help meet permit under all conditions, improve the economics of operating a BNR facility, and understand exactly what is happening within these processes.

They are not easy to use and require not only training, but also knowledge of the kinetics of the biological system. However, these simulators are improving continuously. There is no doubt that modeling and simulation are assuming a prominent role in nutrient removal system design and will be important for operation and control.

REFERENCE

Water Environment Federation (1998) *Biological and Chemical Systems for Nutrient Removal.* Special Publication; Water Environment Federation: Alexandria, Virginia.

Chapter 7

Sludge Bulking and Foaming

Introduction	233	Troubleshooting Sludge Bulking Problems	243
Microscopic Examination	234	Foaming Problems and Solutions	243
Filamentous Bulking	236	Stiff White Foam	247
Process Control for Filamentous Bulking Problems	239	Excessive Brown Foam and Dark Tan Foam	247
Dissolved Oxygen	239	Dark Brown Foam	247
Nutrient Balance	240	Very Dark or Black Foam	247
pH	241	Filamentous Foaming	247
Chemical Addition	241	Conclusion	252
Chlorination	241	References	252
Biological Selectors	242		
Nonfilamentous Bulking	243		

INTRODUCTION

Foaming and bulking are perhaps the most common operating problems with biological nutrient removal (BNR) processes. Because most BNR processes are an extension of the activated sludge process, in many cases, standard activated sludge troubleshooting guidelines associated with bulking and foaming can apply to BNR problems.

Operators must know and recognize problems associated with filamentous organisms. For example, if there is clear supernatant above a poorly settling sludge, it means that the settling process is being hindered by filamentous organisms. Operators should also know how to identify various types of microorganisms, what effect each has on the system, and how to control them. The most important tool an operator has is a microscope.

MICROSCOPIC EXAMINATION

By examining mixed liquor with a microscope, much can be determined about the BNR process. The presence of various microorganisms within the mixed liquor floc can quickly indicate good or poor treatment. It is not necessary to be a skilled microbiologist or to be able to identify or count individual species. Instead, it is important to recognize significant groups of microorganisms, such as the following:

- Filamentous bacteria (Figure 7.1);
- Protozoa (amoebas, flagellates, and free-swimming or stalked ciliates) (Figure 7.2); and
- Rotifers (Figure 7.3).

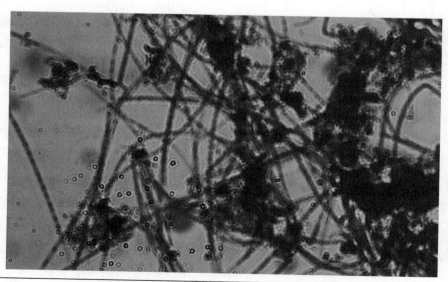

FIGURE 7.1 Filamentous bacteria.

Sludge Bulking and Foaming 235

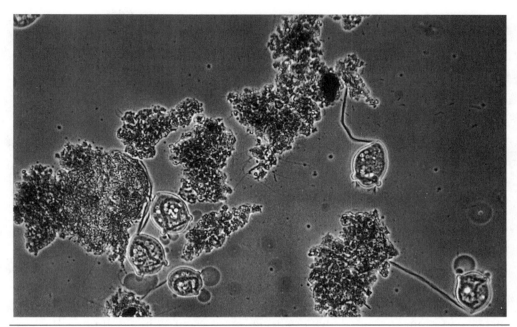

FIGURE 7.2 Stalked ciliates.

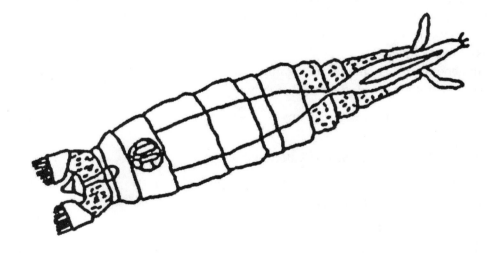

FIGURE 7.3 Rotifers.

Through the microscope, operators can observe the presence or absence of various filamentous organisms, protozoa, rotifers, and other organisms and also the floc structure itself. All of these observations allow operators to make informed decisions on process conditions and process control alternatives.

The identification of filamentous microorganisms is important and can help narrow the investigative process during troubleshooting. Knowing whether a filamentous microorganism is *Nocardia*, *Microthrix*, *Thiothrix*, Type 021N, or others is important in pinpointing an activated sludge settling problem and the corrective action. The specific type of filamentous microorganism will be an indication of whether the problem is caused by a low dissolved oxygen (DO) concentration, nutrient-deficient waste, a low food-to-microorganism ratio (F/M), or other conditions.

FILAMENTOUS BULKING

The presence of some filamentous organisms in activated sludge is important because they add the backbone to the floc structure, thus helping sludge to settle in the final clarifiers, producing a clear effluent. However, as the filamentous organisms increase in number or if particular species are present, solids will not settle well, which results in high effluent suspended solids and possible permit violations.

A variety of influent and plant conditions can encourage the growth of filamentous organisms that cause filamentous bulking, such as low DO, low or high F/M, insufficient nutrients, sulfides, or low pH values. Table 7.1 is a general guide that relates bulking by specific filamentous organisms to various plant conditions (Jenkins et al., 2004).

In general, once filamentous organisms are identified, specific operational controls can be implemented, such as the following:

- Improving the treatment environment, which may include the use of a selector;
- Chlorinating the return activated sludge (RAS);
- Adding nutrients, such as nitrogen, phosphorus, and iron;
- Correcting the DO concentration in the biological reactor; and/or
- Correcting a pH condition.

It is important to understand that not all bulking conditions are caused by filamentous organisms. For example, the presence of a cloudy effluent above a poorly

TABLE 7.1 Filament types as indicators of conditions causing activated sludge bulking (Jenkins et al., 2004). Copyright © 2004 From *Manual on the Causes and Control of Activated Sludge Bulking, Foaming, and Other Solids Separation Problems*, 3rd ed. by Jenkins et al. Reproduced with permission of Routledge/Taylor & Francis Group, LLC.

Cause	Filamentous organism
Low DO concentration	*S. natans*
	Type 1701
	H. hydrossis
Low F/M	Type 0041
	Type 0675
	Type 1851
	Type 0803
Elevated low molecular weight organic acid concentration	Type 021N
	Thiothrix I and II
	N. limicola I, II and III
	Type 0914
	Type 0411
	Type 0961
	Type 0581
	Type 0092
Hydrogen sulfide	*Thiothrix* I and II
	Type 02IN
	Type 0914
	Beggiatoa spp.
Nutrient deficiency	Type 02IN
Nitrogen	*Thiothrix* I and II
Phosphorus	*N. limicola* III
	H. hydrossis
	S. natans
Low pH	Fungi

settling sludge indicates viscous bulking, which is caused by organic loading outside design parameters, overaeration, or the presence of toxics.

Classic sludge bulking in the clarifier is illustrated in Figure 7.4.

If filamentous microorganisms are present when observing mixed liquor under a microscope and bulking is occurring, typical causes include the following:

- Low DO concentrations in biological reactors;
- Insufficient nutrients;
- Too low or variable pH;
- Widely varying organic loading;
- Industrial wastes with high biochemical oxygen demand (BOD) and low nutrients (nitrogen and phosphorus), such as simple sugars or carbohydrates (i.e., food processing waste);

FIGURE 7.4 Troubleshooting guide on settleability tests: (a) good settling sludge (effluent quality problem because of settling), (b) bulking sludge, (c) clumping/rising sludge (denitrification), (d) cloudy effluent, (e) ash on surface, and (f) pinpoint floc and stragglers.

- High influent sulfide concentrations that cause the filamentous microorganism Thiothrix to grow and produce filamentous bulking;
- Very low F/M, allowing *Nocardia* predominance;
- Massive amounts of filaments present in influent wastewater or recycle streams; and
- Insufficient soluble five-day BOD (BOD_5) gradient (generally measured as F/M in a series of biological reactors or cells).

Depending on the filament, operators can implement short-term solutions and evaluate long-term solutions for future implementation. Short-term solutions involve treating the symptoms (changing influent feed points, changing RAS rates, adding settling aids, and chlorinating); and long-term solutions involve treating the cause (adding a selector, controlling mixed liquor pH, controlling influent septicity, adding nutrients, changing aeration rates, and changing F/M).

PROCESS CONTROL FOR FILAMENTOUS BULKING PROBLEMS

If a moderate to large number of filamentous microorganisms are present, the most important thing is to identify the type of filament. If the plant staff does not have the equipment or expertise, outside help should be used. The next thing is to identify the cause of the filament growth. The following sections describe ways of identifying the causes.

DISSOLVED OXYGEN. Measure the DO concentration at various locations throughout the biological reactor.

(1) If the typical DO concentration throughout the reactor is less than 0.5 mg/L, there is insufficient DO in the biological reactor. Increase aeration until DO increases to 1.5 to 3 mg/L throughout the tank.

(2) If DO concentrations are nearly zero in some parts of the reactor, but are higher in other locations, the air distribution system on a diffused air system may be off balance or the diffusers in an area of the reactor may need to be cleaned. Balance the air system and clean the diffusers. If a mechanically aerated system is used, increase aerator speed or raise the overflow weirs.

(3) If DO concentration is low only at the head of the tank, which is being operated in the plug-flow pattern, consider changing to the step-feed or complete-mix flow patterns or using tapered aeration, if possible. However, a low DO concentration only at the head of the reactor may not be a problem, as long as adequate reaction time is available in the rest of the reactor.

NUTRIENT BALANCE. Calculate the ratio of BOD_5 to nitrogen (use total Kjeldahl nitrogen expressed as nitrogen [N]), BOD_5 to phosphorus (expressed as phosphorus [P]), and BOD_5 to iron (expressed as iron [Fe]). Generally, an adequate nutrient balance in an activated sludge system is 100:5:1 (BOD:N:P). Refer to Appendix A of Activated Sludge (WEF, 2002) for procedures on checking nutrient levels and calculating how much nutrient to add if there is a problem. In general, anhydrous ammonia is used to add nitrogen, trisodium phosphate is used to add phophorus, and ferric chloride is used to add iron. The ratio of BOD_5 to required nutrients changes with solids retention time (SRT). For example, high SRTs produce less sludge and result in lower nitrogen and phosphorus requirements. Typically, 3 to 5 mg/L of nitrogen (3 mg/L at high SRTs and 5 mg/L at low SRTs) and 1 mg/L of phosphorus is needed for every 100 mg/L of BOD_5. Typically, a nutrient-deficient waste will not occur with domestic wastewater. However, if a plant is receiving industrial discharge, such as from a canning company, a nutrient-deficient waste could occur.

(1) If there are insufficient nutrients in the wastewater, two things can happen. Filamentous bacteria will predominate or take over the mixed liquor suspended solids (MLSS). Organic materials will only be partially converted to end products; that is, there will be insufficient BOD_5 or chemical oxygen demand removal.

(2) If excess nutrients are added to the wastewater to overcompensate for a nutrient deficiency, a large fraction of these nutrients may not be incorporated to the MLSS and may, therefore, pass into the effluent.

(3) Nutrients, if required, should be added at the head of the biological reactor. Mixed liquor settleability should be carefully observed to see if it is improving. If settleability improves, the nutrient dose can be reduced by 5% per week until settleability decreases. Then the dose should be increased by 5% and the settleability observed.

Nutrients are expensive and should be carefully applied. For example, adding excess ammonia may create a nitrification demand. The nutrient dosage should be increased as BOD concentrations increase, which takes into account the effects of the additional microorganism growth that will occur. If settleability does not improve readily, nutrient dosing should be continued until the actual problem is identified and solved, because the problems that are causing poor settleability may be interrelated.

pH. Some filamentous bulking can occur because of low pH values. As stated in Chapter 3, nitrification destroys alkalinity and reduces the biological reactor pH. If a plant is only nitrifying or if there is very limited influent alkalinity, it may be necessary to raise the MLSS pH by adding sodium bicarbonate ($NaHCO_3$), caustic soda (NaOH), or lime ($Ca[OH]_2$) at the head of the biological reactor.

Adding $NaHCO_3$, NaOH, or $Ca(OH)_2$ to raise pH is expensive. Mixed liquor settleability should be closely monitored to observe changes to ensure that the chemical added is effective. If no improvement occurs within 2 to 4 weeks, and nothing else in the process has changed in the meantime, then chemical addition should be stopped.

CHEMICAL ADDITION. *Chlorination.* Many plants use chlorination to control filamentous bulking. It must be done carefully so as not to harm the floc-forming bacteria and negatively affect treatment capabilities. Chlorination should not be performed randomly. There should be a good procedure set for its use.

One approach is to set a target sludge volume index (SVI) (or other measure of settleability) for satisfactory operation of secondary clarifiers and solids processing units and chlorinate only when the target is significantly and consistently exceeded (Jenkins et al., 1993, 2004). This approach treats the symptom, not the underlying problem.

Location of the chlorine application point is critical. The point should be located where there is excellent mixing, where the sludge is concentrated, and where the wastewater concentration is at a minimum (to reduce unwanted reactions with the chlorine) (Richard, 1989). The three common application points are in the RAS stream, directly in the biological reactor at each aerator, and in an installed sidestream that recirculates mixed liquor within the biological reactor.

Chlorine dose and the frequency at which organisms are exposed to chlorine are the two most important parameters. The dose is adjusted so that concentrations are lethal at the floc surface, but not within the floc. The chlorine dose should be based on the solids inventory in the process (biological reactors plus clarifiers). This is called the overall chlorine mass dose, and effective dosages are in the range of 1 to 10 g/kg·d mixed liquor volatile suspended solids (MLVSS) (1 to 10 lb chlorine/d/1000 lb MLVSS). The dose must be accurately measured, preferably by using a chlorinator or hypochlorite pumps dedicated to this purpose. Initially, start at a lower dose and gradually increase the dose until it is effective (Richard, 1989).

Frequency of exposure is a measure of how often the entire solids inventory is subjected to an excessively high chlorine dosage. Required frequency of exposure depends on relative growth rates of filamentous and floc-forming organisms and the effectiveness of each dose. While the frequency of exposure is plant-specific, three times or more per day were reported as sufficient (Jenkins et al., 1993, 2004), and success has been reported at frequencies as low as once per day (Richard, 1989). The actual dose applied at each exposure should be simply the total calculated dose per day divided by the number of exposure events.

Control tests should be performed during chlorination to assess the effects of chlorine on both filamentous and floc-forming organisms. The tests should measure settleability (i.e., SVI), effluent quality (turbidity), and activated sludge quality (microscopic examination) (Jenkins et al., 1993, 2004). An adequate chlorine dose should start to improve settleability within 1 to 3 days (Richard, 1989). A turbid, milky effluent and a reduction in BOD removal are signs of over chlorination, although a small increase in effluent suspended solids and BOD concentrations are normal during chlorination for bulking control. The microscopically visible effects of chlorine on filaments include the following (in order):

- Intracellular sulfur granules (if present) disappear,
- Cells deform and cytoplasm shrinks, and
- The filaments break up and dissolve.

Chlorine does not destroy the sheath of sheathed filaments, and this causes poor sludge settling until they are wasted from the system. Chlorination should be stopped when only empty sheaths remain, not continued until the SVI falls (Richard, 1989). Adding chlorine beyond this point may over chlorinate. Because the effects of chlorine on filaments can be detected microscopically before settleability improves, microscopic examinations can provide an early indication of filament control.

Biological Selectors. A uniform low concentration of soluble BOD, as found in completely mixed biological reactors, is considered by some to encourage the growth of some types of filamentous microorganisms that compete with floc formers to cause low F/M bulking. One technique suggested to control F/M filaments is use of a selector zone for the wastewater and RAS. In this selector zone, a high-growth-rate environment is provided to promote the growth of floc-forming organisms at the expense of filamentous organisms.

The use of biological selectors (compartmentalizing or dividing the aeration basin into discrete cells or zones) is another way of controlling bulking sludge. Biological nutrient removal systems are generally not only designed (compartmentalized) to improve sludge settlability by controlling the growth of filamentous bacteria, but also may include recovery of oxygen or alkalinity through denitrification or other steps.

Nonfilamentous Bulking. If few or no filamentous microorganisms are present, first determine if the F/M ratio is higher or lower than normal. The presence of small, dispersed floc is characteristic of an increased F/M. If the F/M is higher by 10% or more, then decrease the wasting rate to reduce the F/M. Once the F/M is decreased, dispersed floc should disappear and the effluent quality should improve, but it may take a period of two to three SRTs.

TROUBLESHOOTING SLUDGE BULKING PROBLEMS

Table 7.2 summarizes the various problems and their suggested solutions.

FOAMING PROBLEMS AND SOLUTIONS

The presence of some foam (or froth) on the biological reactor is normal for the activated sludge process. Typically, in a well-operated process, 10 to 25% of the reactor surface will be covered with a 50- to 80-mm (2- to 3-in.) layer of light tan foam.

Under certain operating conditions, foam can become excessive and affect operations. Three general types of problem foam are often seen: stiff white foam, brown foam (greasy dark tan foam and thick scummy dark brown foam), and a very dark or black foam (Figure 7.5).

If stiff white foam is allowed to build up excessively, it can be blown by the wind onto walkways and plant structures and create hazardous working conditions. It can also create an unsightly appearance, cause odors, and carry pathogenic microorganisms. If greasy or thick, scummy foam builds up and is carried over with the flow to

TABLE 7.2 Troubleshooting guide for bulking sludge (WEF, 2002).

Observation	Probable cause	Necessary check	Remedies
1. Clouds of billowing homogeneous sludge rising and extending throughout the clarifier. Mixed liquor settles slowly and compacts poorly in settleability test, but supernatant is fairly clear.	A. Improper organic loading or DO concentration.	1. Check and monitor trend changes which occur in the following: a. Decrease in MLVSS concentration b. Decrease in MCRT c. Increase in F/M d. Change in DO concentrations e. Sudden SVI increase from normal, or decrease in sludge density index	1. Decrease WAS rates by not more than 10% per day until process approaches normal operating parameters. 2. Temporarily increase RAS rates to minimize solids carryover from clarifier. Continue until normal control parameters are approached. 3. DO concentration throughout biological reactor should be greater then 0.5 mg/L, preferably 1 to 3 mg/L.
	B. Filamentous organisms.	1. Perform microscopic examination of mixed liquor and return sludge. If possible, try to identify type of filamentous organisms, either fungal or bacterial.	1. If no filamentous organisms are observed, refer to Probable cause "A" above.
		2. If fungal organism is identified, check industries for wastes that may cause problems.	2. Enforce industrial waste ordinance to eliminate wastes. Also see Remedy 4 below.
		3. If bacterial organisms are identified, check influent wastewater and in-plant sidestream flows returning to process for massive filamentous organisms.	3. Chlorinate influent wastewater at 5- to 10-mg/L dosages. If higher dosages are required, use extreme caution. Increase dosage in 1- to 2-mg/L increments. 4. Chlorinate RAS at 2 to 3 g/kg (lb/d/1000 lb) MLVSS. 5. Optimized operational performance or upgrading of other in-plant unit processes will be required if filamentous organisms are found in sidestream flows.

(continued)

TABLE 7.2 Troubleshooting guide for bulking sludge (WEF, 2002) *(continued)*.

Observation	Probable cause	Necessary check	Remedies
	C. Wastewater nutrient deficiencies.	1. Check nutrient concentrations in influent wastewater. The BOD-to-nutrient ratios should be 100 parts BOD to 5 parts total nitrogen to 1 part phosphorus to 0.5 parts iron.	1. If nutrient concentrations are less than average ratio, field tests should be performed on the influent wastewater for addition of nitrogen in the form of anhydrous ammonia, phosphorus in the form of trisodium phosphate and/or iron in the form of ferric chloride.
		2. Perform hourly mixed liquor settleability tests.	2. Observe tests for improvement in sludge settling characteristics with the addition of nutrients.
	D. Low DO concentrations in biological reactor.	1. Check DO concentrations at various locations throughout the reactor.	1. If average DO concentration is <0.5 mg/L, increase airflow rate until the DO concentration level increases to between 1 and 3 mg/L throughout the reactor.
			2. If DO concentrations are nearly zero in some parts of the reactor, but 1 mg/L or more in other locations, balance the air distribution system or clean diffusers.
	E. pH in biological reactor is less than 6.5.	1. Monitor plant influent pH.	1. If pH is less than 6.5, conduct industrial survey to identify source. If possible, stop or neutralize discharge at source.
			2. If the above is not possible, raise pH by adding an alkaline agent such as caustic soda or lime to the biological reactor influent.

(continued)

TABLE 7.2 Troubleshooting guide for bulking sludge (WEF, 2002) *(continued)*.

Observation	Probable cause	Necessary check	Remedies
		2. Check if process is nitrifying because of warm wastewater temperature or low F/M loading.	1. If nitrification is not required, increase WAS rate by not more than 10% per day to stop nitrification.
			2. If nitrification is required, raise pH by adding an alkaline agent such as caustic soda or lime to the aeration influent.

FIGURE 7.5 Sludge bulking in clarifiers.

the secondary clarifiers, it will tend to build up behind the influent baffles and create additional cleaning problems. It can also plug the scum-removal system. Table 7.3 summarizes the various problems and their suggested solutions.

STIFF WHITE FOAM. Stiff white billowing foam, indicating a young sludge (low SRT), is found in either new or underloaded plants. This means that the MLSS concentration is too low (low activated sludge solids inventory), and the F/M is too high.

EXCESSIVE BROWN FOAM AND DARK TAN FOAM. Foam, which appears excessively brown, is generally associated with plants operating at low-loading ranges. Plants designed to nitrify and operating in the nitrifying mode will typically have low-to-moderate amounts of chocolate brown foam.

DARK BROWN FOAM. Thick, scummy dark brown foam indicates an old sludge (long SRT) and can result in additional problems in the clarifier by building up behind influent baffles and creating a scum disposal problem. Biological nutrient removal plants tend to run at long SRTs, so this is a common problem associated with that process.

VERY DARK OR BLACK FOAM. The presence of very dark or black foam indicates either insufficient aeration, which results in anaerobic conditions, or that industrial wastes such as dyes and inks are present.

FILAMENTOUS FOAMING. Three filamentous organisms can cause activated sludge foaming: *Nocardia* sp. (most common), *Microthrix parvicella* (less common), and Type 1863 (rare) (Jenkins et al., 1993, 2004; Richard, 1989). These three types of filamentous organisms can be distinguished from one another through microscopic examination and staining. *Nocardia* foaming seems to be the most common and severe.

Filamentous organisms can produce a stable, viscous, brown foam on the biological reactor surface that can shift over onto the clarifier surface and may even escape in the effluent, violating permit limits. Foaming can range from being a nuisance to being a serious problem. In cold weather, foam may freeze and have to be removed manually with a pick and shovel. In warm weather, it often becomes odorous.

Samples of the foam and mixed liquor should be microscopically examined to determine if the foaming is, in fact, attributable to filamentous growths (Jenkins et

TABLE 7.3 Troubleshooting guide for foaming sludge (WEF, 2002).

Observation	Probable cause	Necessary check	Remedies
1. White, thick, billowing or sudsy foam on biological reactor surface.	A. Overloaded biological reactor (low MLSS concentration) because of process startup. Do not be alarmed: this problem usually occurs during process startup.	1. Check biological reactor BOD loading kg/d (lb/d) and kg (lb) MLVSS in biological reactor. Calculate F/M to determine kg/d (lb/d) MLVSS inventory for current BOD loading.	1. After calculating the F/M and kg MLVSS needed, if the F/M is high and the MLVSS inventory is low, do not waste sludge from the process or maintain the minimum WAS rate possible if wasting has already started.
		2. Check secondary clarifier effluent for solids carryover. Effluent will look cloudy.	2. Maintain sufficient RAS rates to minimize solids carryover, especially during peak flow periods.
		3. Check DO concentration in biological reactor.	3. Try to maintain DO concentrations between 1.0 and 3.0 mg/L. Also be sure that adequate mxing is being provided in the biological reactor while attempting to maintain DO concentrations.
	B. Excessive sludge wasting from process causing overloaded biological reactor (low MLSS concentration).	1. Check and monitor for trend changes, which occur in the following: a. Decrease in MLVSS concentration b. Decrease in MCRT c. Increase in F/M d. DO concentrations maintained with lower air rates e. Increase in WAS rates	1. Reduce WAS rate by not more than 10% per day until process approaches normal control parameters. 2. Increase RAS rate to minimize effluent solids carryover from secondary clarifier. Maintain sludge-blanket depth of 0.3 to 0.9 m (1 to 3 ft) from clarifier floor.
	C. Highly toxic waste, such as metals or bacteriocide, or colder wastewater temperatures, or severe temperature variations resulting in reduction of MLSS concentrations.	1. Take MLSS sample and test for metals and bacteriocide and temperature	1. Reestablish new culture of activated sludge. If possible, waste sludge from process without returning it to other in-plant systems. Obtain seed sludge from another plant, if possible.

(continued)

TABLE 7.3 Troubleshooting guide for foaming sludge (WEF, 2002) *(continued)*.

Observation	Probable cause	Necessary check	Remedies
		2. Monitor plant influent for significant variations in temperature.	2. Actively enforce industrial-waste ordinances.
	D. Hydraulic washout of solids from secondary clarifier.	1. Check hydraulic residence time in biological reactor and surface overflow rate in secondary clarifier.	
	E. Improper influent wastewater and/or RAS flow distribution causing foaming in one or more biological reactors.	1. Check and monitor for significant differences in MLSS concentrations among multiple biological reactors.	1. MLSS and RAS concentrations and DO concentrations among multiple reactors should be consistent.
		2. Check and monitor primary effluent and/or RAS flowrates to each biological reactor.	2. Modify distribution facilities as necessary to maintain equal influent wastewater and/or RAS flowrates to biological reactors.
2. Shiny, dark-tan foam on biological reactor surface.	A. Biological reactor approaching underloaded (high MLSS concentration) condition because of insufficient sludge wasting from the process.	1. Check and monitor for trend changes, which occur in the following: a. Increase in MLSS concentration b. Increase in MCRT c. Decrease in F/M d. DO concentrations maintained with increasing air rates. e. Decrease in WAS rates	1. Increase WAS rate by not more than 10% per day until process approaches normal control parameters and a modest amount of light-tan foam is observed on biological reactor surface.
			2. For multiple reactor operation, refer to Observation 1, Probable cause "E" of this table.

(continued)

TABLE 7.3   Troubleshooting guide for foaming sludge (WEF, 2002) *(continued)*.

Observation	Probable cause	Necessary check	Remedies
3. Thick, scummy dark-tan foam on biological reactor surface.	A. Biological reactor is critically underloaded (MLSS concentration too high) because of improper WAS control program.	1. Check and monitor for trend changes, which occur in the following: a. Increase in MLVSS concentration b. Increase in MCRT c. Decrease in F/M	1. Increase WAS rate by not more than 10% per day until process approaches normal control parameters and a modest amount of light-tan foam is observed on biological reactor surface 2. For multiple tank operation refer to Observation 1, Probable cause "E" of this table.
4. Dark-brown, almost blackish sudsy foam on biological reactor surface. Mixed liquor color is dark-brown to almost black. Detection of septic or sour odor from biological reactor.	A. Anaerobic conditions occurring in biological reactor.		

al., 1993, 2004; Richard, 1989). Because *Nocardia* foaming is the most common type experienced in BNR plants, the following discussion focuses on identification, causes, and control of *Nocardia*.

The analysis of *Nocardia* foaming involves two factors.

(1) The factors that allow *Nocardia* to grow in activated sludge, and
(2) The conditions that cause foaming (Jenkins et al., 1993, 2004; Richard, 1989).

Nocardia growth is typically associated with warmer temperatures, grease, oil, and fats present in treated wastes and longer SRTs (typically more than nine days), although it has been frequently encountered at SRTs of two days, at cold temperatures and without the presence of grease, oil, and fat.

It is important for operators to understand that, because the foam sits on the surface of the biological reactors, it typically has a longer SRT than the underlying MLSS, and calculations of retention time may be incorrect (Richard, 1989).

Nocardia foaming seems to involve the hydrophobic (water repellant), waxy nature of the *Nocardia* cell wall, which tends to cause flotation under aeration (Richard, 1989). *Nocardia* cells in the mixed liquor concentrate in the foam and continue to float even after they die.

Plants prone to *Nocardia* foaming often, but not always, receive oil and grease wastes (for example, from restaurants without grease traps); have poor or no primary scum removal; recycle scum rather than remove it from the plant; and have biological reactor configurations, such as submerged wall cut outs or effluent gates that trap foam.

The best way to deal with *Nocardia* foaming is to try to prevent the conditions that encourage *Nocardia* growth; however, that is not always possible because exact cause-and-effect relationships have not been completely established. Once established, *Nocardia* foaming can be extremely difficult to eliminate because of the following reasons:

(1) Foam is difficult to break with water sprays.
(2) Chlorinating RAS, although often helpful, does not eliminate *Nocardia* because most of it is in the floc and not exposed to chlorine.
(3) Increased wasting has its limitations because of the following:
    - Foam is not wasted with the waste activated sludge (WAS).
    - Even if foam and scum are removed from the process, they can cause problems in downstream units, such as digesters, and also can be recycled

with decant or supernatant to the activated sludge process.
- Reducing the SRT to fewer than nine days (the classic cure for *Nocardia* growth and foaming) may be inadequate and may result in the loss of nitrogen removal. Numerous *Nocardia* species are involved in foaming. Of the two most dominant, one is slow-growing and the other is fast-growing. The SRT needed to control *Nocardia* foaming may depend on which species is involved (Jenkins et al., 1993, 2004; Richard, 1989).

One proactive approach and one being recommended by most design consultants for dealing with *Nocardia* is to provide surface-wasting facilities to remove foam and scum from the surface of the biological reactors or a convenient channel. The foam and scum should then conveyed to the thickening process (with the WAS) and removed from the system. Foam must be removed and separated so that it will not recycle through the plant.

Another method is to provide a fine contained spray of a highly concentrated chlorine solution (0.5 to 1.0%) directly to the surface of the biological reactor where the foam is located; however, because the foam can be very thick and viscous, the chlorine may not penetrate very far into the foam layer.

CONCLUSION

Filamentous bulking and foaming are two major operating problems associated with BNR facilities. They not only negatively affect effluent quality, but they create serious housekeeping and odor problems. It is important for operators to understand what causes these problems to occur and what works best at your plant to control them. It is important to perform microscopic examinations every day and to determine if conditions are occurring that might cause filaments to grow.

REFERENCES

Jenkins, D.; Richard, M.; Daigger, G. T. (1993) *Manual on the Causes and Control of Activated Sludge Bulking and Foaming*, 2nd ed.; Lewis Publishers: Chelsea, Michigan.

Jenkins, D.; Richard, M.; Daigger, G. T. (2004) *Manual on the Causes and Control of Activated Sludge Bulking, Foaming, and Other Solids Separation Problems*, 3rd ed.; CRC Press: Boca Raton, Florida.

Richard, M. (1989) *Activated Sludge Microbiology.* Water Pollutution Control Federation: Alexandria, Virginia.

Water Environment Federation (2002) *Activated Sludge,* 2nd ed.; Manual of Practice No. OM-9; Water Environment Federation: Alexandria, Virginia.

Chapter 8

Chemical Addition and Chemical Feed Control

Introduction	256	*Long Creek Wastewater Treatment Plant, Gastonia, North Carolina*	269
Carbon Supplementation for Denitrification	257	Volatile Fatty Acid Supplementation for Biological Phosphorus Removal	271
Methanol Addition	257	Acetic Acid	273
Methanol Addition to Activated Sludge Biological Nutrient Removal Processes	259	Alternate Chemical Volatile Fatty Acid Sources	274
Methanol Addition to Tertiary Denitrification Processes	263	Case Study: McDowell Creek Wastewater Treatment Plant, Charlotte, North Carolina	274
Methanol Feed Control	265	*Effect of Dewatering Filtrate*	275
Manual Control	265	*Optimization of Chemical Dosages*	275
Flow-Paced Control	265	Alkalinity Supplementation	277
Feed-Forward Control	265	Alkalinity	277
Feed-Forward and Feedback with Effluent Concentration Control	266	Alkalinity Supplementation	279
Alternate Carbon Sources for Denitrification	266	*Sodium Hydroxide*	279
Case Studies	268	*Calcium Hydroxide*	281
Havelock Wastewater Treatment Plant, Havelock, North Carolina	268	*Quicklime*	281
		Magnesium Hydroxide	282

Sodium Carbonate	282	*Example 3*	287
Sodium Bicarbonate	282	Step 1	287
Alkalinity Considerations	282	Step 2	287
Usable Alkalinity	282	Phosphorus Precipitation	288
Volatile Fatty Acids and Other Alkalinity	283	Iron Compound Chemical Addition	290
Alkalinity Measurement	284	Aluminum Compound Chemical Addition	297
High-Purity-Oxygen Activated Sludge	284	Lime Addition	303
Phosphorus Precipitation	285	Other Options for Chemical Precipitation of Phosphorus	305
Practical Examples	285	Chemical Feed Control	306
Example 1	285	Case Study: Northwest Cobb Water Reclamation Facility, Cobb County, Georgia	306
Step 1	285		
Step 2	285		
Example 2	286	Chemical Feed System Design and Operational Considerations	309
Step 1	286		
Step 2	286	References	311

INTRODUCTION

As discussed in previous chapters, successful operation of a biological nutrient removal (BNR) system may require the addition of chemicals. For example, a readily biodegradable carbon substrate (or soluble biochemical oxygen demand [BOD]/organics) is needed for efficient denitrification. Short-chain volatile fatty acids are needed for biological phosphorus removal. In some areas of the country, where water is naturally soft, alkalinity supplementation may be necessary to maintain the right pH conditions throughout the aeration tank for nitrification. Finally, if chemical phosphorus removal is used, either for polishing after a biological phosphorus removal process or as the main means of phosphorus removal, metal salts and polymer may be needed. This chapter will focus on the basic system and oper-

Chemical Addition and Chemical Feed Control

ating requirements for each, including dosage calculation examples and case studies of wastewater treatment plants (WWTPs) using chemicals to augment their nutrient removal process. The next four sections focus on specific chemicals and their use in the BNR process. General chemical feed system storage, safety, and operational considerations are included at the end of the chapter.

CARBON SUPPLEMENTATION FOR DENITRIFICATION

As discussed in earlier chapters, to achieve BNR through nitrification and denitrification, the ammonia-nitrogen in the wastewater must first be nitrified or converted to nitrate and nitrite. In denitrification, the nitrate is then used as the oxygen source for oxidation of simple carbon compounds through cellular respiration. Therefore, in a denitrification process where the objective is to remove nitrate, a readily biodegradable carbon source must be available. The carbon substrate requirements can be met by the influent wastewater soluble BOD, from cell mass decay in endogenous denitrification or by a supplemental carbon source. This section will focus on the use of supplemental chemicals to provide the necessary carbon source for denitrification, in the absence of sufficient available soluble organic material.

METHANOL ADDITION. Methanol is the most commonly used and best documented carbon substrate for denitrification. Methanol can be supplemented to the activated sludge BNR system in the first anoxic zone in a two- or three-stage BNR configuration and can be dosed to the same location or the postanoxic zone in a four- or five-stage BNR configuration. Methanol has been the substrate of choice for plants operating tertiary denitrification filters. Methanol also can be dosed to other tertiary fixed-film denitrification processes, including tertiary denitrifying biological aerated filters (BAFs) and the postanoxic section of a moving bed biofilm reactor (MBBR) arranged in a four-stage BNR configuration.

Although methanol has generally been widely available and relatively cost-effective, there are some handling concerns. Methanol is highly flammable, with a flash point of 12°C (54°F). This must be taken into consideration in design and handling to meet National Fire Protection Association (NFPA) and other code requirements (NFPA, 2001). A summary of chemical information for methanol is provided in Table 8.1.

Although the specific local code requirements govern, methanol storage tanks, piping, and appurtenances must be of metal construction when inside buildings as a result of flammability considerations. Similarly, close attention must be paid to

TABLE 8.1 Properties of 100% methanol (Merck Index, 1991).

Chemical formula	CH_3OH
Molecular weight	32.04 g/mole
Specific gravity	0.7915
Density	0.79 kg/L (6.6 lb/gal)
Flash point	12°C (54°F)
Freezing point	-117°C (-179°F)
Description	Colorless, odorless liquid

building classification and NFPA 820 requirements (NFPA, 2001), which may specify a Class 1 Division 1 classification requiring that the related electrical systems be rated as explosion-proof. Where feasible, methanol feed systems are generally located outdoors and away from other systems. In some cases, nonsparking materials of construction are required.

To reduce the risk associated with buildup of flammable vapors in the storage tank headspace, some installations have elected to use an inert gas blanket in the storage tank headspace. More recently, floating covers have been used in methanol storage tanks, thus maintaining the actual headspace at a minimum and at significantly lower cost than an inert gas system (similar to an anaerobic digester floating cover). Pressure relief valves and flame arrestors also are typically installed on the methanol storage tanks. With the exception of the flammability-related precautions, methanol feed system components are similar to those of other liquid chemical feed systems.

There have been a number of studies over the years looking at denitrification kinetics, carbon requirements, and cell yields. McCarty et al. (1969) studied the effects of a number of substrates on denitrification. Based on this research, methanol requirements can be quantified using eq 8.1. It is noted that nitrate (NO_3-N), nitrite (NO_2-N), and dissolved oxygen all exert a methanol demand. While dissolved oxygen (DO) should be negligible within the anoxic zone, there is generally some DO carried into the anoxic zone in the mixed liquor recycle from the aerobic zone. Long-term operating experience with denitrification filters has shown that approximately 3 mg/L methanol is required for each milligram per liter of nitrate denitrified.

$$\text{Methanol requirement} = 2.47\ NO_3\text{-}N + 1.53\ NO_2\text{-}N + 0.87\ DO \tag{8.1}$$

Each gram of methanol contains 1.5 g of chemical oxygen demand (COD), as determined by balancing the equation for oxidation of methanol (CH_3OH) to carbon dioxide (CO_2) and water (H_2O), as follows:

$$CH_3OH + 3/2\ O_2 \rightarrow CO_2 + 2H_2O \tag{8.2}$$

In this equation, 1.5 mol oxygen (or 1.5 mol × 32 g/mol = 48 g) are required to oxidize 1 mol methanol (or 1 mol × 32 g/mol = 32 g). Taking the ratio of the oxygen requirement to methanol oxidized results in 48 g/32 g or 1.5 g of COD/g methanol oxidized.

It is noted methanol is a single carbon compound that is not taken up by phosphate-accumulating organisms (PAOs) as a substrate for biological phosphorus removal. Therefore, if a BNR system requires substrate for denitrification and volatile fatty acid (VFA) for biological phosphorus removal, methanol cannot be used to meet both objectives. Chemical addition for VFA supplementation is discussed in a later section.

Methanol Addition to Activated Sludge Biological Nutrient Removal Processes. If the presence of carbon substrate (or soluble BOD) in the influent wastewater is not sufficient for complete denitrification in the anoxic zones of the activated sludge BNR system, methanol can be added directly to the anoxic zones to enhance the process. If the nitrified mixed liquor recycle rate is high enough that there is excess nitrate in the first anoxic zone, it may be beneficial to add methanol to that location, as shown in Figure 8.1. The methanol can be fed directly to the anoxic zone, where it will be thoroughly mixed with the wastewater and mixed liquor solids by the anoxic zone mixer. Alternately, the methanol can be fed to the BNR process influent wastewater. This dosage location is much simpler from a chemical feed system equipment standpoint; however, depending on the individual plant layout, it may not be as efficient from a process standpoint (for example, if it is added to the plant influent and if significant turbulence occurs over primary effluent weirs or in flow splitter boxes upstream of the BNR system, the methanol will be stripped from the liquid). In addition, the presence of entrained air will result in some of the methanol being consumed for aerobic respiration.

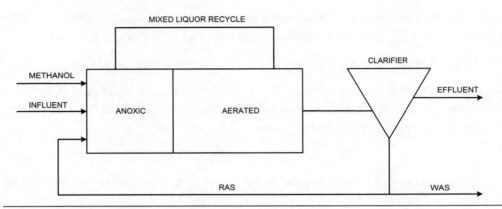

FIGURE 8.1 Methanol addition to first anoxic zone.

If the plant has a four- or five-stage activated sludge process, additional nitrate removal can be achieved in the post-anoxic zone (second anoxic zone) through endogenous denitrification. However, if the volume of the post-anoxic zone is small, it may be necessary to add a supplemental carbon source to increase denitrification rates above those of endogenous respiration. A schematic is shown in Figure 8.2. In this case, it is necessary to add the chemical directly to the post-anoxic zone; if it is added upstream from that location, the methanol will be oxidized in the aerated zone.

In most cases, a manual or flow-paced methanol feed control is adequate for activated sludge systems. Methanol control schemes are discussed in further detail later in this section and in Chapter 13.

If methanol is being added to the first anoxic zone of the activated sludge BNR process, the methanol supplementation requirement can be calculated as follows:
In metric units:

$$(Q_{in} + Q_{ras} + Q_{mlr}) \times (NO_3\text{-}N_{in} - NO_3\text{-}N_{out}) \times$$
$$(3 \text{ mg methanol/mg } NO_3\text{-N removed}) \times$$
$$(1000 \text{ L/m}^3) \times (1 \text{ kg}/1\,000\,000 \text{ mg}) = \text{methanol required, kg/d} \qquad (8.3)$$

In U.S. customary units:

$$(Q_{in} + Q_{ras} + Q_{mlr}) \times (NO_3\text{-}N_{in} - NO_3\text{-}N_{out}) \times$$
$$(3 \text{ mg methanol/mg } NO_3\text{-N removed}) \times 8.34 = \text{methanol required, lb/d} \quad (8.3)$$

Where

Q_{in} = BNR system influent flow (m³/d [mgd]),
Q_{ras} = return activated sludge flowrate (m³/d [mgd]),
Q_{mlr} = nitrified mixed liquor recycle flow rate to anoxic zone (m³/d [mgd]),
$NO_3\text{-}N_{in}$ = nitrate in anoxic zone (after denitrification using influent BOD as the only carbon source) (mg/L) and
$NO_3\text{-}N_{out}$ = target effluent nitrate from the anoxic zone (after additional denitrification) (mg/L).

It is important to note that operating history has shown that, depending on the WWTP and the BNR system design, in practice, the 3-to-1 ratio of methanol to nitrate may not be sufficient to achieve the desired additional denitrification in the first anoxic zone of an activated sludge system. Operators may find, through experience, that the methanol dosage rates calculated in the above example may not be adequate. In this case, the 3-to-1 dosage ratio should be gradually increased and a dosage ratio developed that reflects the actual performance at the treatment plant.

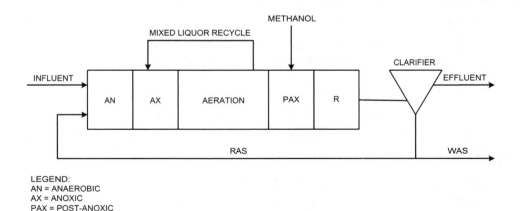

FIGURE 8.2 Methanol addition to post-anoxic zone.

The reason for the sometimes significant difference in the theoretical dosage compared to the actual required dosage remains a topic for additional research. There is evidence to suggest that bacteria that are able to degrade methanol efficiently under anoxic conditions (i.e., through denitrification) need a sufficient anoxic solids retention time (SRT) to dominate over those that degrade methanol more efficiently under aerated conditions. In high-rate activated sludge systems with relatively small anoxic volumes, excess methanol dosage may result in methanol breakthrough to the aerobic zone and selection of a separate population of methanol degrading bacteria that grow best under aerated conditions only (Purtshert and Gujer, 1999). It has been suggested that approximately a 2-days anoxic SRT is needed to avoid selecting for methanol degraders that cannot denitrify efficiently. This issue has been observed with methanol addition to the first anoxic zone, but has not generally been a problem with methanol addition to the post-anoxic zone of a four- or five-stage BNR process.

In later work, pilot testing in New York City showed that, when dosing acetate to the anoxic zones in a step-feed BNR system, the specific denitrification rate was up to 7 times higher than observed with methanol (Carrio et al., 2002). The observed methanol-to-nitrate ratios ranged from 8:1 to 15:1 (with methanol reported as COD). Under controlled conditions, a dosage ratio closer to 8:1 COD/NO_3-N was required; in converting from COD to methanol (1.5 g COD/g methanol), the corresponding ratio would be approximately 5.4 g methanol/g NO_3-N.

If methanol is added to the post-anoxic zone of the BNR process, the following can be used to estimate the required dosage rate:

In metric units:

$$(Q_{in} + Q_{ras}) \times (NO_3\text{-}N_{in} - NO_3\text{-}N_{out}) \times (3 \text{ mg methanol/mg } NO_3\text{-N removed}) \times (1000 \text{ L/m}^3) \times (1 \text{ kg}/1\,000\,000 \text{ mg}) = \text{methanol required, kg/d} \quad (8.4)$$

In U.S. customary units:

$$(Q_{in} + Q_{ras}) \times (NO_3\text{-}N_{in} - NO_3\text{-}N_{out}) \times (3 \text{ mg methanol/mg } NO_3\text{-N removed}) \times 8.34 = \text{methanol required, lb/d} \quad (8.4)$$

Where

Q_{in} = BNR system influent flow (m³/d [mgd]),
Q_{ras} = return activated sludge flowrate (m³/d [mgd]),
NO_3-N_{in} = nitrate entering post anoxic zone (aeration zone effluent nitrate) (mg/L), and
NO_3-N_{out} = target effluent nitrate from the post-anoxic zone (after denitrification) (mg/L).

Methanol Addition to Tertiary Denitrification Processes. For tertiary denitrification systems, such as denitrification filters, denitrifying BAFs, and MBBR systems with a post-anoxic fixed-film zone, the supplemental carbon source is vital for operation of the system. This is because denitrification is located downstream from the main aeration process, and essentially all soluble BOD in the influent wastewater has been removed. For denitrification filters and BAFs, the methanol is added to the nitrified secondary effluent wastewater at a location upstream from the denitrification influent, as shown in Figure 8.3. For an MBBR system, where the media are arranged in a four-stage BNR configuration including a post-anoxic zone, the methanol would be added directly to the post-anoxic zone, as shown in Figure 8.4.

Proper control over the methanol dosage is a very important component for tertiary denitrification systems. Overfeeding wastes chemical and could increase the BOD of the denitrification system effluent. The risk of increased BOD may not be a critical issue for WWTPs with moderate BOD limits, but for plants with BOD limits of approximately 5 mg/L or lower, this becomes an important consideration. Underfeeding the carbon source reduces the amount of nitrate removed, and the plant may not achieve the desired effluent nitrate or total nitrogen (TN) concentration.

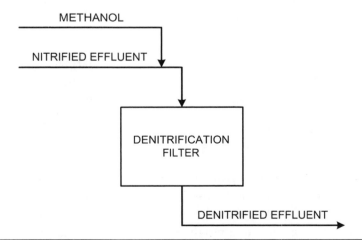

FIGURE 8.3 Methanol addition to denitrification filters.

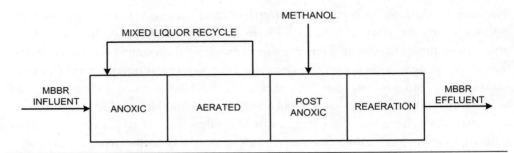

FIGURE 8.4 Methanol addition to MBBR.

Methanol feed requirements can be estimated using eq 8.1, as shown below, for a denitrification filter influent nitrate of 5 mg/L, target effluent nitrate of 1 mg/L, filter influent nitrite of 0.2 mg/L, and filter influent DO of 4 mg/L.

$$C_m = 2.47\, NO_3 + 1.53\, NO_2 + 0.87\, DO \quad \text{(from eq 8.1)}$$
$$C_{m'}\, mg/L = 2.47\,(5-1) + 1.53\,(0.2-0) + 0.87\,(4-0)$$
$$C_{m'}\, mg/L = 13.6\, mg/L$$

After determining the concentration of methanol required, the methanol feed rate must be calculated. For a WWTP flowrate of 22 710 m³/d (6.0 mgd [4166 gpm]) and a required methanol concentration of 13.6 mg/L, the methanol feed rate would be calculated as follows:

Example in metric units:

22 710 m³/d × 13.6 mg/L × 1000 L/m³ × (kg/1 000 000 mg) × (1 L methanol/0.7915 kg) = 390 L/d = 16 L/h methanol

Example in customary units:

(6 mgd × 13.6 mg/L × 8.34 × 1 gal methanol/6.6 lb = 103 gal/d methanol [4.3 gal/h])

As discussed earlier, operating history has shown that the ratio of 3 g methanol to 1 g of NO_3-N to be removed is a close approximation for tertiary denitrification systems. At times, increased methanol requirements may be observed; however, to avoid significant BOD breakthrough to the effluent, the methanol dosage should not exceed four times the amount of nitrate present in the filter influent.

Methanol Feed Control. There are a number of alternatives for control of methanol dosages to BNR processes. These include manual control, automatic flow-paced control, automatic feed-forward control using flow and influent nitrate concentration, and automatic feed-forward and feedback control using flow, influent nitrate concentration, and effluent nitrate concentration.

MANUAL CONTROL. For manual control of chemical dosing, all pumping rate adjustments and sampling are performed manually. Based on a certain level of nitrate removal, the operator manually calculates the methanol dose required and the corresponding pumping rate and manually sets the methanol feed pump. Based on periodic monitoring of actual methanol consumption and the effluent NO_3-N and NO_2-N concentrations, the chemical feed rate may be adjusted up or down. This does not provide a high level of control or the ability to tightly optimize chemical feed dosage, but is simple and, in many cases, adequate, especially if the effluent TN (and allowable NO_X-N) concentrations are not stringent. In the case of methanol addition to the preanoxic (first anoxic) zone of an activated sludge BNR system, the process is relatively forgiving, because a minor overdose of methanol will simply result in bleedthrough to the next aerated zone, where it will be oxidized (rather than passing through to the effluent and causing permit violation on BOD). However, for tertiary denitrification processes, there is risk of increase in the effluent BOD if methanol is overdosed and if nitrate and nitrite concentrations are low.

FLOW-PACED CONTROL. Flow-paced control represents the simplest level of automatic control. Based on the average anoxic zone (or tertiary denitrification process) influent nitrate concentration and the required level of nitrate removal, the operator manually calculates the average methanol dose required and the corresponding average methanol pumping rate. The control system is then set to modulate the pumping rate up and down with fluctuations in wastewater flow. Generally this should only apply to dry weather operation, and methanol dosages should not be increased above that required to meet the daily maximum hour of the dry weather diurnal curve. This level of automatic control is still relatively simple, but provides an increased ability to optimize the chemical dosages. It should be adequate, in many cases, for methanol addition to an activated sludge BNR system, but may not be a high enough level of control to consistently prevent overdose and increased effluent BOD in tertiary denitrification processes.

FEED-FORWARD CONTROL. A feed-forward control scheme, where denitrification influent nitrate concentrations are measured and used in combination with flow to continuously vary the methanol feed rate, offers the next level of automatic control. Because the methanol dose is based on both wastewater flow and concentration, it is feasible to operate in this mode during wet weather events and dry weather. This means of control was discussed in the *Manual for Nitrogen Control* (U.S. EPA, 1993), with the caveat that reliance on online monitoring systems for measurement of wastewater nitrate concentrations was a weak link. While still an area of development, advancements in online instrumentation for monitoring of nutrient concentrations have made this a feasible and reliable means of control with the selection of the right instruments.

FEED-FORWARD AND FEEDBACK WITH EFFLUENT CONCENTRATION CONTROL. This represents the most complex level of chemical feed control. It is currently offered as a patented package known as TetraPace from Tetra Process Technologies (Tampa, Florida) as a control enhancement to their denitrification filter process. The system is designed to operate continuously in the flow-paced or feed-forward (flow and influent concentration) modes, with a correction at discrete intervals to achieve an effluent concentration setpoint. This system was developed to offer tight control of methanol dosage to allow WWTPs to meet low nitrate levels without increasing the effluent BOD or total organic carbon. As with the feed-forward control, reliability of online nitrate monitoring was initially the weak link; however, since that time, this mode of control has been successfully applied at a number of denitrification filter installations.

ALTERNATE CARBON SOURCES FOR DENITRIFICATION. Although methanol is the most common, there are a number of alternate chemicals that can be used to provide supplemental carbon for denitrification, including ethanol, denatured alcohol, acetic acid, and sodium acetate (Elmendorf, 2002). Ethanol is readily available, but is intoxicating when consumed and is heavily taxed and regulated, making it a less favorable carbon source than some of the other options. Denatured alcohol is ethanol with added substances to make it unfit for human consumption. Acetic acid also is a good carbon source and is used as substrate for denitrification at several WWTPs in the United States. However, the cost of acetic acid has historically been significantly higher than that of methanol. Sodium acetate also can be a good option, but it comes in a powdered form, requiring handling facilities for dry chemical and mixing into solution.

In addition to the purchase cost considerations, methanol may not be as efficient for denitrification as other substrates, making other substrates possibly more attractive if anoxic zones are limited in volume. For example, McCarty et al. (1969) investigated the effect of a number of carbon substrate sources on the denitrification rate. Among those tested, acetic acid resulted in higher denitrification rates than methanol. Others have observed similar results, including Carrio et al. (2002), as discussed earlier.

There is one potential disadvantage associated with the use of acetic acid to enhance denitrification. Addition of acetic acid to the post-anoxic zone of a five-stage BNR system can result in release of stored phosphorus from the mixed liquor suspended solids (MLSS). In general, this should not be too problematic if the post-anoxic zone detention time is not too long, if some residual NO_3-N remains, and if sufficient aeration volume is available for uptake of the released phosphorus in the reaeration zone. The addition of acetic acid to denitrification filters downstream of a biological phosphorus removal process also could result in some release of phosphorus from the remaining biological solids in the clarifier effluent feed to the filters. In most cases, acetic acid should still achieve the desired result if dosed at these locations, but could be an issue if both the nitrogen and phosphorus limits are very low.

Because of the expense of adding pure chemicals, a number of WWTPs also have looked to industrial waste product sources of supplemental carbon. There are a number of possibilities, including sugar wastes, molasses, and waste acetic acid solution from pharmaceutical manufacturers. There are several important considerations, however. First, the industrial source must be a "clean" source, relatively free of nutrients, metals, and other contaminants (both dissolved and debris). The consistency of the carbon concentration is also important. If the dosage location is the anaerobic or anoxic zone of the activated sludge BNR system, some variation in the readily biodegradable carbon content of the industrial source can be tolerated in the process without adverse effects. However, because of the location at the end of the treatment process, only a very clean and consistent source can be used for tertiary denitrification filters. Another consideration is availability. The industry's production schedule may have seasonal or other variation, depending on demand for the primary product, and the waste product may not be available in the required quantity on a consistent schedule. In this case, there may still be considerable savings achieved by using the industrial waste source, but the plant should have the capability to dose another chemical also.

CASE STUDIES. *Havelock Wastewater Treatment Plant, Havelock, North Carolina.* The Havelock WWTP in eastern North Carolina is rated for 7200 m³/d (1.9 mgd) monthly average flow. The plant facilities include influent pumping; screening; grit removal; an activated sludge system, consisting of two sets of aeration basins operating in series; final clarifiers; deep-bed denitrification filters; UV disinfection; and mechanical postaeration. Solids handling facilities include aerobic digestion, gravity belt thickening, and a sludge storage basin (Figure 8.5). Polyaluminum chloride is added to the clarifier influent for phosphorus precipitation, and methanol is added to the filter influent to provide a carbon source for denitrification.

Havelock is required to meet an effluent TN limit of 9700 kg/a (21 400 lb/yr), or 3.7 mg/L TN at the design flow. The activated sludge system is operated with an MLSS setpoint of 5000 to 6000 mg/L (corresponding to an SRT of 25 to 30 days). The plant has operated very well in this mode, with effluent ammonia concentrations consistently lower than 0.2 mg/L. The influent nitrate concentration to the filters averages approximately 12 mg/L. Since denitrification in the filters began in 1998, Havelock has consistently met its TN limit.

The most significant challenge with respect to meeting the nitrogen limit was initially related to methanol feed control. In 1998, the National Pollutant Discharge

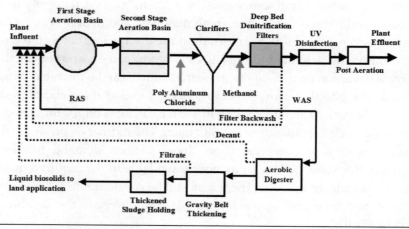

FIGURE 8.5 Havelock WWTP schematic.

Elimination System (NPDES) permit had specified a carbonaceous BOD (CBOD) limit of 3 mg/L monthly average. This CBOD limit was the most stringent in the state and would have been difficult for any plant to meet. The addition of methanol to the filters for denitrification made it even more challenging.

At first, the methanol feed was controlled through flow pacing. While the system performed well, it was sometimes difficult to balance the requirements for very low CBOD with denitrification. In particular, it was difficult to closely match the carbon requirement under varying nitrate concentrations.

As a result, an online nitrate analyzer was installed. The control algorithm was first modified to include feedback based on the effluent nitrate concentration and was subsequently modified to incorporate feed-forward and feedback controls based on flow and influent and effluent nitrate (patented TetraPace system). This has enhanced the operation and reliability of the denitrification process without risk of methanol overdose by more closely matching the methanol feed to the demand. It has also reduced variations in effluent nitrate. The plant staff estimates that methanol consumption has been reduced by approximately 30%.

Havelock now has over five years of operation with the denitrification filters, with excellent results. The effluent TN averaged 3 mg/L in 2001 and 2002. Approximately 2 mg/L of the effluent TN is in the form of NO_x-N. Currently, the main operations challenge is related to high flows during storm events. During peak flow events, the filters are typically taken out of denitrification mode and are operated for filtration only.

Long Creek Wastewater Treatment Plant, Gastonia, North Carolina. The City of Gastonia, North Carolina, owns and operates two BNR plants: the Long Creek WWTP and Crowders Creek WWTP. Both plants have monthly average effluent TN limits of 6 mg/L during April through October, and both facilities add a 20% waste acetic acid solution from a pharmaceutical manufacturer to enhance denitrification. Gastonia WWTPs historically have received significant industrial discharges, and these currently average approximately one-third of the total wastewater flow. This case study will focus on the Long Creek WWTP.

Long Creek WWTP is rated for 60 600 m^3/d (16 mgd) maximum monthly flow (approximately 49 000 m^3/d [13 mgd] average flow) and currently averages approximately 26 500 m^3/d (7 mgd). The plant has preliminary treatment facilities, influent pumping, primary clarifiers, four first-stage activated sludge BNR trains operating in the anaerobic/anoxic/oxic (A^2/O) mode, a common post-anoxic and reaeration basin, secondary clarifiers, filters, chlorine disinfection, and post aeration. Waste activated sludge (WAS) is thickened in a dissolved air flotation thickener, and the thickened WAS and primary sludge are directed to anaerobic digesters. Supernatant is drained from the digesters to thicken the solids concentration, and the liquid residuals are land-applied.

To remove sufficient nitrate to meet the effluent TN limit of 6 mg/L, acetic acid is dosed to the junction box where the first-stage BNR effluent comes together from the four A^2/O trains and enters the post-anoxic zone (Figure 8.6). To ensure consistent compliance with the limit, Gastonia targets an effluent TN of 4 to 5 mg/L. In past years, this has required nitrate removal to as low as 1 mg/L, as the plant had a residual soluble organic total Kjeldahl nitrogen (TKN) concentration of approximately 4 mg/L (even though completely nitrified to very low ammonia concentrations). In 2002 and 2003, discontinuation of some of the industrial discharge resulted in the reduction of the residual soluble TKN to more typical levels of approximately 1.5 mg/L, easing up the requirement to denitrify completely.

During 2003, the seasonal average NO_x-N concentration entering the post-anoxic zones was approximately 5.6 mg/L. Grab samples taken in the fourth post-anoxic zone averaged 2.3 mg/L NO_x-N, showing a total removal of 3.3 mg/L NO_x-N.

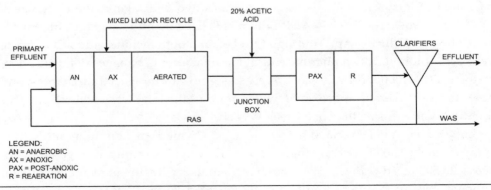

FIGURE 8.6 Long Creek WWTP schematic.

During this time, the entire post-anoxic zone volume (4010 m^3 [1.06 mil. gal]) was in service with a nominal detention time of approximately 3.4 hours. The acetic acid dosage was varied as needed based on sampling data, and also was increased at times when digester supernatant flows (and the corresponding return of ammonia to the head of the plant) were high. Although it has not resulted in any operating difficulties, there appears to be some phosphorus release taking place in the post-anoxic zones, sometimes on the order of several milligrams per liter of orthophosphorus (OP). Profile data suggest that much of this is taken back up in the reaeration zones and in the clarifiers. The plant has the capability to add alum to the clarifier influent screw pump wetwell; however, recently, no alum addition has been needed to meet the monthly average effluent TP limit of 1 mg/L.

Control of the acetic acid system is manual. The plant staff regularly takes grab samples through the different zones of the BNR basins to monitor process performance. These data, in combination with composite plant effluent samples, are examined, and periodic adjustments made as needed. The manual operation has worked well. However, in an effort to optimize the system operation and reduce chemical costs, the city has been considering flow pacing of the chemical pumps (during dry weather operation) and installation of an online nitrate monitor. Online monitoring would provide the operators with a more continuous picture of what is happening in the basin throughout the day as flows and loads fluctuate, and would also enable the number of grab samples to be reduced. This optimization effort is ongoing at both Gastonia plants.

VOLATILE FATTY ACID SUPPLEMENTATION FOR BIOLOGICAL PHOSPHORUS REMOVAL

In biological phosphorus removal, VFAs are taken up in the anaerobic zone PAOs, and stored phosphorus is released, while excess phosphorus is taken up under aerobic conditions (accumulation of poly-β-hydroxybuterate) (Fuhs and Chen, 1975). For proliferation of PAOs and efficient biological phosphorus removal performance, a good source of VFA is required. The VFAs may be present in the influent wastewater. Naturally occurring VFA sources include (1) raw wastewater, where collection systems have long detention times and/or multiple pump stations; and (2) breakdown of more complex organic compounds in the anaerobic zones of the BNR process. However, if the naturally occurring VFA content is insufficient, a supplemental source is needed. This VFA source can be added to the anaerobic zones, as shown in Figure 8.7.

For biological phosphorus removal, VFAs are required, and research suggests that a mix of acetic and propionic acids are optimal. Plants located in warm climates and with long collection systems may have sufficient VFA in the influent wastewater. If VFA supplementation is needed (i.e., added to the existing influent soluble BOD, some of which will ferment to usable VFA products within the anaerobic zones), the full-scale observed supplemental chemical dosage requirements are typically on the order of 5 to 10 mg/L VFA per mg/L phosphorus removed.

The effect of a number of organic substrates on biological phosphorus removal was researched by Abu-garrah and Randall (1991). Ratios of phosphorus uptake per COD used and COD used per milligram per liter of phosphorus removed are summarized in Table 8.2. This work suggests that acetic acid is the most effective chemical substrate for biological phosphorus removal enhancement.

TABLE 8.2 Effect of organic substrate on enhanced biological phosphorus removal.[a,b]

Substrate	mg/L phosphorus uptake[c] / mg/L COD used	mg COD used[d] / mg P removed
Formic acid	0	Infinity
Acetic acid	0.37	16.8
Propionic acid	0.10	24.4
Butyric acid	0.12	27.5
Isobutyric acid	0.14	29.1
Valeric acid	0.15	66.1
Isovaleric acid	0.24	18.8
Municipal wastewater	0.05	102[e]

[a] An SRT of 13 days was used for all experiments.
[b] Source: Abu-garrah and Randall (1991).
[c] Total phosphorus uptake in aerobic zone.
[d] COD used and phosphorus removed in total system.
[e] Value obtained with highly aerobic wastewater.

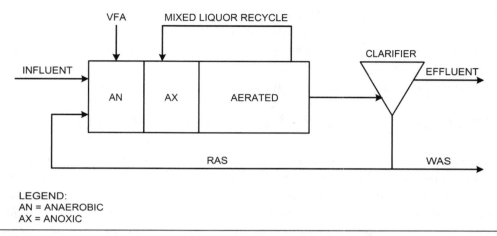

FIGURE 8.7   VFA addition for biological phosphorus removal.

ACETIC ACID. Acetic acid (CH_3COOH) is the most common VFA dosed for enhanced biological phosphorus removal. Acetic acid is commonly delivered as glacial (approximately 100% solution), 84 and 56% solution. Although not as volatile as methanol, glacial acetic acid has a relatively low flash point (40°C [104°F]) and has the added complication of a 17°C (62°F) freezing point. Similar to methanol, meeting code requirements for a flammable liquid is an important component of design (unless very dilute solutions are used, nearing the properties of water), and measures must be taken to avoid freezing. Because of the flammability concerns, systems designed for use of high concentrations of acetic acid generally have similar requirements to those of methanol systems. Storage tanks, piping, and appurtenances should be constructed of metal materials. Because of the corrosivity of acetic acid, type 316 stainless steel is typically used. If glacial acetic acid is used in warm climates, it may be necessary to consider an inert gas blanket or floating cover because of the relatively low flash point. Similar to methanol, all applicable NFPA, U.S. Department of Transportation (Washington, D.C.), Occupational Safety and Health Administration (Washington, D.C.), and local code requirements must be confirmed.

As mentioned above, one additional handling problem associated with acetic acid solutions greater than approximately 85% is that the freezing point is well above that of water. Glacial acetic acid has a freezing point close to 17°C (62°F), requiring continuous heating under most climate conditions. However, as acetic acid is diluted with water, the freezing point decreases, and all dilutions below 85% have a freezing point at or below that of water.

A summary of acetic acid chemical properties is given in Table 8.3.

ALTERNATE CHEMICAL VOLATILE FATTY ACID SOURCES. In addition to acetic acid, there are other options for VFA supplementation. Because of the expense of adding pure chemicals, some WWTPs also have looked to industrial waste product sources of supplemental carbon. There are a number of possibilities, including sugar wastes, molasses, and waste acetic acid solution from a pharmaceutical manufacturer. It is important to ensure that these sources are free of contaminants and debris. Other considerations for the use of alternate chemical VFA sources are the same as discussed in the Alternate Carbon Sources for Denitrification section.

CASE STUDY: MCDOWELL CREEK WASTEWATER TREATMENT PLANT, CHARLOTTE, NORTH CAROLINA. The McDowell Creek WWTP is owned and operated by Charlotte-Mecklenburg Utilities. In 1999, the plant was

TABLE 8.3 Properties of acetic acid (Celanese Chemicals, 2000; Hoechst Celanese Chemical Group, 1988).

Chemical formula	CH_3COOH		
	COD equivalent, mg/L = 1.07 × acetic acid, mg/L		
Molecular weight	60.05 g/mol		
Description	Colorless liquid, strong vinegar odor		
Solution strength	Glacial (100%)	56%	20%
Specific gravity	1.051	1.0615	1.026
Density, kg/L (lb/gal)	1.05 (8.76)	1.06 (8.85)	1.03 (8.56)
Flash point, °C (°F)	42.8 (109)	63.3 (146)	80.6 (177)
Freezing point, °C (°F)	16.6 (61.9)	-23 (-9.4)	-6.5 (20.3)

upgraded to BNR to reduce nutrient loads to Mountain Island Lake, one of the city of Charlotte's drinking water supplies. McDowell can operate in one of several three-stage BNR processes for total nitrogen, and normally operates in the University of Cape Town (UCT) mode. Other treatment units include preliminary treatment facilities, primary clarifiers, deep-bed denitrification filters, UV disinfection, and cascade aeration. Before August 2002, primary sludge and WAS were cosettled before anaerobic digestion. The digested sludge is dewatered with belt filter presses.

Since BNR operation began in 1999, the plant staff has experimented with the different BNR modes. The TN limit of 10 mg/L was easily met, but achieving reliable biological phosphorus removal was initially more difficult. McDowell has several operating challenges related to phosphorus removal process performance, including a relatively "fresh" wastewater (no VFA) and high phosphorus levels in the dewatering filtrate stream. However, using an optimized combination of chemical addition, dewatering filtrate equalization, and a unique supplemental carbon source (as discussed later), this plant has maintained an average of less than 0.2 mg/L effluent phosphorus for the past three years.

Effect of Dewatering Filtrate. As discussed above, McDowell operates anaerobic digesters for biosolids stabilization and solids reduction. The filtrate produced during dewatering of the digested sludge is high in both ammonia and orthophosphorus. Because dewatering is carried out on a weekday shift basis, this resulted in very high phosphorus loads to the BNR process during the day. To equalize the phosphorus load, the filtrate is directed to an equalization basin and pumped back to the plant over a 24-hour period. Phosphorus concentrations in the dewatering filtrate average 75 to 150 mg/L, depending on the operation, resulting in an additional 14 to 41 kg/d (30 to 90 lb/d) in the plant influent (1 to 2 mg/L OP).

Optimization of Chemical Dosages. McDowell has very little influent VFA, and it is necessary to supplement VFA to the anaerobic zones. However, even with significant acetic acid addition, the BNR process effluent TP could creep above 1 mg/L at times. The plant experimented with chemical dosing locations and found, on the basis of operations reliability and cost, that addition of alum to the primary clarifiers to reduce the phosphorus load from the filtrate and cosettled WAS is the most economical solution.

To further reduce chemical costs, Charlotte-Mecklenburg Utilities decided to test a soft drink bottling waste sugar water as an alternate VFA source. From November 2000 to February 2001, sugar water addition was piloted in one full-scale treatment train, while acetic acid addition was continued in the second treatment train. Nearly equal operating results were observed for both chemicals. McDowell has continued to use the waste sugar water as a supplemental carbon source, with excellent results. Currently, the available sugar water supply does not quite meet the total VFA requirement at the plant, so the balance is made up with acetic acid as needed. The net savings is approximately $150,000 per year (Goins, 2002).

In summary, the McDowell Creek WWTP overcame its phosphorus removal challenges and consistently produces an effluent of excellent quality. Although the plant operates filters, it is noted that there is no chemical addition to the filter influent, and the secondary effluent and final effluent quality are similar for all parameters, including phosphorus. The McDowell effluent total phosphorus concentration has averaged less than 0.2 mg/L for several years (Figure 8.8). It is noted that there are many values less than or equal to the laboratory method detection limit of 0.1 mg/L; consequently, the plant effluent phosphorus is likely frequently below this value.

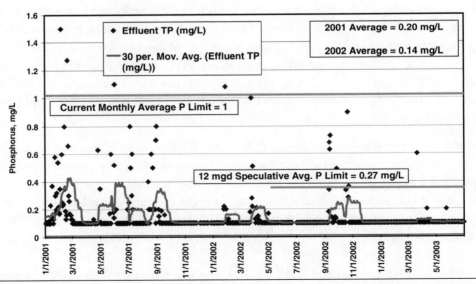

FIGURE 8.8 McDowell Creek WWTP effluent phosphorus (2001 through 2003).

The McDowell plant is currently being expanded to a capacity of 45 000 m^3/d (12 mgd). Speculative permit limits issued in early 2003 reflect a dramatic decrease in the effluent phosphorus limit, reducing it from 1 mg/L monthly average to 0.27 mg/L monthly average during summer months and 0.32 mg/L monthly average during the winter season. Because of the plant's excellent performance with respect to phosphorus removal, no additional phosphorus removal processes are being added.

ALKALINITY SUPPLEMENTATION

Alkalinity is a measure of a wastewater's acid neutralizing capacity. Alkalinity and pH are closely related and are of importance in BNR facilities. The nitrification process consumes alkalinity, which, in turn, causes the wastewater pH to drop. As the pH drops, the rate of nitrification can decrease and stop at a pH of approximately 6. Alkalinity supplementation may be needed to support nitrification in some wastewater treatment facilities or to support nitrification at certain times of the year.

The second concern for alkalinity is its role in biological phosphorus removal processes. Sometimes the influent alkalinity is measured to assess if enough acetic and propionic acids are present to support biological phosphorus removal by using the method to determine the volatile acid concentration anaerobic digesters.

In some nutrient removal facilities, iron or aluminum salts are used to chemically precipitate phosphorus. The chemical precipitation process consumes alkalinity as a result of the acidic nature of the chemicals used. If the nutrient removal facilities include nitrification, alkalinity consumption for both nitrification and phosphorus precipitation must be accounted for to determine if alkalinity supplementation is necessary.

ALKALINITY. Alkalinity is a general parameter because many different ions contribute to alkalinity. In wastewater, alkalinity is generally associated with bicarbonate, although other ions are also present in significant concentrations. The alkalinity test, *Standard Methods* (APHA et al., 1998) Method 2320, is an acid titration. Sulfuric acid is added until the pH is reduced to 4.5. The amount of acid added is reported in terms of calcium carbonate and is defined as alkalinity.

Alkalinity has also been defined as the buffering capacity of a wastewater. Buffering generally applies to the ability of water to resist a change in pH from acid or base addition.

The most common buffer encountered in wastewater treatment is the carbonate family. The carbonate family includes carbonic acid (H_2CO_3), bicarbonate (HCO_3^-), and carbonate (CO_3^{-2}). Carbonic acid is also measured as dissolved carbon dioxide. When alkalinity is measured, it does not account for all three of the carbonate family of compounds. As sulfuric acid is added in the alkalinity titration, the hydrogen ion added converts carbonate into bicarbonate and bicarbonate into carbonic acid. Nothing happens to any carbonic acid in solution. Carbonic acid is formed, but it is not measured in the alkalinity test.

There are several key issues regarding alkalinity that must be understood before implementing an alkalinity supplementation system. These include the following:

(1) If a wastewater has a high pH, it does not necessarily have a high alkalinity. The alkalinity could be 50 mg/L as calcium carbonate ($CaCO_3$), but the pH could range anywhere from 5 to 10. The reverse is also true. The pH could equal 7, but the alkalinity could range anywhere from 5 to over 500 mg/L as $CaCO_3$. Alkalinity is the buffer against changing pH and, therefore, does not set the initial value.

(2) As nitrification proceeds and alkalinity is destroyed, bicarbonate is converted to carbonic acid. As far as the convention for measuring alkalinity is concerned, and if we assume that all of the alkalinity is bicarbonate, alkalinity has been destroyed. However, in buffer chemistry, bicarbonate has been converted to carbonic acid. Nothing has been destroyed, as the chemicals are still present in the wastewater; bicarbonate has merely been converted to another chemical form.

When sodium hydroxide (NaOH) or lime is added into an aeration basin, the pH does not necessarily increase because of a very important chemical relationship. As nitrification occurs, an acid is formed, and bicarbonate is converted in carbonic acid. When sodium hydroxide is added, the hydroxide ion reacts with carbonic acid to convert it back to bicarbonate, and this reaction prevents the pH from rising too high because of the addition of a strong base.

(3) When nitrification occurs, the references always say that alkalinity is destroyed. Based on the alkalinity definition of the acid titration to a pH of 4.5, then alkalinity is destroyed. However, in reality, no mass is destroyed; carbonate and bicarbonate are merely converted to a new form. As the pH in the titration is lowered or approaches 4.5, all carbonate species are con-

verted to carbonic acid, yielding the apparent destruction of alkalinity. In fact, alkalinity has been converted to acidity. Acidity is defined as the amount of caustic expressed as calcium carbonate required to raise the pH to 8.3. Therefore, between pH 8.3 and 4.5, both alkalinity and acidity coexist. When alkalinity is destroyed, acidity is created. A simple analogy is the image of the scales of justice with alkalinity on one pan and acidity on the other. As the pH is lowered, alkalinity is removed from the alkalinity pan and moved to the acidity pan. Conversely, when acidity is destroyed, alkalinity is formed.

ALKALINITY SUPPLEMENTATION. When alkalinity supplementation is required, there are several chemicals that can be used. The preferred chemical is influenced by local conditions, local chemical prices, and operator preferences. Chemicals that can be used for alkalinity supplementation include the following:

- Sodium hydroxide (caustic soda) (NaOH);
- Calcium hydroxide (lime) [$Ca(OH)_2$];
- Calcium oxide (quick lime) (CaO);
- Magnesium hydroxide [$Mg(OH)_2$];
- Sodium carbonate (soda ash) (Na_2CO_3); and
- Sodium bicarbonate ($NaHCO_3$).

The data in Table 8.4 show a comparison of mass use between the various alkalinity supplements. Table 8.4 only presents a comparison of mass use; an economic comparison is very site-specific and dependent on market conditions and shipping distance. Each chemical is discussed separately in the following paragraphs.

Sodium Hydroxide. Sodium hydroxide (commonly called caustic or caustic soda) is commonly used for alkalinity supplementation because of its ease of handling. Sodium hydroxide (caustic) is not the lowest cost chemical; however, compared to calcium hydroxide (lime), it is much easier to use, and the annual maintenance costs for the caustic storage and feed system is much lower. Many utilities believe that the ease of handling for sodium hydroxide (caustic) far outweighs the higher cost of the chemical.

Sodium hydroxide can be purchased at a 50% by weight solution strength or, in some locations, it is available at 20 or 25% by weight solution strengths. It is classi-

TABLE 8.4 Comparison of mass use per calcium carbonate equivalent.

Chemical	mg of chemical/ mg of $CaCO_3$
Sodium hydroxide (NaOH)	0.8
Calcium hydroxide [$Ca(OH)_2$]	0.74
Quick lime (CaO)	0.56
Magnesium hydroxide [$Mg(OH)_2$]	0.58
Sodium carbonate (Na_2CO_3)	1.06
Sodium bicarbonate ($NaHCO_3$)	0.84

fied as a strong base and, if overdosed, can raise the pH much higher than expected. Dilute sodium hydroxide solutions must be freeze-protected to less than 0°C. A 50% by weight sodium hydroxide solution freezes at approximately 12.8°C, so the bulk storage tank and piping must be heated and insulated. Once the liquid temperature drops below 12.8°C, sodium hydroxide will begin to crystallize out of solution. It is very difficult to redissolve crystallized sodium hydroxide. Local chemical suppliers should be contacted to obtain freeze protection curves and other pertinent chemical data for sodium hydroxide.

If sodium hydroxide is delivered to the site at 50% strength and then diluted on-site with plant or potable water, scaling will occur at the point of mixing. The local pH will rise well above pH 10, and calcium carbonate will form immediately; consequently, the dilution system should be designed so that the section of piping where the mixing occurs can be easily cleaned. Dual mixing sections may ensure that one chemical delivery line is always operable.

The point of caustic addition to the WWTP is also vulnerable to scaling. If caustic is added in a pipe with an in-line diffuser, the diffuser will scale, and the pipe will also plug with scale over time. If the caustic is added in a return activated sludge (RAS) line, the high flowrate in the RAS line may protect the line from scaling; however, the diffuser, if one is used, will still scale.

Calcium Hydroxide. Calcium hydroxide or hydrated lime is also commonly used for alkalinity supplementation. Hydrated lime has been slaked by the manufacturer and is sold as a dry material. Hydrated lime must be slurried before use, but does not have to be reslaked. The slurry mixing tank can be operated at room temperature. The slurry tank will be susceptible to scaling if plant water or potable water is used, as the hardness in these waters will precipitate.

Calcium hydroxide is typically prepared in a 3 to 5% by weight slurry. A small fraction of the calcium hydroxide will dissolve and raise the pH to approximately 12. Calcium hydroxide precipitates above a pH of 12. The pH is high enough to be dangerous to the operators. Calcium hydroxide slurry solids settle easily, so the slurry is often conveyed to the point of application in a trough. Pinch valves are sometimes used in a pumped slurry recycle loop to control that rate of application.

Calcium hydroxide is often preferred over quicklime because slaking is not needed. Scaling is a major concern and causes many hours of hard labor to remove scale from feed lines and pumps. The slurry is also very abrasive, and this abrasion also creates more maintenance problems by eroding piping and pump impellers.

Calcium hydroxide should be added at a point of turbulence. The slurry solids may settle and accumulate in the process. Settled solids dissolved very slowly, if at all. Depending on the calcium hydroxide dose, the local pH at the point of application may be raised high enough for scale to form. The scaling can cause problems in the pipelines between processes because of a loss of capacity.

Quicklime. Quicklime is calcium oxide. Slaking is a process where the calcium oxide is mixed with water and allowed to react with wate,r according to the following reaction:

$$CaO + H_2O = Ca(OH)_2 \tag{8.5}$$

The slaking reaction is exothermic-it releases heat. Slaker operating temperatures are typically between 120 and 180°C. The elevated temperature that occurs during slaking increases the amount of scale formed. Calcium solubility decreases as temperature rises.

Slaking is a messy, labor-intensive process. Because it is so labor-intensive to keep the equipment serviceable, it is logical to assume that most operators do not like the process. So, why would a utility consider using quicklime for alkalinity supplementation? Hydrated lime is typically approximately $11 to 16.5/metric ton ($10 to 15/ton) more expensive compared to quicklime.

Comparing hydrated lime and quicklime on a cost-per-ton basis is very deceiving. On a pound-per-pound basis, hydrated lime and quicklime are not equal. According to eq 8.5, 1 mol hydrated lime equals 1 mol quicklime. On a weight basis, it takes 0.599 kg (1.32 lb) of hydrated lime to equal 0.5 kg (1 lb) of quicklime. Even before considering the cost-per-ton price, it takes 32% more hydrated lime compared to quicklime. Because hydrated lime typically costs more than quicklime, a life-cycle cost comparison may show that quicklime is more economical to use, even when the operating and maintenance costs associated with slaking are taken into account.

Magnesium Hydroxide. Magnesium hydroxide is another base that is gaining acceptance for use as an alkalinity supplement. The costs are regionally driven; therefore before selecting magnesium hydroxide for long-term use, delivery costs should be investigated. Magnesium hydroxide is generally sold as a slurry, but a dry product can also be purchased. One benefit of magnesium hydroxide is that it will not raise the pH above approximately 10.5, which correlates to its precipitation point range of pH 10.2 and 10.5. While this chemical limitation does prevent producing a very high pH at the point of application, it does not completely eliminate potential scaling. Scaling will begin above a pH of 8.

For plants that have anaerobic digestion, magnesium hydroxide use may incur hidden costs. Struvite is a magnesium ammonium phosphate precipitate that forms in anaerobic digesters. An increased magnesium concentration may cause more struvite to form. Before recommending magnesium hydroxide for alkalinity supplementation, digester operations should be reviewed to determine if an elevated magnesium concentration will complicate struvite control.

Sodium Carbonate. Sodium carbonate or soda ash can be used for alkalinity supplementation. Its use is not widespread because other chemicals are easier to handle and cost less. Soda ash is generally available as a dry product, so it must be dissolved on-site.

Sodium Bicarbonate. Sodium bicarbonate can be used for alkalinity supplementation. Its use is not widespread because other chemicals are easier to handle and cost less. Sodium bicarbonate is generally available as a dry product, so it must be dissolved on-site.

ALKALINITY CONSIDERATIONS. *Usable Alkalinity.* When alkalinity is measured, the pH is dropped to 4.5. However, at a pH of 4.5, some VFA alkalinity is

One solution to this dilemma is to vent the final cell in the activated sludge process. If the final cell of the HPOAS reactor is vented with outside air, excess accumulation of carbon dioxide is vented to the atmosphere, and the partial pressure of carbon dioxide is lowered. With this modification, the carbon dioxide actually vents out of solution, thereby raising the pH. The treated effluent pH rises and recovers alkalinity. Recovered alkalinity is also returned to the first reaction cell by the RAS recycle.

Phosphorus Precipitation. When iron salts or aluminum compounds are used to precipitate phosphorus, alkalinity is consumed. Both iron and aluminum react with phosphorus to precipitate and form iron phosphate ($FePO_4$) or aluminum phosphate ($AlPO_4$). If excess iron or aluminum is added, it forms its respective hydroxide precipitate [$Fe(OH)_3$ and $Al(OH)_3$].

At a neutral pH, approximately one-half the phosphate is present as $H_2PO_4^-$ and one-half as HPO_4^{-2}. When precipitated, phosphate must give up an average of 1.5 mol hydrogen/mol phosphorus precipitated. This is equivalent to 2.46 mg $CaCO_3$ removed/mg phosphorus removed.

When excess alum or iron is added to the aeration basin, additional alkalinity consumption must be accounted for in the formation of hydroxide precipitates. For aluminum, 5.56 mg $CaCO_3$ is consumed per milligram of aluminum precipitated as aluminum hydroxide. For iron, 2.69 mg $CaCO_3$ is consumed per milligram of iron precipitated as ferric hydroxide.

PRACTICAL EXAMPLES. The following are a series of practical examples using different chemicals under differing conditions.

Example 1. A 38 000-m^3/d (10-mgd) WWTP has to nitrify to meet strict NPDES permit limits. The effluent alkalinity has been averaging 47 mg/L as $CaCO_3$, and it is desired to keep the effluent alkalinity at 100 mg/L as $CaCO_3$. How much additional hydrated lime [$Ca(OH)_2$] must be added? Alkalinity supplementation will be made using hydrated lime. Lime purity is 98%.

STEP 1. Calculate alkalinity needs. Alkalinity to be added equals 100 mg/L desired $CaCO_3$ − 47 mg/L actual $CaCO_3$ in the effluent = 53 mg/L alkalinity as $CaCO_3$.

STEP 2. Calculate hydrated lime feed rate. The problem states that the plant flow is 38 000 m^3/d (10 mgd). The mass of alkalinity to be added in kilograms per day (pounds per day) equals the flow in cubic meters per day times the alkalinity needed in milligrams per liter divided by 1000 (flow in million gallons per day times 8.34 times the alkalinity needed in milligrams per liter). Convert the mass of alkalinity into a hydrated lime feed rate as kilograms per hour (pounds per hour). From the data in Table 8.4, use 0.74 as the conversion factor from $CaCO_3$ to $Ca(OH)_2$.

38 000 m^3/d × 53 mg/L alkalinity as $CaCO_3$ × (1000 L/m^3) × (kg/1 000 000 mg) = 2014 kg/d $CaCO_3$(10 mgd × 8.34 × 53 mg/L alkalinity as $CaCO_3$ = 4420 lb/d $CaCO_3$)

Hydrated lime = 2014 kg/d $CaCO_3$ × 0.74 = 1490 kg/d hydrated lime (4420 lb/d $CaCO_3$ × 0.74 = 3271 lb/d)

Purchased product = 1490 kg/d hydrated lime × 1/0.98 product purity = 1520 kg/d purchased hydrated lime (3271 lb/d hydrated lime × 1/0.98 product purity = 3338 lb/d)

Convert to hourly rate = 1520 kg/d purchased material / (24 h/d) = 63.3 kg/h (3338 lb/d purchased material / 24 h/d = 139 lb/h).

Example 2. A 38 000-m^3/d (10-mgd) WWTP has to nitrify to meet strict NPDES permit limits. The effluent alkalinity has been averaging 47 mg/L as $CaCO_3$, and it is desired to keep the effluent alkalinity at 100 mg/L as $CaCO_3$. This facility wants to use quick lime. How much additional quick lime (CaO) must be added? Lime purity is 90%.

STEP 1. Calculate alkalinity needs. Alkalinity to be added equals 100 mg/L desired $CaCO_3$ − 47 mg/L actual $CaCO_3$ in the effluent = 53 mg/L alkalinity as $CaCO_3$.

STEP 2. Calculate quick lime feed rate. The problem states that the plant flow is 38 000 m^3/d (10 mgd). The mass of alkalinity to be added in kilograms per day (pounds per day) equals the flow in cubic meters per day times the alkalinity needed in milligrams per liter divided by 1000 (million gallons per day times 8.34 times the alkalinity needed in milligrams per liter). Convert the mass of alkalinity to a quicklime feed rate as kilograms per hour (pounds per hour). From the data in Table 8.4, the conversion factor from $CaCO_3$ to CaO is 0.56.

38 000 m^3/d × 53 mg/L alkalinity as $CaCO_3$ × (1000 L/m^3) × (kg/1 000 000 mg) = 2014 kg/d $CaCO_3$

(10 mgd × 8.34 × 53 mg/L alkalinity as $CaCO_3$ = 4420 lb/d $CaCO_3$)

Quick lime = 2014 kg/d $CaCO_3$ × 0.56 = 1128 kg/d quicklime (4420 lb/d $CaCO_3$ × 0.56 = 2475 lb/d quick lime)

Purchased product = 1128 kg/d quick lime × 1/0.9 product purity = 1253 kg/d purchased quick lime (2475 lb/d quick lime × 1/0.9 product purity = 2750 lb/d).

Convert to hourly rate = 1253 kg/d purchased material / 24 h/d = 52.2 kg/h purchased product (2750 lb/d purchased material / 24 h/d = 115 lb/h).

Example 3. This is a 38 000-m³/d (10-mgd) WWTP having to nitrify to meet strict NPDES permit limits. The effluent alkalinity has been averaging 47 mg/L as $CaCO_3$, and it is desired to keep the effluent alkalinity at 100 mg/L as $CaCO_3$. This facility wants to use sodium hydroxide. How much additional sodium hydroxide (NaOH) must be added? Alkalinity supplementation will be made using 50% by weight sodium hydroxide. Sodium hydroxide density is 1.53 kg/L(12.76 lb/gal).

STEP 1. Calculate alkalinity needs. Alkalinity to be added equals 100 mg/L desired $CaCO_3$ − 47 mg/L actual $CaCO_3$ in the effluent = 53 mg/L alkalinity as $CaCO_3$.

STEP 2. Calculate sodium hydroxide feed rate. The problem states that the plant flow is 38 000 m³/d (10 mgd). The mass of alkalinity to be added in kilograms per day (pounds per day) equals the flow in cubic meters per day times the alkalinity needed in milligrams per liter divided by 1000 (million gallons per day times 8.34 times the alkalinity needed in milligrams per liter). Convert the mass of alkalinity to a quicklime feed rate as kilograms per hour (pounds per hour). From the data in Table 8.4, use 0.8 as the conversion factor from $CaCO_3$ to NaOH.

38 000 m³/d × 53 mg/L alkalinity as $CaCO_3$ × (1000 L/m³) × (kg/1 000 000 mg) = 2014 kg/d $CaCO_3$ (10 mgd × 8.34 × 53 mg/L alkalinity as $CaCO_3$ = 4420 lb/d $CaCO_3$)

Sodium hydroxide = 2014 kg/d $CaCO_3$ ×0.8 = 1611 kg/d sodium hydroxide (4420 lb/d $CaCO_3$ × 0.8 = 3536 lb/d)

Purchased product = 1611 kg/d sodium hydroxide × 1/0.5 product weight % = 3222 kg/d sodium hydroxide solution (3536 lb/d sodium hydroxide × 1/0.5 product weight % = 7072 lb/d).

Convert to a daily volume = 3222 kg/d sodium hydroxide solution / 1.53 kg/L = 2106 L/d = 2.1 m³/d of 50% by weight NaOH solution (7072 lb/d sodium hydroxide solution/12.76 lb/gal = 554 gal/d)

Convert to hourly feed rate = (2106 L/d) / (24h/d) = 87.8 L/h = 0.087 m³/h of solution (554 gal/d solution / (24 h/d) = 23 gal/h).

PHOSPHORUS PRECIPITATION

The basic principle of chemical phosphorus removal is precipitation followed by sedimentation. Phosphorus precipitation is the transformation of soluble phosphorus to a particulate form and the removal by sedimentation of these particles together with any phosphorus already present as an insoluble particulate. The following cations typically are used for the precipitation of phosphorus from wastewater:

- Iron,
- Aluminum, and
- Calcium.

Under the right conditions, all three cations form insoluble precipitates with orthophosphate. Orthophosphate is the primary phosphorus species affected by chemical removal; however, it is important to note that influent particulate phosphorus can solubilize, so attention must be paid to the total influent phosphorus concentration. The following three parameters are of particular importance for design, operation, and analysis of chemical phosphorus removal:

- Minimum achievable phosphate concentration,
- Effects of pH, and
- Dose requirements.

To determine the appropriate chemical use for phosphorus removal, the overall treatment process should be evaluated. A target effluent phosphorus concentration may be the objective of chemical addition, but the downstream biological demand for phosphorus should also be considered when removing phosphorus in the primary clarifiers. The biological process will suffer from nutrient deficiency if too much phosphorus is removed with chemical addition.

To minimize sludge production related to chemical phosphorus removal, consideration should be given to chemical addition across the primary clarifiers to reduce the soluble phosphorus concentration to levels equal to the phosphorus demands associated with downstream biological process (i.e., secondary treatment) (Figure 8.9). The overall reduction in phosphorus from the secondary treatment occurs when solids are wasted from the system. More chemical can be added in the biological process for chemical polishing of phosphorus at the final clarifier should additional phosphorus removal be required. This overall treatment scheme for phosphorus removal will yield efficient use of chemical, minimize overall sludge production, and make maximum use of the biological uptake of phosphorus for nutrient demand.

Plant personnel should obtain a specification data sheet or certified analysis for any chemicals to be used in the process to assess if the increased load of chemical impurities on the treatment plant is acceptable. This is particularly important for land disposal of biosolids or in water reclamation facilities. It is generally not necessary to use high-purity chemicals in chemical feed application, as technical grade from a reputable manufacturer is sufficient. Pickle liquor from some industrial sources has a higher probability for containing metal contaminants, so it is important to obtain specifications for the delivered chemical to ensure these contaminants do not have an adverse effect on the plant operation or permit.

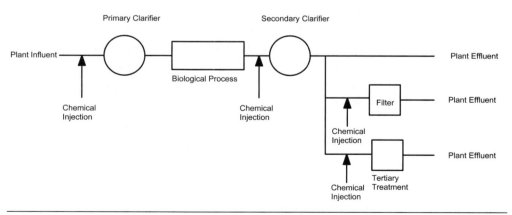

FIGURE 8.9 Chemical phosphorus removal dosing locations.

The subsequent sections will discuss the different chemicals available, the process of precipitation, and sample calculations.

IRON COMPOUND CHEMICAL ADDITION. There are four iron compounds that can be used for phosphorus precipitation. These include ferrous chloride ($FeCl_2$), ferrous sulfate [$Fe(SO_4)$], ferric chloride ($FeCl_3$), and ferric sulfate [$Fe_2(SO_4)_3$]. Ferric chloride is the most widely used and most widely available of the four chemicals mentioned.

Ferric and ferrous chemicals can be added before the primary clarifier, and the precipitate that is formed will settle in the primary clarifier. Performance of iron addition is highly dependent on reaction time. It can take 5 to 15 minutes to achieve a complete reaction. There should be a mixed floc zone for the iron to react and form an insoluble precipitate. If a floc zone is not available, then the chemical should be added further upstream to ensure that proper retention time is achieved. Ferric ions can also be added before the final clarifiers. The ferric precipitate will form upstream of the clarifier and then either be wasted from the system or returned to the aeration basin in the return activated sludge. The precipitate returned to the aeration basins will result in an increase in the basin solids concentration and must be taken into account in the design and operation of the aeration basins.

Ferrous chemicals can also be added before the aeration basin. The ferrous ions will oxidize to ferric ions and then form a precipitate. It should be noted that additional oxygen demand will be created in the aeration basin to oxidize ferrous ions to ferric ions. Ferrous ions should never be added to the final clarifier, as excess soluble iron will appear in the final clarifier effluent. If ferrous ions are in the final clarifier effluent, it will consume chlorine, foul UV systems, add total suspended solids to the plant effluent as the ferrous ions oxidize, and add color in the plant effluent.

Both ferrous (Fe^{2+}) and ferric (Fe^{3+}) ions can be used in the precipitation of phosphorus. With Fe^{3+}, the reaction can be written as follows:

$$Fe^{3+} + PO_4^{3-} \rightarrow FePO_4 \tag{8.6}$$

The reaction between ferrous ions and phosphate ions can be written as follows:

$$3\ Fe^{2+} + 2\ PO_4^{3-} \rightarrow Fe_3(PO_4)_2 \tag{8.7}$$

From the chemical reaction equation, it will take 1 mol ferric ion to react with 1 mol phosphate, and thus a 1:1 mol ratio of Fe:P. Because the molecular weight of iron is 55.85, and the molecular weight of phosphorus is 30.97, the equivalent weight ratio of Fe:P is 1.8:1, as shown in the following calculation:

(55.85 g Fe/1 mol Fe) / (30.97 g P/1 mol P) = 1.8/1 or 1.8:1

A larger amount of iron is required in actual situations than the chemistry of the reaction predicts. With Fe^{2+}, the situation is more complicated and not fully understood. Using the chemical reaction equation from eq 8.7, the mole ratio of Fe:P would be 3:2. However, experimental results indicate that, when Fe^{2+} is used, the mole ratio of Fe:P will be essentially the same as when Fe^{3+} is used, especially when the ferrous iron is added to the aeration basin, which allows ferrous to oxidize to ferric. Table 8.5 indicates the stoichiometric mole and weight ratio for each chemical mentioned.

Bench-, pilot-, and full-scale studies have shown that considerably higher than stoichiometric quantities of chemical typically are necessary to meet phosphorus removal objectives as a result of competing hydroxide and sulfide reactions. Sulfide will compete with phosphate initially for the ferric or ferrous ion; therefore, if a plant has significant levels of sulfide, then the iron dosing must be higher for the same amount of phosphorus removed. This concept is especially important when using chemical phosphorus removal across the primary clarifiers. As the dose rate of iron increases significantly to remove more phosphate, then the sulfide becomes a smaller part of the overall iron consumption in the treatment process.

When excess iron is added, the iron reacts to form a ferric hydroxide precipitate, as shown in eq 8.8. Alkalinity is consumed as both the iron phosphate and ferric

TABLE 8.5 Mole and weight ratios for iron addition.

Chemical	Molecular weight	Mole ratio (chemical / P)	Weight ratio (chemical / P)
Ferric chloride ($FeCl_3$)	162.35	1:1	5.24:1
Ferric sulfate [$Fe_2(SO_4)_3$]	399.7	0.5:1	6.45:1
Ferrous chloride ($FeCl_2$)	126.85	1.5:1	6.14:1
Ferrous sulfate [$Fe(SO_4)$]	151.85	1.5:1	7.36:1

hydroxide precipitates are formed. Essentially all of the iron added will precipitate as iron sulfide, iron phosphate, or ferric hydroxide.

$$Fe^{3+} + 3OH^- \rightarrow Fe(OH)_3 \tag{8.8}$$

When ferrous iron is used, iron precipitates according to eq 8.8, when the ferrous is eventually oxidized to ferric iron. When ferrous is added to a primary clarifier, ferric hydroxide is not formed from excess iron addition until the wastewater enters the activated sludge basin and ferrous reacts with oxygen to form ferric. This is also why ferrous should not be used in the final clarifiers for phosphorus removal, as excess ferrous or unreacted ferrous will carry over into the disinfection system to consume chlorine and form a precipitate (contribute to effluent total suspended solids [TSS]). Furthermore, if a UV disinfection system is used, iron will interfere with UV absorbance and foul the lamp sleeves, increasing the frequency of lamp cleaning.

All iron solutions mentioned are acidic. Ferric or ferrous compounds contain substantial amounts of free sulfuric acid or hydrochloric acid; thus, the acidic nature of the chemical will neutralize alkalinity in the water and suppress the pH of the process water. Furthermore, the removal of phosphorus will also remove alkalinity and lower the wastewater pH through the formation of phosphorus and hydroxide precipitates.

Iron compounds are most effective for phosphorus removal at certain pH values. For Fe^{3+}, the optimum pH range is 4.5 to 5.0. This is an unrealistically low pH, not typically attained in most municipal wastewaters. For Fe^{2+}, the optimum pH is approximately 8; however, good phosphorus removal can be obtained between pH 7 and 8. The addition of lime or sodium hydroxide may be necessary to ensure that the pH does not decrease dramatically with the addition of ferrous salts. Where the water is aerated following Fe^{2+} addition, the use of a base may not be necessary.

Jar tests should be performed to validate chemical dosage for phosphorus removal and to assess alkalinity destruction and pH depression from dosing of iron-based chemicals. The jar testing will determine requirements for additional chemical feed that may be necessary to maintain the required alkalinity for nitrification.

A design dose curve for chemical phosphorus removal using ferric ions was developed using literature and pilot-plant data (Figure 8.10). The literature data includes laboratory-scale test data and data from sites that are operating with chemical phosphorus removal (Luedecke et al., 1988). The data set includes results for a range of pH values (mostly 6.5 to 7.5), temperatures, and wastewater characteristics.

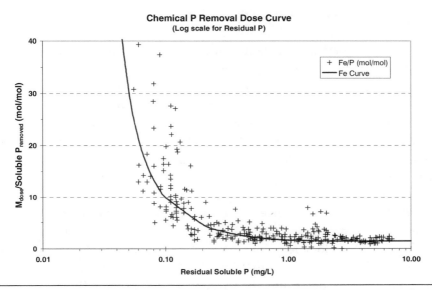

FIGURE 8.10 Ratio of Iron (Fe^{3+}) dose to phosphorus removed as a function of residual soluble orthophosphate concentration (Luedecke et al., 1988, and data from the Blue Plains Wastewater Treatment Plant, Washington, D.C.).

The effluent soluble phosphorus concentration is labeled "residual soluble P" on the logarithmic x-axis. The molar ratio for metal ion dose to soluble phosphorus removed is labeled "M_{dose}/soluble $P_{removed}$ (mol/mol)" on the y-axis. Note that the curves apply to the soluble portion of the phosphorus only.

The curve used to fit the data is based on the following equation:

$$y = \frac{a}{\left(1 + b \times e^{-cx}\right)} \tag{8.9}$$

Where

x = residual soluble phosphate (mg/L),
y = mole iron required per mole soluble phosphate removed,
a = 1.48,
b = -1.07, and
c = 2.25.

In general, the data at lower residual soluble P concentrations is more scattered. Because of this variability in the dose response found in the literature, bench-scale jar testing is recommended at each plant to determine the actual molar dose required to reach the targeted effluent soluble phosphorus. In addition, it is improbable that iron can be used to achieve effluent soluble phosphorus concentrations much below 0.10 mg/L. Even at this concentration, a molar dose of 12.0 mol/mol may be required. Dosing greater than 40 mol/mol of iron will not push the soluble phosphorus much below 0.08 mg/L. This is the practical solubility limit for ferric phosphate at typical wastewater pH values. Table 8.6 includes general information for the different chemicals mentioned in this section.

The molar dose for phosphorus precipitation is based on the desired final effluent soluble phosphorus concentration rather than the starting phosphorus concentration. For example, to meet a 0.5 mg/L soluble phosphorus concentration requires a 2.27 mole ratio of ferric ion to phosphorus or a weight ratio of 4.1 g Fe^{3+}/g P. To remove 2.5 mg/L P (from 3 to 0.5 mg/L) requires an iron dose of 10.25 mg/L Fe^{3+}. It is important to determine if a value greater than the influent soluble phosphorus concentration should be used because of the potential of solubilization of the particulate phosphorus, which would increase the soluble phosphorus concentration above the measured influent concentrations.

TABLE 8.6 Chemical properties.

Chemical	Weight percent in commercial solutions	Specific gravity*
Iron (Fe)	N/A	N/A
Phosphorus (P)	N/A	N/A
Ferric chloride ($FeCl_3$)	37 to 47%	1.39 to 1.53
Ferric sulfate [$Fe_2(SO_4)_3$]	43 to 50%	1.5 to 1.611
Ferrous chloride ($FeCl_2$)	28 to 35%	1.32 to 1.38
Ferrous sulfate [$Fe(SO_4)$]	19%	1.25

* To convert specific gravity to density, multiply by 8.34 to get lb solution/gal (lb/gal × 0.1198 = kg/L).

As noted previously, using dual application points may yield the optimum operating point with respect to chemical dose and sludge production. Using the same concentrations as above, if the phosphorus concentration were to be reduced to 1 mg/L in the primary clarifier, with additional iron added to the aeration basin to achieve a final effluent of 0.5 mg/L soluble phosphorus, less iron will be used. The iron dose to get a 1 mg/L soluble phosphorus out of the primary clarifier requires a molar ratio of 1.67:1 or a weight ratio of 3 g Fe^{3+}/g P. Therefore, to remove 2 mg/L phosphorus across the primary clarifier requires an iron dose of 6 mg/L. To remove the remaining 0.5 mg/L of soluble phosphorus across the secondary treatment system would require a molar ratio of 2.27:1 or a weight ratio of 4.1 g Fe^{3+}/g P, which equates to an iron dose of 2.05 mg/L Fe^{3+}. The total iron dose to meet a residual soluble phosphorus concentration of 0.5 mg/L is 6 + 2.05 = 8.05 mg/L Fe^{3+}, as opposed to 10.25 mg/L Fe^{3+} if all of the phosphorus is removed at one time. There is a 20% savings in chemical use and a reduction in the overall chemical sludge production. It should be noted that this example has neither take any credit for the phosphorus, which would be removed biologically across the secondary treatment system, nor has it accounted for potential solubilization of particulate phosphorus. Actual dosages must be fine-tuned in the field to account for these issues. A similar relationship exists when aluminum is used for phosphorus precipitation.

A sample calculation is provided to calculate the dose required to precipitate soluble phosphorus. To determine the dose of ferric chloride required, assume the following conditions:

- Influent plant flowrate is 38 000 m³/d (10 mgd),
- Soluble phosphorus influent concentration to the plant is 3.0 mg/L, and
- A residual primary effluent phosphorus concentration of 1 mg/L is required.

The amount of soluble phosphorus to be removed in kilograms per day (pounds per day) is as follows:

$$P = (3 \text{ mg/L} - 1 \text{ mg/L}) \times 38\,000 \text{ m}^3/\text{d} \times 1000 \text{ L/m}^3 \times (1 \text{ kg}/1\,000\,000 \text{ mg})$$
$$= 76 \text{ kg/d} ([3 \text{ mg/L} - 1 \text{ mg/L}] \times [10 \text{ mgd}] \times 8.34 = 166.8 \text{ lb/d})$$

The amount of soluble phosphorus removed in kilogram-moles per day (pound-moles per day) is as follows:

P = 76 kg/d / (30.97 kg/kg-mol P) = 2.454 kg-mol P/d ([166.8 lb/d] / ([30.97] lb/lb-mol P = 5.39 lb-mol P/d)

The amount of ferric ions required in moles of iron/moles of phosphorus removed to achieve residual soluble phosphorus concentration of 1 mg/L is as follows:

$Y = 1.48/(1-1.07e^{(-2.25)(1)})$

$Y = 1.67$ mol Fe^{3+}/mol soluble phosphorus removed

Amount of ferric ions required in kilograms per day (pounds per day):

Fe^{3+} = (1.67 kg-mol Fe^{3+}/kg-mol P) × (2.454 kg-mol P/d) × (55.85 kg Fe^{3+}/kg-mol) = 228.8 kg/d (1.67 lb-mol Fe^{3+}/lb-mol P) × (5.39 lb- mol P/d)(55.85 lb Fe^{3+}/lb-mol) = 502.7 lb Fe^{3+}/d)

Dosage of 100% ferric chloride required in kilograms per day (pounds per day):

$FeCl_3$ = (228.8 kg/d Fe^{3+}) × (162.2 kg $FeCl_3$/55.85 kg Fe^{3+}) = 664.5 kg/d 100% ferric chloride ([502.7 lb Fe^{3+}/d] × [162.2 MW $FeCl_3$/55.85 MW Fe^{3+}] = 1460 lb/d)

Then calculate the volume of 37% ferric chloride solution to be added per day:

Volume = (664.5 kg/d) / (0.37 kg $FeCl_3$/kg solution) / 1.34 kg/L = 1340 L/d = 1.34 m³/d

(1460 lb/d / [0.37 lb $FeCl_3$/lb solution] / [1.34 sp gr × 8.34] = 353 gal/d)

The feed rate of 37% ferric chloride required to reduce the soluble phosphorus from 3.0 mg/L to 1 mg/L is 1.34 m³/d (353 gal/d).

There will be a significant quantity of sludge produced when chemical addition is added to the process to remove phosphorus. This is further discussed in Chapter 10. Other issues affecting the decision to use an iron-based chemical include the following:

- Total dissolved solids (TDS) will increase in the treatment system;
- Iron will be present in the final sludge, which will be a benefit for using the sludge as a soil amendment;
- Overdosing of iron may result in effluent iron concentrations, which adversely affects UV disinfection performance and maintenance.

Ferric or ferrous compounds are acidic, so storage and handling issues are of concern. Fiberglass-reinforced plastic (FRP) or polyethylene tanks can be used to store ferric chloride, ferrous chloride, ferric sulfate, or ferrous sulfate. Recommended metering pumps include peristaltic, solenoid, or diaphragm types. Carrier water should be avoided if possible; the chemical will react with the carrier water and cause plating in the chemical feed lines. If it is necessary to add carrier water for mixing or dilution, then it should be added as close to the injection point as possible, to minimize the plating effects. The pump heads should be polyvinyl chloride (PVC). Piping, valves, and fittings should be PVC or chlorinated polyvinyl chloride (CPVC). Personnel should wear personnel protective equipment (PPE) when handling chemicals. The PPE should include, but not be limited to, gloves, respirators, goggles, aprons, and face shields, and should be worn when working or handling any iron salt solutions.

ALUMINUM COMPOUND CHEMICAL ADDITION. There are three aluminum compounds that are used in the wastewater industry for phosphorus removal. These include aluminum sulfate, sodium aluminate, and polyaluminum chloride. Sodium aluminate is typically used for process water that requires additional alkalinity, and polyaluminum chloride is used when enhanced solids removal is also a treatment objective; however, the most common chemical is aluminum sulfate, which is more commonly known as alum. Aluminum ions can combine with phosphate ions to form aluminum phosphate, as follows:

$$Al^{3+} + (PO_4)^{3-} \rightarrow AlPO_4 \tag{8.10}$$

The above equation indicates that it will take 1 mole of aluminum ion to react with 1 mole of phosphate, and thus a 1:1 mole ratio of Al:P. Because the molecular weight of aluminum is 26.98 and the molecular weight of phosphorus is 30.97, the weight ratio of Al:P is 0.87:1, as shown in the following calculation:

(26.98 g Al/1 mol Al) / (30.97 g P/1 mol P) = 0.87/1 or 0.87:1

With aluminum sulfate, there are two moles of aluminum; therefore, the stoichiometric mole ratio for the chemical is 0.5:1, and the weight ratio is {342 g $Al_2(SO_4)_3$ × 0.5}/30.97 g P = 5.52 :1. Similarly, the other chemical stoichiometric weight ratios and mole ratios are provided in Table 8.7.

TABLE 8.7 Mole and weight ratios for phosphorus removal using aluminum compounds.

Chemical	Molecular weight	Mole ratio (chemical/P)	Weight ratio (chemical/P)
Aluminum sulfate [$Al_2(SO_4)_3 \cdot 14H_2O$]	594	0.5:1	9.59:1
Sodium aluminate ($Na_2O \cdot Al_2O_3 \cdot 3H_2O$)	218	0.5:1	3.52:1
Polyaluminum chloride ($AlCl_3$)	133.5	1:1	4.31:1

A larger amount of aluminum is required for actual operation than the chemistry of the reaction predicts. Jar tests of the process water should be performed to determine the appropriate amount of chemical to be used to treat the process water.

The pH of the process water will affect the solubility of aluminum phosphate. Stumm and Morgan (1970) state that the solubility of $AlPO_4$ is pH-dependent and varies (Table 8.8).

The optimum pH for removal of phosphorus is in the range 5.5 to 6.5. At a pH greater than 6.6, aluminum can be effective, but to achieve the same level of phosphorus removal would require a dosage higher than the stoichiometric dose.

TABLE 8.8 Solubility of aluminum phosphate (Stumm and Morgan, 1970).

pH	Approximate solubility of aluminum (mg/L)
5	0.03
6	0.01 (minimum solubility)
7	0.3

Addition of alum will lower the pH of wastewater because alum solutions are acidic and aluminum precipitants consume alkalinity. The extent of pH reduction will depend principally on the alkalinity of the wastewater; the higher the alkalinity, the lower the reduction in pH for a given alum dosage. Most wastewaters contain sufficient alkalinity, so that even large alum dosages will not lower the pH below approximately 6.0 to 6.5. In exceptional cases of low wastewater alkalinity, pH reduction may not be so great that addition of an alkaline substance, such as sodium hydroxide, soda ash, or lime, will be required. However, if the plant also nitrifies, alkalinity consumption by phosphorus precipitation must be added to the nitrification alkalinity demand to evaluate the overall effect on the system.

Bench-, pilot-, and full-scale studies have shown that considerably higher than stoichiometric quantities of alum generally are necessary to meet phosphorus removal objectives. A competing reaction, responsible for the pH reduction mentioned above, at least partially accounts for the excess alum requirement. It occurs as follows:

$$Al^{3+} + 3OH^- \rightarrow Al(OH)_3 \tag{8.11}$$

Therefore, sludge that is generated will include aluminum hydroxide and aluminum phosphate.

Sodium aluminate can also serve as a source of aluminum for the precipitation of phosphorus. The chemical formula for sodium aluminate is $Na_2Al_2O_4$ or $NaAlO_2$. One commercial form is the granular trihydrate, which may be written $Na_2O \cdot Al_2O_3 \cdot 3H_2O$ and which contains approximately 46% Al_2O_3 or 24% Al. In contrast to alum, which reduces pH, a rise in pH may be expected on addition of sodium aluminate to wastewater.

A design dose curve for chemical phosphorus removal using aluminum ions was developed using literature and pilot-plant data (Figure 8.11). The literature data includes laboratory-scale test data and data from sites that are operating with chemical phosphorus removal (Gates et al., 1990). The full data set includes results for a range of pH values (mostly 6.5 to 7.5), temperatures, and wastewater characteristics. The effluent soluble phosphorus concentration is labeled "residual soluble P" on the logarithmic x-axis. The molar ratio for metal ion dose to soluble phosphorus removed is labeled "M_{dose}/soluble $P_{removed}$ (mol/mol)" on the y-axis. Note that the curves apply to the soluble portion of the phosphorus only. The curve used to fit the data is based on the following equation:

$$y = \frac{a}{\left(1 + b \times e^{-cx}\right)} \tag{8.12}$$

Where

x = residual soluble phosphate in mg/L,
y = moles aluminum required per mole soluble phosphate removed,
a = 0.8,
b = -0.95, and
c = 1.9.

In general, the data at lower residual soluble phosphorus concentrations are more scattered. Because of this variability in the dose response found in the literature, bench-scale jar testing is recommended at each plant to determine the actual molar dose required to reach the targeted effluent soluble phosphorus. Aluminum compounds can be used to produce a residual soluble phosphorus concentration 0.05 mg/L. At this point, the curve becomes steeper, causing the amount of aluminum required to increase significantly with each minor decrease in the target residual soluble phosphorus concentration.

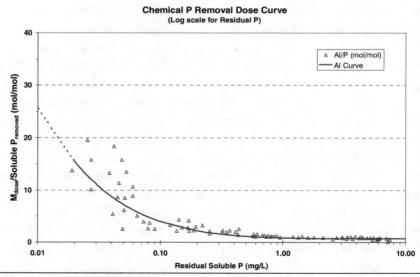

FIGURE 8.11 Ratio of aluminum (Al^{3+}) dose to phosphorus removed as a function of residual orthophosphate concentration (Gates et al., 1990).

Table 8.9 includes the molecular weight for the different chemicals mentioned in this section, which is necessary to calculate the dosage required:

A sample calculation is provided to calculate the dose required to precipitate soluble phosphate. Consideration should be given to whether there is a potential for particulate phosphorus to solubilize and increase the soluble phosphorus concentration from the measured value. To determine the dose of alum required, assume the following conditions:

- Influent plant flowrate is 38 000 m^3/d (10 mgd),
- Soluble phosphorus influent concentration to the plant is 3.0 mg/L, and
- A residual soluble phosphorus concentration of 1 mg/L is required.

The amount of soluble phosphorus to be removed in kilograms per day (pounds per day) is as follows:

P = (3 mg/L − 1 mg/L) × 38 000 m^3/d × 1000 L/m^3 × (1 kg/1 000 000 mg) = 76 kg/d ([3 mg/L − 1 mg/L] × [10 mgd] × 8.34 = 166.8 lb/d)

The amount of soluble phosphorus removed in kilogram-moles (pound-moles) per day is as follows:

P = (76 kg/d) / (30.97 kg/kg-mol P) = 2.454 kg-mol P ([166.8 lb/d] / [30.97 lb/ lb-mol P] = 5.39 lb-mol P/d)

TABLE 8.9 Chemical properties.

Chemical	Molecular weight	Weight percent in commercial solutions	Specific gravity
Aluminum	26.97	N/A	N/A
Phosphate	30.97	N/A	N/A
Aluminum sulfate	594	48.5%	1.335
Sodium aluminate	218	20%	1.46
Polyaluminum chloride	133.5	51%	1.4

The amount of aluminum ions required in moles aluminum/moles phosphate removed is as follows:

$Y = 0.8/(1-0.95e^{(-1.9)(1)})$

$Y = 0.93$ mol Al^{3+}/mol soluble P removed

The amount of aluminum ions required in kilograms per day (pounds per day) is as follows:

Al^{3+} = (0.93 kg-mol Al^{3+}/kg-mol P) × (2.454 kg-mol P/d) × (26.98 kg/kg-mol Al^{3+}) = 61.5 kg/d Al^{3+} ([0.93 lb-mol Al^{3+} /lb-mol P] × (5.39 lb-mol P/d) × [26.98 lb/lb-mol Al^{3+}] = 135 lb Al^{3+}/d)

The dosage of 100% aluminum sulfate required in kilograms per day (pounds per day) is as follows:

Alum = 61.5 kg Al^{3+}/d × (594 kg alum/ 26.97 kg Al^{3+}/2) = 678.4 kg/d alum ([135 lb Al^{3+}/d] × [594 lb alum/ 26.97 lb Al^{3+}/2] = 1487 lb/d alum)

Assuming 48.5% alum is available, the m^3/d (gal/d) of alum feed would be as follows:

Volume = (678.4 kg/d alum) / 0.485 / (1.335 kg/L) = 1048 L/d = 1.048 m^3/d alum

([1487 lb/d]/ 0.485 / [1.335 × 8.34 lb/gal] = 275 gal/d alum)

The alum required per day to reduce the soluble phosphorus from 3.0 to 1 mg/L is 1.048 m^3/d (275 gal/d). This example was based on a single application point; however, similar to the discussion in the Iron Compound Chemical Addition section, the optimum operating mode may involve dual application points (i.e., chemical addition to the primary clarifiers and to the secondary treatment process), as a result of the savings in chemical and reduction in sludge production.

There will be a significant amount of sludge produced when chemical is added to the process to remove phosphorus. The quantity of sludge and handling considerations are discussed in Chapter 10.

Two other issues should be considered when aluminum is added to the treatment process. Aluminum has no advantage as a soil amendment; therefore, if the final sludge is blended with soil, there may be additional concerns with aluminum content in the amendment. Aluminum sulfate and, to a lesser extent, sodium aluminate and polyaluminum chloride, will increase the TDS in the system. The effects of the increase of the TDS on the treatment process and effects at the discharge point should be identified. Discharging high TDS can be of concern where reclaimed water is used for irrigation purposes.

Aluminum compounds are mildly acidic, so storage and handling issues are of concern. Fiber-glass-reinforced plastic or polyethylene tanks can be used to store any of the aluminum compounds. Recommended metering pumps include solenoid, peristaltic, and diaphragm types. Carrier water should be avoided, if possible, as it will result in a higher pH, and aluminum hydroxide will precipitate, causing plating in the chemical feed lines. If it is necessary to add carrier water for mixing or dilution, then it should added as close to the injection point as possible to minimize the plating effects. The pump heads should be PVC. Piping, valves, and fittings should be PVC or CPVC. Personnel should wear PPE when handling chemicals. The PPE should include, but not be limited to, gloves, respirators, goggles, aprons, and face shields, and should be worn when working or handling any aluminum salt solutions.

LIME ADDITION. The use of lime to treat raw wastewater dates back to the mid-1800s. Lime is fed by either slaking quicklime or adding water to hydrated lime, and then the slurry is fed to the application point. Lime has been used to increase alkalinity, remove phosphorus, and improve removal efficiencies across primary clarifiers. Because of the chemistry involved, using lime for phosphorus removal is only applicable at the primary clarifiers. Some studies have shown that total phosphorus removals of 80% or more with concurrent reduction in BOD of 60 to 70% can be achieved across the primary clarifiers. With the addition of large quantities of lime, the primary effluent pH was increased to 9.5 or 10; however, no adverse effects were observed on the downstream biological process, as the carbon dioxide generated by biological activity served to correct the pH to near neutral levels.

There are two methods of adding lime to remove phosphorus from the process water. The methods are identified as low- and high-lime treatment. Low-lime treatment can provide 80% phosphorus removal on a consistent basis. By adding downstream tertiary filtration and facilities for addition of metal coagulants and polymer, the total effluent phosphorus can be reduced to less than 1.0 mg/L.

The high-lime treatment offers an even higher level of efficiency across the primary clarifiers. With flocculant aids and filtration, it can consistently produce final phosphorus levels below 1.0 mg/L without the use of additional treatment methods.

The advantage of low-lime addition in primary treatment stems from two fundamentals: the chemical law of mass action and the need for phosphorus in biological systems.

Lime precipitates more phosphorus during initial stages of the reaction than when the pH has been elevated and phosphorus concentrations are quite low. This effective performance by lime typically continues through the first 75% of phosphorus reduction, which occurs before pH 10 when lime demands for softening reactions increase. Overall lime demands become much higher when the process is operated to reduce phosphorus to very low levels.

One advantage of the low-lime process is the opportunity to use existing primary clarifiers for phosphorus removal. Capital expenditure, therefore, could be relatively small. The process requires equipment for feeding and flash mixing lime. Flocculation generally occurs in the inlet zone of the clarifier, but separate facilities may be preferred for this process. The elevated pH of primary effluent is reduced by recarbonation as a result of carbon dioxide produced in biological metabolism in secondary treatment. The low-lime treatment system may not be appropriate upstream of trickling filter systems, unless it has been determined that high effluent pH is tolerable, because recarbonation is minimal in trickling filter units.

High-lime treatment is defined as the addition of sufficient lime in primary facilities to achieve a pH of 11. Recarbonation with stored carbon dioxide may be necessary, and facilities for this would typically be provided. Recovery of lime by recalcination would frequently be included in high-lime systems.

High-lime treatment is applicable when effluent quality requirements include special provisions, such as softening (for reuse), low levels of soluble compounds of metals, improved virus removal, or reliable and consistent reduction of phosphorus below 1.0 mg/L, without supplemental metal addition. Some form of tertiary solids removal, such as filtration, would typically be used in these plants.

High-lime treatment would typically be combined with other biological, physical, or chemical processes to provide an overall system of advanced waste treatment. If biological treatment is included, the primary clarifier phosphorus residual should be high enough to meet metabolic requirements. High-lime treatment may result in a net increase in plant effluent levels of TDS and alkalinity.

Calcium ion reacts with phosphate ion in the presence of hydroxyl ion to form hydroxyapatite. This material has a variable composition; however, an approximate equation for its formation can be written as follows, assuming, in this case, that the phosphate present is hydrogen phosphate ion (HPO_4^{2-}):

$$3\ HPO_4^{2-} + 5\ Ca^{2+} + 4\ OH^- \rightarrow Ca_5(OH)(PO_4)_3 - 3\ H_2O \qquad (8.13)$$

The reaction is pH-dependent. The solubility of hydroxyapatite is so low, however, that even at a pH as low as 9.0, a large fraction of the phosphorus can be removed. In lime treatment of wastewater, the operating pH may be predicated on the ability to obtain good suspended solids removal rather than on phosphorus removal.

Although it is possible to calculate an approximate lime dose for phosphorus removal, this is generally not necessary. In contrast to iron and aluminum salts, the lime dose is largely determined by other reactions that take place when the pH of wastewater is raised. Some of these reactions are discussed later. Only in waters of very low bicarbonate alkalinity would the phosphate precipitation reaction consume a large fraction of the lime added.

Two other issues should be considered when lime is added to the treatment process. Relatively high calcium concentrations in the process water can inhibit volatile suspended solids destruction in digesters. Also, high calcium content in the final sludge may not be advantageous to certain soils, if the sludge is ultimately used for soil amendments.

Lime is either gravity fed or pumped to the point of application. Materials of construction for lime systems are carbon steel or PVC. Personnel should wear personal PPE when handling chemicals. The PPE should include, but not be limited to, gloves, respirators, goggles, aprons, and face shields, and should be worn when working or handling any chemical solids or slurries.

The forms of lime available are hydrated lime ($CaOH_2$) or quicklime (CaO). Each has advantages and disadvantages as discussed previously in the Alkalinity Supplementation section.

OTHER OPTIONS FOR CHEMICAL PRECIPITATION OF PHOSPHORUS. Alternative chemicals available to precipitate phosphorus include magnesium hydroxide and polymer. Magnesium hydroxide raises the pH to precipitate phosphorus and therefore would yield similar results as lime addition, but the chemical handling issues are not as significant as lime. Magnesium hydroxide is received in a liquid form and can be used similar to ferric or aluminum addition. Tank materials of construction would be FRP or polyethylene. Piping, valves, and fittings would be PVC or other compatible plastic material.

Polymer is regarded as an enhancement to metal salt addition. Polymer used alone would require a tremendous amount of chemical; the costs would prohibit polymer from being used as a stand-alone phosphate removal system. Polymer can be added to iron- or aluminum-based chemicals to enhance the removal of phosphorus from the process water. Metal salts and polymer can be added to the primary clarifier and has the added benefit of better TSS removal and particulate BOD removal.

Typical polymer systems would require stainless steel or FRP storage or aging tanks. PVC piping, valves, and fittings would be required for polymer service. Because polymers are typically sensitive to shear and higher viscosity than most metal salts, progressive cavity pumps would be recommended for polymer service.

Polymer does not have the chemical handling issues associated with most other chemicals. Generally, slips and falls are the most common hazards when handling polymer solution.

CHEMICAL FEED CONTROL. To monitor and control chemical phosphorus removal, total phosphorus should be measured at the plant effluent, and orthophosphate should be measured at the plant influent and the biological process influent and effluent. This can be done using laboratory analyses or using an online monitoring system. Operators can then make adjustments to chemical dosage based on the sampling results. Options for control of chemical dosages for phosphorus removal include manual, flow-paced, feed forward, and feed forward and feedback with effluent concentration control, similar to those discussed earlier in the Methanol Addition section of this chapter and in Chapter 13.

CASE STUDY: NORTHWEST COBB WATER RECLAMATION FACILITY, COBB COUNTY, GEORGIA. The Northwest Cobb Water Reclamation Facility (WRF), located northwest of Atlanta, Georgia, is owned and operated by the Cobb County Water System (CCWS). The plant was expanded from 15 000 to 30 000 m^3/d (4 to 8 mgd) in 1998 and consists of an influent pumping station, bar screens, aerated grit chambers, primary clarifiers, activated sludge system, secondary clarifiers, traveling bridge filters, UV disinfection, and cascade postaeration. Primary and waste activated sludge can be aerobically digested and dewatered with belt filter presses.

The current operation for solids handling includes using a section of the aerobic digesters for sludge holding and dewatering the raw sludge for landfill disposal.

Northwest Cobb WRF has the capability to operate in several biological nutrient removal configurations. After commissioning of the 30 000-m^3/d (8-mgd) expanded plant, operation for biological phosphorus removal was tested. However, it was necessary to add chemical to the dewatering filtrate and/or at the plant headworks for odor control reasons. Because metal salts, such as ferrous chloride, were being used for this purpose, significant phosphorus precipitation also was being achieved making biological phosphorus removal unnecessary. Therefore, Northwest Cobb WRF operates the activated sludge system for BOD removal and nitrification only and removes phosphorus through chemical precipitation using ferrous chloride and/or alum addition. A schematic and typical operating profile of phosphorus concentrations are shown in Figure 8.12.

Northwest Cobb WRF has had excellent performance over the past five years, including the Association of Metropolitan Sewerage Agency (Washington, D.C.) gold awards in 2001 and 2002. By feeding the ferrous chloride to the plant headworks with alum trim to the secondary clarifiers, effluent phosphorus concentrations have averaged well below 0.2 mg/L for the past several years. Ferrous chloride and alum dosages are typically targeted at approximately equal volumes, but are varied

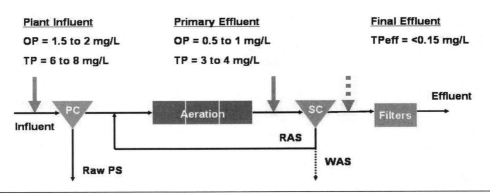

FIGURE 8.12 Northwest Cobb WRF phosphorus profile.

depending on activated sludge settling characteristics and the current cost of ferrous chloride and alum. Average chemical dosages for the past several years are summarized in Table 8.10.

During 2002, Cobb County implemented a supplemental sampling program to provide additional characterization of the influent wastewater and decide whether the plant expansion to 45 000 m^3/d (12 mgd) would incorporate chemical or biological phosphorus removal. These additional data revealed the following:

- The influent readily biodegradable COD is relatively low, at 12% of total COD (typical value is approximately 20%).
- There is VFA material in the influent wastewater or primary effluent. The VFAs are the carbon substrate used in biological phosphorus removal.
- The influent OP/TP ratio is approximately 0.2 (OP averaged less than 2 mg/L). Typically, the OP represents at least one-half of the TP. With this low OP/TP ratio, Northwest Cobb has lower-than-average soluble phosphorus and higher-than-average particulate phosphorus concentrations. With a low influent OP, good TSS (and particulate phosphorus) removal in the primary clarifiers, and some OP removal resulting from the ferrous chloride addition, very little phosphorus remains in the primary effluent for biological phosphorus removal.

TABLE 8.10 Northwest Cobb WRF chemical dosages for phosphorus removal.

Year	Average flow, m^3/d (mgd)	Average effluent phosphorus, mg/L	Chemical	Dose, m^3/d (gpd)
2001	23 000 (6.2)	0.13	Ferrous chloride[a]	0.946 (250)
			Alum[b]	1.00 (265)
2002	24 000 (6.4)	0.16	Ferrous chloride[a]	1.51 (400)
			Alum[b]	0.3 (75)
2003	27 000 (7.1)	0.08	Ferrous chloride[a]	0.757 (200)
			Alum[b]	1.63 (430)

[a]Ferrous chloride strength is 1 lb/gal as Fe^{2+} (1 lb/gal = 0.1198 kg/L).
[b]Alum strength is 49% as alum.

Northwest Cobb WRF plans to continue operation for nitrification with chemical phosphorus removal. Because of the low influent OP and lack of VFAs, conditions for biological phosphorus removal are not favorable, and chemical addition had the lowest present worth cost for phosphorus removal for the expansion to 45 000 m^3/d (12 mgd).

As part of two recent revisions to the NPDES permit, the effluent phosphorus limit for Northwest Cobb WRF was reduced from 0.6 to 0.23 mg/L monthly average. To optimize and ensure reliable operation at this low phosphorus limit and to obtain additional information on plant performance, CCWS installed a Chemscan (Applied Spectrometry Associates, Inc., Waukesha, Wisconsin) online analyzer in April 2003 to obtain real-time measurements of effluent phosphorus and ammonia. In combination with tracking of chemical dosages through the plant supervisory control and data acquisition system, the plant staff can follow phosphorus trends and make adjustments more quickly, if needed. The plant is operated for a target effluent phosphorus concentration of 0.1 mg/L. Because much of the historical data shows effluent phosphorus concentrations at the method detection limit of 0.1 mg/L, the laboratory lowered the method detection limit to 0.05 mg/L in November 2002. Since that time, effluent phosphorus concentrations below 0.1 mg/L have frequently been recorded (Figure 8.13).

CHEMICAL FEED SYSTEM DESIGN AND OPERATIONAL CONSIDERATIONS

All chemicals, whether gas, solid, or liquid, require a feeding system to accurately and repeatedly control the amount applied. Effective use of chemicals depends on accurate dosages and proper mixing. The effectiveness of certain chemicals is more sensitive to dosage rates and mixing than that of others. The design of a chemical feed system must consider the physical and chemical characteristics of each chemical used for feeding, minimum and maximum ambient or room temperatures, minimum average, and maximum wastewater flows, minimum average, and maximum anticipated dosages required, and the reliability of the feeding devices.

Operators and maintenance personnel should be aware of the hazards and characteristics of the chemicals that fed at the plant. Resources for design and operation of chemical feed systems include the following:

- Material data safety sheets;
- Chemical supplier chemical technical specifications;

310 Biological Nutrient Removal (BNR) Operation in Wastewater Treatment Plants

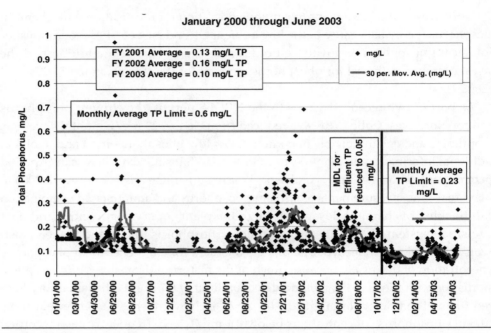

FIGURE 8.13 Northwest Cobb WRF effluent phosphorus (2000 through 2003).

- Water Environment Federation, *Operation of Municipal Wastewater Treatment Plants*, Manual of Practice No. 11 (WEF, 1996);
- Water Environment Federation, *Protecting Workers from Exposure to Chemical and Physical Hazards at Wastewater Treatment Plants* (WERF, 1999); and
- Water Environment Federation, *Biological and Chemical Systems for Nutrient Removal* (WEF, 1998).

REFERENCES

Abu-gharrah, Z. H.; Randall, C. W. (1991) The Effect of Organic Compounds on Biological Phosphorus Removal. *Water Sci. Technol.*, **23**, 585.

American Public Health Association; American Water Works Association; Water Environment Federation (1998) *Standard Methods for the Examination of Water and Wastewater,* 20th ed.; American Public Health Association: Washington, D.C.

Carrio, L.; Abraham, K.; Stinson, B. (2002) High-Rate Total Nitrogen Removal with Alternative Carbon Sources. *Proceedings of the 75th Annual Water Environment Federation Technical Exhibition and Conference* [CD-ROM], Chicago, Illinois, Sept 28–Oct 2; Water Environment Federation: Alexandria, Virginia.

Celanese Chemicals (2000) Product Description and Product Handling Guide: Glacial Acetic Acid, Corpus Christi, Texas. Celanese Chemicals: Dallas, Texas.

Elmendorf, H. A. (2002) Carbon Augmentation for Biological Nutrient Removal Using External Carbon Sources. *Proceedings of Preconference Seminar: Carbon Augmentation for BNR of the 75th Annual Water Environment Federation Technical Exhibition and Conference* [CD-ROM], Chicago, Illinois, Sep 28–Oct 2; Water Environment Federation: Alexandria, Virginia.

Fuhs, G. W.; Chen, M. (1975) Microbiological Basis of Phosphate Removal in the Activated Sludge Process for the Treatment of Wastewater. *Microbiol. Ecol.*, **2** (2), 119–138.

Gates, D. D.; Luedecke, C.; Hermanowicz, S. W.; Jenkins, D. (1990) Mechanisms of Chemical Phosphorus Removal in Activated Sludge with Al(III) and Fe(III). *Proceedings of the 1990 Specialty Conference on Environmental Engineering;* American Society of Civil Engineers: Reston, Virginia, 322.

Goins, P. (2002) A Sweet Alternative. *Proceedings of Preconference Seminar: Carbon Augmentation for BNR; of the 75th Annual Water Environment Federation Technical Exhibition and Conference* [CD-ROM], Chicago, Illinois, Sep 28–Oct 2; Water Environment Federation: Alexandria, Virginia.

Hoechst Celanese Chemical Group (1988) Technical Bulletin, Physical Properties of Aqueous Acetic Acid Solutions, Customer Technical Service Laboratory, Corpus Christi, Texas. Hoechst Celanese Chemical Group: Bishop, Texas.

Luedecke, C.; Hermanowicz, S. W.; Jenkins, D. (1988) Precipitation of Ferric Phosphate in Activated Sludge: A Chemical Model and its Verification. *Water Sci. Technol.*, **21** (352).

McCarty, P. L.; Beck, L.; Amant, P. St. (1969) Biological Denitrification of Wastewaters by Addition of Organic Materials. *Proceedings of the 24th Industrial Waste Conference*; Purdue University: Lafayette, Indiana.

Merck Index (1989) A Summary of Physical and Chemical Properties of Methanol, Gasoline (BTEX) and Benzene; Merck Publishing Group; Merck & Co., Inc.: Rahway, New Jersey.

National Fire Protection Association (2001) *Standard System for the Identification of the Hazards of Materials for Emergency Response,* Standard No. 704; National Fire Protection Association: Quincy, Massachusetts.

Purtschert, I.; Gujer, W. (1999) Population Dynamics by Methanol Addition in Denitrifying Wastewater Treatment Plants. *Water Sci. Technol.*, **39** (1), 43–50.

Stumm, W.; Morgan, J. J. (1970) *Aquatic Chemistry.* Wiley & Sons: New York.

U.S. Environmental Protection Agency (1993) *Nitrogen Control Manual;* EPA-625/R-93-010; Office of Research and Development, U.S. Environmental Protection Agency: Washington, D.C.

Water Environment Federation (1998) *Biological and Chemical Systems for Nutrient Removal.* Special Publication; Water Environment Federation: Alexandria, Virginia.

Water Environment Federation (1996) *Operation of Municipal Wastewater Treatment Plants,* 5th ed., Manual of Practice No. 11; Water Environment Federation: Alexandria, Virginia.

Water Environment Research Federation (1999) *Protecting Workers from Exposure to Chemical and Physical Hazards at Wastewater Treatment Plants.* Water Environment Research Federation: Alexandria, Virginia.

Chapter 9

Sludge Fermentation

Overview of Fermentation Processes 314	Unified Fermentation and Thickening Process 329
Function and Relationship to Biological Nutrient Removal Process 314	Primary Sludge Fermentation Equipment Considerations 329
Hydrolysis 317	Sludge Collector Drives 329
Acidogenesis 317	Primary Sludge Pumping 330
Acetogenesis 318	*Fermented Sludge Pumping* 330
Methanogenesis 318	*Sludge Grinders or Screens* 330
Primary Sludge Fermentation 319	Fermentate Pumping 330
Return Activated Sludge Fermentation 320	Mixers 331
Primary Sludge Fermenter Configurations 322	Scum Removal 331
	Odor Control and Covers 331
Activated Primary Sedimentation Tanks 322	Corrosion and Protective Coatings 332
Complete-Mix Fermenter 326	Instrumentation 332
Single-Stage Static Fermenter 327	*Flow Measurement* 332
Two-Stage Complete-Mix/Thickener Fermenter 328	*Oxidation–Reduction Potential* 332
	Level Measurement 333
	Sludge Density Meters 333
	pH Meters 333

Headspace Monitoring	333	Kalispell, Montana	338
Return Activated Sludge Fermentation	333	South Cary Water Reclamation Facility, North Carolina	341
Configuration	333	*Plant Description*	341
Equipment	334	*Fermentation Process Description*	341
Control Parameters	335		
Case Studies	337	*Operating Parameters*	343
Kelowna Wastewater Treatment Plant, Canada	337	References	345

OVERVIEW OF FERMENTATION PROCESSES

FUNCTION AND RELATIONSHIP TO BIOLOGICAL NUTRIENT REMOVAL PROCESS. As presented earlier in Chapter 4, the secondary influent volatile fatty acid (VFA) chemical oxygen demand-to-phosphorus ratio (COD:P) has significant bearing on the selection and proliferation of the enhanced biological phosphorus removal (EBPR) organisms, namely phosphate-accumulating organisms (PAOs). In cases where the plant influent does not contain sufficient VFAs because of reduced fermentation of raw wastewater in the collection system resulting from cold temperatures or steep (well-aerated) or short collection systems (short detention time), primary settling of particulate organic material further reduces the food sources for the PAOs. At plants where nitrogen and/or phosphorus removal is of importance, supplementation of VFA material may be necessary to ensure the biological nutrient removal (BNR) process achieves the desired effluent quality.

Plant recycles from downstream processes, such as sludge thickening, digestion, or biosolids dewatering, contain high concentrations of ammonia-nitrogen and phosphorus as a result of biological and physical biomass breakdown and lysis reactions that take place in the sludge processing operations (Figure 9.1). Although these recycle streams also contain some COD, the COD:P and the COD-to-nitrogen (COD:N) ratios are generally low enough to increase the nitrogen and phosphorus loads to the BNR process without increasing the biodegradable COD content to the extent necessary to remove the additional nutrient load. At plants where chemicals

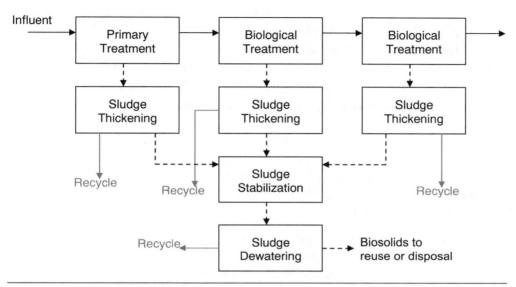

FIGURE 9.1 A typical WWTP operation schematic.

are added to the solids thickening and dewatering units, the organic and nutrient loads in recycle streams may be lower, because the metal precipitates of nitrogen and phosphorus species tend to keep the nutrients in the solids fraction, preventing resolubilization. Plant operators must monitor the sludge and recycle streams to ensure stable operation of the BNR process.

For an EBPR system to produce effluent soluble phosphorus concentrations of 1 mg/L or below, it has been shown that the EBPR influent must contain total COD-to-total phosphorus (TCOD:TP) ratios of 45:1 or higher. In cases where this cannot be sustained as a result of low plant influent organic material content or increased nutrient load to secondary treatment from plant recycle streams, fermentation of the waste sludges may be practiced as a means of returning "valuable" VFAs to the BNR process, where they contribute to the influent COD loading.

For some wastewater treatment plants (WWTPs), where nitrogen removal (i.e., nitrification/denitrification sequencing) is practiced, influent TCOD-to-total Kjeldahl nitrogen (TCOD:TKN) ratios may not be high enough to support denitrification to desired levels, and carbon augmentation may be necessary. In cases where postdenitrification is used, using fermenter supernatant for carbon supplementation may be counterproductive, because this fermentate also contains nitrogen and phosphorus. For predenitrification systems, such as modified Ludzack-Ettinger (MLE) or anaerobic/anoxic/oxic configurations, fermented sludge supernatant can be a feasible carbon augmentation option, depending on the overall nutrient and COD mass balances of the system, including all the fermenter, recycle, and influent flows and loads.

The use of an external VFA source from industry can be a viable option for some WWTPs, where suitable industrial wastewater VFA sources are readily available. Volatile fatty acids, such as acetic acid, can also be purchased through a chemical supplier. However, fermenting sludge for on-site VFA generation has a number of benefits, including the following:

(1) The plant gains independence from outside VFA sources. If an industrial source of VFA is used and the industry has intermittent operation or waste generation schedules, the BNR performance at the WWTP could be hampered when the VFA waste is not available. In addition, the operating cost associated with on-site generation of VFA is often lower than the cost of chemicals.

(2) On-site sludge fermentation shortens the BNR process anaerobic zone detention time that is traditionally used at plants without fermenters. The fermenter supernatant introduced to the anaerobic zones of EBPR systems (or anoxic zones, in the case of denitrification) contains VFA ready for microbial uptake, allowing the time allotted for hydrolysis reactions to be reduced. This decreases the anaerobic zone volume requirements of the BNR process.

(3) On-site sludge fermentation further stabilizes the waste sludges before the main sludge stabilization process, which consists of sludge digestion at a great majority of WWTPs. Operation of a sludge fermenter before digestion would decrease the load on the digesters to some extent. This benefit would vary from plant to plant, depending on the plant influent characteristics and the operation of the liquid-solids separation processes.

There are three different means of sludge breakdown that can be used in fermenters.

(1) Biological. Anaerobic bacteria are selected and grown in fermenter tanks. The extracellular enzymes of these organisms function to break down (i.e.,

hydrolyze) the complex organic material into smaller molecules. This fermentation step is also the "acid digestion" step of traditional anaerobic sludge digestion. To maximize VFA production, the fermenter solids retention time (SRT) must be short enough to maintain the anaerobic biological activity at acid-digestion level and avoid the subsequent conversion to methane that would typically take place in an anaerobic digester.

(2) Chemical. Through addition of acid or alkaline addition, sludge pH is adjusted to break down organic material. This type of hydrolysis is not desirable for carbon augmentation of BNR systems, because the pH extremes (2 or 14) used in chemical hydrolysis do not necessarily result in VFA formation, but lead to the break down of organic material and lysis of microbial cells.

(3) Thermal. Similar to chemical hydrolysis, thermal manipulation of waste sludges does not typically lead to VFA formation. These two methods are typically applied to sludge processing and digestion enhancement.

For the purposes of this chapter, only biological hydrolysis will be considered as a viable fermentation method for secondary treatment carbon augmentation.

The transformation of complex particulate organic material found in wastewater sludges to biogas under anaerobic conditions is mediated by several groups of microorganisms. Gujer and Zehnder (1983) and van Haandel and Lettinga (1994) described the following four distinct phases in the anaerobic digestion process: (1) hydrolysis, (2) acidogenesis, (3) acetogenesis, and (4) methanogenesis. These phases are discussed in more detail in the following sections.

Hydrolysis. Complex organic matter is converted into lower-molecular-weight dissolved compounds. The process requires the mediation of exoenzymes that are excreted by fermentative bacteria. Proteins are degraded to amino acids, carbohydrates are transformed into soluble sugars, and lipids are converted into long-chain fatty acids and glycerine. In practice, hydrolysis can be the rate-limiting step in anaerobic digestion, particularly at lower temperatures.

Acidogenesis. Dissolved compounds generated by hydrolysis are taken up by fermentative bacteria and excreted as simple organic compounds, such as VFAs, alcohols, and lactic acid; and mineral compounds, such as carbon dioxide, hydrogen, ammonia, and hydrogen sulphide gas. The process is carried out by a diverse group of fermentative bacteria, most of which are obligate anaerobes. However, some facultative bacteria can also metabolize organic matter via the oxidative pathway.

Acetogenesis. The products of acidogenesis are converted mainly to acetate, propionate, hydrogen, and carbon dioxide. Approximately 70% of the COD originally present in the sludge is converted to acetic and propionic acids, and the remainder of the electron donor capacity is concentrated in the formed hydrogen.

Methanogenesis. Methane is produced from acetate by acetotrophic bacteria or from the reduction of carbon dioxide by hydrogenotrophic bacteria.

A schematic representation of the above four conversion processes, with the percent COD involved in each transformation, is shown in Figure 9.2. The first three conversion processes are generally known as acid fermentation, and the fourth process is referred to as methanogenic fermentation. Figure 9.2 clearly shows that acetate is an important intermediate compound and the key carbonaceous substrate for the methanogenic bacteria. While acid fermentation takes place at an oxidation-reduction potential (ORP) greater than -300 mV, methane fermentation takes place at ORPs

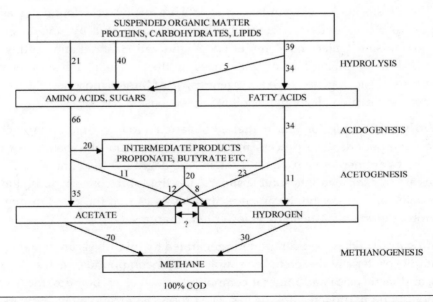

FIGURE 9.2 Four phases of anaerobic digestion (Gujer and Zehnder, 1983; Van Haandel and Lettinga, 1994).

below -550 mV. In a primary sludge fermenter that is used to augment the supply of readily biodegradable carbon to enhance operation of the BNR process, the growth of methane-forming organisms must be avoided to optimize VFA production.

PRIMARY SLUDGE FERMENTATION. The first use of a dedicated primary sludge fermenter for on-site VFAs production was in the design of the Kelowna BNR plant in Canada in the late 1970s (Barnard, 1977). An existing sludge digester was converted to a gravity thickener and used for acid fermentation to augment the fermentation taking place in the anaerobic zone of the process. In the early years of operation, the VFA-rich fermenter supernatant was returned to the inlet to the primary clarifiers, and it was difficult to determine the important role of the fermenter in the performance of the plant. Oldham and Stevens (1984) rerouted the pipework to allow direct discharge of the fermenter supernatant to the anaerobic zone of the bioreactor, as was the original intent of the designers. During the course of a subsequent optimization study, the importance of fermenter supernatant was clearly demonstrated by switching the fermenter discharge from one bioreactor module to the other. Removal of the supernatant stream from one module soon resulted in a large increase in the effluent phosphorus concentration from that module. Efficient phosphorus removal was only restored once the supernatant discharge was reintroduced to the module. When the fermenter supernatant stream was evenly divided between the two modules, effluent phosphorus concentrations of below 0.25 mg/L were achieved consistently at the plant.

Barnard (1984) and researchers at the City of Johannesburg, South Africa (Osborn et al., 1986), allowed a primary sludge inventory to build up in the primary clarifiers at BNR plants in South Africa by recycling primary sludge to the inlet end of the primary clarifiers-either directly or through elutriation tanks. Fermentative conditions were allowed to develop in the sludge layer at the bottom of the primary clarifiers, so that some of the complex organics in the sludge were hydrolyzed to VFAs and other soluble carbonaceous compounds through acid fermentation. These substrates entered the BNR bioreactor with the primary effluent. Barnard (1984) referred to this concept as activated primary sedimentation tanks.

Rabinowitz and Oldham (1985) and Rabinowitz et al. (1987) found that phosphorus removal in a pilot-scale BNR process having a primary clarifier and a completely mixed sidestream fermenter was 100% greater than a similar process treating raw wastewater with comparable characteristics, without primary clarification and sludge fermentation.

Carrio et al. (2002) found that the specific denitrification rates were significantly greater using primary sludge fermentate and sodium acetate instead of methanol as a supplementary carbon source for denitrification in a pilot-scale step-feed process in New York City. Further, the study found that no acclimatization period was required for fermentate and acetate addition, but that slightly higher biomass yields were observed than with methanol addition. The COD:N requirement for acetate or fermentate was approximately 50% less than that for methanol. It was found that only select denitrifying microorganisms can use methanol reliably in cases where the anoxic zone retention time is short. Therefore, if methanol is used only as a backup, denitrification is expected to be impaired, whereas the use of fermenter supernatant is an immediately available backup.

In the past 20 years, primary sludge fermenters have been used in BNR plants in Canada, the United States, Europe, South Africa, Australia, and New Zealand. Several facilities having primary sludge fermenters are reported to be unable to meet their effluent phosphorus limits biologically when the fermenters are taken out of service.

RETURN ACTIVATED SLUDGE FERMENTATION. Although it has not been as widely used as primary sludge fermentation, fermentation of a portion of the return activated sludge solids can be a viable option for internal production of VFAs. This fermentation option could be used at any activated sludge BNR plant, but it particularly applies to WWTPs that operate secondary treatment processes only and do not have primary clarifiers.

The level of VFA production that can be achieved specifically through fermentation of the return activated sludge (RAS) is not well-documented in the literature. However, the viability of the source has been shown by several researchers and operating evidence. For example, in the Phostrip process (Biospherics, Inc., Beltsville, Maryland), the influent wastewater went directly to the aeration basin. The RAS was directed to an anaerobic thickener (stripper) for release of phosphorus, and the supernatant treated with lime in a separate clarifier. Fermentation of RAS solids took place in the anaerobic thickener, according to Fuhs and Chen (1975), thus providing a VFA source for uptake by PAOs and driving the biological phosphorus removal portion of this process. Because all of the RAS was fermented, there was too much secondary release of phosphorus, which necessitated the treatment with lime of the supernatant.

In the late 1980s, the patented Orange Water and Sewer Authority process, in which primary sludge is fermented and added as a VFA source directly to the RAS in a sidestream anaerobic zone (nutrification zone), was developed. The RAS from this anaerobic zone is then directed to the mainstream activated sludge system. In subsequent research, the sidestream biological phosphorus removal process configuration was further developed to include a second sidestream consisting of fermentation of a portion of the RAS (Lamb, 1994). This patented process differed from the fermentation of RAS within the Phostrip process in that the fermented RAS was sent directly to a sidestream anaerobic zone rather than to the mainstream activated sludge system aeration zone, and there was no lime treatment. This process option has been used at several WWTPs in North Carolina.

As an alternative to fermenting a portion of the RAS, fermentation of mixed liquor solids also has been used. In the original four-stage Bardenpho pilot-plant work conducted by Barnard in 1972, a small portion of the mixed liquor solids was inadvertently sent to a "dead zone," where acid fermentation of the biomass produced VFA that returned to the post-anoxic (second anoxic) zone, resulting in the release and uptake of phosphorus by the PAOs. Under this operating condition, the total influent phosphorus of 8 mg/L was reduced to less than 0.2 mg/L. When the dead zone was disconnected from the second anoxic zone, the apparent phosphorus removal was dramatically reduced, with effluent phosphorus concentrations of approximately 4 mg/L (Barnard, 1974, 1976, and 1985).

The capability of RAS fermentation to produce a product high in VFAs has been indirectly shown through Fothergill and Mavinic's (2000) work with autothermal thermophilic aerobic digestion (ATAD) of waste activated sludge. Pilot testing was conducted to investigate the effect of feeding secondary sludge, as a mixture with primary sludge, to an ATAD system as a potential source of supplemental VFA for BNR processes. Under anaerobic aerated conditions (oxygen restricted environment where -500< ORP<-200 mV), the results showed that the net production of VFA increased with higher percentages of secondary sludge compared to primary sludge only.

One disadvantage of fermenting activated sludge solids is the high nutrient content of the fermentate. In a biological phosphorus removal system, the PAOs will release phosphorus in the fermentation zone, producing a high-phosphorus supernatant (similar to that of the anaerobic reactor of the Phostrip system). In addition, ammonia also would be released back into the main process. However, depending on the circumstances, the overall benefit of the additional VFA source can outweigh any negative effect from the return streams.

PRIMARY SLUDGE FERMENTER CONFIGURATIONS

Four principal primary sludge fermenter configurations were described in detail, with their advantages, disadvantages, and typical design criteria, by Barnard (1994) and Rabinowitz (1994). A fifth primary sludge fermenter configuration also has been developed and patented (Baur, 2002a). Process schematics of each of the five primary sludge fermenter configurations are presented in Figures 9.3 to 9.7. A short description of each process, with typical design criteria, operating considerations, and advantages and disadvantages, are presented in this section.

The two principal control parameters for the operation of primary sludge fermenters are the fermenter SRT and hydraulic retention time (HRT). The fermenter SRT is controlled by adjusting the solids inventory and the sludge wastage rate. By increasing the fermenter SRT, the growth of slower growing fermentative organisms is favored, and more complex molecules and higher acids are produced. Conversely, decreasing the SRT favors the growth of faster growing organisms, resulting in simpler biochemical pathways and the production of acetic acid and, to a lesser extent, propionic acid. The ratio of VFA produced per volatile suspended solids (VSS) added to a fermenter has a fairly broad range, from 0.05 to 0.3 g VFA/g VSS added.

The fermenter HRT is controlled by adjusting the primary sludge and elutriation water (added to wash and separate the released soluble VFAs from the particulate matter as an overflow stream) pumping rates. Increasing the HRT increases the available time for the conversion of solubilized substrates to VFAs. However, too long an HRT results in the production of complex molecules and higher acids.

The key parameter for monitoring the performance of primary sludge fermenters is the VFA concentration in the fermenter supernatant. This is best measured by gas chromatography or high-performance liquid chromatography, as these methods provide accurate information about the concentration of individual VFAs present. The distillation method is a reasonable method for measuring the total VFA concentration, but tends to be inaccurate at concentrations lower than 100 mg/L. The concentration of soluble COD in the fermenter supernatant also provides a reasonable indication of the VFA concentration. The ORP in the sludge blanket can indicate the level of the anaerobic activity in the fermenter and whether optimal conditions for acid fermentation are being maintained.

ACTIVATED PRIMARY SEDIMENTATION TANKS. This is the simplest type of primary sludge fermenter and was proposed by Barnard (1984). Primary sludge from the primary clarifier is recycled to the inlet of the clarifier, either directly or through an elutriation tank, so that a fermenting sludge inventory is allowed to build up on the clarifier floor. A fraction of the sludge withdrawn from the primary clarifiers is wasted to the sludge-handling system. The major advantages of this fermenter are its simplicity and the fact that no additional unit processes are required. Although this type of fermenter has been used successfully to enhance the biological phosphorus removal characteristics of BNR processes, it has several disadvantages, as follows:

(1) It results in a high solids loading rate to the primary clarifiers, which often leads to solids loss over the clarifier weirs and a resultant additional solids loading to the BNR process.
(2) It is difficult to control the sludge age of the fermenting sludge mass, and these fermenters have a tendency to promote methane and sulphide formation in the sludge mass in warmer climates, leading to reduced VFA yields and odor and corrosion problems. For example, Osborn et al. (1986) reported that, in a large wastewater plant with four activated primary tanks in Johannesburg, South Africa, the entire contents of each tank must be wasted every three to four days to avoid methane and sulphide formation in the clarifiers.
(3) The VFAs produced cannot be discharged directly to the BNR process, but must be conveyed in the primary effluent. The opportunity exists, therefore,

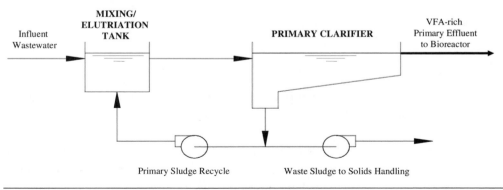

FIGURE 9.3 Activated primary sedimentation tanks.

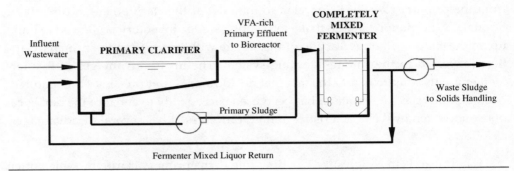

FIGURE 9.4 Complete-mix fermenter.

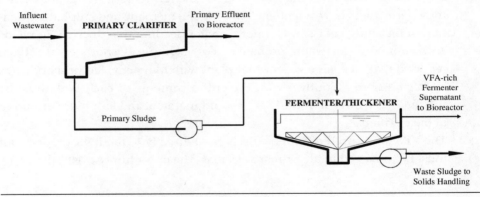

FIGURE 9.5 Single-stage static fermenter.

for the VFAs to be stripped or aerobically metabolized because of air entrainment in the passage between the primary clarifiers and the BNR process.

(4) Recycling the primary sludge can lead to a buildup of fibrous material and plastics in the sludge mass, which can cause blockages in the primary sludge pumps and pipework. This problem is generally solved by continuously screening the entire sludge flow at some point in the recycle loop.

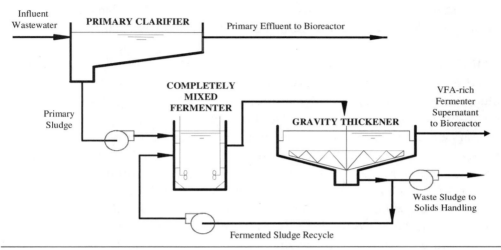

FIGURE 9.6 Two-stage fermenter/thickener.

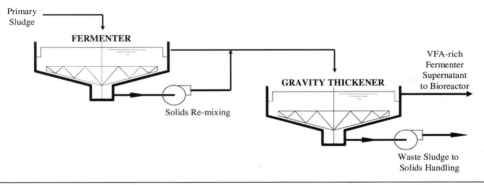

FIGURE 9.7 Unified fermentation and thickening fermenter.

Activated primary sedimentation tanks are generally designed on the basis of the sludge inventory required to achieve a given sludge age, or SRT. Sludge ages used are typically between two and four days. The wastage rate is chosen to allow the sludge mass blanket in the primary clarifier to build up to a given height above the clarifier floor (i.e., 1.5 or 2 m). Primary sludge recirculation rates are typically approximately 5 to 10% of the average dry weather flowrate to the plant. The sludge removal mechanisms must be able to cope with the higher solids inventory in the tank.

COMPLETE-MIX FERMENTER. This fermenter is similar in concept to the activated primary sedimentation tanks and was proposed by Rabinowitz et al. (1987). Sludge from the primary clarifiers is pumped into a completely mixed tank, where acid fermentation occurs. The tank overflow is returned by gravity to the inlet of the primary clarifiers, where it is mixed with the incoming wastewater. Surplus primary sludge is wasted from the fermenter. The fermenter HRT is determined by the tank volume and the sludge recycle rate. The fermenter SRT is determined by the sludge wastage rate. The major advantage of this fermenter over the activated primary sedimentation tank is that it is possible to more accurately control the sludge age, and, consequently, the degree of methane formation and sulphide generation is significantly reduced. Disadvantages with this fermenter are similar to the activated primary tanks (i.e., excessive solids losses over the primary clarifier weirs as a result of the high solids loading rates; and the loss of some of the VFAs produced in the fermenter, as a result of aerobic activity and stripping in the passage through the primary clarifiers). Operating difficulties reported with these fermenters are excessive solids losses over the primary clarifier weirs, "roping" of fibrous material as a result of vortexing action of the mixers in the complete-mix tanks, formation of a stable scum blanket in the tanks, and blockages of the outlet pipework and primary sludge pumps. The use of inline grinders, chopper pumps, or fine screens on the primary sludge line is recommended to either screen out or macerate the sludge mass continuously.

Complete-mix fermenters are typically designed to handle between 4 and 8% of the average dry weather flowrate to the plant. Units are sized to provide an HRT of between 6 and 12 hours, an SRT of 4 to 8 days, and a solids concentration of between and 1 and 2%. Fermenter VFA concentrations of between 300 and 500 mg/L have been reported in the fermenter itself, with the fermenter adding between 15 and 30 mg/L of VFA (as acetic acid) to the primary effluent entering the bioreactor. Mixing energy provided by the mixers should be sufficient to prevent solids deposition on the tank floor and the formation of a stable scum layer on the surface, but not so great as to cause vortexing and excessive air entrainment in the fermenter. The use of slow-speed mixers equipped with variable speed drives that impart between 8 and 10 W/m^3 (40 to 50 hp/mil. gal) into the liquid is recommended.

In addition to the conventional complete-mix fermenter, some plants operate modified versions of this configuration. For example, in Penticton, British Columbia, the fermenter has been operated partially as an upflow anaerobic sludge blanket reactor. The fermenter was mixed each day for a short period, but was unmixed the remainder of the time. Primary sludge is pumped into the sludge blanket. This resulted in thicker sludge to the digesters, improved fermentation, and reduced solids loadings to the primary clarifiers.

SINGLE-STAGE STATIC FERMENTER. This fermenter is a gravity thickener with an increased side water depth to allow for the storage of a fermenting sludge mass on the thickener bottom. This fermenter configuration was used in the design of the BNR facility for Kelowna, Canada, and its use was first documented by Oldham and Stevens (1984). Primary sludge is pumped into a center well and allowed to settle and thicken in the unit. Thickened primary sludge is drawn off from the bottom of the fermenter, at a solids concentration of 5 to 8%, and wasted to the sludge handling system. The wastage rate is generally based on controlling the fermenter SRT by maintaining a given sludge blanket height or a sludge inventory in the fermenter. A major advantage of this fermenter is that the VFA-rich fermenter supernatant can be discharged directly to the anaerobic zone of the BNR process, thus allowing for optimal use of this substrate source in the EBPR mechanism. This also allows the use of BNR process configurations in which the anaerobic zone is not at the head end of the process (i.e., those that include a preanoxic zone in which the return activated sludge is denitrified either by endogenous respiration alone or in conjunction with a small portion of primary effluent before entering the anaerobic zone). In more recent static fermenter designs, a source of elutriation water (either primary or final effluent) is typically fed to the unit with the primary sludge to flush out the VFAs produced and inhibit methane and sulphide formation.

Primary sludge is typically pumped into the fermenter at a rate of between 4 and 8% of the average dry weather flow to the plant. The loading rate to the unit is typically approximately 25 to 40 kg/m^2·d, which is significantly lower than the solids loading rate generally used for gravity thickeners. Side water depths of between 3.5 and 5 m are used so that the required sludge inventory can be maintained. A high-torque sludge scraper mechanism is required to cope with the high sludge inventory generally maintained in these units. Sludge ages of between 4 and 8 days are typically used, depending on the fermenter temperature. There is some VFA in the thickened sludge, which is discharged to the sludge handling system, but the fraction decreases with increasing sludge thickness. When the sludge is thickened to a solids concentration of 6 to 8%, the fraction of VFA being pumped to the sludge handling process is relatively low. Recycling of the thickener underflow to the inflow to elutriate the VFA has been suggested, but experience indicates that the degree of thickening is then reduced, and the benefit may be limited.

TWO-STAGE COMPLETE-MIX/THICKENER FERMENTER. This fermenter configuration combines the positive features of the complete mix fermenter and the single-stage static fermenter and consists of a complete-mix tank and a gravity thickener in series. It has been suggested that the gravity thickener can also be replaced by a thickening centrifuge. The first full-scale application was in Rotorua, New Zealand. The first large full-scale application of a two-stage complete mix/gravity thickener fermenter was at the Bonnybrook WWTP in Calgary, Canada, and is described by Fries et al. (1994). Primary sludge is pumped into the completely mixed tank, and overflows by gravity to the thickener. Most of the sludge inventory is stored in the complete mix tank. Thickened sludge from the thickener bottom is recycled to the complete-mix tank, and a portion of the thickened primary sludge is wasted to sludge handling to maintain the desired fermenter SRT. The VFA-rich thickener supernatant is conveyed directly to the anaerobic zone of the bioreactor. As in the case of the static fermenter, a source of elutriation water (either primary or final effluent) is typically fed into the gravity thickener section with the complete-mix tank outlet to flush out the VFAs produced and inhibit methane and sulphide formation.

These fermenters are typically designed to operate at an SRT of 4 to 8 days, with a solids concentration of between 1.5 and 2% in the complete-mix tank. The HRT of the complete-mix tank is typically between 12 and 24 hours. Loading rates and side water depth for the gravity thickeners should be those recommended by the Water Environment Federation® (formerly the Water Pollution Control Federation [1980]) (i.e., 100 to 150 kg/m^2·d [20 to 30 lb/d/sq ft] and 3.5 to 4.0 m [11 to 13 ft], respectively). One problem in the design and operation of these units is that the recycling of the sludge between the units significantly increases the solids loading to the thickeners. For this reason, the primary sludge pumping rate is typically lower than that used in other fermenters, typically 2 to 4% of the average dry weather flow to the plant. The thickened sludge recycle rate from the thickener to the complete-mix tank is typically 50% of the primary sludge pumping rate. Mixing energy provided to the complete-mix tank is the same as that provided in the complete-mix fermenters (i.e., slow-speed mixers that impart 8 to 10 W/m^3 [40 to 50 hp/mil. gal] to the fermenter liquid).

UNIFIED FERMENTATION AND THICKENING PROCESS. The unified fermentation and thickening (UFAT) process consists of two thickeners in series (Figure 9.7). The first is operated as a fermenter, and the settled solids and supernatant are recombined and directed to the second thickener. Elutriation water can be added to the thickener to condition the solids and improve settling. The VFA-rich supernatant from the second thickener is directed to the BNR process, while the settled solids are sent to solids processing. The UFAT process was developed and patented by Clean Water Services (Hillsboro, Oregon) and is used at the Durham Advanced Wastewater Plant (Tigard, Oregon) (Baur, 2002a and 2002b).

One of the advantages of this configuration is the ability to further control the fermenter SRT by varying the solids pumping rate from the bottom of the fermenter. The fermenter itself is unmixed, and operates such that the fermenting solids are stratified, with the sludge layer at the bottom having the longest SRT. The VFA in the sludge blanket is elutriated by recombining the fermented sludge with the fermenter overflow before directing the combined stream to the thickener. The thickener is operated as needed to meet downstream solids processing requirements. Operating data from the UFAT fermenter at the Durham facility show supernatant VFA concentrations of 250 to 350 mg/L after the first fermentation stage, and 400 to 550 mg/L after the thickening stage (Baur, 2002b).

PRIMARY SLUDGE FERMENTATION EQUIPMENT CONSIDERATIONS

There are a number of unique considerations associated with fermentation equipment; the purpose of this section is to discuss the major equipment types and related operating issues.

SLUDGE COLLECTOR DRIVES. In activated primary tanks, static fermenters, and in the thickener tank of a two-stage fermenter, deeper sludge blankets are maintained compared to those of a conventional primary clarifier or gravity thickener. With the deeper blanket, an increased solids inventory is being held in the tank, and high sludge solids concentrations generally are observed. For successful operation, the collector drives must be designed to accommodate the additional torque load. To avoid corrosion problems, the collector drive and mechanism must have a corrosion resistant coating, or be constructed of stainless steel.

PRIMARY SLUDGE PUMPING. The primary sludge flow and percent solids being pumped can vary significantly, depending on the type of fermenter being used. If an activated primary tank is used, the primary sludge stream will generally have a relatively high solids content (approximately 4 to 5% total solids) because of the depth of the sludge blanket being maintained within the clarifier. In this case, most facilities use a positive displacement type pump, such as a progressive cavity or plunger pump.

If a static fermenter or completely mixed fermenter are being used, primary sludge is generally pumped at higher rates (approximately 1 to 5% of the total plant flow), resulting in primary sludge solids concentrations lower than 0.5%. Although a progressive cavity pump could be used for this application, with the less concentrated sludge stream, a recessed impeller centrifugal pump should perform the work with lower maintenance requirements.

Fermented Sludge Pumping. Fermented sludge from an activated primary, static fermenter, or thickener tank of a two-stage fermenter may range in solids concentration from 2 to 10% total solids and often higher than 6%. Pumping of very concentrated sludge streams can be problematic for any WWTP, and the fermented sludge is no exception. To ensure that there is no "rat-holing" in the fermenter anaerobic sludge blanket, the pumps must be able to accommodate the lower sludge withdrawal rates and higher solids concentrations. In addition, sufficient capacity must be provided to allow increases in sludge wasting if necessary to avoid excessive anaerobic activity within the fermenter. Positive displacement pumps, such as progressive cavity and rotary lobe, are generally used for pumping of the thickened, fermented sludge.

Sludge Grinders or Screens. For pumping of primary sludge and fermented sludge, grinders (or possibly chopper-type pumps for pumping of a dilute primary sludge stream to a complete-mix or static fermenter) are often used to grind up rags, plastics, and other debris that might otherwise get caught in the sludge pump or piping. Alternately, the sludge flow can be screened to remove the debris. Because of the likelihood of debris and thick sludge to cause clogging of the lines, incorporation of grinding or screening is a critical component for ensuring that the fermenter can be operated successfully.

FERMENTATE PUMPING. The VFA-rich fermentate can be returned to the influent or primary effluent wastewater before entering the BNR process, or it can be sent directly to the BNR process anaerobic zones. If the fermentate is sent directly to the anaerobic zones, fermentate pumps, piping, and valves are generally needed. However, the ability to control the fermentate addition and avoid stripping of VFA materials over primary clarifier and other weirs is considered, by many, to be worth the additional cost. Centrifugal nonclog pumps with adjustable frequency drives are generally used for this purpose. Because the fermentate still contains solids, some prefer a recessed-impeller centrifugal pump rather than the closed-impeller type.

MIXERS. In the complete-mix and two-stage fermenter/thickener configurations, primary sludge is fermented in a completely mixed tank. The mixing energy should be sufficient to prevent solids deposition on the tank floor and the formation of a stable scum layer on the surface, but not so great as to cause vortexing and excessive air entrainment in the fermenter. The use of slow-speed mixers equipped with variable speed drives that impart between 8 and 10 W/m^3 (40 and 50 hp/mil. gal) into the liquid is recommended (Rabinowitz and Abraham, 2002).

SCUM REMOVAL. Radial scum skimming and collection should be provided on static fermenters and the thickener tank of a two-stage fermenter. Scum can be collected and sent to a scum concentrator or to the digestion process with the remainder of scum collected elsewhere at the facility. Although fermenter scum can be sent back to the head of the plant by gravity, this results in rehandling of the scum at the primary clarifiers, which may be the solution, in some cases, but is not optimal. Scum buildup also can be problematic in complete-mix fermenters. Mixing within the complete mix fermenter should be intensive enough to allow entrainment of scum within the main fermenter contents. One option might be to use a pumped mixing system with discharge nozzles at the fermenter surface to break up the scum, similar to that used by some WWTPs in anaerobic digesters.

ODOR CONTROL AND COVERS. Fermenter units should be covered and the headspace air scrubbed in a chemical scrubber system to control odors. Two- and three-stage chemical scrubbers are commonly used. An alternate means of air treatment would be to return it through the diffused aeration system in the BNR reactor. Typical headspace venting is three to six air changes per hour. Similar to conventional gravity thickener operation, odors can be minimized in static fermenters and two-stage fermenter/thickeners by adding dilution, or elutriation water at the thickener influent. Common elutriation water sources include primary or secondary effluent.

Fermenter covers should be designed to reasonably minimize the headspace and thereby reduced the volume of air to be treated in the odor control system. The covers are subject to the corrosive nature of carbon dioxide (CO_2) and hydrogen sulfide (H_2S) and the associated reduced pH, and it is important that they be constructed of a material that is corrosion resistant to handle a pH of less than 2.0. Low-profile aluminum geodesic dome covers are a good alternative. In a static fermenter or thickener, the collector drive motor and gear box can be mounted above the dome, allowing access from the outside. Stainless steel or protective coatings should be considered for the odor control blowers/fans and the air piping.

CORROSION AND PROTECTIVE COATINGS. As discussed earlier in this chapter, CO_2 gas produced during fermentation dissolves in condensed moisture in the headspace and produces corrosive carbonic acid. In addition, sulfate reduction activity in the fermenter results in H_2S formation, also a common cause of corrosion. To avoid corrosion problems, surfaces from a few meters (several feet) below the liquid level and surfaces above the liquid level in the fermenter should receive a protective coating to withstand reduced pH conditions as low as 1.0. Surfaces maintained below the liquid level should be less susceptible to corrosion, although coating of the entire tank might be considered for a new installation. The pH in the liquid is likely to range from 5.5 to 6.5.

INSTRUMENTATION. *Flow Measurement.* Magnetic flow measurement is typically used on the primary sludge feed to the fermenter, fermented sludge flow, and the fermentate flow. The ability to measure these flows allows the operator to perform a flow balance around the fermenter and provides information needed to make SRT calculations. Measurement of the fermented sludge flow is also important, with respect to control of downstream digestion and dewatering operations. Measurement of the supernatant flow assists with quantifying the VFA sent to the BNR basins and may aid in BNR system operations decisions.

Oxidation–Reduction Potential. The ORP can be used as an indicator of fermenter performance. Acid-forming bacteria can proliferate at ORP values of approximately -300 mV, while significant decreases in ORP to -600 mV or more can indicate the onset of methanogenesis, which adversely affects the resulting VFA available for use in the BNR process. However, it has been noted at several installations that, in practice, an actual correlation between VFA production and ORP was not observed, and use of the ORP meters was not helpful for operations (Oldham and Abraham, 1994).

Level Measurement. It is difficult for operators to check the actual tank level visually because fermenters are covered, and level measurement should be provided in the fermenter tank. Ultrasonic level measurement has been used for this purpose.

Sludge Density Meters. It can be difficult to obtain samples to measure fermenter solids. Sludge density meters can be helpful on the primary sludge feed line, fermenter waste line, and in the fermenter itself. This information can be used in combination with flow to perform mass-balance calculations for use in checking the fermenter SRT and overall VSS destruction. Although less convenient, sludge samples can also be manually collected from each location and analyzed in the laboratory for the same purpose.

pH Meters. Although every plant will be slightly different as a result of varying conditions of wastewater alkalinity, pH can be used as an indicator of fermenter performance. If pH meters are installed and the operating pH conditions monitored, a pH "signature" can be developed. If fermenter pH trends outside the normal operating range, the operator has an indication that conditions are changing in the fermenter (i.e., increased pH might signify the onset of methanogenesis) and that an operational adjustment should be made (Oldham and Abraham, 1994).

Headspace Monitoring. There is a potential for production of H_2S and methane gas within the fermenter, and headspace H_2S and methane monitors are typically recommended. Depending on the plant and the operation, fermenter headspace H_2S concentrations can exceed 200 ppm. An alarm should be initiated if the headspace gas in the fermenter reaches 5% of the lower explosive limit (LEL) for safety reasons and to allow the plant staff time to initiate changes to prevent the onset of methanogenesis.

RETURN ACTIVATED SLUDGE FERMENTATION

Fermentation of a portion of the RAS can be used for generation of the needed VFA source for operation of a BNR process. Although this fermentation option could be used at any activated sludge BNR plant, it particularly applies to WWTPs that operate secondary treatment processes only and do not have primary clarifiers. There are several variations on the configuration for this fermentation option; this section will focus on the operation of the sidestream RAS fermentation process that is currently being used at several plants in North Carolina.

CONFIGURATION. In the patented sidestream biological phosphorus removal process using RAS fermentation (Lamb, 1994), RAS from the secondary clarifier is first directed to a sidestream anaerobic zone. A portion of the RAS from the anaerobic zone is directed (typically pumped) to a sidestream fermentation zone. Effluent from the sidestream fermentation zone is sent back to the anaerobic zone as the VFA source, as shown in Figure 9.8. The sidestream RAS fermentation zone is similar in appearance to any anaerobic or anoxic zone in the mainstream BNR process. As an alternate to the sidestream biological phosphorus removal configuration, the fermentation zone effluent could be sent directly to the anaerobic zone of the main BNR process.

EQUIPMENT. Equipment for the sidestream fermentation process includes the following:

- Mixers for the fermentation zone. These may consist of submersible propeller mixers, jet mixing, vertical turbine mixers, or other means of unaerated mixing.
- Solids recycle pump. The solids recycle pump directs a portion of the RAS to the fermentation zone. Depending on the hydraulics of the system, a second pump might be needed to return the fermentate to the designated location. A

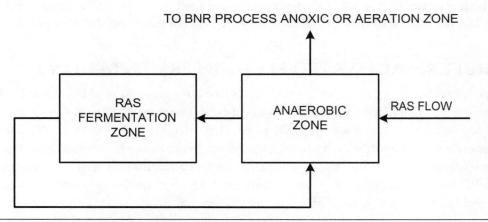

FIGURE 9.8 Schematic of sidestream biological phosphorus removal process with RAS fermentation.

submersible pump is most commonly used. The use of variable frequency drives provides the capability to adjust the RAS flow to the fermentation zone.

- Odor control. Odors generated by the sidestream fermentation zones are generally not problematic. Known installations consist of typical open-top zones partitioned in concrete tanks (Figure 9.9).

CONTROL PARAMETERS. Similar to primary sludge fermentation systems, the control parameters consist of HRT and SRT. In the completely mixed sidestream RAS fermentation zone, HRT and SRT are equal. Depending on the actual VFA production and process results, the HRT might be varied from a few hours to approximately two

FIGURE 9.9 South Cary Water Reclamation Facility sidestream RAS fermentation zone.

days. At this time, there is not much available information on the control systems required for this type of operation, although several plants use these methods successfully. During development of the sidestream biological phosphorus removal process with RAS fermentation (Lamb, 1994), full-scale testing included an RAS sidestream preanoxic zone HRT of 2 hours, RAS sidestream anaerobic zone HRT of 2 hours, and an RAS fermentation zone HRT of 66 hours, or nearly 3 days. Approximately 6% of the total RAS flow was diverted to the fermentation zone. The plant effluent phosphorus concentrations averaged less than 0.3 mg/L as TP. However, fermentation zone VFA data are not available.

Suggested process control parameters for RAS fermentation include regulating the pumping rate to the fermentation zone, allowing adjustment of the HRT. The phosphorus release in the anaerobic zones should be monitored as a measure of effectiveness, allowing identification of the pumping rate that maximizes VFA production.

Sampling recommendations for tracking the effectiveness of the fermentation zone performance are provided in Table 9.1. After process changes, these can be measured daily or at least several times a week until stabilized, at which time, intermittent sampling should be adequate.

TABLE 9.1 Sidestream RAS fermentation zone sampling parameters.

Parameter	Process considerations
RAS, TSS, VSS, filtered COD	Upstream from fermentation zone to establish baseline conditions
Fermentation zone TSS, VSS	Examine TSS and VSS destruction in fermentation zone
Fermentation zone filtered COD, VFA	Examine soluble COD and VFA production in the fermentation zone
Fermentation zone NH_3-N and OP	Examine phosphorus and ammonia release in fermentation zone
Anaerobic zone OP	Examine phosphorus release in the anaerobic zones of the BNR process

CASE STUDIES

KELOWNA WASTEWATER TREATMENT PLANT, CANADA. In the early 1990s, the Kelowna WWTP was upgraded and expanded to a capacity of 40 ML/d, and the original single-stage static fermenter (converted digester) was replaced by two 15.2-m diameter static fermenters (converted secondary clarifiers from the original conventional activated sludge plant, Figure 9.10). The original five-stage Bardenpho process was replaced by a modified three-stage BNR process, which consists of two larger, 14-cell trains and two smaller, 7-cell trains, all operated in parallel. Each of the four trains has an anaerobic, anoxic, and aerobic zone. The primary effluent base flow is discharged to the anaerobic zone, with the higher flows being bypassed to the anoxic zone. The VFA-rich fermenter supernatant is split between the anaerobic cell at the head end of each of the four trains. Mixed liquor from the four bioreactor trains flows into five secondary clarifiers. Effluent from the secondary clarifiers

FIGURE 9.10 Kelowna static fermenters.

is polished on five dual-media (sand/anthracite) tertiary filters and disinfected in a medium pressure UV disinfection system before discharge to the environmentally sensitive Okanagan Lake.

In 2003, the average effluent TP concentration from the plant, based on 7-day composite samples, was 0.11 mg/L. The corresponding average effluent orthophosphorus (OP) concentration, based on 24-hour composite samples, was 0.03 mg/L. The average annual effluent TKN and nitrate-nitrogen (NO_3-N) concentrations, based on 24-hour composite samples, were 1.80 and 2.78 mg/L, respectively (Carey, 2004).

The two static primary sludge fermenters are 15.2 m in diameter and have a side water depth (SWD) of 3.25 m. The depth at the center of the units is 5.8 m. Primary sludge is pumped to the fermenters in a relatively dilute form, at a constant rate, that is approximately 5% of the average influent flowrate to the plant. The scraper mechanisms have a relatively high tip speed of 0.3 m/s. Because of the high solids inventory, material tends to mat in front of the rake, suppressing the gentle mixing for which the pickets are intended. The mechanism is equipped with an automatic reversing drive to allow the rotation to be occasionally reversed for short periods to dislodge the collected mats. The drive, mounted on the center platform, is accessed by a half-bridge, which spans the tank's radius. The sludge blanket height is maintained by adjusting the amount of sludge removed daily and is controlled on the basis of maintaining a fermenter supernatant VFA concentration between 150 and 250 mg/L and keeping the rake mechanisms torque at below 40% of the maximum. During the summer months, when a shorter fermenter SRT is required, a 1.1-m sludge blanket height is maintained. During the colder winter months, the sludge blanket height is increased to approximately 1.8 m. Thickened fermenter sludge is withdrawn from the sludge hopper at an average concentration of approximately 6.0% (as dry solids) and is macerated using inline grinders upstream of the sludge pumps. The sludge is then blended with the thickened waste activated sludge from the BNR process and centrifuge dewatered to a solids content of approximately 20%. The fermenters are covered by flat, low-profile FRP covers, and the headspace foul air is treated in a dedicated chemical scrubber/biofilter arrangement.

KALISPELL, MONTANA. The 11 ML/d (3.1-mgd) wastewater treatment plant at Kalispell, Montana, consists of a modified University of Cape Town process with a two-stage fermenter. Kalispell discharges to Ashley Creek, a tributary to the sensitive Flathead Lake system in northern Montana. The existing BNR reactor was designed with eleven cells to permit varying the size of the anaerobic, anoxic, and aerobic zones, in response to seasonal fluctuations in loads and temperatures caused by wet weather and snowmelt. In addition, the last two aerobic cells, constituting up to 30% of the bioreactor volume, can be bypassed to avoid secondary release of phosphorus and ammonia resulting from endogenous conditions.

Operating challenges at Kalispell include the very low winter wastewater temperatures and high flows during the spring months resulting from snow melt and increased rainfall. Since commissioning of the BNR and fermenter systems in 1992, the plant has consistently maintained effluent phosphorus concentrations of less than 0.4 mg/L, including averaging less than 0.15 mg/L TP for the past five years (1998 through 2002) (Emrick and Abraham, 2002).

The fermenter consists of a completely mixed fermentation tank followed by a gravity thickener (Figure 9.11). Primary sludge is pumped to the fermenter on a timed basis throughout each day. The pumps are set to five cycles per hour, and pump 4.8 minutes each cycle. Primary sludge solids concentrations are typically approximately 5000 mg/L. The complete-mix stage of the fermenter is operated at a 4- to 5-day SRT. Good operation and control is achieved at a target total suspended solids (TSS) concentration of approximately 12 000 mg/L in the compete-mix tank. The fermenter sludge recycle solids concentration averages approximately 20 000 mg/L. Thickened fermented sludge is transferred to the digesters with air diaphragm pumps. Plant effluent nonpotable water (NPW) is added to the thickener influent to help elutriate the VFAs. The fermentate is pumped to the BNR system first anaerobic cell using recessed impeller centrifugal pumps. The fermenter has operated well and has consistently produced VFA concentrations between 200 mg/L (winter operation) and 450 mg/L (summer) (Emrick, 2004a; Natvik et al., 2003). The solids content of the thickener supernatant averages less than 100 mg/L (Emrick, 2004b).

To break up any scum buildup in the complete-mix stage of the fermenter system, the plant operates a scum-buster external pumping system. The fermenter recirculation pumps are chopper-type pumps to grind up debris. At one point, the plant staff tried using a grinder, but found more reliable operating results with the chopper pump.

One aspect of the fermenter operation that has been of concern is the corrosive environment. Severe corrosion of the concrete has occurred above the water line over the years, necessitating repairs. When originally constructed nearly 15 years ago, the fermenter had been coated with a coal tar epoxy, but this coating was unable to with-

FIGURE 9.11 Kalispell complete-mix fermenter and thickener.

stand the extremely corrosive environment at the water line and above. The gravity thickener stage was recoated with a more resistant coating in 1995, and the same was done with the complete-mix tank several years after that. The plant staff has found that stainless steel and plastic coatings likely have the longest design life within the fermenter.

The two-stage fermenter has worked very well, but there is a lot of tankage and associated pumps and equipment, including chopper pumps, air diaphragm pumps, mixers, scum buster, and supernatant pumps. Kalispell is currently planning future conversion to a static fermenter when the plant expands to 24 600 m^3/d (6.5 mgd). The existing fermenter would be converted, and the project also would include construction of an additional static fermenter tank to meet future capacity needs. This change would serve to simplify the operation and the quantity of equipment and is expected to result in similar VFA production.

SOUTH CARY WATER RECLAMATION FACILITY, NORTH CAROLINA.
Plant Description. The 48 400-m^3/d (12.8-mgd) South Cary Water Reclamation Facility, North Carolina, has operated its BNR process since 1999. The treatment processes include preliminary treatment, activated sludge BNR system, secondary clarifiers, deep-bed filters, UV disinfection, and postaeration. Waste activated sludge is gravity-belt thickened, aerobically digested, and land-applied. The South Cary Water Reclamation Facility currently discharges to Middle Creek, a tributary to the Neuse River. Because of nutrient impairment of the Neuse River, WWTP discharges were given annual nitrogen mass limits. Under the current capacity, South Cary must meet a limit corresponding to approximately 4.6 mg/L total nitrogen (TN).

Before expansion and upgrade to BNR, the plant consisted of two completely mixed activated sludge basins, originally designed to meet effluent biochemical oxygen demand and ammonia-nitrogen (NH$_3$-N) limits of 5 and 1 mg/L, respectively. These basins were circular in shape and were modified for BNR by constructing a total of 16 zones within each basin. A third basin, identical to the existing two basins, was constructed to meet capacity requirements. Many of the zones have the capability to be operated in multiple modes (i.e., anaerobic, anoxic, aerobic), as shown in Figure 9.12. This provides South Cary the capability to operate in a number of BNR modes (including MLE and three-, four-, or five-stage BNR configurations), as needed, to optimize performance or meet changing process needs.

Fermentation Process Description. A unique aspect of the South Cary BNR process is the capability to operate a "sidestream" fermentation process to enhance biological phosphorus removal. These sidestream fermentation options consist of sending the RAS to dedicated mixed zones, which are capable of being operated as anoxic (for endogenous RAS denitrification), for fermentation (where all or a portion of the RAS is allowed to ferment, producing VFAs to enhance the BNR process), and as anaerobic (for phosphorus release and VFA uptake in biological phosphorus removal). These RAS fermentation options and their interface with the main BNR system flowsheet are shown in Figure 9.13. South Cary currently operates in a four-stage Bardenpho configuration and achieves effluent total nitrogen concentrations between 2 and 3 mg/L and TP concentrations of less than 0.5 mg/L. All RAS flow is directed to an anoxic cell, to allow the nitrate to deplete, and then to the anaerobic cells. The RAS flowrates average approximately 45% of the influent flow during the summer and 55% of the influent flow during the winter. Approximately 3% of the total RAS flow is sent from the anaerobic zones to fermentation. The fermented RAS is sent back to the anaerobic zone (Figure 9.13c).

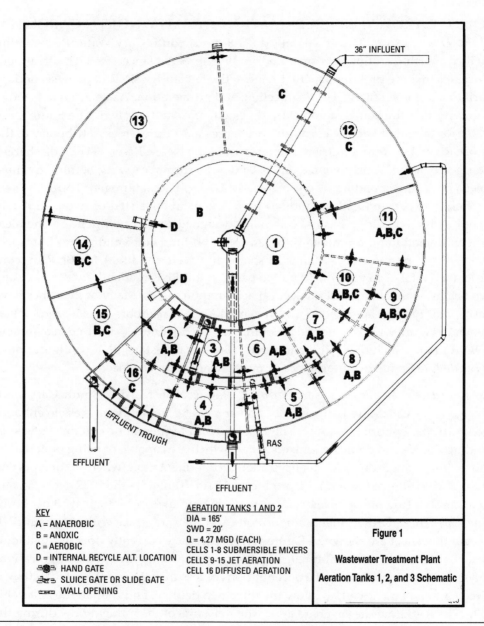

FIGURE 9.12 Plan view of South Cary Water Reclamation Facility BNR system (Stroud and Martin, 2001).

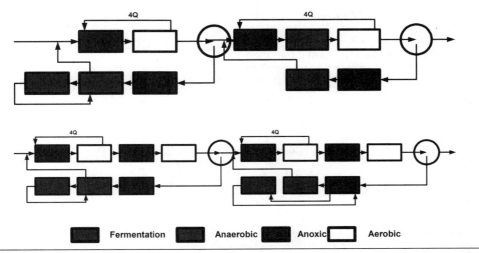

FIGURE 9.13 Biological nutrient removal process configurations with sidestream RAS fermentation (Stroud and Martin, 2001): (a) MLE with sidestreams, (b) three-stage BNR, (c) Bardenpho A, and (d) Bardenpho B.

Operating Parameters. The general operating parameters at South Cary are summarized in Table 9.2.

The RAS fermentation process has worked well, and the plant has had excellent performance. At this time, specific testing is not conducted in the RAS fermentation zones. Relative to the overall BNR process operation, South Cary has observed the best performance when considering the following operating parameters:

- Keeping dissolved oxygen levels in the aerobic zones at 1 to 2 mg/L to help prevent oxygen bleed through to the secondary anoxic zones.
- The best TN results are achieved when a small quantity of ammonia remains in the effluent.
- Scum generation intensifies with changes in season (hot weather to colder, and cold to hot).
- Settled sludge in the clarifiers is less compact than in the basic nitrification system.
- Keeping the solids concentrations within the desired range is more critical to the overall BNR process than in basic nitrification systems.

TABLE 9.2 Summary of operating parameters at the South Cary Water Reclamation Facility.

Typical average plant influent parameters	
Flow	20 000 m^3/d (5.3 mgd)
BOD	160 to 190 mg/L
TSS	200 to 300 mg/L
NH$_3$-N	23 mg/L
TKN	36 mg/L
TP	6.7 mg/L
Number of BNR basins in service	2
Mainstream BNR process	
Anoxic zone	Zone 1 (nitrate recycle is sent from zone 13)
Aerobic zone	Zones 10, 9, 11, 12, 13, and 14
Post-anoxic zone	Zone 15
Reaeration zone	Zone 16
Typical mixed liquor suspended solids concentration	2500 to 2800 mg/L
Sidestream RAS fermentation process	
Anoxic zone	Zones 4 and 5 (endogenous denitrification of remaining nitrate in the RAS)
Anaerobic zone	Zones 6, 9, and 2 (phosphorus release and VFA uptake for biological phosphorus removal)
Sidestream fermentation zone	Zones 7 and 8 (RAS flow pumped from third anaerobic zone to first fermentation zone. Fermented RAS flow from second fermentation zone is directed to the first RAS anaerobic zone).
Typical RAS TSS	5000 to 5800 mg/L

REFERENCES

Barnard, J. L. (1974) Cut P and N without Chemicals. *Water Wastes Eng.*, Part 1, **11** (7), 33–36; Part 2, **11** (8), 41–43.

Barnard, J. L. (1976) A Review of Biological Phosphorous Removal in Activated Sludge Process. *Water SA*, **2** (3), 136–144.

Barnard, J.L. (1977) Letter Report to Knight and Peisold Consulting Engineers Documenting Design Recommendations for Upgrade of the Kelowna Plant to BNR; P. G. J. Meiring & Partners, Pretoria, South Africa; August 9.

Barnard, J. L. (1984) Activated Primary Tanks for Phosphate Removal. *Water SA*, **10** (3), 121–126.

Barnard, J. L. (1985) The Role of Full Scale Research in Biological Phosphate Removal. *Proceedings of the University of British Columbia Conference on New Directions and Research in Waste Treatment and Residuals Management*, Vancouver, Canada, June 23–28; University of British Columbia: Vancouver.

Barnard, J. L. (1994) Alternative Prefermentation Systems. Preconference Seminar on Use of Fermentation to Enhance Biological Nutrient Removal of the 7th Annual Water Environment Federation Technical Exposition and Conference, Chicago, Illinois, Oct 15–19; Water Environment Federation: Alexandria, Virginia.

Baur, R. (2002a) Unified Fermentation and Thickening Process. U.S. Patent 6,387,264, May 14.

Baur, R. (2002b) Achieving Low Effluent Phosphorus Concentrations in Activated Sludge Effluent: Chemical vs. Biological Methods. Preconference Seminar on Carbon Augmentation for BNR of the 75th Annual Water Environment Federation Technical Exhibition and Conference, Chicago, Illinois, Sep 28–Oct 2; Water Environment Federation: Alexandria, Virginia.

Carey, S., City of Kelowna, British Columbia (2004) Personal Communication Regarding Primary Sludge Fermenter Design, Operation and Performance Data; May 3.

Carrio, L.; Abraham, K.; Stinson, B. (2002) High Rate Total Nitrogen Removal with Alternative Carbon Sources. *Proceedings of the 75th Annual Water Environment Federation Technical Exhibition and Conference* [CD-ROM], Chicago, Illinois, Sep 28–Oct 2; Water Environment Federation: Alexandria, Virginia.

Emrick, J.; Ken, A. (2002) Long Term BNR Operations-Cold in Montana! *Proceedings of the 75th Annual Water Environment Federation Technical Exhibition and Conference* [CD-ROM], Chicago, Illinois, Sep 28–Oct 2; Water Environment Federation: Alexandria, Virginia.

Emrick, J., City of Kalispell, Montana (2004a) Personal Communication Regarding Primary Sludge Fermenter Operation at the City of Kalispell WWTP; April 20.

Emrick, J., City of Kalispell, Montana (2004b) Personal Communication Regarding Primary Sludge Fermenter Operation at the City of Kalispell WWTP; July 29.

Fothergill, S.; Mavinic, D. S. (2000) VFA Production in Thermophilic Aerobic Digestion of Municipal Sludges. *J. Environ. Eng.*, **126** (5), 389–396.

Fries, M. K.; Rabinowitz, B.; Dawson, R. N. (1994) Biological Nutrient Removal at the Calgary Bonnybrook Wastewater Treatment Plant. *Proceedings of the 67th Annual Water Environment Federation Technical Exposition and Conference*, Chicago, Illinois, Oct 15–19; Water Environment Federation: Alexandria, Virginia.

Fuhs, G. W.; Chen, M. (1975) Microbiological Basis of Phosphate Removal in the Activated Sludge Process for the Treatment of Wastewater. *Microbiol. Ecol.*, **2** (2), 119–138.

Gujer, W.; Zehnder, A. J. B. (1983) Conversion Process in Anaerobic Digestion. *Water Sci. Technol.*, **15** (8/9), 127–167.

Lamb, J. C. (1994) Wastewater Treatment with Enhanced Biological Phosphorus Removal and Related Purification Processes. U.S. Patent 5,288,405, February 22.

Natvik, O.; Dawson, B.; Emrick, J.; Murphy, S. (2003) BNR "Then" vs. 'Now," A Case Study-Kalispell Advanced Wastewater Treatment Plant. *Proceedings of the 76th Annual Water Environment Federation Technical Exhibition and Conference* [CD-ROM], Los Angeles, California, Oct 11–15; Water Environment Federation: Alexandria, Virginia.

Oldham, W. K.; Abraham, K. (1994) Overview of Full-Scale Fermenter Performance, Use of Fermentation to Enhance Biological Nutrient Removal. Preconference Seminar on Use of Fermentation to Enhance Biological Nutrient

Removal of the 67th Annual Water Environment Federation Technical Exposition and Conference, Chicago, Illinois, Oct 15–19; Water Environment Federation: Alexandria, Virginia.

Oldham, W. K.; Stevens, G. M. (1984) Initial Operating Experiences of a Nutrient Removal Process (Modified Bardenpho) at Kelowna, British Columbia. *Can. J. Civ. Eng.*, **11**, 474–479.

Osborn, D. W.; Lotter, L. H.; Pitman, A. R.; Nicholls, H. A. (1986) Enhancement of Biological Phosphate Removal by Altering Process Feed Composition. WRC Report No 137/1/86; Water Research Commission: Gezina, South Africa.

Rabinowitz, B. (1994) Criteria for Effective Primary Sludge Fermenter Design. Conference Seminar on Use of Fermentation to Enhance Biological Nutrient Removal of the 67th Annual Water Environment Federation Technical Exposition and Conference, Chicago, Illinois, Oct 15–19; Water Environment Federation: Alexandria, Virginia.

Rabinowitz, B.; Abraham, K. (2002) Primary Sludge Fermentation. Preconference Seminar on Carbon Augmentation for BNR, Water Environment Federation Annual Conference and Exposition of the 75th Annual Water Environment Federation Technical Exhibition and Conference, Chicago, Illinois, Sep 28–Oct 2; Water Environment Federation: Alexandria, Virginia.

Rabinowitz, B.; Oldham, W. K. (1985) The Use of Primary Sludge Fermentation in the Enhanced Biological Phosphorus Removal Process. *Proceedings of the ASCE/CSCE International Conference on New Directions and Research in Waste Treatment and Residuals Management,* Vancouver, British Columbia, June 23–28; American Society of Civil Engineers: Reston, Virginia; Canadian Society for Civil Engineers: Montreal, Quebec.

Rabinowitz, B.; Koch, F. A.; Vassos, T. D.; Oldham, W. K. (1987) A Novel Design of a Primary Sludge Fermenter for Use with the Enhanced Biological Phosphorus Removal Process. *Proceedings of the IAWQ Specialized Conference on Biological Phosphate Removal from Wastewaters,* Rome, Italy, October; International Association on Water Quality: London.

Stroud, R.; Martin, C. (2001) South Cary Water Reclamation Facility's Nutrient Removal Modifications and Reduction Success, Town of Cary, North Carolina. *Proceedings of the 74th Annual Water Environment Federation Technical Exposition and Conference* [CD-ROM], Atlanta, Georgia, Oct 13–17; Water Environment Federation: Alexandria, Virginia.

van Haandel, A. C.; Lettinga, G. (1994) *Anaerobic Sewage Treatment: A Practical Guide for Regions with a Hot Climate.* Wiley & Sons: Chichester, United Kingdom.

Water Pollution Control Federation (1980) *Sludge Thickening.* Manual of Practice No. FD-1; Water Pollution Control Federation: Washington, D.C.

Chapter 10

Solids Handling and Processing

Introduction	350	Sidestream Management Alternatives	365
Issues and Concerns	351	Recycle Equalization and Semitreatment	366
Influent Load Variations	351	*Equalization*	366
Influent Amenability to Biological Nutrient Removal	351	*Solids Removal*	366
Mean Cell Residence Time	352	*Aeration*	366
Struvite Formation	352	*Operational Issues*	367
Sludge Production	353	Sidestream Treatment	367
Nutrient Release	353	*Nitrogen Removal*	367
Release Mechanisms	353	*Stand-Alone Sidestream Treatment and Full-Centrate Nitrification*	368
Sources of Secondary Release	354		
Primary Clarification	354	*Operational Issues with Separate Recycle Nitrification Process*	370
Final Clarification	354		
Thickening	354	Influent Solids	370
Stabilization	355	Struvite	371
Dewatering	355	Alkalinity Feed	371
Estimating Recycle Loads	357	Aeration Efficiency	372
Eliminating or Minimizing Recycle Loads	360	Reactor Configuration	372
Sidestream Management and Treatment	362		

Foaming	372	Struvite Control Alternatives	387
Separate Recycle Treatment—Nitrogen Removal (Nitrification and Denitrification)	372	*Phosphate Precipitating Agents*	388
		Dilution Water	389
		Cleaning Loops	389
Single Reactor System for High Activity Ammonium Removal over Nitrite	372	*Hydroblasting*	389
		Equipment and Pipe Lining Selection	389
ANAMMOX	373	*Magnetic and Ultrasonic Treatment*	389
Ammonia Stripping	374		
Combination Sidestream Treatment and Biological Nutrient Removal Process—Return Activated Sludge Reaeration	374	*Lagoon Flushing*	391
		Controlled Struvite Crystallization (Phosphorus Recovery)	391
		Facility Design	391
Formation of Struvite and Other Precipitates	376	*Process Design*	392
Struvite Chemistry	377	Case Studies	395
Biological Nutrient Removal and Struvite	380	Conclusion	395
		References	396
Areas Most Susceptible to Struvite Formation	382		

INTRODUCTION

The sludge wasted from an enhanced biological phosphorus removal (EBPR) process contains approximately 4 to 10% phosphorus on a dry-weight basis. Of this, approximately 2% (dry-weight basis) represents the metabolic requirements, which is organically bound in all microbes. The remaining is a result of enhanced phosphorus uptake by phosphorus-accumulating organisms (PAOs), which is stored as polyphosphate volutin granules, an unstable inorganic compound containing magnesium. The nitrogen content of the sludge is approximately 8 to 12% (dry-weight basis). Consequently, biological nutrient removal (BNR) sludges will need to be handled and processed with caution to ensure that the resulting recycle streams do not overload the mainstream process, causing regulatory noncompliance with respect to effluent nutrient levels. This chapter outlines the key issues related to recycle streams, including sources, estimation, and management of return loads.

ISSUES AND CONCERNS

The key issues and concerns associated with handling and processing of BNR sludges are briefly described in the following sections.

INFLUENT LOAD VARIATIONS. The return streams containing the released nitrogen and phosphorus are typically blended with the plant influent and recycled through the bioreactor. These sidestreams are not always continuous and, in many facilities, occur intermittently. Biologically mediated processes-in particular, the BNR process reactions-are extremely sensitive to influent load variations. While the average recycle loading may not be significant, the short-term peak loads imposed by return streams can overwhelm the BNR system. For example, if dewatering operations occur over one shift, five days per week, the recycle loading could potentially be four times the loading generated by a 24 hour-per-day, 7 day-per-week operation. The complex microbial consortium has a limited ability to quickly respond to influent variations by self-adjusting itself. The period of acclimation is directly influenced by mean cell residence time (MCRT), mixed liquor suspended solids (MLSS), and the magnitude and duration of peak loads. Within limits, higher MCRT and MLSS enhance microbial diversity and system robustness, while extremely high and persistent loadings can be catastrophic.

INFLUENT AMENABILITY TO BIOLOGICAL NUTRIENT REMOVAL. Commonly used measures of influent amenability to BNR are based on the amount of rabidly biodegradable substrate available to sustain EBPR and denitrification (nitrogen removal). As a first approximation, minimum biochemical oxygen demand-to-total Kjeldahl nitrogen (BOD:TKN) and BOD-to-total phosphorus (BOD:TP) ratios of 3:1 and 20:1, respectively, may be used to assess the site-specific availability of an adequate carbon source for BNR. This determination should be made on the influent to the BNR bioreactor and should include all major recycle loads. Municipal wastewater of primarily domestic origin typically contains sufficient carbon substrate for

BNR, if excess BOD removal does not occur in the primary clarifiers. Return streams, which are characteristically low in BOD and relatively high in nitrogen and phosphorus, can depress the BOD:TP and BOD:TKN ratios in the bioreactor influent. In addition, if solids capture is not optimized in the EBPR sludge thickening and dewatering operations, significant phosphorus-rich solids would be returned to the primary clarifier and mixed with substrate-rich primary sludge. This can potentially result in secondary phosphorus release, if anaerobic conditions prevail in the sludge blanket. This released phosphorus will lower the bioreactor influent BOD:TP ratio. It should be noted that the minimum ratios indicated above are a first approximation of the nutrient removal capability. Even if the ratios reveal that adequate rapidly biodegradable substrate is available, other influent characteristics and operating factors could compromise the ability of the system to achieve reliable BNR.

MEAN CELL RESIDENCE TIME. Mean cell residence time determines the solids inventory that must be maintained in the system for reliable system performance. Typically, BOD and phosphorus removal can be accomplished at relatively low MCRT (2 to 4 days) under most wastewater temperature conditions. Generally, longer MCRTs are required for nitrification, and they are strongly linked to wastewater temperature. In the winter, microbial activity is lower as a result of colder temperatures. Hence, a higher solids inventory (higher MCRT) is required to maintain the same level of nitrification. If solids capture is not optimized during sludge thickening and dewatering operations, the resulting recycle streams will contain elevated solids levels, which could cause nitrification inhibition as a result of a lower active fraction in the bioreactor.

STRUVITE FORMATION. When EBPR sludge is anaerobically digested, struvite ($MgNH_4PO_4$) formation can occur because all of the required ingredients (namely ammonia, phosphate, magnesium, and pH increase) are encountered in the digester. Ammonia is a byproduct of anaerobic stabilization, while magnesium is released when the internally store volutin granules are degraded during the phosphorus release mechanism. Increase in pH is caused by a release of carbon dioxide (CO_2) from solution as a result of turbulence in pumps, centrifuges, and pipe bends. An in-depth discussion on struvite formation is presented in the Formation of Struvite and Other Precipitates section. While struvite is the most important chemical precipitate formed in the digester environment, other compounds, such as brushite ($CaHPO_4 \cdot 2H_2O$) and vivanite [$Fe_2(PO_4)_3 \cdot 8H_2O$], may also form if favorable conditions are encountered. The formation of these compounds as a result of operating conditions represents another phosphorus removal mechanism (inherent chemical precipitation).

SLUDGE PRODUCTION

Sludge production estimate in biological processes is based on the actual growth rate adjusted for the decay rate. Process MCRT also plays a key role in waste sludge production. According to Randall et al. (1992), conversion to BNR has an effect on solids generation because of the following:

- Differences in PAO decay rates under aerobic, anoxic, and anaerobic conditions;
- Differences in decay rates between PAOs and nonPAOs; and
- Differences in yield coefficients under anoxic and aerobic conditions.

Inclusion of nitrogen removal has shown a 5 to 15% reduction in waste sludge production (Sen et al., 1990; Waltrip, 1991). On the other hand, implementation of EBPR may increase sludge production at the same MCRT because of the lower decay rate of PAOs. This has been attributed to bacterial predators preferring organisms that do not contain stored phosphorus granules (Wentzel et al., 1989).

NUTRIENT RELEASE

RELEASE MECHANISMS. Two mechanisms are implicated in the release of stored phosphorus in EBPR systems (Chaparro and Noguera, 2002). Primary release, which is always accompanied by carbon uptake and storage, occurs in the anaerobic environment. Primary release is a prerequisite for biological phosphorus removal and is promoted in the BNR treatment scheme. Secondary phosphorus release is not associated with substrate uptake and can occur in the anaerobic zone, when volatile fatty acids (VFAs) are depleted; in the anoxic zone, when nitrates are depleted, converting it to an anaerobic zone; or in the aerobic zone, as a result of cell lysis (Barnard and Fothergill, 1998; Stephens and Stensel, 1998). These conditions are encountered when the three zones are oversized or when the bioreactor is underloaded (e.g., nights and weekends). Another secondary release mechanism is cell lysis, which results in the release of both stored and organically bound (metabolic) phosphorus.

Because there is no concomitant energy storage during secondary release, subsequent aerobic uptake of the released phosphorus may not be possible, and elevated effluent phosphorus levels could result. In the case of nitrogen, the main release mechanism is cell lysis.

SOURCES OF SECONDARY RELEASE. A number of traditional solids management strategies can increase the nutrient loading on the wet-stream nutrient removal processes. The following sections present a description of these solids management strategies and the potential effect on the nutrient loading.

Primary Clarification. When primary and BNR sludges are cosettled in the primary clarifier, phosphorus-rich waste activated sludge (WAS) is brought into contact with substrate-rich primary sludge in an anaerobic environment within the sludge blanket. Under these operating conditions, stored phosphorus is released into the bulk liquid, thereby increasing the phosphorus load to the bioreactor. Chaparro and Noguera (2002) reported that a primary sludge-to-WAS ratio of 1:1 (by volume) provided the optimum balance between carbon substrate (VFA) and the polyphosphate content of WAS, resulting in the highest phosphate release.

Final Clarification. Similar to primary clarifiers, anaerobic conditions could manifest in final clarifiers operated with deep sludge blankets, leading to phosphorus release. Some facilities do not waste sludge over the weekend and holidays, thereby allowing the sludge blanket to increase in the final clarifiers. The amount released is influenced by the phosphorus content of the sludge, depth of sludge blanket, dissolved oxygen (DO), nitrate levels within the sludge blanket, and wastewater temperature.

Thickening. Gravity thickening of EBPR waste sludge is likely to trigger phosphorus release as a result of anaerobic conditions within the sludge blanket. The rate and extent of phosphorus release is increased if co-thickening of primary and waste activated sludge is practiced. Pitman (1999) reported that comixing of primary and WAS in a gravity thickener (GT) caused the phosphorus level in the thickener to exceed 100 mg/L. A survey conducted by Pitman et al. (1991) indicated that WAS thickening using dissolved air flotation (DAF) produced liquors with 0.2 to 10 mg/L orthophosphorus, while the use of gravity belt thickeners (GBTs) resulted in 10 to 20 mg/L orthophosphorus in the filtrate. The lower observed phosphorus release during DAF thickening may be attributed to aerobic conditions.

Stabilization. One of the objectives of sludge stabilization is volatile solids destruction. Some of the unit processes that achieve this objective (such as digestion) also solubilize organic nutrients, which are recycled to the mainstream BNR process through decanting and dewatering operations. Pitman et al. (1991) reported that anaerobic digestion of BNR sludges can release up to 130 mg/L of phosphorus and 1000 mg/L of nitrogen. In another study, approximately 60% of the phosphorus removed in the EBPR process was released during anaerobic digestion (Murakami et al., 1987). It should be noted that the actual recycle load will depend on how much of the solubilized phosphate and ammonia are chemically precipitated as struvite, brushite, and vivanite in the anaerobic digester environment. Table 10.1 provides information on recycle streams generated by commonly used sludge stabilization processes (Jeyanayagam and Husband, 2002).

Conversion of existing mesophilic (approximately 35°C [95°F]) anaerobic digestion to thermophilic (approximately 55°C [131°F]) operation to maximize existing digester capacity or potentially achieve Class A biosolids should also consider the increase in the recycle loads caused by the increased volatile solids destruction.

Dewatering. Depending on the sludge stabilization process used, dewatering can either precede (e.g., thermal drying) or follow (e.g., digestion) the stabilization process. With respect to recycle loads, the latter would be the most critical. Dewatering operation itself does not cause significant nutrient release, but generates recycle streams containing nutrients that are released in upstream processes. The following are some of the key effects of the recycle stream from dewatering operations:

- As discussed previously, most often, dewatering operations are not continuous. While the average 24-hour recycle loading may not be significant, the intermittent loading could overwhelm the main BNR process.

- Poor solids capture during thickening and dewatering operations will increase the recycle of nutrient-laden solids. This can potentially release phosphorus when mixed with influent wastewater and settled in the primary clarifier.

- Poor solids capture can increase the solids loading to the biological treatment system if primary settling is not provided. These "junk" solids will take up bioreactor space and reduce the MCRT to values less than that required to meet process goals, such as nitrification.

- Some polymers used in sludge dewatering have been found to inhibit nitrification when recycled to the bioreactor.

TABLE 10.1 Recycle loads from BNR sludge stabilization processes and their effects (Jeyanayagam and Husband, 2002).

Stabilization process	Constituents recycled	Potential effects on BNR process	Potential effects on Effluent
Anaerobic digestion	TP	Lower BOD:TP Substrate limited	Increased TP
ATAD	TKN	Lower BOD:TKN Substrate limited Reduced denitrification Reduced alkalinity and aeration credits Increased aeration requirements resulting from increased TKN load Reduced nitrification resulting from inadequate aerobic volume	Increased ammonia-nitrate-nitrogen, and TN
Aerobic digestion	TP	Lower BOD:TP Substrate limited	Increased TP
	Nitrate	Reduced denitrification resulting from increased nitrate load to BNR process Reduced alkalinity and aeration credits Reduced bioreactor anaerobic volume resulting from recycled nitrate Reduced VFA and phosphorus uptake	Increased nitrate-nitrogen and TN Increased TP

ESTIMATING RECYCLE LOADS

Return streams from BNR sludge operations typically contain BOD, TSS, nitrogen, and phosphorus. The characteristics of these recycle streams cannot be generalized because they exhibit wide variability and are strongly influenced by many site-specific factors, including the following:

- Plant influent characteristics,
- Mainstream unit processes used and their performance,
- Solids stream unit processes used and their performance,
- Nitrogen and phosphorus content of solids,
- Operating schedule, and
- Extent of other removal mechanisms (struvite formation).

The recycle BOD and TSS loads resulting from most conventional sludge processing operations are modest and do not generally pose a challenge. Table 10.2 provides typical concentration ranges for BOD and TSS in BNR sludge processing sidestreams (U.S. EPA, 1987).

Recycle nitrogen loads are best characterized by measuring the flow and analyzing the parameters of interest. However, measuring existing recycle phosphorus levels before incorporating EBPR will not reflect the additional phosphorus that will be in the recycle stream. The following guidelines may be used in estimating recycle nitrogen and phosphorus loads:

- A mass balance based on actual operating data should be used to predict recycle loads. Care should be taken to ensure that the true operating conditions are reflected (i.e., 3, 5, or 7 days per week).
- Mass balances should be developed for average conditions and adjusted to assess the effect of operating schedules and peak conditions. Both present and future scenarios should be examined. It should be noted that a mass balance provides a "snapshot" of operating conditions and cannot be used to examine the dynamic behavior of the system.
- Nutrient release during digestion is related to volatile solids reduction. Nitrogen release is approximately 8 to 10% of volatile solids destruction.
- Depending on the phosphorus content of the solids, the phosphorus release from EPBR sludges is approximately 4 to 8% of volatile suspended solids

TABLE 10.2 Biochemical oxygen demand and TSS levels in sludge processing sidestreams (adapted from U.S. EPA, 1987).

Source	Recycle stream	BOD_5, mg/L	TSS, mg/L
Thickening			
Gravity thickening	Supernatant	100 to 1200	200 to 2500
DAF	Subnatant	50 to 1200	100 to 2500
Centrifuge	Centrate	170 to 3000	500 to 3000
Stabilization			
Aerobic digestion	Decant	100 to 2000	100 to 10 000
Anaerobic digestion	Supernatant	100 to 2000	100 to 10 000
Composting (static pile)	Leachate	2000	500
Wet air oxidation	Decant liquor	3000 to 15 000	100 to 10 000
Incineration	Scrubber water	30 to 80	600 to 10 000
Dewatering			
Belt filter press	Filtrate	50 to 500	100 to 2000
Centrifuge	Centrate	100 to 2000	200 to 20 000
Sludge drying beds	Underdrain	20 to 500	20 to 500

(VSS) reduction. During anaerobic digestion, if the conditions are right, some of the released ammonia and phosphate will be chemically precipitated as phosphate complexes (e.g., struvite).

- The average soluble five-day biochemical oxygen demand (BOD_5) concentration in digestion processes is in the range of 50 to 100 mg/L (Daigger, 1998).
- The mass of nitrogen and phosphorus in the recycle streams is distributed between the liquid and solids components. In the absence of site-specific

recycle stream characteristics, the mass balance presented in Figure 10.1 (adapted from Daigger, 1998) may be used. For example, the phosphorus recycle load (Ptotal, recycle) may be estimated as follows:

$$P_{total,\ recycle} = (P_{sol,\ recycle}) + (P_{part,\ recycle})$$
$$(P_{sol,\ recycle}) = (Q_{recycle}/Q_{in}) \times (P_{sol,\ in})$$
$$(P_{part,\ recycle}) = (100 - S_{capture}) \times (P_{part,\ in})$$
$$(P_{part,\ in}) = (X_{in} \times P_{content})$$

Where

$P_{sol,\ recycle}$ and $P_{part,\ recycle}$ = soluble and particulate fraction of the recycle phosphorus (kg/d);

$Q_{recycle}$ and Q_{in} = influent and recycle flows (mgd) (mgd $\times$ 3785 = m^3/d);

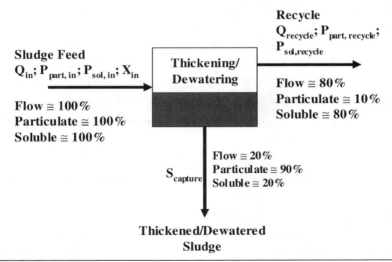

FIGURE 10.1 Partitioning of flow and particulate and soluble fractions thickening and dewatering (adapted from Daigger, 1998).

$P_{sol, in}$ and $P_{part, in}$ = soluble and particulate fraction of the influent phosphorus (kg/d);
X_{in} = influent solids load (kg/d);
$S_{capture}$ = solids capture (%); and
$P_{content}$ = phosphorus content of solids (%).

ELIMINATING OR MINIMIZING RECYCLE LOADS

By implementing appropriate solids handling and processing practices, recycle loads can be eliminated or minimized at the source. Ideally, primary sludge and EBPR waste sludge should be handled and processed separately, as shown in Figures 10.2 and 10.3 (Jeyanayagam and Husband, 2002). However, this may not always be practical. The following is a list of design and operational approaches that are likely to eliminate or minimizing recycle loads:

- Blending the primary and WAS sludges should be moved as far downstream as possible in the sludge treatment train. Following blending, the sludge should be processed quickly before significant release could occur.

- Maintaining a shallow sludge blanket in primary clarifiers may prevent nitrogen and phosphorus solubilization. It will also improve clarifier performance, particularly during high wet-weather-induced flows. However, this will result in a relatively thin underflow (<1.5%). Sen et al. (1990) found that, at the Bowie Wastewater Treatment Plant (WWTP) in Maryland, maintaining the sludge blanket at least (1 m [3.3 ft]) below the overflow weir allowed only a fraction of the released phosphorus to escape with the supernatant.

- Secondary release in final clarifiers is a common problem at many BNR facilities, particularly in the warmer months, when septic conditions are readily established in the sludge blanket. This may be avoided by implementing an effective wasting strategy and maintaining a shallow blanket. Doing so will eliminate anoxic conditions and the potential for floating sludge and subsequent sludge blanket washout resulting from denitrification. Some facilities maintain a high DO concentration in the bioreactor effluent to minimize the potential for denitrification within the sludge blanket. However, maintaining a high DO at the end of the bioreactor may be not always viable, because the internal recycle necessary for denitrification in the bioreactor will recycle

Solids Handling and Processing 361

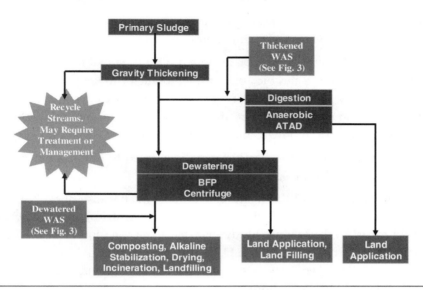

FIGURE 10.2 Primary sludge management options (Jeyanayagam and Husband, 2002).

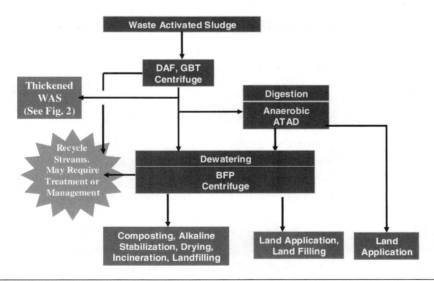

FIGURE 10.3 Waste activated sludge management options (Jeyanayagam and Husband, 2002).

oxygen to the anoxic zone and reduce denitrification. Also, this practice does not represent an energy efficient operation.

- The BNR bioreactor should be designed with the flexibility to waste sludge from the end of the aerobic zone of the bioreactor to keep the sludge "fresh" and minimize the likelihood of releasing phosphorus downstream. In addition, the biomass wasted from the aerobic zone contains the highest phosphorus content (Rabinowitz and Barnard, 2002).
- Sludge wasted from the clarifier may be aerated to inhibit and delay phosphorus release during subsequent processing.
- Gravity thickening of EBPR sludge will most likely cause phosphorus release and should be avoided. Gravity belt thickening, centrifugation, or DAF can be used for WAS thickening. Sludge wasted from the bioreactor may be more economically thickened with a DAF unit because of the dilute solids concentration. If co-thickening is practiced, it should be completed immediately after blending to avoid phosphorus release.
- The type of sludge treatment process used will determine recycle stream characteristics. Composting, thermal drying, and advanced alkaline stabilization produce minimal recycle loads. Anaerobic digestion and autothermal thermophilic aerobic digestion (ATAD) will release higher nutrient concentrations.
- Both thickening and dewatering should be optimized with proper polymer dose to maximize solids capture.

SIDESTREAM MANAGEMENT AND TREATMENT

Plant solids process recycle streams (sidestreams) have been recognized as a source of operational problems and as having the potential to negatively influence effluent quality. Typically, plant operators were concern with the suspended solids recycle loadings and odorous compounds. At plants requiring nitrification or removal of nutrients (nitrogen and phosphorus), these sidestream liquors can represent a significant nutrient loading on the liquid stream processes. The quality of the recycle loads described in the Estimating Recycle Loads section demonstrates the high nutrient concentration of these sidestreams. The highest nutrient recycle loading is generated from the dewatering of anaerobically digested sludge or thermally conditioned sludge, where nutrients are released from microorganisms and a byproduct of volatile solids destruction. The ammonia and soluble phosphorus concentration in a typical mesophilic digester will be approximately 800 and 1200 mg/L, depending on the influent feed concentration of VSS and VSS destruction within the digester. Nutrient concentrations in the dewatering liquid can increase by as much as 50 to 75%, if thermophilic digestion is used. Aerobic digesters can generate high oxidized nitrogen (nitrite and nitrate) depending on their operating scheme. For the purpose of this section, the discussion will be related to digested sludge thickening or digested sludge dewatering recycle streams.

Depending on the sludge dewatering equipment, the recycle stream concentration may be more dilute, but the recycle load not any different. One example is a belt filter press (BFP), where a significant portion of the dewatering flow is spray water. While the nutrient recycle load remains the same, the filtrate concentration is much lower than dewatering the sludge with a centrifuge.

Concentrated nutrient-laden sidestream can increase nutrient loadings by 15 to 50% to the liquid stream units, depending on the sludge stabilization process and whether the facility manages outside sludges. The effect of these higher loadings, even on an equalized basis, will present difficulties, such as inadequate alkalinity for nitrification and inadequate readily biodegradable matter (RBOM) to support denitrification and biological phosphorus removal. Facilities must increase the size of their main stream process, separate treatment, or provide a combination of sidestream treatment and main plant treatment.

An important consideration in any design is to understand when these highly concentrated nutrient rich sidestreams are generated. As shown in Figure 10.4, the net increase in ammonia loading to a BNR system is only 3 mg/L higher (15% of the entire load) when equalized over 24 hours, 7 days per week. However, when sludge dewatering system is operated only 5 days per week for 8 hours per day, the resulting influent concentration to the BNR system is increased from a 3 mg/L to approximately 13 mg/L during the middle of the day. This is a tremendous increase loading on the aerobic and anoxic processes. These recycle streams can cause the following:

- Increased oxygen demand. The aeration equipment may not be adequate to manage the higher oxygen demand. This can cause low DO concentrations, which may result in the growth of poorly settling microorganisms (filamentous); breakthrough of ammonia through the aerobic reactor (nitrification); and, in the case of dual nutrient removal process, secondary release of biologically removed phosphorus.

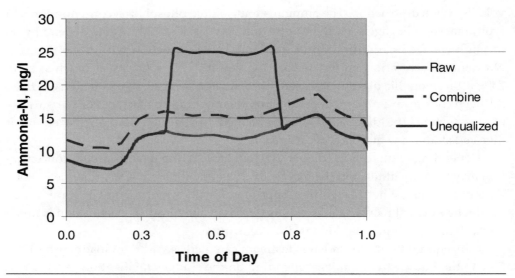

FIGURE 10.4　Influent loading to a BNR system.

- Higher energy costs. With adequate aeration capacity, the resulting oxygen demand peaks during the portion of the day when power utilities will charge their highest energy rates.
- Imbalance of RBOM. As noted earlier, optimal performance of nutrient removal requires a certain ratio of RBOM and nutrient-whether for biological phosphorus removal or denitrification. These spike loadings of nutrients can result in a poor RBOM-to-nutrient ratio and result in an overall higher daily nutrient discharge effluent quality.
- pH and alkalinity. The alkalinity available in anaerobically digested sludge recycle streams is approximately 50% of the amount needed for complete nitrification. Complete nitrification will require 7.1 mg/L (as calcium carbonate [$CaCO_3$]) alkalinity, whereas the typical digested sludge will only have approximately 50 to 60% of the alkalinity required to fully nitrify the ammonia in the centrate. While, on average, there can be adequate alkalinity in the raw

wastewater to allow the mixture of recycle streams and raw wastewater to be completely nitrified, spike loading of recycle streams can cause depressed pH in the nitrification process.

SIDESTREAM MANAGEMENT ALTERNATIVES. Recognizing that sidestream processes can dramatically affect the influent nutrient loading to the nutrient removal processes, plant staff must carefully weigh methods to reduce their effect on other plant processes that will negatively affect effluent quality. The first step that should be taken is to characterize the recycle streams from the various sludge management units. Quantification of the recycle loads will help identify those streams that can have the most effect on the liquid stream process units and need to be controlled and treated. Plant staff should conduct in-plant stream sampling and analysis to define the flow and concentration of the various solids handling system recycle streams. These criteria should include the following:

- Flow and hours of operation;
- Total and soluble COD;
- Total and soluble carbonaceous BOD_5;
- Total and volatile suspended solids;
- Total solids and total volatile solids;
- pH;
- Alkalinity;
- Nitrogen series (TKN, ammonia, nitrite, and nitrate); and
- Phosphorus (TP, orthophosphorus).

As noted earlier, the thickening or dewatering of anaerobically digested sludge will have the greatest effect on the nutrient removal processes. While operating these units for 24 hours per day is the natural first response by a design engineer to resolve spike loadings during the day, it is likely not the most cost-effective operating scheme for many small- and medium-sized facilities. Many of these facilities do not require operating sludge dewatering equipment 24 hours per day. Also, staffing these units during off hours represents a financial burden, and neighbors do not favor truck traffic 24 hours per day, which forces on-site storage of dewatered cake, creating additional issues. Other more cost-effective steps are available. This discussion highlights the importance of considering the operation of the sludge dewatering facilities during process design.

RECYCLE EQUALIZATION AND SEMITREATMENT. *Equalization.* Equalization of concentrated sidestreams is the first alternative facilities consider in lieu of modifying solids management operations. The equalization facility should be sized to provide the plant staff the ability to control the amount and time period of discharge of these sidestreams to the main plant liquid stream. The equalization tank can be sized to contain the expected total volume of recycle flows (including miscellaneous water, such as belt press spray water, which can be significant volume) to maximize the flexibility of when sidestreams are recycled to the plant. Some facilities have found it advantageous to recycle at night, when the lowest plant flows occur. This will require that drains from the concentrated sidestreams be isolated and directed to the equalization tank.

Solids Removal. Because there can be significant suspended solids and residual polymer in the sidestream liquors, the solids settling rate can be very high. In addition to solids, a large amount of hair and other fibers may be present that can foul and clog equipment downstream. Also, during the startup and wash down of dewatering equipment, large amounts of sludge can be discharged. Field measurements of centrifuge dewatering recycle streams (centrate) at three facilities have indicated that approximately 95% of the solids settle to the bottom of a 1-L cylinder in 10 minutes. To take advantage of this high settling rate and reduce the solids recycle loads back to the wet stream processes, a small settling tank before the equalization tank or provisions in the equalization tank to allow solids to settle and be removed (to sludge thickening devices or sludge holding tanks) should be considered. As noted earlier, plant staff should understand the characteristics of the recycle streams and determine the overall solids loading being imposed on the liquid stream units. If a separate solids settling tank is not deemed viable, then sending the equalized recycle stream to primary settling tanks should be considered.

Aeration. Another partial treatment scheme in combination or separate from equalization tank is aerating the sidestream. Operation at facilities have shown that aerating the highly concentrated and warm dewatering liquors from anaerobically digested sludge can reduce the ammonia loading by approximately 50%, as a result of a combination of nitrification and some offgassing of ammonia. As noted earlier, the characteristics of digested sludge recycle stream has approximately 50% of the alkalinity required for nitrification. This becomes a limiting factor on the degree of ammonia removal. The following is required to achieve 50% ammonia removal:

- Detention time. Detention time must be sufficient for nitrifiers to reproduce. If the liquor temperature is in the 30°C range, detention times of 2 to 4 days are sufficient. The solids retention time (SRT) will be equal to the hydraulic retention time (HRT).
- Nitrifiers. There will be a need to seed the system with some nitrifiers.
- Complete mix. Plug-flow systems will not maintain the nitrifier population in the system.
- Oxygen. Sufficient DO must be provided to allow the ammonia oxidizers to work.

Operational Issues. Because of the high concentration of ammonia and sulfide, recycle stream can be odorous. Depending on the particular locality of a facility and characteristics of the recycle stream, unacceptable odors may occur. Another issue that will be of concern when aerating digested sludge is the potential to form struvite. The reduction in carbon dioxide in digested sludge recycle liquors can cause struvite to form. See the Formation of Struvite and Other Precipitates section for more information.

Bench-scale testing can be performed in the plant laboratory by aerating sludge dewatering liquor and batch feeding the system to develop some idea of the potential nitrification rates.

SIDESTREAM TREATMENT. Some facilities have taken proactive steps to perform some or complete treatment of their highly concentrated nutrient streams. This section has been divided between those treatment processes that manage nitrogen and those that achieve phosphorus removal.

Nitrogen Removal. Highly concentrated and warm liquors from the thickening or dewatering of anaerobically digested sludge can have the greatest effect on the wet stream treatment processes. For the purposes of this discussion, the sidestream quality will be 200 mg/L BOD_5, 500 to 1000 mg/L TSS, 800 mg/L ammonia-nitrogen, and 1000 mg/L TKN. There have been a number of processes that have proven successful on pilot-or full-scale facilities. Effluent quality will vary, depending on the operating strategy. The history of separate sidestream treatment for nitrogen treatment is a relatively new and ongoing process. Some of the combination sidestream treatment and BNR process will be easier to incorporate to some facilities than others and will expand by the time this book is published. However, the basic principles and theory of each described process will assist the reader in understanding the benefit and operating challenges for each process scheme.

Stand-Alone Sidestream Treatment and Full-Centrate Nitrification. As described earlier, aerating warm, concentrated sidestream liquors in a completely mixed tank can achieve significant nitrification under the right operating strategy. However, to achieve full nitrification, additional alkalinity will be required. Separate centrate treatment in an activated sludge system has been tested using sequential batch reactor and flow through a conventional activated sludge system. Depending on the detention time, SRT, and chemical feed system, partial or complete nitrification of the ammonia can be achieved. The principles of nitrification are the same as reported in Chapter 3; however, unlike most wastewaters, the sludge recycle stream can be much warmer than typical wastewater (28 to 35°C). This allows for a shorter SRT. Proposed design criteria for separate activated sludge nitrification are given in Table 10.3 (Mishalani and Husband, 2001).

Successful full-scale operations of a separate digested sludge dewatering liquor constructed in 1994 at the Roundhill facility in West Midlands, England, were

TABLE 10.3 Separate centrate nitrification design criteria (Mishalani and Husband, 2001).

Hydraulic retention time	1 to 2 days
Solids retention time	3 to 5 days
Clarifier overflow rate	400 to 800 gpd/sq ft[a]
Oxygen demand	1 lb oxygen/lb[b] BOD_5 and 4.6 lb oxygen/lb ammonia
Alkalinity demand	7.1 lb as $CaCO_3$/lb TKN influent minus 90% of the sludge dewatering alkalinity

[a]gpd/sq ft x 0.040 74 = $m^3/m^2 \cdot d$.
[b]lb/lb x 1000 = g/kg.

reported. This facility was designed to operate six months during the winter, to allow the main plant to achieve effluent limits. Wastewater temperatures in the main plant can be approximately 8°C. The sidestream treatment unit consists of a primary clarifier, aeration tank (50 hours), and final settling tank, operating at a SRT of 14 days or greater. A second facility designed for the Minworth facility also achieves greater than 95% ammonia oxidation, with only 15 hours of detention time and final settling tanks. The sidestream facility is only operated six months of the year, during the winter. During the warmer months, the main plant is able to achieve its effluent ammonia concentration. This reduces ammonia loading on the BNR facility and thus the aerobic detention time required in the main plant to achieve full nitrification. Waste sludge is discharged to the effluent and the main plant activated sludge facility.

A significant amount of oxidized nitrogen is present and must be managed. At the Minworth facility, anoxic zones were created in the main plant to form an anoxic selector to that improved MLSS settleability and reduced nitrate concentration. Another method to manage this high oxidized nitrogen stream is to discharge this stream to the plant headworks. Studies conducted in Phoenix, Arizona (Carrio et al., 2003), have shown that recycling this material upstream of primary settling tanks resulted in significant reduction of nitrate (approximately 5 mg/L as nitrate-nitrogen) and reduced aqueous sulfide by 2 mg/L. This provided additional benefit to the separate nitrogen oxidation system. Because the primary settling tanks were operating with "no sludge blankets," no problem with sludge floating in the primary settling tanks was observed. Another option is to discharge this highly concentrated nitrate to an anoxic zone in the main plant BNR.

Another benefit that may be obtained through separate nitrification is the use of the WAS from the centrate nitrification process as a seed source for nitrifiers to the main plant nitrification reactors. A patented process by M2T Technologies, Lotepro Environmental Systems & Services (Lotepro) (State College, Pennsylvania) called In-Nitri (Kos et al., 2000), uses the WAS from the separate nitrification reactor to supplement the main plant activated sludge system with nitrifiers-referred to as bioaugmentation. The claimed benefit of the In-Nitri process, illustrated in Figure 10.5, is that the nitrifier seed will allow facilities to operate at lower SRT (lower MLSS) and more quickly recover from the high flow or cold temperature that reduce the SRT in the main activated sludge facility. The viability of the nitrifiers grown in the warm recycle stream in colder wastewater (main plant BNR) has been questioned by some, but research conducted on laboratory-scale systems by the University of Manitoba

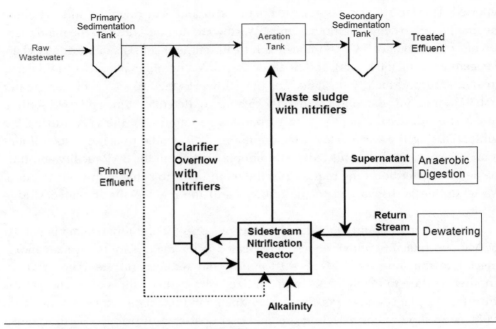

FIGURE 10.5 In-Nitri sidestream nitrification process.

(Winnepeg, Manitoba, Canada) indicated a slight decrease in nitrification versus theory when taking temperature correction into account. Increased nitrification was demonstrated in a pilot study conducted in the Southwest, when WAS from a separate centrate nitrification reactor was feed to a high-purity-oxygen activated sludge system.

Operational Issues with Separate Recycle Nitrification Process. *INFLUENT SOLIDS.* Optimal influent solids concentrations range from 200 to 1000 mg/L. When influent solids increase to higher than 1000 mg/L, the ability to operate at a SRT of 5 days becomes more difficult as clarifier solids loading rates increase. While loss of solids over the clarifier weir is not as great a concern with the sidestream treatment process because the effluent is being recycled to the treatment plant, excess solids can reduce the overall system SRT. This may lead to the loss of complete nitrification by washing out the nitrite oxidizers (nitrobacter). While oxidizing ammonia to nitrite has many advantages, there is concern that nitrite entering the main liquid stream may bleed through to the effluent chlorination system. Nitrite will increase chlorine demand and can cause effluent fecal coliform exceedances.

Studies have also shown that, with too few solids (such as adding a settling basin before the activated sludge process), nitrifiers will be dispersed with no solids to weigh them down. Accordingly, the effective SRT can be reduced to the HRT. Designers have suggest installing a presettling basin with the ability to bypass a portion of the solids recycle stream to maintain an influent solids concentration of 500 mg/L, to serve as a ballast for the nitrifiers.

STRUVITE. As noted earlier, without some positive control on the formation of struvite (when present), the aeration of the centrate can cause carbon dioxide to blow off and increase the potential to form struvite. With a complete-mix aeration tank, if the influent is rapidly mixed, thereby reducing the ammonia concentration, the potential to form struvite is reduced. Also the use of ferric chloride or other precipitant of phosphorus has been successful in reducing struvite formation. Plant staff can test the potential of struvite formation by aerating samples of the recycle stream in a bucket and examining the aeration stone for buildup of struvite.

ALKALINITY FEED. Depending on the aeration devices and conditions, caustic or sodium bicarbonate, or both, are suitable alkalinity feed sources. While caustic is often preferred because of the ease in handling and smaller storage requirement versus bicarbonate, it is associated with the following certain process disadvantages:

- Poor pH buffering. Overdosing is a concern. If overdosing does occur, the pH can increase to >9, where nitrifier will be destroyed.
- No inorganic carbon addition. Nitrifier requires inorganic carbon to grow. Sodium bicarbonate ($NaHCO_3$) will add inorganic carbon, where caustic (NaOH) adds none. When using caustic, the process will rely on carbon dioxide generated during BOD reduction and dissolution of carbon dioxide from the atmosphere.

The cost for alkalinity is the largest operational cost for separate recycle stream nitrification.

AERATION EFFICIENCY. Testing conducted on one demonstration facility nitrifying anaerobically digested sludge centrate recycle treatment process for nitrification indicated that the aeration transfer efficiency alpha factor ranged from 0.7 to 1.0, depending on the type of fine-bubble aeration device. Both ceramic and membrane fine-bubble aeration devices showed higher-than-normal alpha values in the centrate nitrification tank.

REACTOR CONFIGURATION. It is preferred that the tank be completely mixed versus a plug-flow reactor for a number of reasons, including the following:

- pH control. This avoids the need for numerous alkalinity feed points to maintain the pH in suitable range along the tank.
- Uniform aeration demand.
- Lower ammonia concentrations will reduce the potential for inhibiting ammonia oxidizers as a result of excess un-ionized ammonia. This will also reduce the potential to strip ammonia (odors).
- Reduced potential for nitrifier washout if the aeration tank detention time is greater than the nitrifier growth rate (minimum SRT).

FOAMING. During initial startup of these sidestream reactors, excess foaming can occur. Foaming can also occur during periods when high levels of polymer feed into the plant. This foam is typically very light and subjected to blowing because of wind. Controlling the foam by means of spray water has been successful. Provisions for wasting the foam should be provided.

Separate Recycle Treatment—Nitrogen Removal (Nitrification and Denitrification).

Options for nitrification and denitrification of recycle streams are similar to the processes described in Chapter 3. Of course, the volume, temperature, and wastewater constituents will result in nonconventional sizing of the tankage and chemical feed system. The recycle stream has a relatively low RBOM versus ammonia concentration; thus, supplemental RBOM must be added (such as methanol or acetic acid). Because of the high concentration of ammonia, there will be a need for 16 to 36 hours aerobic detention time. Stand-alone processes include those described in the following sections.

Single Reactor System for High-Activity Ammonium Removal Over Nitrite.

Single reactor system for high activity ammonium removal over nitrite (SHARON) is a patented process. This process achieves both ammonia oxidation and nitrogen removal-essentially total nitrogen (TN) removal. The goals for this process are to operate a single or multistage reactor (with high recycles) to achieve ammonia oxidation to nitrite and denitrification using an external source of RBOM. This process takes advantage of the high temperature of anaerobically digested sludge recycle stream.

By operating at a minimal SRT, the ammonia oxidizers (nitrosomonas) which have a shorter reproduction rate than the nitrite oxidizers (nitrobacter), will survive in the system, but the nitrite oxidizers will be washed out. This will reduce the amount of oxygen required, because oxidizing ammonia to nitrite versus nitrate saves 33% of the oxygen demand. Also, subsequent denitrification can be achieved with 40% less RBOM because nitrite is being reduced. The other benefit is that denitrification produces alkalinity. Some supplemental alkalinity addition is required; however, compared to a centrate nitrification process, the alkalinity addition requirement is approximately 80 to 90% less. Projected effluent quality from the SHARON process is approximately 100 mg/L of ammonia and oxidized nitrogen compounds. Unlike other separate centrate nitrification processes, no seeding (bioaugmentation) benefit is claimed by the patent holders for solids from the SHARON process discharged to the main plant BNR process.

ANAMMOX. Another novel process that is currently being tested is called the ANAMMOX (anaerobic ammonia oxidation) process-a novel patent biological process. Paques BV (Balk, Netherlands) holds the rights to marketing. This process consists of partial nitrification and subsequent conversion of ammonia in the presence of nitrite to nitrogen gas under anoxic conditions, with the nitrite as the electron acceptor (Mulder et al., 1995). This autotrophic process reportedly can save 40% oxygen and needs no organic carbon source for denitrification. This process is still being investigated, and full-scale tests are currently underway. In this process, 55 to 60% of the ammonium is oxidized to nitrite (partial nitrification), and the remaining ammonium is oxidized with nitrite in an anoxic reactor system (Siegrist et al., 2003).

This process requires careful control of the process to washout nitrite oxidizers so only nitrite is formed, and the ratio of nitrite to ammonia must be approximately 1 to 1.3. Other process requirements include using a sequential batch reactor with a HRT of 0.5 days and an SRT of 15 to 20 days. The principle advantage is that autotrophic denitrification of the nitrite eliminates the need for an external carbon source. This reportedly can save approximately 40% of the operating cost over other separate sidestream biological nitrogen removal process.

Ammonia Stripping. Ammonia stripping can be achieved by raising the pH of the liquor so that the aqueous ammonia is existing in equilibrium with its gaseous counterpart in accordance with Henry's law. When the pH exceeds 9.5, un-ionized ammonia prevails and can be stripped from the sidestream liquid. This requires a large amount of air to achieve high ammonia removal. Also, the gas stream should be captured to eliminate odor problems. This requires condensing the offgas and capturing the ammonia. Stripping via pH can cause scaling problems and is subject to freezing during cold weather.

A second option is to steam strip the ammonia from the recycle stream. Studies conducted by the New York City Department of Environmental Protection (Gopalakrishnan et al., 2000) indicated that steam stripping is more cost-effective than hot air stripping and can achieve from 70 to 90% removal. The pH does not have to be adjusted (from 7 to 7.5) in a steam stripper, thereby reducing scaling issues. Pilot-scale studies indicated that steam stripping of centrate (anaerobically digested sludge recycle liquor) could be a feasible alternative for long-term operations. Pretreatment included 2-mm screens and solids settling to prevent clogging of the reactor and heat exchanger resulting from the organic material, including hair, in the centrate stream. The stripped ammonia was sent to a distillation column, where the gas stream is cooled and a pure ammonia solution is formed. Clogging did occur in heat exchangers, again as a result of the organic material in the centrate and calcium and magnesium. Pretreatment requirements include fine screening and suspended solids settling.

COMBINATION SIDESTREAM TREATMENT AND BIOLOGICAL NUTRIENT REMOVAL PROCESS—RETURN ACTIVATED SLUDGE REAERATION.

Facilities faced with nitrification requirements have isolated the solid recycle stream and discharged these highly concentrated streams to sludge regeneration tanks and/or sludge reaeration zones of their facilities. The principle is based on directing the highly concentrated sidestream to the return activated sludge (RAS) reaeration zone. This zone has a relatively long detention time and high concentration of MLSS. As shown in Figure 10.6, the Prague WWTP (Praha, Czech Republic) discharges anaerobically digested sludge dewatering liquors to a separate RAS aeration tank (regeneration tank). The plant reportedly uses the first half of the regeneration tank to nitrify, and the second half of the regeneration tank to denitrify by mixing. The facility reports excellent effluent quality and improved MLSS settleability.

The New York City Department of Environmental Protection has also piloted and conducted a full-scale demonstration of separate RAS regeneration systems with

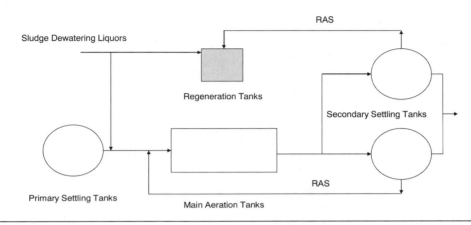

FIGURE 10.6 Flow scheme of Prague wastewater treatment plant.

alkalinity addition and some primary effluent to increase nitrification rates in their facilities. This process uses the nitrifiers grown in the main plant as the source of nitrifiers to oxidize the ammonia in the sludge recycle, by taking advantage of the longer detention time in the RAS operations (Figure 10.7).

A similar process is the bioaugmentation batch enhanced (BABE) process (Salemi et al., 2003). This process potentially reduces the concern about how nitrifiers grown in the separate reactor not being acclimated to the conditions in the main plant BNR process. This process relies on RAS from the main plant as the source of nitrifiers in the separate nitrification reactor. The concept is to use the same nitrifiers that were successfully nitrifying in the main plant to nitrify the recycle ammonia. Data, including fluorescence in situ hybridization (FISH) analysis, were used to identify the specific microorganisms performing nitrification and confirmed that the nitrifiers grown in this process were viable in the main plant BNR reactors.

All of these processes take advantage of the benefit of using nitrifying organisms from the main plant BNR process (RAS). Whether the process is incorporated to a separate RAS regeneration tank or in the RAS reaeration pass of a step-feed activated sludge process, the overall philosophy is the same: maximize nitrification of the side stream liquors using RAS in the main plant. Site-specific conditions will define whether supplemental alkalinity is required.

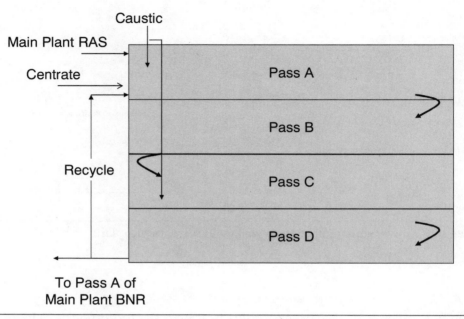

FIGURE 10.7 Separate centrate nitrification in four-pass system.

FORMATION OF STRUVITE AND OTHER PRECIPITATES

Struvite deposits are quite common and found at most municipal WWTP that operates anaerobic digesters (Figures 10.8 to 10.13). Fortunately, in most cases, the associated problems do not exceed some required regular maintenance. However, problems can range from clogged valves, frozen valves, and failing instrumentation, to the virtual destruction of major equipment.

For BNR plants, the potential for struvite formation is significantly higher than for conventional plants. That is, PAOs compensate the negative charge of accumulated phosphate with uptake of positively charge metal ions, preferably magnesium and potassium. The latter are released alongside phosphate during anaerobic digestion. This results in a significant digester magnesium loading increase.

FIGURE 10.8 Struvite deposit from a 0.3-m (12-in.) lagoon decant line, which broke loose and got caught in a check valve at Columbia Boulevard Wastewater Treatment Plant, Portland, Oregon.

Because of this relationship between EBPR and magnesium uptake, struvite control should be part of the operation strategy for BNR plants with anaerobic WAS stabilization.

STRUVITE CHEMISTRY. Struvite, also known as magnesium ammonium phosphate, is a crystal that forms in a liquid environment with concentrations of its constituents high enough to reach supersaturation. That is, when the product of ion concentrations exceeds the solubility product constant (Ksp).

$$Mg^{2+} + NH_4^+ + PO_4^{3-} + 6H_2O \rightarrow MgNH_4PO_4(H_2O)_6 \tag{10.1}$$

FIGURE 10.9 Struvite deposits found in 75-mm (3-in.) magnetic flowmeter on dewatering centrate line. The thicker the deposit on the flowmater, the bigger the error because the meter measures more flow than is actually there because of the reduced diameter.

As the concentration of the reactants (magnesium [Mg^{2+}], ammonium [NH_4^+], and phosphate anion [PO_4^{3-}]) further increases, the reaction proceeds faster, and precipitation accelerates. The reaction is reversible, and struvite can be dissolved when the liquid environment surrounding the crystals is below maximum solubility. A number of factors contribute to the level that struvite will form or potentially deposit on pipe and equipment surfaces. Figure 10.14 illustrates the relationship between the maximum solution product and pH. For conditions relevant to municipal wastewater, the solubility of struvite increases with decreasing pH, and vice versa.

FIGURE 10.10 Image of struvite deposit in 25-mm (1-in.) pipe to a streaming current meter, which controlled dewatering polymer feed. Inaccurate measurement by the streaming current meter, as a result of the deposits, can result in polymer overdosing or reduced dewatering performance.

The struvite crystal forms in two stages: nucleation and crystal growth. The nucleation step is the limiting step in the reaction, which is when the three reactants first form crystal embryos (nuclei).

Once a nuclei has formed, the crystal continues to grow, limited now only by the availability of substrate and crystal growth kinetics. The limiting nucleation step can be bypassed by introducing seeds. Suspended solids in sludge can serve as seeds, a fact that explains the observation that struvite deposits are more common in parts of the solids processing system that convey or hold process streams with low values of

FIGURE 10.11 Close-up of recovered struvite deposit (20× magnification).

suspended solids (i.e., dewatering filtrate or lagoon decant). Note that surface roughness, even on a microscale, also has a crystal seeding effect. Thus, in an environment free of suspended solids, surface roughness can represent a major factor regarding struvite deposits and its prevention.

BIOLOGICAL NUTRIENT REMOVAL AND STRUVITE. Enhanced biological phosphorus removal is established when bacteria are subjected to a series of anaerobic and aerobic environments while providing adequate substrates (VFA, etc.). Under aerobic conditions, these bacteria store phosphorus in the form of polyphosphates beyond their metabolic requirements, which results in a low effluent phosphorus concentration. Under anaerobic conditions, the bacteria use the energy in the polyphosphates to accumulate organic substrate, resulting in a release of phosphate

FIGURE 10.12 Photograph of recovered struvite crystals from anaerobic digester at Durham Advanced Wastewater Treatment Plant, Tigard, Oregon (200× magnified).

in the anaerobic zone. The biochemistry and kinetics of these reactions have been described by Wentzel et al. (1991).

During the uptake of phosphorus and formation of polyphosphates, the bacteria also accumulate the negative charge of phosphate during the formation of polyphosphate ($nH_2PO_4^-$). To maintain electroneutrality, positive metal ions, such as magnesium (Mg^{2+}) and potassium (K^+), are also accumulated. A correlation between magnesium and phosphate has been confirmed and, on average, the observed magnesium uptake ranges between 0.2 and 0.4 mg magnesium/mg phosphorus.

During anaerobic digestion, phosphorus stored by PAOs is released back into solution. Maintaining electroneutrality, this release requires the release of previously stored positive metal ions. This release is proven to be at the same molar ratio as the uptake. In addition, further phosphate and ammonia are released as a result of the

FIGURE 10.13 Struvite deposits in a 75-m (3-in.), rubber-lined, 90-deg elbow; the buildup occurred during a two-month material testing period.

breakdown of other organic material in the digestion process. The total release of stored phosphorus and cell phosphorus can range between 60 and 100%.

AREAS MOST SUSCEPTIBLE TO STRUVITE FORMATION. Commonly, struvite problems only become apparent after equipment failed or pipes clog, creating a situation where otherwise simple maintenance can become a costly emergency, potentially requiring process shutdowns and equipment replacement.

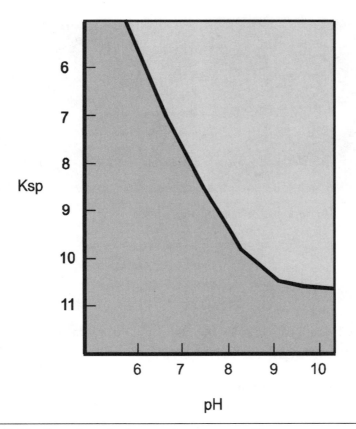

FIGURE 10.14 Struvite solution product versus pH.

Knowing and anticipating where struvite is likely to form becomes a valuable asset in avoiding severe operational problems associated with struvite deposits.

Struvite formation is very pH-sensitive, and, between the pH of 4 and 12, solubility decreases with increasing pH. The pH of digested sludge inside the anaerobic digester is typically near neutral, but begins to increase as soon as it is exposed to atmospheric conditions, which is a result of the CO_2 saturation during anaerobic digestion. This saturation causes dissolved CO_2 to escape into the gas phase as soon as a positive CO_2 concentration gradient between the liquid and gas phase develops. Loss of CO_2 from the liquid matrix causes the pH to increase, as follows:

FIGURE 10.15 Struvite deposits on a belt filter press.

$$HCO_3^- + H^+ \rightarrow CO_2\uparrow + OH^- \qquad (10.2)$$

The same CO_2 loss appears to be taking place at place of local turbulence or cavitations, resulting in increased struvite formation at pipe elbows, mixer blades, valves, and pumps impellers.

Based on the chemistry of struvite formation and observations at many treatment plants, the following locations are most likely be effected by struvite scales:

- Belt filter presses. Despite the significant dilution with spray water, struvite formation on BFPs is very common and can be severe. Deposits, such as shown in Figure 10.15, can reduce the dewatering efficiency by clogging belt and rollers and the operating life of belt and bearings as a result of belt misalignment.

- Dewatering filtrate storage and conveyance system (especially if gravity flows with free discharge). Two factors make dewatering centrate or filtrate conveyance system most susceptible to struvite deposits: (1) exposure to atmospheric conditions and (2) low TSS. The latter becomes apparent when considering that suspended solids provide seeds for crystals to grow on, and the more such seeds are present, the less struvite will grow on pipe and equipment surfaces.

- Dewatering filtrate pumping stations. Filtrate pumping stations are especially plagued with deposits as a result of the turbulence and negative pressure change This is especially true for pumping stations with little or no positive suction head, creating prime conditions where CO_2 is stripped out of solution, causing pH to increase and subsequently increasing struvite formation.
- Digested sludge transfer pumping stations. Very similar to the filtrate pumping station, the negative pressure change in the pump suction causes a CO_2 stripping from the liquid matrix, which raises the pH. Because of the presence of high suspended solids concentration, the amount of deposits is less than observed at filtrate pumping stations.
- Digester decant boxes. Decant boxes have a very high struvite formation potential because they combine exposure of digested sludge to atmospheric conditions with local turbulence. Depending on the design of the decant boxes, this can result in very rapid growth of struvite scales. The most vulnerable are vertical drop pipes below the decant box that have varying sludge levels, resulting in digested sludge cascading into the pipe.
- Long sludge transfer lines. Long transfer lines are especially problematic, because they are difficult to access and clean (Figure 10.16). The length of the line does not necessarily increase struvite formation, but the often intermitted operation of such a long transfer line does. The intermitted operation can also create problems by allowing struvite crystals that have already formed to settle out and compact on the pipe bottom. The resulting grit layer can be very difficult to remove.
- Turbulence causing fittings in digested sludge and dewatering centrate lines. Pipe fittings, such as valves and elbows, create areas of local turbulence that are prone to develop struvite, especially downstream of the anaerobic digester (exposed to the atmosphere).

STRUVITE CONTROL ALTERNATIVES. Figure 10.17 illustrates the basic adjustments for every struvite control alternative; that is, either lowering the pH or reducing the concentration of at least one constituent to move the solution product from crystallized to the dissolved area. Most control methods represent a combination of both.

Figure 10.16 Photograph captured during a camera inspection image of a 10-km (6-mile) sludge transfer line at Eugene Springfield Water Pollution Control Facility (Eugene, Oregon). The photograph shows some struvite deposits on the pipe surface and a layer of struvite grit.

The goal of struvite control is to prevent struvite from depositing and developing to scales that impair the functionality of equipment and facility operation. However, it is important to distinguish between struvite formation and struvite deposits. The formation of struvite is not necessarily bad. In EBPR plants, struvite formation is thought to contribute the phosphorus removal. The formation of struvite as internal crystals that are removed in the sludge handling process does not affect the operation of the plant. Figure 10.18 shows struvite crystals found in digested sludge.

The emphasis of struvite control should be on minimizing struvite deposits; monitoring and maintaining those areas that are susceptible to scaling; and protecting critical equipment, such as heat exchangers, dewatering equipment, or recirculation pumps.

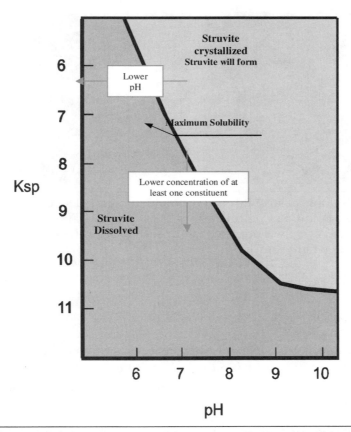

FIGURE 10.17 Struvite solution product versus pH. Upper and lower arrows show direction of struvite control.

Struvite control alternatives for the operator are mostly limited to chemical addition and scale removal. Various chemical agents can be used to control struvite formation. Some facilities have had good experience spraying the belts and rollers with a struvite-suppressing chemical agent offered by several companies. These agents are typically cost-competitive with commonly used phosphorus precipitating agents (ferric and alum). However, they are only recommended to protect or clean specific equipment. For systems with chemical struvite control, phosphorus

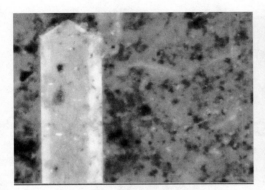 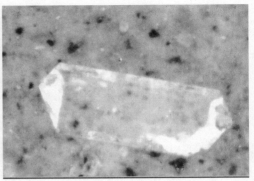

FIGURE 10.18 Example of struvite crystals found in digested sludge at Eugene Springfield Water Pollution Control Facility (200× magnification). Though barely visible to the naked eye, their presence can be seen by sparkling that occurs when held at an angle to a light source.

precipitating chemicals are preferable. Researchers have found that polymer selection can affect the struvite formation potential at dewatering facilities. Cationic polymers can have a disassociating effect on magnesium, thus increasing the load of free magnesium.

Phosphorus Precipitating Agents. When using phosphorus precipitating chemicals, it is important to understand that is not necessary to removal all dissolved phosphorus. The key is to remove enough to lower the solution produce far enough below the maximum solubility, such that no significant amounts of struvite will form downstream of the chemical addition point. As a general rule, 20 to 30% removal efficiency would be sufficient for struvite control downstream of the anaerobic digester discharge.

Because of its effect on secondary treatment, minimizing the recycle phosphorus load by maximizing phosphorus precipitation from the solids stream could be beneficial and cost-effective. Preferably, a phosphorus-precipitating chemical should be added upstream of dewatering, when possible. The effect of the chemical on the dewatering efficiency should be tested before permanently installing such chemical treatment.

Dilution Water. Whenever possible, the addition of dilution water should be the first choice downstream of sludge dewatering. It is very effective at a low cost. Depending on the strength of the dewatering filtrate (or centrate), dilution rates of 25 to 50% can be sufficient to successfully prevent struvite formation. Diluting with other recycle flow, such as WAS thickening filtrate, backwash return of GT overflow would help to minimize the overall plant recycle flow, which cumulatively reduces plant capacity.

Cleaning Loops. Only costly excess chemical addition will completely prevent struvite formation. Such chemical addition rates are economically undesirable. One way to protect critical equipment, such as heat exchangers, pumps, and inaccessible pipe lines, is by the installation of cleaning loops. By installing one injection and one return tap and the necessary isolation valves, the operator can run a cleaning solution through the system to periodically remove struvite scales. Such cleaning solutions are commercially available, or the service could also be contracted out. When selecting a cleaning chemical, the operator must assure that all elements of the cleaning loop are compatible and will not be damaged.

Hydroblasting. Hydroplaning is a very effective struvite removal method; however, because of its cost, it is only recommended in cases where scale removal is only required every other year. The only limitation of hydroblasting is the maximum available distance that can be reached from the entry point. In most cases, the available equipment limits the maximum length to 45 to 90 m (150 to 300 ft).

Equipment and Pipe Lining Selection. Whenever the replacement of pipes or equipment becomes necessary, the replacement should be selected under consideration of its susceptibility to scaling. It is commonly know that pipes with smooth lining are lees susceptible to struvite scales. It is important to understand that the material must be smooth on a microscopic level to be effective. Otherwise, the small impurities illustrated in Figure 10.19 will serve as seeds, from which new crystals can grow, causing the strong physical bond between the pipe and the scale. There are a number of high-purity materials on the market, but not all of them are available as pipe liner. Polyvinylidene fluoride (PVDF) and Harvel LXT® (Harvel Plastics, Inc., Easton, Pennsylvania, www.harvellxt.com) have proved to be very effective. When installing such high-purity material, follow the manufacturers' instructions closely to avoid local impurities.

Magnetic and Ultrasonic Treatment. Over the past decade, a number of products from other industries using magnetic and ultrasonic technologies have crossed over

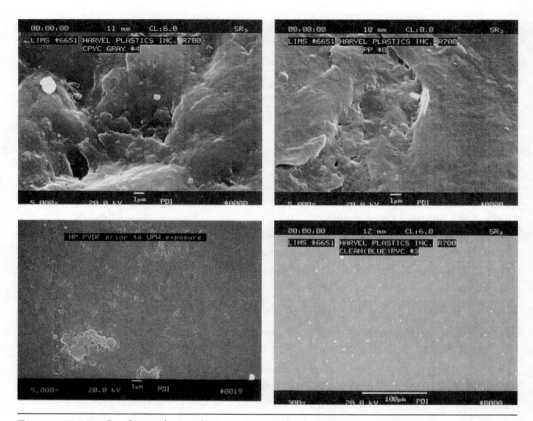

FIGURE 10.19 Surface of pipe linings magnified 5000x. Top left—chlorinated polyvinyl chloride (CPVC); top right—polypropylene; bottom left—PVDF; and bottom right—Harvel LXT® (image courtesy of Harvel Plastics, Inc., Easton, Pennsylvania).

and have been offered as struvite control devices. Most often, such devices would be mounted either to the outside of the pie or a spool piece. To the knowledge of the authors, no such installation has been proven successful to date, which is not to say that improvements and new developments of those technologies will not produce different results in the future.

Lagoon Flushing. Sludge storage lagoons introduce one complicating factor. Concentrating struvite constituents through evaporation increases the struvite formation potential. To minimize the effect of evaporation, the lagoon decant could be periodically or continuously be replaced with secondary effluent. To minimize the effect of recycle phosphorus load, phosphorus-precipitating chemicals could be added to the lagoon decant. When operating this type of lagoon decant flushing, one should anticipate some temporary depreciation in secondary effluent quality (i.e., increased turbidity and TSS).

Controlled Struvite Crystallization (Phosphorus Recovery). Currently still in the emerging technology stages, phosphorus recovery via struvite formation has been proven feasible in full-scale applications. Conceptually, the process is not much different from other metal phosphorus precipitation; only, in this case, the metal is magnesium. The key difference is that the crystallization reaction requires more time and a separation stage to recover the formed crystals. However, if phosphorus recovery is not desired, no separation is necessary. The available technologies either use a fluidized bed reactor or operate in batch mode. In both cases, the pH is adjusted and, if necessary, magnesium is added to optimize the molar ratios for phosphorus removal. The recovered struvite can be used as a fertilizer. However, unlike in Europe or Asia, no real demand exists for recovered struvite, but this is likely to change as the technology becomes more readily available. Without the cost recovery from the either struvite sales or reduced disposal cost, other factors could still make the technology economically viable. Magnesium hydroxide is less expensive than other phosphorus precipitating chemicals, and phosphorus ammonia is also removed, which reduces the nitrogen recycle load. Furthermore, for areas with limitation of phosphorus content in secondary effluent and biosolids, this technology offers an alternative to accomplish both, by extracting the phosphorus from the solids recycle stream.

Facility Design. Operators, from time to time, become involved in the design process of a new facility or facility upgrade on a review and design workshop level. Most operational problems related to facility design are a result of the fact that scaling problems were not anticipated. Following is a list of recommendations that should be followed when designing a solids processing facility that will or may be used to anaerobically digested waste activated sludge:

- Minimize length of sludge and filtrate conveyance system and minimize the number of fittings as much as possible;

- If long pipelines are necessary, provide cleanouts every 76 m (250 ft);
- Provide taps and isolation valves around critical equipment (i.e., pumping stations and heat exchanges);
- Use pinch valves where possible;
- Provide excess filtrate conveyance capacity to allow for dilution water addition;
- Use glass-lined pipes for general piping and ultra-pure pipe liners for critical sections (i.e., pump suction and discharge, and overflow pipes);
- Avoid free discharge of digested sludge into wet wells, which will force CO_2 out of solution and increase the pH; and
- If centrate or filtrate storage is required, provide chemical addition for phosphorus precipitation of pH control.

Process Design. The process design of secondary treatment and biosolids stabilization controls the phosphorus and magnesium mass balance; that is, the amount of phosphorus that enters solids processing and in which form it enters. In cases where EBPR is desired, the options are limited.

The phosphorus transfer into the anaerobic digester (accumulation of phosphorus in WAS) can be minimized by the following:

- Preventing enhanced biological phosphorus uptake.
- Design anaerobic selector to be dominated by glycogen-accumulating organisms (GAOs). Glycogen accumulating organisms typically coexist with PAOs. Certain conditions give GAOs an advantage, which results in reduced phosphorus uptake.
- Avoid anaerobic selector operation.
- Maintain anoxic conditions in anoxic selector, at all times.
- Minimize yield (longer SRT).
- Stabilized WAS by means other than anaerobic digestion.

Figure 10.20 shows a decision support flow chart that can help when reviewing a process design for options to reduce the struvite formation potential.

Solids Handling and Processing 393

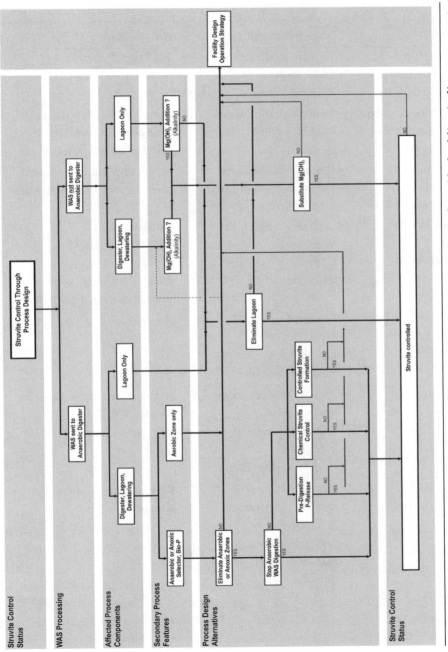

FIGURE 10.20 Struvite control through process design: decision support chart (dashed line indicates current situation at Bozeman Wastewater Treatment Plant, Bozeman, Montana).

TABLE 10.4 Sludge treatment strategies implemented at various BNR facilities.

Plant	Sludge treatment train	Recycle loads
Valrico Advanced Wastewater Treatment Plant, Florida	• Primary sludge: none • WAS: aeration + GBT + lime stabilization	WAS kept "fresh", no significant recycle loads
Bowie WWTP, Maryland (Randall et al., 1992)	• Primary sludge: none • WAS: GT + BFP	4 and 22% of phosphorus removed is recycled in GT and BFP recycle, respectively
Old Maryland City WRF, Maryland (Randall et al., 1992)	• Primary sludge: none • WAS: settled in primary clarifier	43% of phosphorus removed is released in primary clarifier
York River WWTP, Virginia (Randall et al., 1992)	• Primary sludge: GT • WAS: DAF • Codigestion (anaerobic digestion) + GBT	GBT centrate: 170 mg/L TP, 700 mg/L NH_3-N Significant nitrogen and phosphorus removal by struvite and other precipitates
Little Patuxent WRP, Maryland	• Primary sludge: GT • WAS: DAF • Co-dewatering (BFP) + lime stabilization	No significant recycle loads
Kalispell Advanced Wastewater Treatment Plant, Montana	• Primary sludge: primary sludge fermenter + anaerobic digestion • WAS: DAF • Co-dewatering (BFP)	2nd stage anaerobic digestion decant: 30 to 35 mg/L TP BFP filtrate: 3.3 mg/L TP
91st Ave. WWTP Phoenix, Arizona	• Primary sludge: thickening centrifuge • WAS: DAF • Codigestion (anaerobic digestion) + dewatering centrifuge	400 to 600 mg/L ammonia-nitrogen

(continued)

TABLE 10.4   Sludge treatment strategies implemented at various BNR facilities *(continued)*.

Plant	Sludge treatment train	Recycle loads
Kelowna Wastewater Treatment Facility, Canada	• Primary sludge: primary sludge fermenter + thickening centrifuge • WAS: DAF • Co-dewatering (dewatering centrifuge) + off-site composting • Lime treatment of centrate	Centrate quality: <u>Before lime treatment</u> TP = 236 mg/L NH_3-N = 29 mg/L TKN = 94 mg/L Total COD = 1913 mg/L <u>After lime treatment</u> TP = 146 mg/L NH_3-N = 16 mg/L TKN = 69 mg/L Total COD = 1443 mg/L

CASE STUDIES

Examples of sludge management strategies implemented at various full-scale BNR facilities are summarized in Table 10.4. A review of the information illustrates the site-specific nature of recycle stream characteristics.

CONCLUSION

Recycle streams resulting from BNR sludge operations have the potential to impose significant and periodic loading on the main process. The adverse effects of return streams can be controlled or eliminated by proper planning, design, and operation. Failure to reconcile recycle loads could result in regulatory noncompliance.

REFERENCES

Barnard, J. L.; Fothergill, S. (1998) Secondary Phosphorus Release in Biological Phosphorus Removal Systems. *Proceedings of the 71st Annual Water Environment Federation Technical Exposition and Conference,* Orlando, Florida, Oct 3–7; Water Environment Federation: Alexandria, Virginia.

Carrio, L.; Sexton, J.; Lopez, A.; Gopalakrishnam, K.; Sapienza, V. (2003) Ammonia-Nitrogen Removal from Centrate 10 Years of Testing and Operating Experience in New York City. *Proceedings of the 76th Annual Water Environment Federation Technical Exhibition and Conference* [CD-ROM], Los Angeles, California, Oct 11–15; Water Environment Federation: Alexandria, Virginia.

Chaparro, S. K.; Noguera, D. R. (2002) Reducing Biosolids Phosphorus Content from Enhanced Biological Phosphorus Removal Reactors. *Proceedings of the 75th Annual Water Environment Federation Technical Exhibition and Conference* [CD-ROM], Chicago, Illinois, Sep 28–Oct 2; Water Environment Federation: Alexandria, Virginia.

Daigger, G. T. (1998) Recycle Streams-How to Control Impacts on Liquid Treatment Process. *Water Environ. Technol.,* **10** (10), 47–52.

Gopalakrishnan, K.; Anderson, J.; Carrio, L.; Abraham, K.; Stinson, B. (2000) Design and Operational Considerations for Ammonia Removal from Centrate by Steam Stripping. *Proceedings of the 73rd Annual Water Environment Federation Technical Exposition and Conference* [CD-ROM], Anaheim, California, Oct 14–18; Water Environment Federation: Alexandria, Virginia.

Jeyanayagam, S. S.; Husband, J. A. (2002) Handle with Care! Processing BNR Sludges. *Water Environ. Technol.,* **13** (10).

Kos, P.; Head, M. A.; Oleszkiewicz, J.; Warakomski, A. (2000) Demonstration of Low Temperature Nitrification with a Short SRT. *Proceedings of the 73rd Annual Water Environment Federation Technical Exposition and Conference* [CD-ROM], Anaheim, California, Oct 14–18; Water Environment Federation: Alexandria, Virginia.

Mishalani, N. R.; Husband, J. (2001) Nitrogen Removal from Dewatering Sidestreams. *Proceedings of the WEF/AWWA/CWEA Joint Residuals and Biosolids Management Conference,* San Diego, California, Feb 21–24.

Mulder, A.; Van de Graaf, A. A.; Robertson, L. A.; Kuenen, J. G. (1995) Anaerobic Ammonium Oxidation Discovered in a Denitrifying Fluidized Bed Reactor. *Microbiol. Ecol.*, **16**, 177–183.

Murakami, T.; Koike, N.; Taniguchi, N.; Esumi, H. (1987) Influence of Return Flow Phosphorus Load on Performance of the Biological Phosphorus Removal Process. In *Advanced Water Pollution Control Proceedings of the IAWPRC Special Conference*, Rome, Italy, Sep 28–Oct 2; International Water Association: London.

Pitman, A. R.; Deacon, S. L.; Alexander, W. V. (1991) The Thickening and Treatment of Sewage Sludges to Minimize Phosphorus Release. *Water Res.*, **25** (10), 1285–1294.

Pitman, A. R. (1999) Management of Biological Nutrient Removal Plant Sludges-Change the Paradigms? *Water Res.*, **33** (5), 1141–1146.

Rabinowitz, B.; Barnard, J. L. (2000) Sludge Handling for Biological Nutrient Removal Plants. *International Water Association Yearbook 2000*; International Water Association: London.

Randall, C. W.; Barnard, J. L.; Stensel, H. D. (1992) *Design and Retrofit of Wastewater Treatment Plants for Biological Nutrient Removal*. Technomic Publishing Co., Inc.: Lancaster, Pennsylvania.

Salemi, S.; Berends, D. H. J. G.; van der Roest, H. F.; van der Kuij, R. J.; van Loosdrecht, M. C. M. (2003) Full-Scale Application of the BABE® Technology. *Proceedings of the 9th IWA Specialized Conference on Design, Operation, and Economics of Large Wastewater Treatment Plants*, Praha, Czech Republic, Sep 1–4; International Water Association: London.

Sen, D.; Randall, C. W.; Grizzard, T. J. (1990) Modifications of the Maryland City WRF for Biological Nutrient Removal. U.S. EPA report, Maryland City WRF, Chesapeake Bay Program; U.S. Environmental Protection Agency: Washington, D.C.

Siegrist, H.; Rieger, L.; Fux, C.; Wehrli, M. (2003) Improvement of Nitrogen Removal at WWTP Zurich Werdhoelzli After Connection of WWTP Zurich-Glatt. *Proceedings of the 9th IWA Specialized Conference on Design, Operation, and Economics of Large Wastewater Treatment Plants*, Praha, Czech Republic, Sep 1–4; International Water Association: London.

Stephens, H. L.; Stensel, H. D. (1998) Effect of Operating Conditions on Biological Phosphorous Removal. *Water Environ. Res.*, **70**, 360–369.

Waltrip, D. (1991) Modification of York River WWTP to the VIP Process. Paper presented at the Virginia WPCF Conference, Williamsburg, Virginia, September 25; Water Pollution Control Federation: Alexandria, Virginia.

Wentzel. M.; Dold, C. P. L.; Ekama, G. A.; Marais, G. v. R. (1989) Enhanced Polyphosphate Organism Cultures in Activated Sludge Systems. Part I: Experimental Behavior. *Water SA,* **15** (2), 71–88.

Wentzel. M.; Dold, C. P. L.; Ekama, G. A.; Marais, G. v. R. (1991) Kinetics of Nitrification Denitrification Biological Excess Phosphorus Removal Systems—A Review. *Water Sci. Technol.*, **23**, 555–565.

U.S. Environmental Protection Agency (1987) Sidestreams in Wastewater Treatment Plants. U.S. EPA Design Information Report. *J. Water Pollut. Control Fed.*, **59**, 54–59.

Chapter 11

Laboratory Analyses

Nitrogen	400	*Kjeldahl Method*	409
Types	400	Storage of Samples	409
Sampling and Storage	400	Interference	410
Analyses Methods	401	Phosphorus	411
Ammonia-Nitrogen	401	Types	411
Colorimetric Methods	401	Sampling and Storage	411
Titrimetric Methods	402	Analyses Methods	412
Ion Selective Method	402	Digestion Methods for Total Phosphorus Analyses	412
Ion Chromatography	403	Methods for Orthophosphate Analysis	413
Nitrite-Nitrogen	404	Vanadomolydophosphoric Acid Colorimetric Method	413
Colorimetric Method	404		
Ion Chromatography	404		
Nitrate-Nitrogen	404		
Nitrate Electrode Screening Method	405	Ascorbic Acid Method	413
		Ion Chromatography	413
Chromotropic Acid Method	406	Short-Chain Volatile Fatty Acid Analysis	415
Ion Chromatography for Nitrite and Nitrate	407	Analytical Methods for Short-Chain Volatile Fatty Acid Measurement	416
Organic Nitrogen	408	References	417

NITROGEN

TYPES. The common nitrogen species in wastewaters are ammonium ion, free ammonia, nitrite-nitrate, and organic nitrogen. The forms of the nitrogen species and their descriptions were given in Chapter 2.

SAMPLING AND STORAGE. The determination of nitrogen is essential to monitor biological nutrient removal (BNR) plant performance. The most accurate and reliable results are obtained from the fresh samples. If prompt analysis is not possible, sample preservation is required. Approximately 1 mL (a couple of drops) of concentrated sulfuric acid (H_2SO_4) is sufficient enough to stop any biological reaction. A pH value of 1.5 to 2 is typically desired for acid preservation (APHA et al., 1995). Hydrochloric acid (HCl) is not recommended because it reacts with ammonia and removes it. If wastewater contains residual chlorine, it should be removed to avoid its reaction with ammonia. Following sulfuric acid addition, samples should be stored at 4°C. The neutralization of samples with sodium hydroxide (NaOH) or potassium hydroxide (KOH) is required before analysis.

Sample size and location depends on the nitrogen species of interest. For example, in a modified Ludzack-Ettinger (MLE) process (anoxic and aerobic with an internal nitrate cycle), the samples must be collected from raw influent, primary effluent, anoxic zone, aerobic end, and secondary clarifier effluent. The sampling and sample locations for nitrogen species in an MLE process are given Table 11.1

TABLE 11.1 Sampling and sample locations for nitrogen species in an MLE process.

Nitrogen species	Raw influent	Primary effluent	Anoxic zone	Aerobic zone	Secondary effluent
Ammonia	✓		✓	✓	✓
Nitrite and nitrate	✓*		✓	✓	✓
TKN	✓	✓			✓

*In domestic wastewaters, raw wastewater has very negligible nitrite- and nitrate-nitrogen.

ANALYSES METHODS. *Ammonia-Nitrogen.* Ammonia-nitrogen (NH_3-N) is one of the major compounds to assess BNR plant performance. Ammonia-nitrogen concentration is typically less than 1 mg/L in well-operated BNR plant effluents. Method selection for ammonia-nitrogen analyses is influenced by two factors: (1) concentration range of ammonia, and (2) presence of interference (APHA et al., 1995). Ammonia-nitrogen can be measured by the following techniques:

- Colorimetric,
- Titrimetric,
- Ion selective, and
- Ion chromatography.

In addition, an online analyzer is commercially available for ammonia measurements. Online analyzers for nitrogen and phosphorus measurements are given in Chapter 10.

COLORIMETRIC METHODS. The Nessler and Phenate methods are the two most common methods for colorimetric determination of ammonia. In each method, chemical reagents with ammonia produce a distinct color and are measured in a spectrophotometer. The Nessler method (Standard Method 417-B; APHA et al., 1995) is typically used to determine ammonia concentration in treated wastewater effluents with an ammonia-nitrogen concentration of less than 5 mg/L. Turbidity and color interfere with the method and need to be removed by zinc sulfate (APHA et al., 1995). Manual phenate is a highly sensitive method to determine ammonia-nitrogen concentrations up to 0.5 mg/L. High alkalinity (500 mg/L calcium carbonate [$CaCO_3$]), color, and turbidity may interfere with chemical reagents and colorimetric readings. If sample preservation is provided or samples have high alkalinity or color, a distillation step is recommended before analyses (Standard Method 417-A; APHA et al., 1995). The distillation step relies on hydrolysis of an organic nitrogen compound at pH 9.5 with a borate buffer (APHA et al., 1995). The pH was adjusted with 6 N NaOH. A few glass beads are added to the sample (500 mL) and borate buffer (20 mL) mixture to steam out the distillation apparatus until no traces of ammonia are left in distilate. Ammonia-free distilled water must be used during any dilution or preparation of borate buffer. The Phenate method (Standard Method 417-C; APHA et al., 1995) and Nessler method can be applied to determine ammonia concentrations in activated sludge basins and treated effluent. The following procedure is followed:

- Check alkalinity values of samples;
- Immediately filter samples through 0.45-μm membrane filter following sampling;
- If filtrate is turbid, use zinc sulfate to eliminate turbidity;
- Make three different dilutions for direct Phanate method (1/4, 1/6, 1/10) (no dilutions necessary for Nessler method);
- Duplicate the samples;
- Prepare ammonia standard solutions in the range of 0 (blank) to 10 mg/L for effluent and 0 to 40 mg/L for influent;
- Measure absorbance at 400 to 500 nm for standard and samples;
- Develop a standard curve based on standard readings; and
- Calculate ammonia concentration for samples.

The direct Phenate and Nessler method can also be applied to raw wastewater. However, it requires typical dilution of 1/60 to 1/100 for Phenate and 1/6 to 1/10 for Nessler method, respectively (APHA et al., 1995).

TITRIMETRIC METHODS. The titrimetric method for ammonia is used only on samples carried through the primary distillation step described in Standard Method 417-A (APHA et al., 1995). Approximately 0.5 mL (one drop) of mixed indicators containing methyl red and methylene blue is added to the distillated samples and blank. The samples are titrated with 0.02 N H_2SO_4 until a pale lavender color is observed (APHA et al., 1995). The ammonia concentration is calculated by taking the difference between the volume of H_2SO_4 titrated for sample and the volume of H_2SO_4 titrated for blank. Table 11.2 shows the required sample volume as a function of ammonia-nitrogen predicted in the sample (APHA et al., 1995).

ION SELECTIVE METHOD. The ammonia selective electrode method uses a hydrophobic gas-permeable membrane to separate the sample solution from an electrode solution of ammonium chloride. The solution pH is raised to above 11 with a strong base (i.e., NaOH or KOH) to convert nearly all ammonia to free ammonia. Only free ammonia diffuses from the membrane and changes the internal solution pH that is measured by a pH meter with an expanded millivolt scale (typically -700 mV to 700 mV). A chlorine electrode senses the reduced chlorine concentration if ammonia is present in the sample.

TABLE 11.2 Required sample volume for titrimetric method as a function of ammonia-nitrogen concentration in sample (adapted from *Standard Methods*; APHA et al., 1995).

Ammonia-nitrogen in sample, mg/L	Sample volume, mL
5 to 10	250
10 to 20	100
20 to 50	50
50 to 100	25

The ion selective method (Standard Method 417 E; APHA et al., 1995) is applicable to analyses of ammonia-nitrogen in wastewaters within the range of 0.03 to 1400 mg/L. High concentrations of other dissolved ions may interfere with the measurements. The method is not affected by high turbidity and color.

ION CHROMATOGRAPHY. In most wastewater, the pH falls in a neutral region where total ammonia is mostly in ionized form (ammonium ion). At pH value of 7.3, 99% of the total ammonia is in ionized form. Ion chromotography is a rapid and simple method to analyze many ions in wastewater. It eliminates preparation of a hazardous reagent. Wastewater sample is injected to a stream of carbonate-bicarbonate eluent and passed through a series of ion exchangers. Each ion has a unique affinity to a separator column and exhibits a different separation time.

It is accurate to measure ammonium ion in a wide range. If the pH of the wastewater sample is 7.5 or less, pH correction is not necessary. However, pH values of 8.0 to 8.5 require a correction factor of 1.04 and 1.12, respectively. The minimum detection level for ammonia-nitrogen is approximately 0.05 mg/L. However, tailing of peaks occurs when samples contain high ammonia concentrations (>40 mg/L) (Erdal, 2002). Therefore, dilution is recommended if the ammonia concentration is higher than 40 mg/L. The method can directly be applied to measure ammonia concentration in raw wastewater, activated sludge, and industrial wastewaters. The following step must be performed:

- Prepare at least seven standard solutions in the range of low to high (0.2, 1.0, 5.0, 10, 20, 30, and 40 mg/L);

- Use ion free deionized water during standard solution preparation;
- Take 2 to 3 mL volume from each standard solution and inject to the ion chromatographer;
- Prepare a standard curve for ammonia calculations;
- Filter all samples with a 0.45-μm membrane filter;
- Take a 2- to 3-mL sample with a disposable syringe;
- Inject the samples to the ion chromatograph; and
- Calculate the ammonia concentrations based on the standard curve.

Nitrite-Nitrogen. Nitrite ion is seldom found in a well-operated wastewater treatment plant. Plant operators may observe nitrite formation during the startup period of the plant.

COLORIMETRIC METHOD. Colorimetric determination of nitrite relies on the formation of a reddish purple azo dye at pH 2.0 to 2.5 and its photometric measurements at 540 nm. Diazitized sulfanilic acid and N-ethylenediamine dihydrochloride are combined to produce the azo dye (*Standard Methods*; APHA et al., 1995). The method is very suitable to measure low nitrite concentration in microgram levels. The accurate nitrite range is 5 to 50 μg nitrite-nitrogen/L, if a 5-cm light path and a green filter is used. Higher nitrite concentrations require dilution.

Suspended solids, free chlorine, and nitrogen trichloride interfere with the measurements. The presence of ions may also interfere with the color developments. Before analyses, all samples must be filtered through a 0.45-μm membrane filter. Analysis of fresh samples is recommended for accurate results. If sample storage is necessary, filtering and freezing of samples is enough to preserve the samples for a couple of days. Acid preservation is not recommended.

ION CHROMATOGRAPHY. See ion chromatography techniques for nitrate.

Nitrate-Nitrogen. Nitrate is an oxidized form of nitrogen. Nitrate in wastewater treatment is generated through nitrification and destroyed through denitrification. The performance of BNR plants is assessed by the determination of various nitrogen species. Nitrate-nitrogen is an integral part of the effluent nitrogen discharge requirement, where the effluent discharge limit may be based on ammonia only, nitrate-nitrogen, total nitrogen (TN), or total inorganic nitrogen (TIN). Both TN and TIN include nitrate-nitrogen.

Raw domestic wastewaters do not typically contain nitrate. Determination of nitrate with conventional methods is difficult because relatively complex procedures are followed. Some techniques in Standard Methods (APHA et al., 1995) are not suitable for nitrate analysis in wastewaters. For example, the UV spectrometric method is not recommended because the accuracy of the measurements is significantly reduced in the presence of organic materials.

NITRATE ELECTRODE SCREENING METHOD. This method uses a selective sensor for the nitrate ion that develops a potential across a thin, inert membrane (Standard Method 418 B; APHA et al., 1995). The electrode responds only to the nitrate ion in the range 0.2 to 1400 mg/L (APHA et al., 1995). The electrode responds to nitrate activity rather than nitrate concentration. The electrode is sensitive to chloride and bicarbonate ions. Chloride and bicarbonate concentrations exceeding the nitrate concentration by 5 and 10 times, respectively, may interfere with the nitrate. Using a buffer solution containing silver sulfate (Ag_2SO_4) may minimize such interferences. Reducing the pH to 3 eliminates bicarbonate ions. If nitrite is present in wastewaters, sulfamic acid addition will minimize the nitrite interference. The nitrate electrode should be calibrated before each measurement. For calibration and curve generation, concentrated nitrate stock solution is prepared by dissolving sodium nitrate ($NaNO_3$) in deionized water. Standard solutions are prepared in range 0.5 to 20 mg NO_3-N/L (0.5, 1.0, 2.5, 5.0, 10.0, and 20.0 mg/L). Required buffer solutions are described in *Standard Methods* under section 418 B (APHA et al., 1995). The following procedure is recommended:

- Filter samples with a 0.45-μm membrane filter (if prompt sampling is not possible, add 40 mg mercury chloride);
- Add 10 mL of 0.5 mg/L nitrate-nitrogen standard to a 50-mL beaker and add 10 mL buffer solution;
- Stir 2 to 3 minutes with a magnetic stirrer;
- Immerse tip of electrode and record the voltage;
- Remove electrodes from the beaker;
- Rinse and dry;
- Repeat the measurements for the remaining standard solutions (defined in Table 11.3);
- Construct a calibration curve;

TABLE 11.3 Standard solution concentrations and volumes for standard curve generation.

Transfer volume for standard solution, mL	Standard solution concentration, mg/L nitrate-nitrogen
10	0.5
10	1.0
10	2.5
10	5.0
10	10.0
10	20.0

- Transfer 10-mL sample to a 50-mL beaker, and add a 10-mL buffer solution;
- Stir 2 to 3 minutes with magnetic stirrer;
- Immerse electrode and read the voltage; and
- Calculate the nitrate concentration from the curve (both standard solutions and sample have 1/2 dilution, so the readings will directly reflect the nitrate-nitrogen concentrations for samples).

CHROMOTROPIC ACID METHOD. This is a colorimetric method to detect nitrate-nitrogen concentrations in the range 0.1 to 5.0 mg/L (Standard Method 418 D; APHA et al., 1995). In many BNR plants, effluent nitrate-nitrogen concentrations are typically 3 to 10 mg/L (Randall et al., 1992). If the nitrate concentration is unknown in the sample, 1/3 dilution is recommended. Each mole of nitrate ion in sample reacts with one-half of the moles of chromotropic acid to from a yellow product with a maximum absorbance at 410 nm (APHA et al., 1995). The color develops within 10 minutes and remains stable up to 24 hours. A cooling bath is required for samples because acid addition significantly increases the temperature of the sample, which may reduce the accuracy of the results. The presence of chlorine and nitrite interferes with the color development. The addition of sulfite (sodium sulfate) eliminates the interference of chlorine and other oxidizing agents. For this purpose, a sulfate urea agent is prepared by dissolving 5 g urea and 4 g anhydrous sodium sulfide in water. The following steps should be followed:

- Prepare nitrate standards in the range 0.10 to 5.0 mg/L by diluting 0, 1, 5, 10, 25, 40, and 50 mL standard nitrate-nitrogen solution to 100 mL with water;
- Mix samples and place them into a cooling bath at 10 to 20°C;
- Add all reagents defined in *Standard Methods* section 418 D (APHA et al., 1995);
- Add H_2SO_4 to bring volume to approximately 10 mL;
- Cool the samples at room temperature for 45 minutes;
- Read absorbance at 410 nm;
- Filter samples with a 0.45-µm membrane filter;
- Make 1/2 to 1/3 dilutions with deionized water; and
- Repeat the same procedure for the sample.

ION CHROMATOGRAPHY FOR NITRITE AND NITRATE. Ion chromatography is a very rapid, easy, and accurate method to determine nitrite- and nitrate-nitrogen of wastewaters. It eliminates hazardous chemical use. The anions are separated on the basis of their relative affinities for a strongly basic anion exchanger (separator column). The separated anions are directed onto a strongly acidic cation exchanger (suppressor column), where they are converted to their acidic form. The anions, in their acidic form, are measured by conductivity. The quantification is performed based on the peak area for each ion (APHA et al., 1995). The common anions in wastewater treatment are chloride (Cl^-), nitrite (NO_2^-), phosphate (PO_4^{3-}), nitrate (NO_3^-), and sulfate (SO_4^{2-}). Some organic acids have been reported to interfere with the measurements (APHA et al., 1995). However, no interference has been reported to analyze nitrite, nitrate, and phosphate ions by ion choromatography (Erdal, 2002). A typical anion separation is shown in Figure 11.1. The following procedure is recommended for nitrite and nitrate determination by ion chromatography:

- Prepare a series of nitrite and nitrate standards by weighing sodium nitrite and sodium nitrate (see example below) (Use deionized water during standard preparation);

- Take 2 to 3 mL from the combined standard solutions and inject them to the ion chromatographer;
- Repeat standard injection at least twice; Produce a standard curve based on area and concentration relationship;
- Filter samples with a 0.45-μm pore diameter filter (if prompt determination is not possible, freeze the filter samples at -10°C or below);
- Make dilution, if necessary (use deionized water during dilution);
- Inject samples to the ion chromatographer; and
- Calculate nitrite and nitrate calculations from the curve.

Example: Prepare nitrite and nitrate standard solutions using anhydrous $NaNO_2$ and $NaNO_3$. The combined standard should contain nitrite and nitrate nitrogen concentrations of 0.5, 2.5, 5, 10, 15, and 20 mg/L.

- Use a 500-mL volumetric flask to dissolve each compound,
- Prepare 200 mg/L nitrite- and nitrate-nitrogen as a concentrated solution,
- Weigh precisely 492.8 mg $NaNO_2$ and 607.1 mg $NaNO_3$ and place them into a 500-mL volumetric flask,
- Fill one-half of the flask with deionized water,
- Mixed until they are completely dissolved,
- Add water to the 500-mL mark, and
- Use the volumes given in Table 11.4 to prepare the desired standard solutions.

Organic Nitrogen. The organic nitrogen in wastewaters is related to protein and amino compounds. Their decompositions release organic nitrogen to the medium. In a well-operated BNR plant, effluent organic nitrogen concentration may be 0.5 to 2.0 mg/L. The organic nitrogen concentration is important to know, if effluent nitrogen discharge limit includes either total Kjeldahl nitrogen (TKN) or total nitrogen (TN).

Organic nitrogen is determined through the Kjeldahl method. The Kjeldahl method may be performed by removing ammonia in the beginning of the test. The test, including ammonia, refers to Kjehdahl nitrogen, and the nitrogen concentration is equal to ammonia-nitrogen plus organic nitrogen. The organic nitrogen concentration is estimated by taking the difference between TKN and ammonia-nitrogen.

TABLE 11.4 Volume of concentrated nitrite and nitrate solutions and deionized water needed to prepare 100-mL standard solutions in the range 0.5 to 20 mg/L.

Standard solution concentration, mg/L nitrite- or nitrate-nitrogen	Volume added from the concentrated solution, mL	Deionized volume, mL	Final volume, mL
0.5	0.25	99.75	100
2.5	1.25	98.75	100
5.0	2.50	97.50	100
10	5.00	95.00	100
15	7.50	92.50	100
20	10.0	90.00	100

Kjeldahl Method. Two Kjeldahl methods are available to determine organic nitrogen (APHA et al., 1995).

(1) Macro-Kjeldahl method, and

(2) Semi-micro-Kjeldahl method.

The macro-Kjeldahl method is applicable for samples containing a broad range of organic nitrogen. It requires a relatively large volume for the analysis (APHA et al., 1995). The semi-micro-Kjeldahl method is applicable to samples with high concentrations of organic nitrogen. The sample volume should be chosen to contain organic plus ammonia-nitrogen in the range of 0.2 to 2.0 mg. The macro-Kjeldahl method (Standard Method 420 A; APHA et al., 1995) is typically recommended, if enough sample volume is available (Table 11.5).

As can be seen from Table 11.5, a 250- to 500-mL sample size is required, if organic nitrogen is determined in the plant effluent, assuming the plant effluent contains 1 mg/L organic nitrogen.

STORAGE OF SAMPLES. The most reliable results are obtained from fresh samples. If prompt analysis is not possible, preserve the samples with concentrated sul-

TABLE 11.5 Required sample size for various organic nitrogen concentrations (adapted from *Standard Methods*; APHA et al., 1995).

Organic nitrogen in sample, mg/L	Required sample size, mL
0 to 1	500
1 to 10	250
10 to 20	100
20 to 50	50
50 to 100	25.0

furic acid. The pH of the sample must be at least 1.5 to 2.0 for preservation. Following acid addition, samples can be stored in at 4°C. Mercury chloride addition is not recommended because it reduces the ammonia in the sample.

The macro-Kjeldahl method has three major steps, as follows:

(1) Digestion step. Organic materials present in sample are converted to ammonium sulfate during digestion with concentrated sulfuric acid and in the presence of potassium sulfate (K_2SO_4) and mercuric sulfate catalyst ($HgSO_4$) and heat. Free ammonia and ammonium ion are also converted to ammonium sulfate. A mercury ammonium complex is also formed during digestion and decomposed by sodium thiosulfate ($Na_2S_2O_3$). The digestion apparatus should provide a temperature of 360 to 370°C for effective digestion.
(2) Distillation. Addition of a strong base (NaOH) permits the conversion of ammonia into free ammonia, followed by boiling and condensation of ammonia gas. It is absorbed in boric acid.
(3) Ammonia determination. The ammonia is determined colorimetrically or by titration with a standard mineral acid (APHA et al., 1995).

INTERFERENCE. A high nitrate concentration (10 mg/L or higher) may oxidize a portion of released ammonia during digestion. If reduced organic material is present, nitrate may be reduced to ammonia, resulting in positive interference (APHA et al., 1995). A simultaneous reaction will take place between organics and sulfuric acid, if the sample contains a high amount of organics. The reaction will reduce the available acid for digestion and reduces the efficiency of digestion (APHA et al., 1995). The addition of more acid may eliminate such concerns, but may reduce the digestion temperature to less than 360°C. Therefore, the temperature must be closely monitored during digestion.

PHOSPHORUS

TYPES. Nearly all phosphorus in wastewater is in the form of phosphates. These are classified as orthophosphates, condensed phosphates, and organically bound phosphates. Phosphates that respond to the colorimetric test or ion chromatography without preliminary hydrolysis or digestion refer to reactive phosphorus. Most of the reactive phosphorus in wastewater refers to orthophosphate. If acid hydrolysis at 100°C is applied to samples, it converts dissolved and particulate condensed phosphate to soluble orthophosphate. The phosphate fractions that are converted to orthophosphate only by digestion of organic matter refer to organically bound phosphorus (APHA et al., 1995). The second classification may be made according to its physical state. In this category, phosphorus is divided into two broad categories.

(1) Soluble phosphorus, and
(2) Particulate phosphorus.

Filtration through a 0.45-μm pore diameter filter separates soluble forms from particulate forms of phosphorus. Some colloidal phosphorus may be present in the soluble part. Each type can be subdivided into two categories.

(1) Biodegradable, and
(2) Nonbiodegradable.

The sum of the soluble and particulate phosphorus species is equal to total phosphorus (TP).

SAMPLING AND STORAGE. In an enhanced biological phosphorus removal (EBPR) plant, two major phosphorus parameters are of interest. They are orthophosphate and TP. Phosphorus is released through the anaerobic zone and taken up in the aerobic zone. Sample types and locations for phosphorus species in an anaerobic/oxic process are given in Table 11.6.

Prompt analysis is recommended for reliable results. If the soluble form of phosphate is of interest, filter the sample promptly after collection. Store filtered samples at -10°C. If longer storage is desired, add 40 mg/L mercury chloride. If total phosphorus is of interest, add 1 mL of concentrated HCl to unfiltered samples. If the analysis is performed within 48 hours, freezing without acid addition will be sufficient enough to preserve the samples (APHA et al., 1995). Avoid the use of phosphate-containing detergents for cleaning of phosphorus glassware (APHA et al., 1995).

ANALYSES METHODS. Total phosphorus includes all orthophosphates and condensed phosphates, soluble and particulate, and organic and inorganic fractions. An acid digestion method is required to determine TP. Following digestion, all complex phosphorus is converted to the orthophosphate form and is determined by colorimetric, spectrophotometric, or chromatographic methods.

Digestion Methods for Total Phosphorus Analyses. Three digestion methods are given in *Standard Methods* (Standard Method 424 C; APHA et al., 1995). These are as follows:

(1) Perchloric acid digestion,
(2) Sulfuric-nitric acid digestion, and
(3) Persulfate digestion.

TABLE 11.6 Sample types and locations for an anaerobic/oxic EBPR system.

Sample location	TP	Orthophosphate
Raw influent	✓	
Primary effluent	✓	✓
Anaerobic zone		✓
Aerobic zone		✓
Effluent	✓	

The persulfate digestion method is the most common and less tedious method among the three methods.

Persulfate digestion uses concentrated sulfuric acid solution and ammonium pursulfate as a catalyst. One drop of phenolphthalein indicator is added to a 50-mL sample before boiling at 120 to 125°C for 30 to 40 minutes. The cooled samples are diluted with distilled water, and one drop of phenolphthalein is added. Sodium hydroxide is used to neutralize the sample until a faint pink color develops. Phosphorus concentration is then determined by colorimetric, spectrometric, and chromatographic analysis (APHA et al., 1995).

If TP concentration in activated sludge is measured, a significant amount of dilution may be required for the analyses (1/50 to 1/200). High dilution may significantly reduce the accuracy of the measurements.

Methods for Orthophosphate Analysis. *VANADOMOLYDOPHOSPHORIC ACID COLORIMETRIC METHOD.* Orthophosphate reacts with ammonium molybdate under acidic conditions to form molybdophosphoric acid. Vanadium is added to promote vanadomolybdophosphate, which produces a yellow color. The color intensity is proportional to phosphorus in the sample. The color is spectrometrically detected at 400 to 470 nm. The ferric ion typically interferes at low wavelengths. Therefore, a wavelength of 470 nm is recommended for measurements (APHA et al., 1995). A standard curve is prepared in a phosphorus range of 0 (blank) to 20 mg/L. The detection limit of the method is 200 µg/L. If the digestion step is include, the measured phosphorus values will reflect the TP. Unfiltered activated sludge samples requires 1/50 to 1/200 dilution before the digestion step.

ASCORBIC ACID METHOD. Similar to the previous method, in the ascorbic acid method (Standard Method 424 F; APHA et al., 1995), orthophosphates react with ammonium molybdate and potassium antimonyl tartrate in a highly acidic medium to form intensely colored molybdenum blue by ascorbic acid. The developed color should be measured within 30 minutes following the color development at 880 nm. An arsenate concentration as low as 0.1 mg/L interferes with the color development. Hexavalent chromium and nitrite may also interfere with the results. The minimum detectable concentration of phosphorus is approximately 10 µg/L.

ION CHROMATOGRAPHY. Ion chromatography is a very rapid, easy, and accurate method to determine orthophosphate from wastewaters. It eliminates the use of reagent and other chemicals. The orthophosphates ($H_2PO_4^-$ and HPO_4^{2-}) are separated on the basis of their relative affinity for a strongly basic anion exchanger (separator column). The separated orthophosphates are directed to a strongly acidic cation exchanger (suppressor column), where they are converted to phosphoric acid. The quantification is performed based on the peak area under the phosphate ion (APHA et al., 1995). The following procedure is recommended for orthophosphate analysis by ion chromatography:

- Prepare a series of phosphate standards by weighing potassium phosphate dibasic (K_2HPO_4);
- Use deionized water during standard preparation;
- Inject 2 to 3 mL various standard solutions to the ion chromatographer;
- Repeat standard injections for reliability;
- Prepare a standard curve (area versus concentration);
- Filter samples with a 0.45-μm pore diameter filter (if prompt determination is not available, freeze the filter samples at -10°C or below);
- Make dilution, if necessary (use deionized water during dilution);
- Inject samples containing phosphate; and
- Calculate phosphate calculations from the curve.

The ion chromatography method can measure phosphorus concentration in the range 0.1 to 80 mg/L, without peak tailing. At high phosphate concentrations, dilution of samples is required. The ion chromatography method may also be used for TP measurements (Erdal, 2002). Based on the elution order of phosphate in ion chromatography, either persulfate acid or perchloric acid digestion is recommended. According to the separation curve in Figure 11.1, persulfate digestion is recommended, because chloride ions formed through the perchloric acid digestion method can easily interfere with phosphate ions. The digestion method converts complex phosphorus compounds to orthophosphate, which can be analyzed by ion chromatography. Before sample injections, add concentrated NaOH to neutralize the samples at pH 7 to 7.5 and make at least a 1/40 to 1/100 dilution (Erdal, 2002).

The comparisons of digestion and colorimetric versus digestion and ion chromatography showed that ion chromatography was more accurate than any other techniques for determination of phosphates.

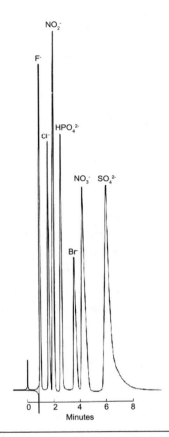

FIGURE 11.1 A typical anion separation in ion chromatography (adapted from *Standard Methods*; APHA et al., 1995).

SHORT-CHAIN VOLATILE FATTY ACID ANALYSIS

Short-chain volatile fatty acids (SCVFAs) are the principal end products of fermentation. The most common form of SCVFAs in domestic wastewater is acetic acid; however, propionic, butyric, valeric, caproic, and heptanoic acids can also be present in wastewater. The concentration and type of SCVFAs in raw wastewater are particularly important because they have a direct effect on biological phosphorus removal and denitrification performance. The effect of the volatile fatty acid (VFA) source on biological phosphorus removal was discussed in Chapter 4.

ANALYTICAL METHODS FOR SHORT-CHAIN VOLATILE FATTY ACID MEASUREMENT

Volatile fatty acids contain negative charges when they are ionized (solubilized) in water. Therefore, some SCVFAs including acetate may be measured through ion chromatography without further derivatization and purification. High-pressure liquid chromatography (HPLC) can also be used to determine VFAs in wastewater. Siegfried et al. (1984) purified the acetate samples with calcium hydroxide and cupric sulfate before analysis in HPLC. The high polarity-free fatty acid product (HP-FFAP) columns are designed primarily for the analysis of organic acids, free fatty acids, or samples that require derivatization and purification. The stationary phase is modified with acid to provide a very inert column that can accommodate the demanding analysis of acids dissolved in water. Short- and long-chain volatile fatty acids (up to 24 carbons) can be analyzed by avoiding time-consuming derivatization.

The SCVFAs can also be analyzed by using gas chromatography (GC). Packed-column GC and capillary GC are robust to determine fatty acids in wastewater and other mixed samples. Fussell and McCalley (1987) successfully determined VFAs in silage through packed-column GC. Highly polar columns are typically used in SCVFA analysis. Hydrogen gas was used as a carrier gas, and detection was carried out by a flame ionization detector. The initial temperature was set at 70°C and held for one minute. Following one minute holding at 70°C, the temperature was increased 10°C per minute to reach a final temperature of 200°C. Figure 11.2 shows an example output of GC run for fatty acid analysis via BP21. Further information can be obtained from following Web sites:

- http://www.supelco.com,
- http://www.sge.com,
- http://www.chem.agilent.com,
- http://www.dionex.com, and
- http://www.usepa.gov.

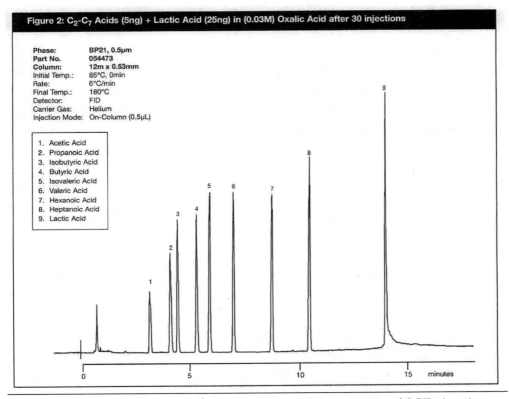

FIGURE 11.2 An example of fatty acid run via BP21 (courtesy of SGE, Austin, Texas).

REFERENCES

American Public Health Association; American Water Works Association; Water Environment Federation (1995) *Standard Methods for the Examination of Water and Wastewater*, 18th ed.; American Public Health Association: Washington, D.C.

Erdal, U. G. (2002) *The Effects of Temperature on EBPR Performance and Microbial Community Structure*. Ph.D. Dissertation; Virginia Polytechnic Institute and State University: Blacksburg, Virginia.

Fussell, R. J.; McCalley, D. V. (1987) Determination of Volatile Fatty Acids and Lactic Acid in Silage by GC. *Analyst*, **112**, 1213.

Randall, C. W.; Barnard, J. L.; Stensel, D. H. (1992) *Design and Retrofit of Wastewater Treatment Plants for Biological Nutrient Removal*. Technomic Publishing Co., Inc.: Lancaster, Pennsylvania.

Siegfried, R.; Ruckemann, H.; Stumff, G. (1984) Eine HPLC-Methode zur bBestimmung Organischer Sauren in Silagen. Landwirtsch, *Forschung*, **37**, 3–4 (in German).

Chapter 12

Optimization and Troubleshooting Techniques

Process Evaluation	422	What	432
Sampling and Testing	422	Where	433
Sampling Locations and Techniques	422	Why	433
		When	433
Sampling Plan	422	How	433
Sample Handling	423	pH	434
In Situ Sampling	424	What	434
Grab Sampling	424	Where	434
Interval Sampling	427	Why	434
Composite Sampling	428	When	434
Sample Location	428	How	434
Mixed Liquor Suspended Solids, Mixed Liquor Volatile Suspended Solids, Return Activated Sludge, and Waste Activated Sludge	430	Alkalinity	435
		What	435
		Where	435
		Why	435
		When	435
What	430	How	436
Where	431	Temperature	436
Why	431	What	436
When	432	Where	436
How	432	Why	436
Settleability and Sludge Volume Index	432	When	437
		How	437

Dissolved Oxygen	437	How	444
What	437	Orthophosphorus	444
Why	437	*What*	444
Where And When	438	*When And Where*	444
How	438	*Why*	444
Oxidation–Reduction Potential	438	*How*	445
What	438	Chemical Oxygen Demand	445
Where	439	*What*	445
Why	440	*Where*	446
When	440	*Why*	446
How	440	*When*	446
Ammonia and Total Kjeldahl Nitrogen	440	*How*	446
What	440	Volatile Fatty Acids	448
Where	441	*What*	448
Why	441	*Where*	448
When	441	*Why*	449
How	442	*When*	449
Nitrite-Nitrogen	442	*How*	449
What	442	Soluble Biochemical Oxygen Demand	450
Where	442	*What*	450
Why	442	*Where*	450
When	442	*Why*	450
How	442	*When*	451
Nitrate-Nitrogen	442	*How*	451
What	442	Nitrification Test	451
Where	442	*What*	451
Why	443	*Where*	451
When	443	*Why*	451
How	443	*When*	452
Total Phosphorus	443	*How*	452
What	443	Denitrification Test	453
When And Where	443	*What*	453
Why	443	*Where*	453

Why	453	*Performance Indicators*	464
When	453	Biological Phosphorus Removal	464
How	454	*Chemical Environment*	464
Biological Phosphorus Removal Potential Test	454	*Solids Retention Time*	465
What	454	*Performance Indicators*	465
Where	454	Optimization and Troubleshooting Guides	465
Why	455	Overview	465
When	455	Optimization and Troubleshooting Guide Format	466
How	455	*Indicator and Observations*	466
Method	455	*Probable Cause*	466
Preservation	455	*Check or Monitor*	466
Analyze	455	*Solutions*	466
Equipment and Supplies	456	*References*	468
Reagents and Chemicals	456	Optimization and Troubleshooting Guides	468
Principle	456	Case Studies	468
Preparation of Samples	457	Wolf Treatment Plant, Shawano, Wisconsin	468
Considerations	457	City of Stevens Point, Wisconsin	491
Procedure	458	City of Dodgeville, Wisconsin	491
Test Evaluation	459	Eastern Water Reclamation Facility, Orange County, Florida	495
Microbiological Activity	460		
What	460		
Where	461		
Why	461		
When	462		
How	462	Wastewater Treatment Plant, Stamford, Connecticut	497
Data Analysis and Interpretation	462	References	499
Nitrification	462		
Chemical Environment	463		
Solids Retention Time	463		
Performance Indicators	463		
Denitrification	463		
Chemical Environment	464		
Solids Retention Time	464		

PROCESS EVALUATION

The first step in using optimization and troubleshooting techniques is to evaluate the process performance. Process evaluation is accomplished through evaluation of existing data and operator observations and, as needed, additional sampling, testing, and data analysis and interpretation.

SAMPLING AND TESTING

Sampling and testing includes determining sampling locations and techniques and selecting appropriate tests that will allow evaluation of process performance.

SAMPLING LOCATIONS AND TECHNIQUES. *Sampling Plan.* Every wastewater treatment facility is unique and will, therefore, require a site-specific sampling plan to monitor the performance of that facility. Operations personnel and facility administrators need to work together to develop a sampling plan that allows a complete understanding of the key operating parameters without creating unneeded effort and excessive expense. Choosing the specific sample locations and parameters to be tested can be an excellent training tool, if done properly. During this exercise, everyone involved should ask themselves the following questions:

- What are our regulatory reporting requirements?
- What are the key cause-and-effect relationships that can affect our process?
- How can we collect the data required to understand these relationships?
- When and how often do these tests need to be collected?
- How will this data be used to manage our facility?
- Will data be available in time to make good operating decisions?
- Do we have the resources needed to collect the data that was identified above?

When determining a sampling plan, it is helpful to start with a copy of the facility flow schematic(s) to illustrate the various potential sample locations at the facility. A sample flow schematic is shown in Figure 12.1. From this schematic, the operators should first identify the location of the samples that are required for permit compliance and create a list of the parameters that need to be collected there. It is important to note that, if this exercise is occurring following a retrofit to an existing facility, much of the current sampling plan may remain, with some additional parameters

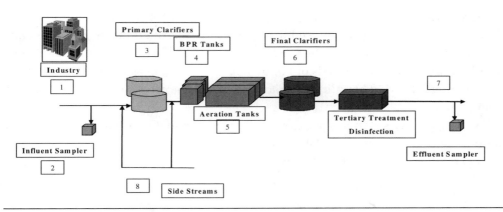

FIGURE 12.1 Flow schematic example.

designed to assist in the monitoring of the new biological nutrient removal (BNR) facilities.

The facility should then build a table summarizing the data that should be collected for each of the sample locations identified on the schematic. On one axis of the table, the operator should list the sample locations, and, on the other, the operator should list the various parameters that will be tracked at the facility. The operator should then start at a sample location and follow the table across, considering how each parameter could be used to monitor the respective sample location. If the operator decides that the parameter for this sample is useful, he should note in the box the interval that it will be sampled and if the sample is required or for process control. Table 12.1 contains an example of a sampling plan that could be prepared for the facility in Figure 12.1.

It is important to understand that implementing a sampling plan during an upset is often too late. Having the sampling plan in place when the process is operating well is critically important when attempting to identify the trigger that may have caused the upset.

Sample Handling. Sample handling should occur in accordance with the regulatory requirements that govern your facility and/or as suggested by a generally accepted resource, such as 40 *CFR* 136 (U.S. Environmental Protection Agency guidelines; Guidelines Establishing Test Procedures for the Analysis of Pollutants, 2005) or *Standard Methods* (APHA et al., 1998). See Part II of 40 CFR 136 (Guidelines Establishing Test Procedures for the Analysis of Pollutants, 2005), or *Standard Methods* for specific information related to handling samples in a manner acceptable for the parameters being tested.

Once a facility's staff has decided the information that they need to collect, they need to decide where the raw data will be collected and how. The raw data will generally be collected in one of the following five forms:

(1) In situ sampling,
(2) Grab sampling,
(3) Interval sampling,
(4) Time composite sampling, and
(5) Flow composite sampling.

IN SITU SAMPLING. In situ samples are those measurements collected directly from the respective tank, channel, or pipe, as shown in Figure 12.2. Examples of in situ measurements include dissolved oxygen (DO), pH, oxidation-reduction potential (ORP), and nitrate. Some in situ measurements, such as DO, are often collected using a field probe. Dissolved oxygen and other parameters can be installed to continuously monitor conditions at a specific point and can be incorporated to the control systems to allow real-time process modifications.

GRAB SAMPLING. A grab sample is a single sample based on neither time nor flow, as shown in Figure 12.3. When collecting grab samples, the volume should be large enough for all laboratory tests to be made. Individual samples are preferably larger than one liter.

Grab samples should also be taken when a situation, such as unusually high concentration, dictates investigation; when sampling tank contents at different depths; when the composite sampler fails to provide enough sample needed for all testing; or when composite sampling is not appropriate for sample type (e.g., fecal coliforms).

For many of the parameters associated with BNR facilities, a grab sample is the best sample type because collecting a composite sample would allow the samples to change with time. This is especially true when hoping to identify current conditions, such as nitrate, phosphorus, and/or ammonia concentrations in various stages of the biological treatment process. Grab samples should be analyzed or preserved immediately.

TABLE 12.1 Example of sampling plan.*

Sample	Sample description	BOD	TSS	S_{bsi}	VFA	pH	Total phosphorus	Ortho-phosphorus	Ammonia	Nitrate	Dissolved oxygen
1	Industry	TC/R/W	TC/R/W			G/R/W	TC/P/W				
2	Raw influent	FC/R/D	FC/P/D	FC/P/W	FC/P/W	G/R/D	FC/P/W				
3	Primary clarifiers	FC/P/D	FC/P/D	FC/P/W	FC/P/W	G/P/D	G/P/3				
4	BPR tanks					G/P/A		G/P/M (profile)	G/P/M (profile)		
5	Aeration tanks		G/P/D MLSS (SVI)			G/P/A		G/P/D			G/P/D
6	Final clarifiers		G/P/D RAS (blanket depth)								
7	Final effluent	FC/R/3	FC/R/3	FC/P/M		FC/R/3	FC/R/3	G/P/D	G/P/W	G/P/W	G/R/3
8	Side-streams	G/P/W	G/P/W				G/P/W		G/P/W	G/P/W	

*Notes:

Type of sample
G - grab
TC - time composite
FC - flow composite

Reason for sampling
P - process control

Interval of sampling
W - weekly
M - monthly
3 - 3 times per week
A - as needed

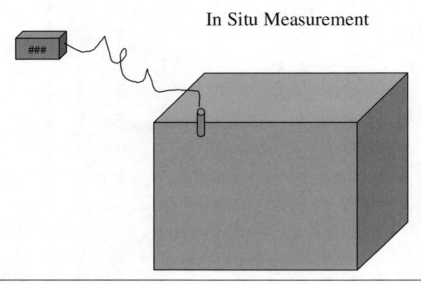

FIGURE 12.2 Illustration of an in situ measurement. The parameter measurement is conducted in the tank or channel of interest.

Some general guidelines for grab sampling of wastewater are as follows:

(1) Samples should be taken at locations where the wastewater is as completely mixed as possible (the sampling locations are described later in this section).
(2) Particles greater than 6.35 mm (0.25 in.) in diameter should be excluded when sampling.
(3) Any floating material growths or other particles that may have collected at a sampling location should not be included when sampling.
(4) If samples are not analyzed immediately, they should be immersed in ice water or refrigerated between 0 and 4°C.
(5) Proper sampling equipment should be provided, and safety precautions should be exercised during all sampling.

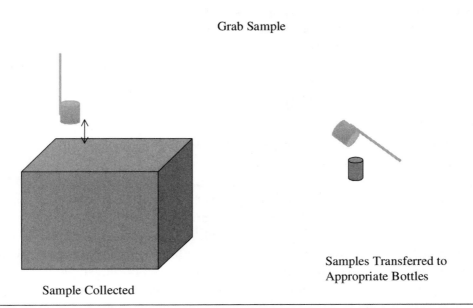

FIGURE 12.3 Illustration of a grab sample. Grab samples can be useful when parameters are time-sensitive and are best when taken from well-mixed or homogeneous solutions.

(6) Collect enough sample in a suitable container. Error can result from attempting to collect small portions for a composite sample. The sample should be taken from a point in the treatment plant process where mixing has created a somewhat homogeneous condition. The nature of wastewater prevents the collection of a relatively small portion that represents the whole. This is particularly true of raw or unsettled wastewater.

INTERVAL SAMPLING. Occasionally, a facility's staff will come to suspect that there are periods of the day that wastewater characteristics change significantly, and they may want to collect data to better understand what is occurring. This may be especially true when significantly affected by a wet industry or sidestream at the facility. An interval sampling event can allow the operations staff to isolate periods of the day by using a sampler equipped with multiple bottles, as shown in Figure 12.4. The sampler can be programmed to segregate sample from various time periods into separate bottles. Interval sampling is generally considered too labor-intensive for daily operation, but should be considered if characterizing a suspected flow stream of high diurnal variability.

COMPOSITE SAMPLING. A composite sample is a combination of individual samples taken at selected time or flow intervals for some specified period to minimize the effect of the variability of an individual sample. Sample portions may be of equal volume or proportional to flow at time of sampling. The volume of a composite sample varies with the number and size of individual samples, but typically varies between 7.5 and 11.4 L (2 and 3 gal). Figure 12.5 shows the time-composite method for creating a composite sample. Individual samples of equal volume are taken at a preselected time interval and are combined into one composite sample.

Figure 12.6 shows one of the two flow-composite methods for creating a composite sample. Individual samples of equal volume may be collected at selected volumetric flow increments, based on a flow meter and combined into one composite sample. Alternatively, individual samples may be collected at a preselected time interval, measured out as a volume proportional to flow at the time of sampling, and combined into one composite sample.

When the flowrate and strength of wastewater are variable, samples must be collected frequently and composited according to flow to be representative. Ideally, a composite sample would be a continuously collected sample, with volumes, at all times, in proportion to rates of flow. Samples taken on raw wastewater will tend to be highly variable, while samples taken on clarifier effluent will tend to be quite uniform because of upstream detention time and the degree of mixing.

Sample Location. It is important that a sample location is selected for any given sample and that all personnel collect the sample at the same place. Sample locations should be selected that allow representative samples to be collected. Additional samples should be collected if there is a possible change in conditions as the flow passes through the tanks. Characteristics of a good sample location include the following:

- The contents are mixed well enough at that location to collect a representative sample, and
- There is accessibility, allowing safe collection of the sample.

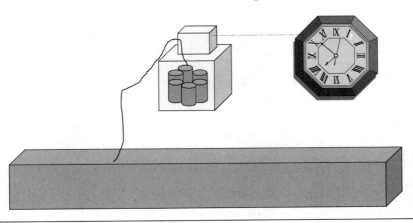

FIGURE 12.4 Illustration of an interval sample. The sampling can be programmed to segregate samples by time periods into separate bottles. This allows a facility to identify diurnal or event-related changes in influent or feed characteristics. The sampler is initiated, and a mechanism advances the bottle, as signaled by a time clock.

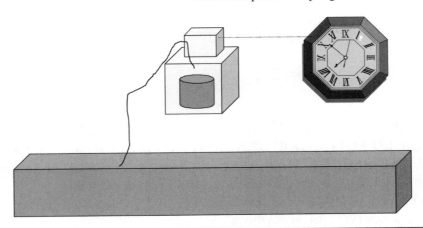

FIGURE 12.5 Illustration of a time-composite sample. Samples are collected by the sampler to meet the programmed interval entered by the operator. These samples can be acceptable in homogenous flows without significant diurnal fluctuations.

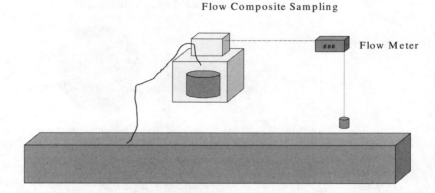

FIGURE 12.6 Illustration of a flow-composite sample. Samples are collected by the sampler to meet the programmed flow interval entered by the operator. These samples require the most equipment but provide the best characterization of streams that vary significantly throughout the day.

MIXED LIQUOR SUSPENDED SOLIDS, MIXED LIQUOR VOLATILE SUSPENDED SOLIDS, RETURN ACTIVATED SLUDGE, AND WASTE ACTIVATED SLUDGE. *What.* The mixed liquor suspended solids (MLSS) test indicates the total suspended solids (TSS) present in the mixed liquor.

The mixed liquor volatile suspended solids (MLVSS) test indicates the volatile suspended solids (VSS) present in the mixed liquor.

The return activated sludge (RAS) suspended solids test indicates the TSS present in the RAS returned from the clarifier to the treatment process.

The waste activated sludge (WAS) suspended solids test indicates the TSS present in the WAS removed from the treatment process to maintain the MLSS concentration at the desired level.

Where. The MLSS and MLVSS test is run on a sample taken from a well-mixed representative location within the process basin. Depending on the process, there may be multiple sample points resulting from varying solids concentrations, such as with the step-feed process and University of Cape Town process.

The RAS test is run on a sample taken from the RAS pipe (or channel or box) in a location that is well-mixed and representative.

The WAS test is run on a sample taken from the WAS pipe (or channel or box) in a location that is well-mixed and representative. When the WAS is pumped from a single hopper, composite sampling may be necessary over the duration of the WAS pumping cycle to accurately determine the mass of solids wasted out of the process.

Why. The MLSS is also used to calculate the sludge volume index (SVI), in conjunction with settleability test results (see the Settleability and Sludge Volume Index section).

The MLVSS result divided by the MLSS result provides the volatile fraction. In typical BNR plants, the volatile fraction is approximately 70 to 80%, depending on the relative concentration of inert material in the process influent and the solids retention time (SRT). Higher SRTs will result in a lower volatile fraction, because volatile material is reduced in the mixed liquor through oxidation of organic material trapped in the biomass by microorganisms and reduction of stored volatiles in the cells of microorganisms. The volatile fraction is a rough indicator of active biomass in suspended-growth treatment processes.

The MLSS and WAS results can be used in conjunction with the volumes of aerobic zones in the treatment process and the WAS flow to calculate the SRT in the aerobic zones. The aerobic SRT is most important for nitrification because the nitrifying organisms grow slower than the other organisms used in BNR. The aerobic SRT needs to be high enough to maintain an adequate population of nitrifying organisms in the process, especially at low temperatures.

The MLSS and RAS results can be used to evaluate whether the RAS rate is appropriate in relation to the process influent flow. For example, if the RAS concentration is two times the MLSS concentration, the RAS flowrate is approximately equal to the flowrate entering the process basin. If the RAS concentration is too close to the MLSS concentration, the RAS flowrate is much higher than the flowrate entering the process basin, and there is high potential for solids overloading occurring at some time during the day at the clarifiers.

When. Samples for MLSS, MLVSS, RAS, and WAS should be pulled several times during the day. These samples can be tested individually, or, if the plant is stable and does not have significant variation in incoming flow, the samples can be composited and then tested.

How. Suspended solids is determined by filtering a known volume of well-mixed sample through a preweighed glass fiber filter, rinsing the solids residue with deionized water to remove dissolved solids, drying for one hour at 103 to 105°C, cooling in a dessicator, weighing the dried residue and filter, and calculating the net weight of the dried residue. The suspended solids test is reported in milligrams per liter. The suspended solids test is Method 2540D in *Standard Methods* (APHA et al., 1998).

Volatile suspended solids is a follow-up to the suspended solids test, where the dried residue and filter from the suspended solids test are placed in a furnace at 450 to 550°C for 20 minutes to burn off volatile material, air cooling for a few minutes, final cooling in a dessicator, weighing the ash residue and filter, and calculating the net weight of the ash residue. The difference between the dried residue and the ash residue is the volatile mass. The VSS test is reported in milligrams per liter. The VSS test is Method 2540E in *Standard Methods* (APHA et al., 1998).

SETTLEABILITY AND SLUDGE VOLUME INDEX. *What.* The settleability measurement is used to indicate the settling and compaction characteristics of mixed liquor in suspended-growth treatment processes at the instantaneous MLSS concentration. The settleability is reported as milliliters per liter, which represents the volume corresponding to the top of the settled solids interface from a one-liter sample of mixed liquor after 30 minutes of settling.

The SVI "normalizes" the settleability using the mixed liquor concentration and is reported as milliliters of settled volume per gram of MLSS. The SVI is used as follows:

- An SVI less than 80 indicates excellent settling and compacting characteristics,
- An SVI from 80 to 150 indicates moderate settling and compacting characteristics, and
- An SVI greater than 150 indicates poor settling and compacting characteristics.

Where. The settleability and SVI tests are conducted on a sample pulled from the end of the aeration basin before clarification.

Why. The settleability and SVI tests verify that the BNR plant has stable operation and may not experience problems with solids carryover at the clarifiers under high flow conditions. It should be noted that clarifier capacity and performance depend on many factors. The SVI test serves as an indicator of how the settleability of the mixed liquor may change over time. The SVI should then be trended against other treatment indices to evaluate how these parameters affect settleability.

When. The solids settling and compacting characteristics can vary during the day as a result of diurnal variations in organic or nutrient loadings. Samples can be pulled manually once (or more, if necessary) per eight-hour shift, tested for settleability and MLSS, and the SVI can be calculated.

How. A sample of mixed liquor is pulled from a well-mixed representative location of the aeration basin upstream of the clarifiers. This sample is gently mixed and placed in a 1-L graduated cylinder with stirring mechanism (or settleometer, calibrated beaker or, less desirable, an unstirred 1-L graduated cylinder). The settled volume is recorded approximately every 5 minutes up to 30 minutes as milliliters per liter. The 30-minute reading is generally used for settleability and SVI. One of the reasons for recording the settled volume every five minutes is that the mixed liquor in some nitrifying plants denitrifies, and solids float up before the 30 minute-period is complete.

The mixed liquor is tested for suspended solids (see the Mixed Liquor Suspended Solids, Mixed Liquor Volatile Suspended Solids, Return Activated Sludge, and Waste Activated Sludge section), and the value is used to calculate SVI, as follows:

$$\text{SVI, mL/g} = ([\text{Settleability, mL/L}]/[\text{MLSS concentration, mg/L}]) \times 1000 \text{ mg/g} \tag{12.1}$$

For example,

Settleability = 200 mL/L

MLSS concentration = 2500 mg/L

SVI, mL/g = ([200 mL/L]/[2500 mg/L]) × 1000 mg/g

= 80 mL/g

The settleability test is Method 2710C, and the SVI calculation is Method 2710D in *Standard Methods* (APHA et al., 1998).

PH. What. pH is a measurement of the acid, neutral, or basic condition of the wastewater. A pH of 7 is neutral. A pH below 7 is acidic, and a pH above 7 is basic. The pH of incoming wastewater is a function of the drinking water supply; acids or bases added through household, commercial, or industrial activity; and buffering capacity (see Alkalinity section). The pH can be changed through chemical addition, byproducts of biological activity, and interaction of the wastewater with air.

Where. The pH should be measured in the wastewater. An alternative is for sample to be pumped past a pH probe, if the sample lines and sample pump are regularly checked to make sure that the sample is representative and not influenced by fouling or plugging of the sample lines.

Why. The pH in the BNR process should be within a narrow range and should be relatively stable for stable operation. If pH is too low or varies significantly, the nutrient removal activity can be slowed to where nutrient limits cannot be met. pH can be used with alkalinity to determine if wastewater characteristics and/or the treatment process are responsible for a pH drop and whether further pretreatment by industrial contributor(s), process modification, or alkalinity addition is needed.

The optimum pH range for nitrifying bacteria is pH 7.5 to 8.6, while an acceptable pH range is 7.0 to 9.0. Nitrifying bacteria can acclimate to a pH in the 6.5 to 7.0 range, with little decrease in activity, if the pH does not vary significantly. The optimum pH for denitrification is between 7 and 8 (WPCF, 1983).

For biological phosphorus removal, the pH should be above 6.5. Although the optimum pH for chemical removal of phosphorus is 5.3 for ferric or 6.3 for alum, chemical removal of phosphorus will occur at higher pH values, with increased chemical dosage. Rarely is the pH of the wastewater adjusted to accomplish chemical removal of phosphorus, unless lime is used.

When. pH is not generally used for control purposes, but can be a quick indicator of a significant industrial constituent in the wastewater entering a BNR plant. Typically, pH in the BNR plant effluent is a compliance parameter. In larger plants, pH is monitored continuously at the plant effluent, and some plants monitor pH in the BNR process.

How. A pH probe is typically used to monitor pH. There are many interferences in wastewater that make pH measurement using pH paper unreliable. The pH test is Method 4500-H+ B in *Standard Methods* (APHA et al., 1998).

ALKALINITY. *What.* Alkalinity is a measurement of the buffering ability of the wastewater and is a rough indicator of carbonate/bicarbonate concentration in the wastewater. Alkalinity can result from hydroxides, carbonates, bicarbonates, and, less commonly, phosphates, silicates, borates, and similar compounds. With sufficient alkalinity, water can take in acid (or base) with no change or only a small change in pH. An alkalinity over 50 mg/L is necessary to prevent pH from dropping sharply when small amounts of acid are introduced to the water by either chemical addition or biological activity. Wastewater alkalinity results from alkalinity in the water supply plus alkalinity added during domestic use (e.g., cleaners and detergents). A typical alkalinity value for raw wastewater is 150 mg/L, but can range between 60 and 250 mg/L, depending on the regional alkalinity in the drinking water.

Where. Samples for alkalinity testing can be taken from the raw wastewater entering a BNR plant and in the various zones of BNR plants.

Why. The raw wastewater alkalinity can indicate if there is insufficient alkalinity to maintain pH under nitrifying conditions. A profile of alkalinity from samples taken through the BNR system (or over time for a sequencing batch reactor [SBR]) can indicate loss of alkalinity through nitrification and production of alkalinity through denitrification. The value of alkalinity testing is in conjunction with pH to identify the cause of low pH problems in a BNR plant.

Alkalinity is most important during the nitrification process, which reduces alkalinity through the metabolism of inorganic carbon by the nitrifying bacteria and produces acid when ammonia is oxidized. The nitrification reaction predicts that, for each gram of ammonia-nitrogen oxidized to nitrate-nitrogen, 7.08 g alkalinity (as calcium carbonate [$CaCO_3$]) are destroyed. When too much alkalinity is destroyed, the pH can decrease to a level outside the narrow pH range for good activity of the nitrifying bacteria.

The denitrification process replenishes alkalinity by destroying acid and producing carbon dioxide. The denitrification reaction predicts that, for each gram of nitrate-nitrogen reduced to nitrogen gas, 3.5 g alkalinity (as $CaCO_3$) are produced. Where alkalinity in the raw wastewater is low, the denitrification step can be performed before or simultaneously with nitrification to reduce the net alkalinity destruction.

When. Alkalinity is not used for process control in most BNR plants, but it is occasionally necessary in areas with low background alkalinity concentrations and/or significant nitrification needs. It is tested when pH problems are experienced, or initially when a BNR process is being designed, to identify whether alkalinity needs to be added. Alkalinity is sometimes used for process control in high-rate BNR plants that are trying to minimize nitrification by operating with low SRT while removing phosphorus only; in this case, the process can be controlled by preventing a decrease in alkalinity that signals the start of nitrification.

How. Alkalinity is a soluble constituent; thus, a sample of the bulk fluid, even if solids are not representative, will be adequate for alkalinity testing. Alkalinity is determined by titrating a wastewater sample of known volume using a standard acid solution. Most commonly, the pH of wastewater is less than 8.3, and the titration down to pH 4.5 indicates primarily bicarbonate alkalinity. The alkalinity concentration is reported in milligrams per liter as $CaCO_3$ equivalents. The alkalinity test is Method 2320B in *Standard Methods* (APHA et al., 1998).

TEMPERATURE. *What.* Temperature of the wastewater in suspended-growth systems is the temperature at which the microorganisms function and oxygen transfer is occurring. In attached-growth systems, the temperature is a function of both wastewater temperature and air temperature. Temperature typically ranges from 3 to 27°C in the United States and as high as 30 to 35°C in areas of Africa and the Middle East (Metcalf and Eddy, 2003). Wastewater temperature arises from the temperature of the drinking water supply and warm water from households, commercial, and industrial water uses. In addition, biological activity adds heat, and heat is lost from surfaces and turbulent areas open to the atmosphere.

Where. Temperature of raw wastewater indicates the temperature of wastewater entering a BNR plant. Temperature at various locations in the process basins of BNR plants indicate the temperatures at which the wastewater treatment and nutrient removal are occurring.

Why. Temperature affects the growth rate and metabolism of microorganisms. Nitrification and denitrification rates are affected by temperature; nitrifying bacteria are sensitive to temperature variations. Low temperatures require longer reaction time and higher SRT to maintain adequate population of slow-growing microorganisms (e.g., nitrifying bacteria) in BNR plants. Temperature also affects oxygen transfer into the wastewater; warmer temperatures reduce the solubility of oxygen in water and, consequently, reduce the oxygen transfer rate.

When. The temperature of wastewater can become critical at the extremes of operating temperature. When the temperature is low, biological activity slows down, and more biomass is needed. During this time, biomass production is also slowed, which requires that a higher SRT be maintained in the BNR plant. When the temperature is high, biological activity is higher, which increases the oxygen demand for treatment. During this time, oxygen transfer is lower as a result of the lower driving force (i.e., the saturation concentration of oxygen is lower in warmer water).

How. Temperature can be measured by thermometer or temperature probe.

DISSOLVED OXYGEN. ***What.*** Dissolved oxygen is a measurement of the oxygen dissolved in a liquid stream. In water, the DO concentration is limited by the maximum equilibrium concentration, or saturation concentration, which depends on the gas used (air or pure oxygen), water temperature, dissolved solids in the water, and elevation (atmospheric pressure). The DO saturation concentration decreases with increasing temperature, increasing dissolved solids concentration, and decreasing atmospheric pressure. In addition, oxygen transfer rates are never 100% efficient, and oxygen uptake occurs with biochemical oxygen demand (BOD) removal and nitrification. Oxygen uptake rates are higher with higher temperature as a result of increased microbiological activity. The critical time for maintaining adequate DO is during high-temperature periods at peak loading. It should be noted that DO, as measured in an activated sludge system, provides the DO concentration in the bulk solution and may not be directly representative of DO concentrations within the floc. The DO within the floc will generally be lower than measured in the bulk solution.

Why. Dissolved oxygen is essential for aerobic microorganisms that are responsible for BOD removal, nitrification, and biophosphorus uptake. In aeration zones, DO concentrations must be sufficient to meet the oxygen requirements of the reactions taking place and should be high enough to achieve the necessary removal rates. Biochemical oxygen demand removal can occur with DO concentrations of 0.5 mg/L or less; however, aerobic activity at low DO can lead to filamentous bulking. In nitrification, a minimum DO concentration of approximately 2.0 mg/L is typically adequate, with nitrification rate decreasing with lower DO concentrations.

Dissolved oxygen should not be present in any significant quantity in the anoxic zones, because denitrification will be inhibited. Similarly, DO should not be present in the anaerobic (fermentation) zone, because volatile fatty acid (VFA) uptake and subsequent biological phosphorus removal will be inhibited.

Where and When. Dissolved oxygen in aeration zones should be monitored several times per day, or continuously if instrumentation is used. Often, DO is used to control aeration to enhance BNR and reduce energy costs. The DO concentrations in anoxic zones, anaerobic zones, and internal recycle or RAS streams are not generally monitored, but can be checked if an upset condition is experienced.

How. The DO is typically measured using the membrane electrode Method 4500-O G in *Standard Methods* (APHA et al., 1998). The equipment includes a probe with an oxygen-permeable, membrane-covered, electrode-sensing element with thermistor for temperature reading and a meter that provides temperature compensation of the sensing element's signal and reads out the DO concentration.

The laboratory method for determining DO, such as azide modification of the iodometric titrimetric method (i.e., Standard Method 4500-O C; APHA et al., 1998), can be used; however, this is time-consuming and impractical for most BNR facilities because of the number of sample points and frequency of sampling.

OXIDATION–REDUCTION POTENTIAL. ***What.*** Oxidation-reduction potential is a measurement of the oxidation or reduction potential of a liquid. A positive ORP value indicates that an oxidation reaction is occurring in solution, and a negative value indicates that a reduction reaction is occurring in solution. The ORP is useful in wastewater treatment, because the various oxidative and reductive reactions can be measured in a spectrum of strengths, from the highest positive values (most oxidative) to the highest negative values (most reductive). The ORP is often used as an indirect measure of chlorine in odor control and disinfection applications. The ORP control setpoints in these applications would generally be greatly positive as a result of the strong oxidative qualities of chlorine.

The ORP can also be used to indicate what type of biochemical activity (i.e., aerobic, anoxic, or anaerobic) is occurring in a BNR facility. In a biological solution, such as activated sludge, an aerobic environment will result in an ORP reading higher than what would be measured in an anoxic environment. An anoxic environment would result in a higher ORP reading than what would be measured in an anaerobic environment (see Figure 12.7).

Figure 12.7 does not attempt to assign ORP values to specific environmental conditions, because the background characteristics of a specific wastewater and the type of ORP probe (silver versus platinum) will affect the readings. For these reasons, it is important that a field correlation is performed before the ORP values are used to control the system. Also, keep in mind that ORP can be influenced by interferences, such as, variation in ORP of incoming wastewater, chemical injection, or recycle streams from solids handling. A field correlation would involve defining the ORP readings under "typical" operating conditions that can be expected when the tank is

- Aerobic (oxygen is present);
- Anoxic (nitrate is present, but oxygen is not present); or
- Anaerobic (no nitrate or oxygen is present).

Once a correlation is made, an operator can measure the ORP to quickly determine if the tank is aerobic, anoxic, or anaerobic. Control systems can be programmed to control recycle rates, aeration, or other devices to meet the desired environmental conditions in the tank.

Where. The ORP should be measured in well-mixed areas in each compartment of the process basin and in the clarifier. If inline meters are used, they should be positioned in the tanks at strategic locations that allow control of the system. This position will vary, depending on the treatment configuration that was selected.

A profile would involve collecting ORP measurements at various locations of the BNR system or in the same location at different times for a SBR. Readings should be

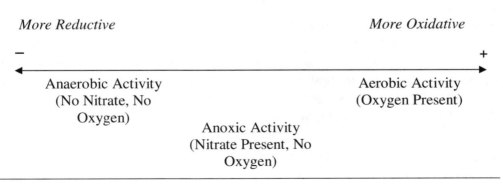

FIGURE 12.7 Relative ORP readings.

taken from the same locations each time to develop a meaningful database of ORP information.

Why. The ORP is a quick, simple measurement that can indicate a condition consistent with aerobic, anoxic, or anaerobic activity in a specific process area. A handheld meter can be used to quickly check a single location or to develop an ORP profile through the process. If your facility has varying background conditions as a result of highly variable wastewater ORP, injection of chemicals that influence ORP, or return streams from solids handling at certain times of day, the ORP can provide a general indication, but should not be relied upon with specific ORP values as control limits or indicators because of the effect of those activities on the ORP readings.

The intent of a profile would be to determine if the tanks (or times) that are designated to be aerobic, anoxic, or anaerobic are meeting their intended conditions. The ORP can also indicate the presence of an unusual wastewater discharge, if the discharge has a strong oxidizing or reducing characteristic.

When. For profiles to be useful, they need to be collected on a periodic basis when the system is performing well. Comparative data should be collected when the system is not performing well. If there is a change in the environments, additional investigation into why the changes have occurred may assist in identifying corrective actions that could help reestablish good treatment. This is likely most useful in biological phosphorus removal (BPR) systems for confirming that the anaerobic zone is truly anaerobic. It should be noted that low nighttime loadings may cause daily fluctuations that would go undiscovered if only performing this profile during the day. For this reason, it may be important to run a nighttime profile, if staffing allows.

How. Handheld meters can be used to profile treatment by dropping the weighted probe into a well-mixed area and obtaining a reading on the ORP meter. The ORP can also be measured using inline measurement, as discussed in the Oxidation-Reduction Potential section of Chapter 13. The ORP measurements, equipment care, and calibrations should be performed as described by the manufacturer of your equipment. You will need to know the type of probe that you own when comparing your ORP values to ORP values obtained or reported by others with a different probe.

AMMONIA AND TOTAL KJELDAHL NITROGEN. *What.* The nitrogen species of interest are discussed in greater detail in Chapter 11. In wastewater, some of the organic nitrogen is hydrolyzed by microorganisms to ammonia-nitrogen (NH_3-

N) in the collection system and the treatment plant. The total Kjeldahl nitrogen (TKN) is oxidized by microorganisms under aerobic conditions and is taken up in new cell growth by the biomass. The oxidation of NH_3-N requires 4.6 kg of oxygen for each kilogram of NH_3-N. The nitrogen uptake in cell growth results in approximately 2 to 7% total nitrogen (TN) in WAS.

Where. Ammonia-nitrogen and TKN can be measured in the influent wastewater, influent to the BNR process, and clarifier effluent. Where there is significant biomass (e.g., mixed liquor), the concentration of organic nitrogen is high, and microbiological activity can change the concentration of nitrogen species, unless the sample is filtered or measured in situ. The NH_3-N can be measured through the BNR process and has value exiting the aeration zones.

Why. Depending on temperature and pH, NH_3-N in wastewater effluent can have toxic effect on microorganisms, macroinvertebrates, and fish in surface waters; this is often the reason for low NH_3-N effluent limits in BNR plants discharging to surface waters. In addition, NH_3-N, if present in very high concentrations, can inhibit nitrification.

The TKN indicates how much nitrogen is in the unoxidized form and represents a nitrogenous oxygen demand, as TKN is oxidized to NO_3-N. This oxygen demand can cause a DO sag (depressed DO concentration) in surface waters downstream of discharge. Also, algae blooms resulting from nutrient release can severely depress DO, especially at night.

The nitrogen species that are included in TN are TKN (which is the total of NH_3-N and organic nitrogen), nitrite-nitrogen (NO_2-N), and nitrate-nitrogen (NO_3-N). Total nitrogen is often a compliance parameter in advanced wastewater treatment facilities.

When. The TKN in plant influent is typically run at the same frequency as compliance monitoring in plants that have a TN limit; NH_3-N in plant influent can also be run at the same time, primarily to evaluate industrial contributions. The TKN and NH_3-N should also be periodically checked in return streams from solids handling processes to evaluate their effect on BNR process performance.

Ammonia-nitrogen is sometimes monitored at the end of aeration zones with NO_2-N and/or NO_3-N to evaluate or control the level of aeration and evaluate nitrification performance. These tests can be run several times per day or can be continuous, if automated measurement is used. The NH_3-N in the effluent should be checked daily and, if a compliance parameter, in accordance with permit requirements.

How. See Chapter 11 of this manual, 40 CFR 136 (Guidelines Establishing Test Procedures for the Analysis of Pollutants, 2005), and/or *Standard Methods* (APHA et al., 1998) for discussion on the analysis of these parameters.

NITRITE-NITROGEN. *What.* The nitrogen species of interest are discussed in greater detail in Chapter 11. In wastewater, NO_2-N can enter the collection system with industrial discharges and can be generated through partial nitrification at the treatment plant. Nitrite-nitrogen is generally not present in significant concentrations in most treatment plants, but can occur if there is incomplete nitrification.

Where. Nitrite-nitrogen can be measured in the plant influent, aeration zones, clarifier effluent, and final effluent after disinfection. Where there is significant biomass (e.g., mixed liquor), microbiological activity can change the concentration of nitrogen species, unless the sample is filtered or measured in situ.

Why. Nitrite-nitrogen in wastewater effluent can have toxic effect on microorganisms, macroinvertebrates, and fish in surface waters. In addition, NO_2-N, if present in high concentrations, can increase chlorine demand. The presence of significant concentrations of NO_2-N can be used to indicate the need for additional aeration in aeration zone(s).

When. Nitrite-nitrogen can be measured in situ or by using a laboratory test. Generally, the laboratory tests are run with NO_2-N measured with NO_3-N as NO_2+NO_3-N. Nitrite-nitrogen can be measured separately from NO_3-N through the BNR process, if there is concern over incomplete nitrification or high chlorine demand. These tests can be run several times per day or can be continuous, if automated measurement is used.

How. See Chapter 11 of this manual, 40 CFR 136 (Guidelines Establishing Test Procedures for the Analysis of Pollutants, 2005), and/or *Standard Methods* (APHA et al., 1998) for discussion on the analysis of these parameters.

NITRATE-NITROGEN. *What.* The nitrogen species of interest are discussed in greater detail in Chapter 11. In wastewater, NO_3-N can enter the collection system with industrial discharges and can be generated through complete nitrification at the treatment plant.

Where. Nitrate-nitrogen can be measured in the plant influent, aeration zones, clarifier effluent, and final effluent after disinfection. Where there is significant biomass (e.g., mixed liquor), microbiological activity can change the concentration of nitrogen species, unless the sample is filtered or measured in situ.

Why. Nitrate-nitrogen in wastewater effluent can be limited by TN effluent limits or NO_3-N limits for groundwater discharge. The NO_3-N profile through the BNR process can indicate performance of nitrification and denitrification and can be used to control aeration or internal recycle rates and make decisions regarding SRT.

When. Nitrate-nitrogen can be measured in situ or by using a laboratory test. Laboratory tests can use either an ion-specific probe or colorimetric test. Colorimetric tests are run with NO_3-N measured with NO_2-N as NO_2+NO_3-N. If NO_3-N is desired using colorimetric tests, NO_2-N is determined, and then NO_3-N concentration is calculated by subtracting the NO_2-N concentration from the NO_2+NO_3-N concentration.

These tests can be run several times per day or can be continuous, if automated measurement is used. The NO_3-N (or NO_2+NO_3-N) in the effluent should be checked daily and, if a compliance parameter, in accordance with permit requirements.

How. See Chapter 11 of this manual, 40 CFR 136 (Guidelines Establishing Test Procedures for the Analysis of Pollutants, 2005), and/or *Standard Methods* (APHA et al., 1998) for discussion on the analysis of these parameters.

TOTAL PHOSPHORUS. *What.* Phosphorus is a macronutrient needed by all living things to reproduce and grow cell mass. Phosphorus is generally regarded as the limiting factor for algae growth in freshwater streams and lakes. It is for this reason that many wastewater treatment plants have phosphorus effluent limits in an attempt to reduce the total phosphorus (TP) load to the watershed. When algae are produced in streams and lakes, they eventually die and are decomposed by organisms, resulting in a reduction of DO. In addition to the potential immediate effects of low DO, the decomposed algae settle to the bottom of the body of water and contribute to the process of filling the stream or lake, making it shallower.

When and Where. Total phosphorus should be monitored as required by your National Pollutant Discharge Elimination System permit and as often as practical from industry, at the influent, from any sidestream, and at the effluent for process control. It is important that enough data is collected when the facility is running well to form a reliable baseline of information.

Why. Because phosphorus is needed for reproduction and cell growth, there is a need for phosphorus in biological wastewater treatment systems. Using the generally recognized ratio of 100:5:1 (BOD:N:P), which is often used to demonstrate nutrient needs, we can get a relative feel for how much phosphorus is consumed to treat the influent wastewater facility that is not performing BPR. Using this ratio, it can be estimated that 1 mg/L of phosphorus is needed for cell growth for every 100 mg/L of BOD removed. This estimate will vary, based on the type of system, and is offered here only to illustrate that all biological systems will remove some phosphorus. Most municipal wastewater treatment plants will have more than enough phosphorus to satisfy cell growth, and the remainder needs to be removed through treatment enhancements.

Phosphorus can occur in many forms and will typically change forms through secondary wastewater treatment. Among the forms of phosphorus, some will be dissolved, and some will be part of the suspended solids. The soluble fraction is generally all in the orthophosphorus state (see the next section titled Orthophosphorus), and these two forms are often used interchangeably.

How. See Chapter 11 of this manual, 40 CFR 136 (Guidelines Establishing Test Procedures for the Analysis of Pollutants, 2005), and/or *Standard Methods* (APHA et al., 1998) for discussion on the analysis of this parameter.

ORTHOPHOSPHORUS. *What.* Orthophosphorus is a soluble form of phosphorus that is very common in treated secondary effluent. Most of the particulate phosphorus will be contained within the activated sludge and should, therefore, not discharge the facility in the effluent, with the exception of the phosphorus included in the effluent suspended solids. For this reason, the majority of the effluent phosphorus in a facility with low effluent TSS will be in the orthophosphorus form.

When and Where. Test kits can be purchased to quickly determine the orthophosphorus, making this an attractive process control tool to some operators. It should be noted that the orthophosphorus data generated by a test kit do not generally provide a reportable value and should be considered an estimate, because the kits are generally not as precise as the laboratory version. In an effluent with low suspended solids, many operators test their effluent quality, on a frequent basis, with the orthophosphorus and routinely compare it to their TP data. These comparisons allow the operators to determine how reliable their orthophosphorus data is in predicting TP.

Why. Orthophosphorus can also be used within BPR processes to quickly monitor the various stages of the system for the release of phosphorus in the anaerobic zone and the subsequent luxury uptake in the aerobic zone. The solids should be removed by allowing the samples to settle or filtering them and should be analyzed as quickly as possible.

How. If using a test kit, follow the manufacturer's directions for proper analysis. If following a laboratory method for orthophosphorus analysis, see 40 CFR 136 (Guidelines Establishing Test Procedures for the Analysis of Pollutants, 2005) or *Standard Methods* 4500-P (APHA et al., 1998).

CHEMICAL OXYGEN DEMAND. *What.* Chemical oxygen demand (COD) is a measure of the COD of a sample that is used in similar applications as BOD. At approximately three hours, the COD analysis is much faster than the standard five-day BOD (BOD_5) test. Variations of the COD analysis have been developed in an attempt to better characterize the effect of COD on biological systems by isolating certain fractions of the COD that are directly used by the phosphate-accumulating organisms (PAOs). Figure 12.8 shows these influent COD subdivisions, as presented by Ekama et al. (1984).

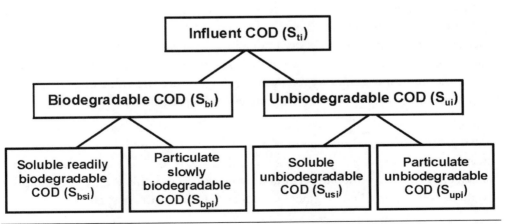

FIGURE 12.8 Division of the total influent COD in municipal wastewater.

The influent COD (S_{ti}) is first subdivided into biodegradable and unbiodegradable fractions. The unbiodegradable fractions will typically pass through the treatment system (S_{usi}) or eventually be wasted from the system (S_{upi}) in the form of particulate matter in the MLSS or primary sludge. The biodegradable fraction is broken into readily biodegradable (S_{bsi}) and slowly biodegradable (S_{bpi}). The S_{bsi} fraction is the key component when dealing with any BNR system evaluation.

Where. Samples for total COD should be collected from the plant influent, just before the influent reaches the secondary treatment process, and the effluent from the secondary treatment process. Other locations may be sampled to quickly evaluate organic contributions to the loading on the secondary treatment process. Samples for soluble COD should be collected from just before the influent reaches the secondary treatment process (S_{bsi}), and the effluent from the secondary treatment process (S_{usi}).

Why. In the anaerobic zone, typically only the S_{bsi} component is susceptible to fermentation to form VFAs within the short retention time (1 to 2 hours). Ekama et al. (1984) found that phosphorus release in the anaerobic zone increased as the influent Sbsi increased. He concluded that a S_{bsi} concentration of at least 25 mg/L was necessary surrounding the PAO in the anaerobic zone for phosphorus release to occur. Other research indicates that a S_{bsi} influent concentration of 60 mg/L is needed for reliable enhanced biological phosphorus removal (EBPR) to occur. Every plant has its own unique wastewater characteristics and S_{bsi} concentrations that correspond to good EBPR performance. Operations staff can develop a database to determine if changes in treatment efficiencies are related to changes in influent wastewater characteristics, such as S_{bsi}. These data can be trended against the effluent quality and other operating parameters to better understand the conditions that affect treatment. A facility can elect to use total COD in their sampling protocol; however, variations in the amount of inorganic oxygen demand may reduce the value of this data. The S_{bsi} analysis may be the preferable parameter for tracking feed to the BPR process.

When. The COD or S_{bsi} data are most useful when they are collected on a regular basis, so that results obtained during good treatment can be compared with results obtained during poor treatment. Sample frequency will likely depend on the resources of each facility, but is recommended to occur at least weekly to monthly. Additional sampling should be conducted when treatment is poor and during seasonal changes.

How. The analysis involves the chemical oxidation of the sample followed either by a titrimetric or color reading of the result to determine the result expressed as milligrams per liter of oxygen (O_2). Because the COD process oxidizes both the inorganic and organic components of the sample, the COD results will typically be greater than the BOD results for the same sample. The BOD results are typically between 50 and 60% of the COD results, but each wastewater will have its own ratio.

The base analysis is the same for both the COD analysis and the S_{bsi} analysis and should be performed based on an approved method, such as 5220 A-D in *Standard Methods* (APHA et al., 1998).

Analysis of S_{bsi} requires two samples of COD to be run: one on the influent to the BNR process and one from the effluent. Both samples receive chemical flocculation and filtration before the COD analysis is conducted. By subtracting the effluent truly soluble COD value from the influent truly soluble COD value, the amount of COD that was not biodegradable will get subtracted from the total. A rapid physical-chemical method for the determination of S_{bsi} in municipal wastewater was developed by Mamais et al. (1993). This method is based on the assumption that the S_{usi} is equal to the truly soluble effluent COD from an activated sludge plant treating wastewater. The method is also described by Park et al. (1997).

The S_{bsi} is determined based on the following equation:

$$S_{bsi} = sCOD - S_{usi} \tag{12.2}$$

Where

S_{bsi} = readily degradable soluble COD,
SCOD = influent truly soluble COD, and
S_{usi} = influent non-readily biodegradable soluble COD.

The additional flocculation step of the Sbsi analysis is described here.

(1) Add 1 mL of 100 g/L zinc sulfate to a 100-mL wastewater sample and mix well with a magnetic stirrer for 1 minute;
(2) Adjust the pH to approximately 10.5 with a 6-M sodium hydroxide (NaOH) solution;
(3) Settle quiescently for a few minutes;
(4) Withdraw clear supernatant (20 to 30 mL) with a pipette and pass through a 0.45-μm membrane filter; and
(5) Measure COD of the filtrate using approved method.

The validity of the flocculation method was assessed by Mamais et al. (1993) by comparing it with the biological method developed by Ekama et al. (1984). Results from four domestic wastewaters demonstrated that the two methods gave virtually identical results.

VOLATILE FATTY ACIDS. *What.* The performance of an EBPR process will vary with the specific VFAs available in the anaerobic zone. The COD consumption during BPR has been estimated by past research to be 50 to 60 mg/L COD per milligram per liter of phosphorus removed from municipal wastewater. Table 12.2, taken from Abu-ghararah and Randall (1991), lists the effect of organic substrate on EBPR. According to this research, acetic acid (HAc) is the most efficient VFA for enhanced BPR.

Fortunately, acetic acid is typically the primary VFA formed from wastewater fermentation, with propionic acid as the secondary VFA. Sufficient VFA production can occur in wastewater collection systems, especially in long forcemains or gravity sewers with long detention times. Several facilities have seen improved reliability of EBPR with the addition of a VFA-laden waste stream, such as a long force main serving an area several miles from the publicly owned treatment works. Fermentation of wastewater or primary sludge involves the reduction of proteins and carbohydrates into, most commonly, acetates, propionates, and butyrates.

Where. The samples should be collected on the influent to the BPR tanks before the introduction of RAS. For plants with primary clarifiers, the sample should be primary effluent; for plants without primary clarifiers, the sample should be the effluent

TABLE 12.2 Effects of organic substrate on EBPR (Abu-ghararah and Randall, 1991).

Substrate	mg/L phosphorus uptake per mg/L COD used	mg COD used per mg phosphorus removed
Acetic acid	0.37	16.8
Propionic acid	0.10	24.4
Butyric acid	0.12	27.5
Isobutyric acid	0.14	29.1
Valeric acid	0.15	66.1
Isovaleric acid	0.24	18.8
Municipal wastewater	0.05	102

from preliminary treatment. Samples should also be collected on the elutriate from prefermentation facilities to determine the performance of these units in generating additional VFAs that are directed to the EBPR anaerobic zone(s).

Why. Enhanced biological phosphorus removal requires the presence of VFA in the anaerobic zone of any BNR wastewater treatment system. When the influent wastewater is weak and not septic, VFA production can potentially be accomplished in the anaerobic zone of the BNR process or outside the BNR system, in a separate process called prefermentation. A facility can elect to continue monitoring VFAs as a process control parameter. This may prove especially useful at BPR facilities that do not have an abundance of VFAs in the influent to the BPR tanks. These facilities can potentially be optimized by tracking the abundance of VFAs and adjusting their processes accordingly.

When. Samples should be taken during times of adequate BPR to obtain a baseline to compare to periods of poor performance. Typically, approximately 40 to 50 mg/L of HAc as COD is desired in the influent to the anaerobic zone; however, this can vary from plant to plant. If the anaerobic zone is larger (greater than 1.5 hours of detention time) then VFA production can potentially be generated within the zone.

How. One method to determine the VFA concentration is to use gas chromatography to measure acetic acid and propionic acid concentrations. These two organic acids are the most prevalent forms in municipal wastewater, and the summation of the two is generally adequate to determine VFA concentrations for EBPR process control. However, this method is difficult to perform, needs special equipment, and would not typically be run in-house, especially at smaller plants. For smaller plants, it would be advisable and likely more cost-effective to send the samples out to a contract laboratory. A sample could be sent to the laboratory, split, and analyzed for VFA and Sbsi to determine if a correlation exists. If a correlation exists, it may be easier to run Sbsi more frequently as a process performance indicator.

A simple method using distillation and titration is contained in *Standard Methods* 5560-organic and volatile acids (APHA et al., 1998). This method, commonly used as a control test for anaerobic digestion, does not appear to give reliable results and gives incomplete and somewhat variable recovery. It should be noted that digester supernatant, although apparently high in VFAs, appears to inhibit EBPR rather than improve performance. It has been hypothesized that this is a result of the presence of hydrogen sulfide and elevated levels of phosphorus in the recycle stream.

An alternative method to determine sufficient fermentation of the influent wastewater is to compare the COD:TP ratio to the BOD_5:TP ratio. If the COD:TP ratio is considerably higher than 40:1 and the BOD_5:TP ratio is considerably lower than 20:1, then the wastewater has likely not undergone sufficient fermentation (Randall et al., 1992).

SOLUBLE BIOCHEMICAL OXYGEN DEMAND. *What.* Soluble BOD (SBOD) is the BOD of the filtrate of a sample. The soluble fraction of the BOD that is loaded to the anaerobic zone of a BPR facility may be more accurate in predicting the response of the PAOs than total BOD would be. This is because the PAOs are able to use only the soluble fraction of the BOD in gaining their competitive advantage while in the anaerobic zone. The SBOD should not replace the BOD_5 for calculating the oxygen demand exerted on the biological system, but may be useful in predicting phosphorus removal success.

Where. Samples for SBOD should be collected just before the influent (primary effluent) reaches the BPR tanks. Soluble phosphorus analysis should also be analyzed on these samples.

Why. The BPR process is dependant on having a readily degradable food source available to the PAOs to remove phosphorus. The greater the ratio of BOD:TP or SBOD:SP, the better chance that the system will be capable of adequately removing enough phosphorus to achieve a high quality effluent. Review of this data over time will allow a facility to determine if variations in effluent quality are the direct result of changes in the amount of food available to the system. It should be noted that the SBOD:SP ratio offers the advantage of being in the correct form to predict how the system will react and is likely a better control parameter than a BOD:TP ratio; however, it may not be as good an indicator as S_{bsi} or short-chain VFAs. The SBOD:SP ratio may be the best indicator for a facility that does not have the equipment and staff needed to run COD or VFA analysis. The key disadvantage of using BOD or SBOD is that it takes five days to get the results; it is, therefore, not a good predictor of the immediate future, but instead could be used to identify long-term trends. Designers may use ratios of 15:1 (SBOD:SP) or greater, and possibly a ratio of 20:1 (BOD:TP), as a threshold when determining if a facility is a good candidate for BPR. If the ratios that are being measured at your facility are at or greater than those values, it is not likely that the facility would be considered BOD-limited. See the optimization/troubleshooting guides in this chapter for additional assistance troubleshooting your facility.

When. These samples should be collected at the normal sample frequency to allow trending of this data. Additional samples should be collected any time there is a significant change in the effluent quality. If it is not practical to perform this analysis that often, this ratio can be used as an additional spot check. It should be noted that this data will not be as valuable if it is recorded too infrequently to properly characterize the wastewater during good operation.

How. A SBOD sample is simply the BOD run on the filtrate of a filtered sample. A sample should be collected and immediately filtered through a 0.45-μm filter into a clean filter flask. The filtrate should then be analyzed for BOD, as described in an approved method, such as Standard Method 5210 (APHA et al., 1998).

NITRIFICATION TEST. ***What.*** The nitrification test described below is intended for suspended-growth systems, that is, an activated sludge process. The purpose of the test is to initially determine a baseline indicator value of nitrification rate for a specific facility and determine the performance of the facility relative to that baseline under the various operating conditions encountered.

Where. The samples for the nitrification test should be collected from the influent to the BNR process before the introduction of RAS and from the RAS.

Why. The performance of the nitrification process depends on the concentration of nitrifiers in the system, the ammonia concentration, and environmental factors (such as wastewater temperature, dissolved oxygen concentration, pH and alkalinity, and the presence or absence of inhibitory compounds in the influent). The nitrification test provides adequate supply of ammonia, dissolved oxygen, and alkalinity. Influent pH, influent wastewater temperature, and biomass concentration (MLSS and MLVSS) are recorded for the test, and the biomass is presumably acclimated to the influent and contains a sufficient proportion of nitrifiers to achieve nitrification. Establishing a baseline indication of nitrification performance is useful at the various temperatures encountered and concentrations of MLSS and MLVSS controlled by plant operations. Periodic use of the test reinforces the baseline indicator values of nitrification rate and will indicate if there is some factor, such as wastewater temperature, inhibitory compounds, or nitrifier population, that is adversely affecting the nitrification process if poor process performance occurs. If the nitrification test shows that adequate nitrification can occur but the plant performance is lacking, then look for conditions related to insufficient DO, inadequate detention time or inadequate mixing, insufficient alkalinity and lowered pH in the process, or intermittent slugs of inhibitory compounds. If the nitrification test shows poor nitrification compared to the baseline, then look at for lowered SRT, insufficient DO, high organic loading, low wastewater temperature, and presence of inhibitory compounds.

When. The test should be performed regularly to maintain baseline information and assist in judging the effectiveness of operational adjustments necessary to maintain nitrification during seasonal changes in wastewater temperature. The test can also be performed in multiple reactors several times a day if slugs of inhibitory compounds are suspected based on observations of full-scale nitrification performance.

How. There are three basic tests that determine nitrification maximum growth rate. These are the Low F/M SBR Method, Washout Method, and High F/M Method as described in the Water Environment Research Foundation publication, *Methods for Wastewater Characterization in Activated Sludge Modeling* (WERF, 2003). The following test is a simplified version of the Low F/M SBR Method and will give a rough indication of nitrification performance but will not be able to provide kinetic information for modeling purposes. To run the nitrification test, a reactor capable of aerating 19 L of mixed liquor will be filled with one-third RAS and two-thirds BNR process influent. A gram of NH_4Cl will be added and the air will be turned on. For a total reactor volume of 19 L, this will provide an additional 13.8 mg/L of NH_3-N, resulting in an initial concentration of approximately 27 to 40 mg/L NH_3-N. Alkalinity will be adjusted with 3 g Na_2CO_3 to provide an additional 150 mg/L alkalinity as $CaCO_3$. A sample will also be taken for MLSS and MLVSS at the beginning and end of the run. Also, the pH and temperature of the mixture should be measured and recorded. The sample will be aerated to maintain the dissolved oxygen at greater then 2 mg/L. A sample will be taken every 15 minutes for a total of 2 hours. The sample bottle will be labeled with the reactor number, date, time, and temperature.

The maximum volume of the eight samples, which are taken over a 2-hour period, will be 200 mL each. The samples will be divided up for the different analyses that will be run in the laboratory. Reserve a volume of up to 20 mL for MLSS and MLVSS analyses. Immediately centrifuge or filter the remaining 180 mL to remove the solids and split the filtrate to provide a volume of 60 mL for NO_2+NO_3-N analysis and 120 mL for alkalinity analysis.

The NO_2+NO_3-N concentrations will be plotted with time. The slope of the line is the production rate of nitrite-N and nitrate-N, and this can be the indicator used to evaluate nitrification process performance. This information can also be used to approximate maximum nitrifier growth rate.

DENITRIFICATION TEST. *What.* The nitrification test described below is intended for suspended-growth systems, that is, an activated sludge process. The purpose of the test is to initially determine a baseline indicator value of nitrification rate for a specific facility and determine the performance of the facility relative to that baseline under the various operating conditions encountered.

Where. The samples for the nitrification test should be collected from the influent to the BNR process before the introduction of RAS and from the RAS.

Why. The performance of the denitrification process depends on the concentration of denitrifiers in the system, the nitrate concentration, and environmental factors (such as wastewater temperature, dissolved oxygen concentration, pH, and the presence or absence of inhibitory compounds in the influent). The denitrification test provides an adequate supply of nitrate and no dissolved oxygen. Influent pH, influent wastewater temperature, and biomass concentration (MLSS and MLVSS) are recorded for the test, and the biomass is presumably acclimated to the influent and contains a sufficient proportion of denitrifiers to achieve denitrification. Establishing a baseline indication of denitrification performance is useful at the various temperatures encountered and concentrations of MLSS and MLVSS controlled by plant operations. Periodic use of the test reinforces the baseline indicator values of denitrification rate and will indicate if there is some factor, such as wastewater temperature, inhibitory compounds, or inadequate denitrifier population, that is adversely affecting the denitrification process if poor process performance occurs. If the denitrification test shows that adequate denitrification can occur but the plant performance is lacking, then look for conditions related to excessive DO, inadequate detention time or inadequate mixing, or intermittent slugs of inhibitory compounds. If the denitrification test shows poor denitrification compared to the baseline, then look for low SRT, high or low pH, low wastewater temperature, and presence of inhibitory compounds.

When. The test should be performed regularly to maintain baseline information and assist in judging the effectiveness of operational adjustments necessary to maintain denitrification during seasonal changes in wastewater temperature. The test can also be performed in multiple reactors several times a day if slugs of inhibitory compounds are suspected based on observations of full-scale nitrification performance.

How. For this test, a reactor capable of mixing 19 L of mixed liquor will be filled with one-third RAS and two-thirds BNR process influent. Three grams of $NaNO_3$ will be added to the mixture. The measured dissolved oxygen should be 0 mg/L. A sample will also be taken for MLSS and MLVSS at the beginning and end of the run. Also, the pH and temperature of the mixture should be measured and recorded. A sample will be taken every 15 minutes for a total of 2 hours. The sample bottle will be labeled with the reactor number, date, time, and temperature.

The maximum volume of the eight samples, which are taken over a 2-hour period, will be 200 mL each. The samples will be divided up for the different analyses that will be run in the laboratory. Reserve a volume of up to 20 mL for MLSS and MLVSS analyses. Immediately centrifuge or filter the remaining 180 mL to remove the solids and split the filtrate to provide a volume of 60 mL for NO_2+NO_3-N analysis and 120 mL for BOD_5 analysis.

The NO_2+NO_3-N concentrations will be plotted with time. The slope of the line is the denitrification rate and can be the indicator used to evaluate the denitrification process performance.

BIOLOGICAL PHOSPHORUS REMOVAL POTENTIAL TEST. *What.* The ratio of organic compounds concentration/phosphorus concentration in plant influent has been used to determine the expected performance of EBPR and to project effluent phosphorus concentrations. However, the reliance on these ratios to predict EBPR potential may exhibit some shortcomings. Research has indicated that plants have been built based on optimistic assumptions and limited data and have proven not to perform as well as initially predicted. This has been the case at some plants where the indicated ratios were favorable for EBPR, but the plant did not remove phosphorus biologically, as desired. Therefore, to assist in the operation and evaluation of these facilities, a bench-scale procedure can be performed to determine if the influent wastewater is amenable to EBPR.

Where. The samples should be collected on the influent to the BPR tanks before the introduction of RAS. For plants with primary clarifiers, the sample should be primary effluent; for plants without primary clarifiers, the sample should be the effluent from preliminary treatment.

Why. The performance of an EBPR unit process is strongly affected by the characteristics of the wastewater influent to the anaerobic zone of the activated sludge system. Biological phosphorus removal cannot be accomplished without specific biodegradable organic substrate. The organic compounds necessary for EBPR are VFAs, such as acetic acid, that are produced from wastewater fermentation. Sufficient wastewater fermentation generally occurs in the collection system, particularly in the summer. Additional fermentation, if needed, has been accomplished at treatment facilities by using specially designed fermenters and, in some instances, increasing the size of the anaerobic zone.

Relationships have been developed for operating plants, such as ratio of total influent BOD to total phosphorus (influent BOD_5:TP), to determine what effluent phosphorus concentrations are achievable. Most data from operating treatment plants indicate that a BOD_5:TP ratio of 20:1 is sufficient to accomplish EBPR (Randall et al., 1992). However, each plant has its own unique influent wastewater characteristics aside from BOD_5 that affect actual TP removal efficiency, such as readily degradable COD and, more specifically, VFAs.

When. The EBPR potential test can be performed when it is suspected that the influent wastewater strength is affecting BPR performance or secondary phosphorus release is occurring in the anaerobic reactor. This test can help confirm that the operation of the facilities is being effected by a low-strength wastewater and give the operator an idea of how to correct the problem. The test is run on a grab influent sample, as the storage of the influent could promote additional fermentation that would not occur in the full-scale system.

How. METHOD. The method is based on a Wisconsin Department of Natural Resources research report (Park et al., 1999).

PRESERVATION. Perform immediately:

- Phosphorus samples preserved using sulfuric acid and refrigerated to 4°C,
- COD samples preserved using sulfuric acid and refrigerated to 4° C, and
- Nitrate samples (if collected) preserved using sulfuric acid and refrigerated to 4° C.

ANALYZE. Run test immediately, 28 days hold time for phosphorus, COD, and nitrate and nitrite samples.

EQUIPMENT AND SUPPLIES.

- Six-paddle stirrer   *Alternatively* two magnetic stirrers
- Stopwatch
- DO meter and probe
- ORP meter and probe (optional)
- pH meter and probe
- Two 1-L graduated cylinders
- Two 2-L jars
- Large-tip pipettes   *Alternatively* turkey basters
- Analytical balance   *Alternatively* preweigh sodium
- Acetate
- Vacuum pump
- Filter flask   *Alternatively* a filter apparatus
- RAS sample containing PAOs
- Wastewater sample
- Filters
- Gloves
- Tweezers

REAGENTS AND CHEMICALS.

- Sodium acetate trihydrate crystal ($CH_3COONa \cdot 3H_2O$): dilute 0.106 g sodium acetate in 100 mL of the 1 L of wastewater designated for the acetate sample. Stir until fully dissolved and recombine with the wastewater sample, for a total volume of 1 L. This is equivalent to 25 mg COD/L after it is mixed with the RAS sample. If using more or less wastewater in the sample, adjust the sodium acetate accordingly. Note: If using anhydrous sodium acetate in place of sodium acetate trihydrate, 0.064 g should be added.
- Sulfuric acid (H_2SO_4): concentrated.

PRINCIPLE. The PAOs that are responsible for BPR require a readily degradable food source to gain a competitive advantage. These PAOs gain an advantage because they can release phosphorus to gain energy in anaerobic conditions and use this energy to uptake readily degradable BOD. If the readily degradable BOD in the wastewater is inadequate, the PAOs will only have a limited advantage and consequently may not be able to remove enough phosphorus to meet effluent TP limits. This screening procedure compares the amount of phosphorus release obtained from samples with and without supplemental acetate added as a BOD source. For both tests, RAS from a successful BPR facility is used to provide PAOs. Comparison between the sample that has been supplemented with acetate and the one that has not will graphically indicate the likelyhood that influent wastewater characteristics could support BPR. The amount of phosphorus released from the "wastewater only" sample can be used to estimate the effluent TP concentration, as described below.

PREPARATION OF SAMPLES.

(1) Collect wastewater samples just upstream of the biological treatment process and upstream of RAS introduction. Wastewater samples should be composited over a 3- to 4-hour period. This provides reasonable composite samples, while minimizing sample holding time. Review historical data for variations in influent volumes and concentrations. Highly variable influent characteristics will require additional testing to properly characterize the facility.

(2) Collect a sample of RAS from a facility that is successfully performing BPR. Mixed liquor suspended solids containing PAOs could also be used. Any sample collected should be from after the aeration tank, so that the PAOs have completed phosphorus uptake. Fill container to the brim and minimize turbulence to minimize incorporation of DO. At least one liter of sample is required for every jar that that is to be tested. Place on ice if the test can not be started within two hours of sampling.

(3) If analysis can not be performed immediately following collection, the samples should be cooled but returned to room temperature before testing.

CONSIDERATIONS.

- The process described simplifies the wastewater to RAS ratio by always using 1:1 by volume. If RAS concentrations are too high or low for anticipated operating conditions, the ratio can be adjusted, but will need to be consistent for both samples. Other points include the following: (1) It is important to use the same ratios for all samples to avoid introducing variability to the test; (2) a mass-balance calculation can be run to determine the best ratio if the solids concentrations are known; and (3) it is better to approximate this mixture than to wait for a TSS analysis to be run.

- Nitrate analysis may be warranted if there is concern for the amount of nitrate interference with the uptake. Nitrate can be analyzed on the same sample that was collected, filtered and preserved for phosphorus analysis, but this will require additional sample and will increase the analytical expense of the test. A combination of nitrite + nitrate (NO_2+NO_3-N) will allow a longer hold time and should provide similar results.
- Chemical oxygen demand should also be analyzed to determine the change in available soluble substrate. The COD can be analyzed on the same sample that was collected, filtered and preserved for phosphorus analysis, but this will require additional sample and will increase the analytical expense of the test.
- Oxidation-reduction potential can also be used to monitor potential nitrate interference, providing real-time data. As oxygen is consumed, the ORP measurement will decrease, reflecting anoxic conditions. The ORP values will decrease again when the nitrate is consumed, and the biological activity in the reactor becomes anaerobic. Recording significant changes in ORP could be used to explain changes in the release rate.

PROCEDURE.

(1) Measure 1 L of influent into both 2-L batch reactors.
(2) Add 25 mg COD/L of sodium acetate to one reactor only (see Reagents and Chemicals section above).
(3) Add 1 L of PAO-containing sludge to each reactor.
(4) Begin mixing the samples at a low rate-high enough to mix the sample, but not so high that oxygen transfer could occur at the surface.
(5) Measure DO (or ORP) to confirm that the samples are oxygen-deficient before starting the stop watch. Monitor DO (or ORP) and pH through the analysis and note significant changes.
(6) Collect and filter samples every 30 minutes for approximately 2 hours, starting at the 0-minute mark for both reactors (Note: time management is critical during this test. A staggered start will allow one person to run more than one reactor at a time, while collecting samples at similar times in the

respective batches. It is also critical that filtration can be performed in the available time. Filtration may be easier if mixing is discontinued for a couple of minutes before sample collection. Sample is then drawn from the clear liquid using a large pipette.). Interval and duration may vary depending on scenario; industrial wastewater with high influent phosphorus should be sampled for at least 3 hours (this should be determined before starting).

(7) Preserve filtered samples for analysis of TP by lowering the pH of samples to less than pH 2 with sulfuric acid (H_2SO_4).

TEST EVALUATION.

(1) Graph the test results of both reactors to compare the release rate with and without the supplemental sodium acetate. Two graphs are provided as examples of possible outcomes. Graph 1 (Figure 12.9a) indicates that there is little potential for BPR with the facility's wastewater characteristics. This is indicated by the lack of a phosphorus release in the sample that does not contain sodium acetate. The PAO organisms proved to be viable when reviewing the data from the batch to which sodium acetate was added, because significant phosphorus release occurred in this reactor. Graph 2 (Figure 12.9b) illustrates a wastewater sample that has potential for BPR. Additional evaluation of this sample is warranted as described in step 2.

(2) Using a ratio of phosphorus excess uptake to phosphorus release of 1.15, the following calculation can be used to predict the effluent phosphorus concentrations based on plant information and operating conditions:

$$(P_{eff}, mg/L) = (P_{inf}, mg/L) - [(P_{release}, mg/L) \times (1.15 - 1.0)] - \{[5 \times (SRT, days) + 90]^{-1} \times (BOD, mg/L)\} \quad (12.3)$$

The last term in the equation represents metabolic phosphorus uptake/removal. A spreadsheet can be developed that graphs the phosphorus releases and calculates an estimated effluent concentration.

(3) The decrease in soluble COD can also be plotted to help determine the ideal anaerobic hydraulic retention time (HRT) for BPR for the wastewater. The soluble COD concentration should decrease rapidly for the first one to two hours and then level off as the COD uptake rate decreases. The point where the COD uptake rate decreases is roughly equal to the ideal anaerobic HRT. An HRT beyond this time could result in secondary release of phosphorus from the mixed liquor, which is undesirable for BPR.

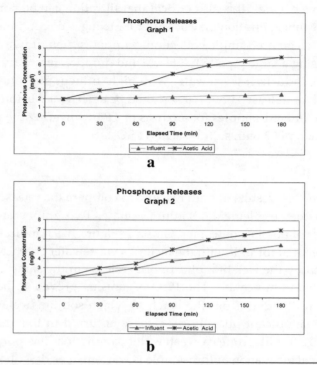

FIGURE 12.9 Examples of possible outcomes of BPR potential test: (a) showing the lack of phosphorus release in the sample that does not contain sodium acetate and (b) showing a sample with potential for BPR.

MICROBIOLOGICAL ACTIVITY. *What.* Monitoring microbiological activity involves the use of a sensor (probe) to detect biological activity by monitoring the strength of a fluorescence signal reflecting from the biomass in an activated sludge system of a wastewater treatment plant. This, in theory, allows the monitoring and control of the biological state or activity in the aeration tanks at low DO concentrations more reliably than a DO probe alone could accomplish. The fluorescence signal is dependent on the concentration of a coenzyme nicotinamide-adenine dinucleotide (NADH) contained within the cells of the biomass in the aeration tanks. The fluorescence signal reflected from the biomass originates from the probe, which emits a UV light with a wavelength of 340 nm. The NADH has a unique property in that it fluoresces back at 460 nm when struck with a UV light of 340 nm. The level of NADH contained in each cell of the biomass changes with the metabolic state of the cells in the system. Under anaerobic conditions (no or low DO concentrations), the concentration of NADH is high, corresponding to a strong signal; and, conversely, when aerobic conditions prevail (higher DO concentrations), the NADH reading is low. Under anoxic conditions, the signal is somewhere in between.

Where. Monitoring microbiological activity can be useful in optimizing the operation of an activated sludge system, operating in a simultaneous nitrification and denitrification mode, detecting influent toxicity or inhibition, detecting a potential slug loading into the plant, and for use with a DO probe for blower aeration automated control.

Why. At a poultry processing wastewater treatment plant in Texas, the aeration tanks were designed and configured to operate as a completely mixed reactor (similar pollutant concentrations and DO levels throughout the tanks). They operate the system based on a timed aeration mode of operation by automatically maintaining DO levels between 0.6 and 0.8 mg/L with an in-tank DO probe and by switching on and off several mechanical aerators in the aeration tanks. Following the aeration phase, the system switches to a timed anoxic (denitrification) phase of operation and automatically maintains lower DO levels (between 0.2 and 0.4 mg/L) by automatically switching off aerators in response to in tank DO levels as measured by the DO probe. The system is then cycled back and forth between the aeration and anoxic phases of operation throughout the day. The NADH probe is used to reduce the time for the anoxic (denitrification) phase of the activated sludge system. When the NADH level changes, it "jumps" past a certain preset level (indicating that denitrification is complete and the system is moving from an anoxic state to an anaerobic state), the system switches to the aeration phase of operation.

Other uses of monitoring microbiological activity include optimizing the existing operation of BNR activated sludge facilities. The NADH probe can be used in a controlled environment to determine the denitrification rate at a wastewater treatment plant to optimize anoxic zone sizes, recycle rates, etc. Early work trying to maintain a constant NADH level in the activated sludge basins of a wastewater treatment plant led to difficulty, because NADH levels can change as a result of more than one outside factor, including the following:

(1) Metabolic state of the biomass, as discussed previously (this is the change in NADH levels that you want to monitor and measure for aeration control).
(2) Biomass concentration in the reactor, more commonly know as changes in MLVSS. The MLVSS concentration can vary, even in a well-run activated sludge facility by varying wasting and return rates, influent organic loadings, etc.
(3) Background concentrations of soluble proteins or combinations of proteins in the wastewater can produce a false signal, indicating higher NADH levels, at times.
(4) The NADH is also sensitive to the pH and temperature of the wastewater.

It can be hard to control the aeration system based on maintaining a constant NADH level, because one is not certain which factor has contributed to changing the NADH level. There has been success using the probe in the fermentation industry where pH, temperature, and biomass concentrations are controlled, and the variable is the metabolic state of the biomass in the system.

When. Typically, this is a continuous online system, providing real-time data and feedback.

How. Online automated instrumentation for monitoring microbiological activity can be purchased from a manufacturer. The limited numbers of facilities contacted that are using this technology indicated that they are able to operate the probe with very little maintenance and that it is a very durable and reliable piece of equipment. They mentioned that they only need to wipe the optical lens on the bottom of the unit once a week, and there is no calibration required.

DATA ANALYSIS AND INTERPRETATION

The analysis and interpretation of data for BNR facilities includes aspects of conventional wastewater treatment. The discussion below is limited to BNR through nitrification and denitrification, and BPR. Detailed interpretation is provided in the Advanced Control section of Chapter 13.

NITRIFICATION. The parameters critical for nitrification are related to operating conditions that favor or indicate the activity of nitrifying organisms. These conditions include the chemical environment, SRT, and performance indicators.

Chemical Environment. The chemical environment is described in part by pH, alkalinity, temperature, DO, ORP, NH_3-N, and COD or BOD. The pH in the aeration basin can greatly affect nitrification. Low pH in the aeration basin can occur when the alkalinity concentration is insufficient to buffer the loss of alkalinity upon nitrification of TKN and tends to inhibit nitrification. The pH and alkalinity of the incoming wastewater affects the pH and alkalinity in the aeration basin, where nitrification occurs. Low temperature or low DO in the aeration tank can slow nitrification rates and, if prolonged, will reduce the population of nitrifiers in the mixed liquor. High organic content relative to nitrogen, typically reported as the COD/TKN or BOD/TKN ratio, will reduce the population of nitrifiers in the mixed liquor. The ORP can be used as a quick indicator of aeration basin conditions-for example, high ORP is an indicator of conditions that favor nitrification, and low ORP would be a reason to look at other indicators, such as DO, COD/TKN or BOD/TKN ratio, or toxicity. For more detail on conditions that affect nitrification, refer to the Nitrification Kinetics section of Chapter 3.

Solids Retention Time. The nitrifying bacteria include species that do not grow as quickly as bacteria that oxidize organic material (COD or BOD). A long enough SRT, as described in the Nitrification Kinetics-Biomass Growth and Ammonia Use section of Chapter 3, is needed for a stable, adequate nitrifier population. The parameters that are used to determine SRT include temperature, MLSS, MLVSS, RAS, WAS, and secondary effluent TSS. With this information, the actual aerobic SRT can be determined and compared to the theoretical SRT necessary for nitrification to occur. Typical minimum ratios for actual aerobic SRT to theoretical SRT range between 1.5 and 2.0.

Performance Indicators. The nitrogen species (i.e., NH_3-N, organic nitrogen, NO_2-N, and NO_3-N) before, within, and after the treatment process indicate nitrification performance. The distribution of nitrogen among the species indicates whether nitrification is complete, partial, or absent. For complete nitrification, the NH_3-N and organic nitrogen in the water, after mixed liquor solids are separated or removed, are expected to be low. High NO_2-N concentration is an indicator of partial nitrification and would be reason to look at indicators of unfavorable chemical conditions, low SRT, low DO in aeration zones, or toxicity. The absence of nitrification would be reason to look at toxicity.

DENITRIFICATION. The parameters critical for denitrification are related to operating conditions that favor or indicate the activity of denitrifying organisms. These conditions include the chemical environment, SRT, and performance indicators.

Chemical Environment. The environment for denitrification is anoxic, or lacking molecular oxygen (i.e., DO). The important parameters are pH, temperature, DO, ORP, NO_3-N, and COD or BOD. While optimum denitrification occurs within a narrow pH range, denitrification is significant unless pH is lower than pH 6 or higher than pH 8. Denitrification can be affected by low temperature as a result of lowered activity of denitrifiers. Denitrification can be inhibited by the presence of DO, which is used preferably by denitrifiers over NO_3-N. The organic concentration, reported as COD or BOD, needs to be adequate for the desired amount of NO_3-N removal. The ORP can be used as a quick indicator of anoxic basin conditions-for example, low ORP is an indicator of conditions that favor denitrification, and high ORP would be a reason to look at other indicators such as DO and COD/TKN or BOD/TKN.

Solids Retention Time. The denitrifying bacteria include many species, including those that can use DO or NO_3-N for metabolism and that grow more quickly than nitrifiers. The parameters that are used to determine SRT include temperature, MLSS, MLVSS, RAS, WAS, and secondary effluent TSS. An adequate SRT and reaction time for denitrification is commonly designed through adequate basin volume for the design MLSS. For attached growth systems, the SRT is accomplished by adequate surface area to maintain the denitrifiers. Low temperature will slow the rate of denitrification. The SRT for denitrification is not generally a concern for combination nitrification and denitrification systems, where the design has adequate anoxic volume because the nitrifier growth rates are more critical than the denitrifier growth rates.

Performance Indicators. The NO_3-N within and after the treatment process indicate denitrification performance. The degree of NO_3-N reduction indicates whether nitrification is complete, partial, or absent. For complete denitrification, the NO_3-N concentration is less than 1 mg/L. The intermediate NO_3-N concentration is an indicator of partial denitrification and would be reason to look at indicators of unfavorable chemical conditions, high DO, low temperature, or toxicity. The absence of denitrification is indicated by no reduction in NO_3-N through the denitrification process and would be reason to look at toxicity.

BIOLOGICAL PHOSPHORUS REMOVAL. The parameters critical for BPR are related to operating conditions that favor or indicate the activity of BPR organisms. These conditions include the chemical environment, SRT, and performance indicators.

Chemical Environment. The environment for BPR is anaerobic. The important parameters are temperature, DO, ORP, NO_3-N, and COD or BOD. The BPR can be affected by low temperature because of inability to achieve anaerobic conditions. The BPR is inhibited by the presence of DO. The organic concentration, reported as COD or BOD, needs to be adequate for the desired amount of BPR (see the Influent Composition and Chemical-Oxygen-Demand-to-Phosphorus Ratio section of Chapter 4). The ORP can be used as a quick indicator of anaerobic basin conditions-for example, low ORP is an indicator of conditions that favor BPR, and high ORP would be a reason to look at other indicators ,such as DO, COD/TP or BOD/TP ratio, or toxicity. In the anaerobic stage, ORP can go too low, resulting in reduction of sulfate to sulfide with associated odors and bulking in the system.

Solids Retention Time. In this case, the aerobic SRT and total system SRT affect BPR rather than a "BPR" SRT. A long SRT results in lower mass removed in WAS, which lowers the amount of TP removed through biomass. A long SRT also will lower the BPR rate, as a result of more endogenous activity of the biomass and less storage capability within the BPR organisms.

Contact time under anaerobic conditions is important to produce VFAs, the preferred food source of BPR organisms. This contact time is determined by the flowrate entering the process and the anaerobic volume provided. In some facilities, VFAs are formed in the collection system, and no contact time or only a short contact time is needed to produce adequate VFAs for BPR.

See the Solids Retention Time and Hydraulic Retention Time section of Chapter 4 for more details.

Performance Indicators. Total phosphorus and orthophosphorus before, within, and after the treatment process indicate BPR performance. The degree of TP reduction indicates whether BPR is complete, partial, or absent. For complete BPR, the TP concentration is less than 1 mg/L. Intermediate TP concentration is an indicator of partial BPR and would be reason to look at indicators of unfavorable chemical conditions, high DO or high NO_3-N, high SRT, excessive recycle of TP, or toxicity. The absence of BPR is indicated by TP reduction only through WAS with non-enhanced TP content of approximately 1% and would be reason to look at toxicity.

OPTIMIZATION AND TROUBLESHOOTING GUIDES

OVERVIEW. The BNR process is upset when indicators show changes from normal. The upset condition may lead to effluent quality that does not meet permit limits for one or more of the following parameters: BOD (or CBOD), TSS, NH_3-N, TN, total inorganic nitrogen, and TP. The first step is to find the problem causing the upset-whether it is loading, aeration, biomass inventory, clarifier operation, internal recycle, pH/alkalinity, or toxicity. To find out, proceed in a logical step-by-step fashion through the possible cause blocks shown in Figure 12.10 and eliminate all of the "NO" answers. This procedure allows the operator or manager to check the whole system to correct problems, not symptoms. Similarly, if the BNR process is performing at less than desired efficiency or reliability, the same procedures can be used to optimize the process.

When the problem is identified, the next step is to go to the appropriate optimization and troubleshooting guide (OTG) to further identify the problem and determine the appropriate response

OPTIMIZATION AND TROUBLESHOOTING GUIDE FORMAT.

The OTGs are arranged in columns, as explained below. The format is similar to that used in the U.S. Environmental Protection Agency's technology-transfer documents and various equipment manufacturers' operation and maintenance manuals (Tillman, 1996).

Indicator and Observations. This shows what has been observed or reported by the operator.

Probable Cause. This shows the most likely cause of the upset or inefficiency.

Check or Monitor. The operator should check the system for specific information to document the current condition and aid in diagnosing the problem. After the solution has been implemented, the operator should perform the listed monitoring to verify process performance improvement until the process is optimized or recovered from upset.

Solutions. The operator should perform one of the suggested solutions addressing the probable cause previously identified. For optimization, only one control parameter should be changed at any one time, and increases or decreases should be limited to 10% of its previous value. For troubleshooting, the operator may perform more than one solution, especially for permit compliance or to alleviate health, safety, or environmental concerns, and has greater latitude in the magnitude of change to remedy the upset. The operator is cautioned to limit the magnitude of any change that could destabilize the process further.

Optimization and Troubleshooting Techniques

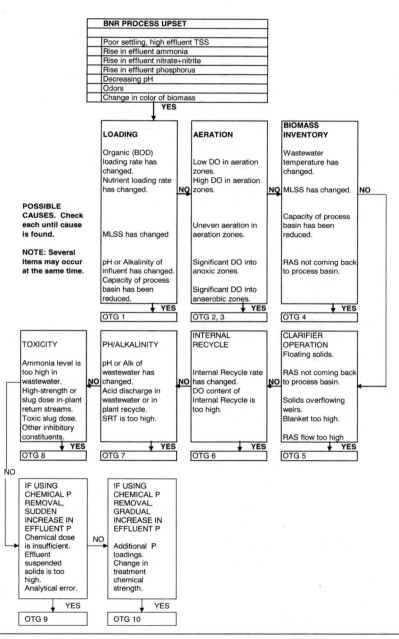

FIGURE 12.10 Decision tree for using optimization/troubleshooting guides.

References. The numbers listed in this column show where additional information is located in either another OTG or in another chapter of this manual.

OPTIMIZATION AND TROUBLESHOOTING GUIDES. Optimization and troubleshooting guides 1 through 10 are provided as Tables 12.3 through 12.12, respectively. Reading across the page of an OTG, follow the numbers and letters. For example, Solution 1a in the "Solutions" column refers to the corresponding Item 1a in the "Probable cause" and "Check or monitor" columns.

Information from a decision-tree-type troubleshooting guide specific to enhanced biological phosphorus removal (EBPR), developed by Benisch et al. (2004), was incorporated to these tables. For troubleshooting EBPR only, this decision tree guide may provide further assistance in identifying and correcting poor process performance.

CASE STUDIES

WOLF TREATMENT PLANT, SHAWANO, WISCONSIN. The Wolf treatment plant serving the Shawano, Wisconsin, area is an 11 360-m^3/d (3-mgd) average daily flow facility that has a 1-mg/L effluent phosphorus limit. The BPR was selected for this facility because significant VFAs were projected as a result of a long forcemain being constructed in this regionalization project (Stinson and Larson, 2003). The BPR was projected to reduce chemical addition rates required to achieve the 1 mg/L effluent limit. The selector conditions achieved in the BPR facility would act to control bulking and foaming associated with excessive filaments in the aeration tanks. A schematic of the process is shown in Figure 12.11.

The flow scheme used at Shawano is a modification to the conventional University of Cape Town (UCT) configuration, but is not the modified UCT. Modeling using BioWin software (developed by EnviroSim Associates, Hamilton, Ontario, Canada) was also conducted during design of this facility. The process allows the operator to select from two RAS locations and adjust the internal recycle rate. The tanks can also be operated in three different anaerobic/oxic (A/O) arrangements. The operations staff has operated in an A/O configuration, with two of the four zones in operation, while a mixer was being repaired. This mode of operation was not as effective as the configuration shown above, but it did allow BPR to continue with some chemical

TABLE 12.3 Optimization/troubleshooting guide 1: loadings.

Indicator or observations	Probable cause	Check or monitor	Solutions	Reference
1.) Raw wastewater odor; dark color in aeration zone. Higher effluent NH_3-N, TKN, or TP than normal.	1a.) Excessive loading of BOD or TKN.	1a.) Check $NO_2 + NO_3$-N at end of process before clarification for nitrification performance.		
		1a1.) If activated sludge, check DO in aeration zone; if rotating biological contactor, check DO in first or second compartment; if trickling filter, check DO in trickling filter effluent.	1a1.) If activated sludge with diffused aeration/mixing, see OTG 2; if activated sludge with mechanical aeration/mixing, see OTG 3. If RBC or trickling filter, try to increase aeration as short-term solution, if possible. Also, see 1a2.	
		1a2.) Check BOD and TKN concentrations in influent to BNR process.	1a2.) Look at plant operations to see if sidestreams from solids handling are causing periodically high loads and adjust operations to even out BOD and TKN loadings to process.	
		1a3.) Check BOD and TKN concentrations in influent to plant.	1a3.) Check for and discourage discharges to collection system that are causing unusually high strength in the influent wastewater.	
	1b.) Effective capacity of process has been reduced.	1b1.) Check for grit deposits at bottom of basins.	1b1.) If activated sludge, increase MLSS to maintain SRT. Clean out accumulated deposits.	
		1b2.) Observe mixing and aeration patterns.	1b2.) If activated sludge with diffused aeration/mixing, see OTG 2. If activated sludge with mechanical aeration/mixing, see OTG 3.	OTG 2 or 3
		1b3.) If attached growth, look for excessive growth or debris that hinders aeration or airflow and causes short-circuiting.	1b3.) Increase sloughing by flooding, increasing recirculation rate, flushing with hose, etc., as applicable. For trickling filter, clean out debris, restore airflow, consider increased recirculation rate. For other attached growth, distribute loading more evenly.	

TABLE 12.3 Optimization/troubleshooting guide 1: loadings *(continued)*.

Indicator or observations	Probable cause	Check or monitor	Solutions	Reference
2.) Pin floc in secondary clarifier effluent, high SVI, sometimes dark tan foam on aeration basin. Higher effluent TP than normal if 2a. occurs.	2a.) Underloading of process (lower flow, BOD, or TKN than design values) causing secondary release of nutrients.	2a.) Compare actual flow, BOD, and TKN values to design values. 2a1.) If multiple trains are in service, calculate loadings if fewer trains are in service. 2a2.) If activated sludge, check MLSS and WAS solids concentrations and WAS flowrate and calculate SRT.	2a1.) If calculated loadings are within design values with fewer trains in service, decrease number of trains in service. 2a2.) If activated sludge, lower MLSS concentration to lower SRT and consider lowering RAS rate.	Chapter 7
	2b.) *Nocardia*-type microorganisms are favored by underloaded condition, influent oil and grease concentration, or plant design.	2b.) See Chapter 7.	2b.) See Chapter 7.	
3.) Loading appears to be uneven (odors, MLSS color, and DO); MLSS concentrations or DO different in different trains. Higher effluent NH$_3$-N or TKN than normal.	3a.) Unequal flow distribution.	3a.) Check flow of process influent to each train. Check NH$_3$-N, NO$_2$+NO$_3$-N at end of process before clarification.	3a.) Adjust flow-splitting devices to provide equal flow to same-sized trains or provide equal loadings to different-sized trains.	
	3b.) Unequal RAS distribution.	3b.) If activated sludge and RAS enters separately, check RAS flow to each train.	3b.) Adjust flow-splitting devices on RAS to provide equal flow to same-sized trains or provide equal loadings to different-sized trains.	
	3c.) Unequal MLSS distribution.	3c.) If process influent and RAS mix and then are split, check for good mixing and check flow to each train.	3c.) Increase mixing if poor mixing is observed before flow distribution. Adjust MLSS flow-splitting devices to provide equal flow to same-sized trains or provide equal loadings to different-sized trains.	

TABLE 12.3 Optimization/troubleshooting guide 1: loadings *(continued)*.

Indicator or observations	Probable cause	Check or monitor	Solutions	Reference
4.) Dark diluted color of MLSS; sometimes white, sudsy foam in aeration zone. Higher effluent NH_3-N or TKN than normal.	4.) MLSS is lower than desired because of excessive wasting.	4.) Check MLSS and DO and check NH_3-N and NO_2+NO_3-N at end of process before clarification.	4.) If MLSS is low, DO in aeration zones is high, NH_3-N is high, and NO_2+NO_3-N is low, decrease wasting to allow MLSS to increase for an SRT adequate for nitrification to occur. See OTG 4.	OTG 4
5.) Low influent pH or low alkalinity. Higher effluent NH_3-N or TKN than normal.	5a.) Low alkalinity in potable water combined with decomposition of wastewater in collection system during warm weather.	5a.) Check influent pH and alkalinity.	5a1.) Decrease SRT by reducing MLSS through increased wasting if nitrification is occurring adequately or if nitrification is not desired. 5a2.) Add alkalinity through chemical addition. See Chapter 8.	Chapter 8
	5b.) Acidic discharge(s) to collection system.	5b.) Check influent pH, alkalinity, and other parameters that could identify source of acidic discharge(s).	5b.) Identify and eliminate source of acidic discharge(s). For short-term solution, see 5a.	
6.) Higher effluent NH_3-N concentration.	6a.) Higher BOD or TKN concentration in influent wastewater.	6a1.) Check BOD and TKN concentrations in influent to BNR process.	6a1.) Look at plant operations to see if sidestreams from solids handling are causing periodically high loads and adjust operations to even out BOD and TKN loadings to process. See Chapter 10.	Chaper 10
		6a2.) Check BOD and TKN concentrations in influent to plant.	6a2a.) Check for and discourage discharges to collection system that are causing unusually high NH_3-N or BOD in the influent wastewater.	
			6a2b.) Check SRT and DO in process; adjust higher, if possible. See OTG 2 or 3 and 4.	OTG 2 or 3, 4

TABLE 12.3 Optimization/troubleshooting guide 1: loadings *(continued)*.

Indicator or observations	Probable cause	Check or monitor	Solutions	Reference
	6b.) Inhibition of nitrification.	6b.) Check pH profile through process, DO in aeration zones, and toxicity.	6b.) For process adjustments, see OTG 7. For toxicity, see OTG 8.	OTG 7, 8
7.) Higher effluent NO_2+NO_3 concentration.	7a.) BOD/TKN ratio has changed.	7a1.) Check BOD and TKN concentrations in influent to BNR process.	7a1.) Look at plant operations to see if sidestreams from solids handling are causing periodically high loads and adjust operations to reduce or even out BOD and TKN loadings to process. See Chapter 10.	Chapter 10
		7a2.) Check BOD and TKN concentrations in influent to plant.	7a2.) If BOD/TKN is low, there may be insufficient BOD for denitrification in anoxic zones.	
		7a2a.) High TKN in influent.	7a2a.) Check for and discourage discharges to collection system that are causing unusually high TKN.	
		7a2b.) Low BOD in influent.	7a2b.) Add carbon source, such as methanol, to increase denitrification. See Chapters 8 and 9.	Chapters 8 and 9
	7b.) Inhibition of denitrification	7b.) Check NO_2+NO_3, DO, and/or ORP profiles through process.	7b.) Optimize the anoxic conditions in anoxic zones; see OTG 2 or 3 for controlling aeration, and OTG 6 for controlling internal recycle.	OTG 2 or 3, 6
8.) Higher effluent TP concentration.	8a.) BOD/TP ratio has changed.	8a1.) Check BOD, soluble BOD, TP, and orthophosphorus in influent to BNR process.	8a1.) Look at plant operations to see if sidestreams from solids handling are causing periodically high loads and adjust operations to reduce or even out TP loadings to process.	Chapter 10
		8a2.) Check BOD, SBOD, TP, and orthophosphorus in influent to plant.	8a2.) If BOD/TP or SBOD/ orthophosphorus is low, there may be insufficient VFAs for biophosphorus removal.	

TABLE 12.3 Optimization/troubleshooting guide 1: loadings *(continued)*.

Indicator or observations	Probable cause	Check or monitor	Solutions	Reference
		8a2a.) High TP in influent.	8a2a.) Check for and discourage discharges to collection system that are causing unusually high TP.	
		8a2b.) Low soluble BOD (low VFAs) in fermentation zone	8a2b.) Add VFAs, such as acetic acid, to fermentation zone to increase biophosphorus removal. See Chapters 8 and 9.	Chapters 8 and 9
		8a3.) Insufficient soluble BOD (VFAs) and additional VFAs unavailable.	8a3.) Use chemical phosphorus removal on short-term basis or, if this is continuing situation, use chemicals and recalculate SRT and wasting based on higher inorganic content of MLSS resulting from chemical solids production. See Chapter 8.	Chapter 8
	8b.) Inhibition of phosphorus release in fermentation zone.	8b.) Check orthophosphorus, NO_2+NO_3, NH_3-N, DO, and/or ORP profiles through process. Observe mixing in fermentation zone.	8b.) Optimize the anaerobic conditions in fermentation zone and ensure that there is no short-circuiting; see OTG 2 or 3 for controlling aeration and OTG 5 for controlling RAS rate. Increase fermentation zone volume, if possible.	OTG 2 or 3, 5
	8c.) Inhibition of biological phosphorus uptake.	8c1.) Check orthophosphorus, DO profiles through process. Observe mixing in aerobic zone(s). Check aerobic HRT.	8c1.) Ensure that there is sufficient HRT and no short-circuiting; see OTG 2 or 3 for controlling aeration. Install baffling, if needed, to block strong currents from inlet to outlet of aeration zones.	OTG 2 or 3
		8c2.) Check influent VFA, and VFA, orthophosphorus profiles through process.	8c2.) If shock load of influent VFA occurs, excessive phosphorus release in fermentation zone may exceed biological phosphorus uptake in the aerobic zone(s). Eliminate excess VFA discharge or equalize VFA load.	Chapter 4

TABLE 12.3 Optimization/troubleshooting guide 1: loadings *(continued)*.

Indicator or observations	Probable cause	Check or monitor	Solutions	Reference
		8c3.) Check pH, temperature, and microorganism population for glycogen accumulating organisms.	8c3.) If other causes of low biological phosphorus uptake have been eliminated from consideration, the treatment plant conditions (low pH, high temperature, etc.) may favor GAOs over phosphorus-accumulating organisms. Consider pH adjustment or chemical phosphorus removal.	Chapter 4
9.) Odors in fermentation zone.	9.) Underloading of process is causing excessive detention time.	9.) Check detention time and ORP in fermentation zone.	9a.) If possible, decrease volume of fermentation zone (e.g., decrease number of basins in service or lower operating water level).	Chapter 4
			9b.) Increase RAS flowrate to return more NO_2+NO_3-N.	
10.) Odors in anoxic zones.	10.) Underloading of process is causing excessive detention time.	10.) Check detention time and ORP in anoxic zones.	10a.) If possible, decrease volume of anoxic zone.	Chapter 3
			10b.) Add some DO to beginning of anoxic zone.	
			10c.) Increase internal recycle to return more NO_2+NO_3-N and DO.	

TABLE 12.4 Optimization/troubleshooting guide 2: aeration/mixing–diffused aeration.

Indicator or observations	Probable cause	Check or monitor	Solutions
1.) Low DO in aeration zones.	1a.) Poor oxygen transfer in aeration zones.	1a.) Check for diffuser problems.	1a.) See 5 and 6 in this guide.
	1b.) Insufficient aeration.	1b.) Check airflow versus calculated airflow for loadings. Check DO profile through process.	1b.) Increase airflow to aeration zones until DO is approximately 2.0 mg/L. If airflow per diffuser is too high and additional aeration zones are available, put additional aeration zones in service.
2.) High DO in aeration zones.	2.) Flow, BOD, and/or TKN are lower than design.	2a.) Check airflow versus calculated airflow for loadings. Check DO profile through process.	2a.) Decrease airflow to aeration zones until DO is approximately 2.0 mg/L. If airflow per diffuser is too low, take aeration zone(s) out of service to maintain at least the minimum recommended airflow per diffuser.
		2b.) Compare actual flow, BOD, and TKN to design values. Calculate required airflow.	2b.) If diffusers allow, reduce the amount of air added by periodically reducing or shutting off air (not more than 1.5 to 2 hours off to prevent odors). Use a combination of on time and either reduced airflow time or off time throughout the day to better match aeration to the oxygen demand.
3.) Significant DO into anoxic zones.	3a.) Too much DO at internal recycle suction.	3a.) Check DO in aeration zone where internal recycle originates.	3a.) For activated sludge process with internal recycle, DO profile can be tapered for low DO at internal recycle suction.
	3b.) Too much turbulence at influent.	3b.) Observe turbulence or splashing at influent.	3b.) Adjust basin levels, use baffles, and/or modify inlet ports to minimize turbulence and introduction of DO to anoxic zones.

(continued)

TABLE 12.4 Optimization/troubleshooting guide 2: aeration/mixing–diffused aeration *(continued)*.

Indicator or observations	Probable cause	Check or monitor	Solutions
4.) Significant DO and/or $NO_2 + NO_3$-N into fermentation zone.	4a.) Too much DO returning with RAS.	4a.) Check DO in RAS. Check RAS concentration and flow.	4a.) Reduce DO in aeration zone immediately upstream of clarifiers. If clarifier operation allows, reduce RAS flow.
	4b.) Too much $NO_2 + NO_3$-N returning with RAS.	4b.) Check $NO_2 + NO_3$-N concentration in RAS. Check RAS concentration and flow.	4b.) If nitrification is not needed, reduce SRT to reduce NO_2+NO_3-N concentration going to clarifiers. If clarifier operation allows, reduce RAS flow.
	4c.) Too much turbulence at influent.	4c.) Observe turbulence or splashing at influent.	4c.) Adjust basin levels, use baffles, and/or modify inlet ports to minimize turbulence and introduction of DO to anaerobic zones.
5.) Poor mixing pattern.	5a.) Diffusers need cleaning or repair.	5a.) Visual observation of mixing pattern.	5a.) Bump or chemically clean in-place diffusers; take basin out of service and manually clean diffusers, replace broken diffusers, or periodically replace all diffusers if beyond expected service life.
	5b.) If blowoffs present, air is releasing through blowoffs.	5b.) Check for excessive diffuser headloss or blowoff malfunction.	5b.) Clean or replace diffusers as in 5a, and/or repair blowoff system.
	5c.) Diffusers not installed at same elevation.	5c.) Check that diffusers are installed level and at same elevation.	5c.) Take basin out of service and adjust diffuser support systems and/or laterals so that they are level and that diffusers are at the same elevation.

(continued)

TABLE 12.4 Optimization/troubleshooting guide 2: aeration/mixing–diffused aeration (*continued*).

Indicator or observations	Probable cause	Check or monitor	Solutions
6.) No or too little turbulence.	6a.) Too few blowers in operation.	6a.) Check number of blowers in operation and control system.	6a.) Increase number of blowers in operation or adjust controls to bring on more blowers if automatically controlled.
	6b.) Blower malfunction.	6b.) Compare airflow at each blower to its performance curve(s).	6b.) Diagnose and repair/replace malfunctioning parts. Check setting of inlet air vanes and valves.
	6c.) Dirty inlet filter.	6c.) Check inlet air pressure between filter and blower.	6c.) Replace inlet filters; clean filters if washable.
	6d.) Valves need adjustment.	6d.) Check airflow to basin.	6d.) Adjust manual valves to better distribute air.
	6e.) Aeration control system needs adjustment.	6e.) Compare control parameter values to set point(s).	6e.) If insufficient air is provided under automatic operation, adjust controls to increase airflow when setpoints are maintained.
	6f.) Air rate too low for proper operation of diffusers.	6f.) Compare airflow divided by number of diffusers to acceptable low value.	6f.) Increase airflow to provide at least minimum recommended airflow per diffuser.
7.) Excessive turbulence over entire basin.	7a.) Too many blowers in operation.	7a.) Check DO in aeration zones and number of blowers in operation.	7a.) If DO is above desired levels, decrease airflow. Stay within range of recommended airflow per diffuser as in 7c.
	7b.) Aeration control system needs adjustment.	7b.) Compare control parameter values to setpoint(s).	7b.) If excess air is provided under automatic operation, adjust controls to decrease airflow when setpoints are maintained.
	7c.) Air rate too high for proper operation of diffusers.	7c.) Compare airflow divided by number of diffusers to acceptable high value.	7c.) Decrease airflow to provide, at most, the maximum recommended airflow per diffuser.

TABLE 12.5 Optimization/troubleshooting guide 3: Aeration/mixing–mechanical aeration.

Indicator or observations	Probable cause	Check or monitor	Solutions
1.) Low DO in aeration zones.	1a.) Poor oxygen transfer in aeration zones.	1a.) Check for aerator problems.	1a.) See 5, 6a, and 7 in this guide.
	1b.) Insufficient aerators in service.	1b.) Check total horsepower in service versus calculated horsepower based on design oxygen transfer. Check DO profile through process.	1b.) Increase number of aerators in service until DO is approximately 2.0 mg/L. If additional aeration zones are available, put them in service.
2.) High DO in aeration zones.	2.) Flow, BOD, and/or TKN are lower than design.	2a.) Check total horsepower in service versus calculated horsepower. Check DO profile through process.	2a.) Decrease level in basin to lower impeller submergence until DO is approximately 2.0 mg/L. If DO is still too high, take aerator out of service, use low speed if two-speed aerators, or take an aeration zone out of service to maintain minimum impeller submergence.
		2b.) Compare actual flow, BOD, and TKN to design values. Calculate required aeration horsepower.	2b.) If aerators allow, periodically reduce the speed of aerator to reduce oxygen transfer but maintain mixing. If aerator is able to resuspend solids, consider shutting off aerator(s) (not more than 1.5 to 2 hours off to prevent odors). Use a combination of on time and either reduced speed time or off time throughout the day to better match aeration to the oxygen demand. For large aerators, on–off operation may be too hard on gear reducers; consider soft-start of motors if on–off operation is used.

(continued)

TABLE 12.5 Optimization/troubleshooting guide 3: aeration/mixing–mechanical aeration *(continued)*.

Indicator or observations	Probable cause	Check or monitor	Solutions
3.) Significant DO into anoxic zones.	3a.) Too much DO at internal recycle suction.	3a.) Check DO in aeration zone where internal recycle originates.	3a.) For activated sludge process with internal recycle, locate internal recycle suction where DO is lowest and adjust submergence, as in 2a. If $NO_2 + NO_3$-N concentration in effluent is low, reduce internal recycle rate.
	3b.) Too much turbulence at influent.	3b.) Observe turbulence or splashing at influent.	3b.) Adjust basin levels, use baffles, and/or modify inlet ports to minimize turbulence and introduction of DO to anoxic zones.
4.) Significant DO and/or $NO_2 + NO_3$-N into fermentation zone.	4a.) Too much DO returning with RAS.	4a.) Check DO in RAS. Check RAS concentration and flow.	4a.) Reduce DO in aeration zone immediately upstream of clarifiers. If clarifier operation allows, reduce RAS flow.
	4b.) Too much $NO_2 + NO_3$-N returning with RAS.	4b.) Check $NO_2 + NO_3$-N concentration in RAS. Check RAS concentration and flow.	4b.) If nitrification is not needed, reduce SRT to reduce $NO_2 + NO_3$-N concentration going to clarifiers. If clarifier operation allows, reduce RAS flow.
	4c.) Too much turbulence at influent.	4c.) Observe turbulence or splashing at influent.	4c.) Adjust basin levels, use baffles, and/or modify inlet ports to minimize turbulence and introduction of DO to anaerobic zones.
	5a.) Water level too low for mechanical aerator impeller.	5a.) Visual observation of waves in basin, surging noise; check impeller submergence level.	5a.) Increase level in basin to provide recommended minimum impeller submergence; repair leaks in weirs or gates that allow water level to drop below desired level at low flows.
5.) Surging noise and waves in aeration zone.	5b.) Basin design prone to wave formation at aerator.	5b.) Visual observation of waves reflecting off surfaces and creating standing wave.	5b.) Adjust aerator speed or position, if possible, or modify basin using baffles to reduce the standing wave effect.

(continued)

TABLE 12.5 Optimization/troubleshooting guide 3: aeration/mixing–mechanical aeration *(continued)*.

Indicator or observations	Probable cause	Check or monitor	Solutions
6.) Motor overload.	6a.) Water level too low for mechanical aerator impeller.	6a.) Visual observation of waves in basin, surging noise; check impeller submergence level.	6a.) See 5a in this guide.
	6b.) Water level too high for mechanical aerator impeller.	6b.) Check impeller submergence level, especially at peak flow.	6b.) Decrease level in basin to limit impeller submergence to recommended maximum at peak flow.
7.) Vibration, reduced splashing, low DO.	7a.) Impeller fouled with debris.	7.) Check for vibration, visual observation of impeller, visual observation of splash pattern.	7a.) Take aerator out of service and remove debris from impeller.
	7b.) Impeller fouled with ice.		7b.) Take aerator out of service and remove ice from impeller, install manufacturer-approved shields that keep splash confined to basin and minimize ice formation.

TABLE 12.6 Optimization/troubleshooting guide 4: biomass inventory.

Indicator or observations	Probable cause	Check or monitor	Solutions	Reference
1.) Wastewater temperature is low, DO is higher. Higher effluent TKN than normal.	1.) Cold weather is slowing down biomass activity.	1.) Check wastewater temperature in process.	1.) Increase SRT to continue nitrification, if required. If nitrification is not required during cold weather, then adjust process to discourage nitrification (lower SRT, less air, and lower MLSS and RAS flowrate).	
2.) MLSS is low, dark diluted color of MLSS.				OTG 1
2a.) Higher effluent TKN than normal.	2a1.) MLSS is lower than desired because of excessive wasting.	2a1.) Check MLSS and DO and check NH_3-N and $NO_2 + NO_3$-N at end of process before clarification.	2a1.) Decrease wasting to allow MLSS to increase for an SRT adequate for nitrification to occur. Select a better time to obtain more consistent MLSS concentration and/or use moving average of seven or more days on which to base wasting. Institute operational limits on the amount of wasting that can occur in a single day.	Chapter 3
	2a2.) Aerobic zone is too small.	2a2.) Check actual SRT corresponding to stable operation versus SRT needed for nitrification.	2a2.) Add another train if train is available, enlarge aerobic zone, or consider integrated fixed-film/activated sludge system modification. For latter, see Chapter 3.	OTG 1
2b.) If using a biophosphorus removal process without nitrification, higher effluent TP than normal.	2b.) MLSS is low to inhibit nitrification and is too low for the temperature to maintain biophosphorus removal population.	2b.) Check MLSS and DO and check orthophosphorus at end of process before clarification.	2b.) Decrease wasting as in 2a1. If nitrification occurs and there is insufficient aeration capacity to maintain stable operation, operate at low MLSS and implement chemical phosphorus removal until temperature increases to where biophosphorus population is restored; see Chapter 8.	Chapter 8

(continued)

TABLE 12.6 Optimization/troubleshooting guide 4: biomass inventory *(continued)*.

Indicator or observations	Probable cause	Check or monitor	Solutions	Reference
3.) MLSS is high, pin floc in secondary clarifier effluent, high SVI, sometimes dark tan foam on aeration basin. Higher effluent TKN or TP than normal.	3a.) High MLSS resulting from reduced wasting or seasonally higher temperatures.	3a1.) If multiple trains are in service, calculate loadings if fewer trains are in service.	3a1.) If calculated loadings are within design values with fewer trains in service, decrease number of trains in service.	
		3a2.) Check MLSS and WAS solids concentrations and WAS flowrate and calculate SRT.	3a2.) Lower MLSS concentration to lower SRT and consider lowering RAS rate.	
	3b.) *Nocardia*-type microorganisms are favored by underloaded condition, influent oil and grease concentration, or plant design.	3b.) See Chapter 7.	3b.) See Chapter 7.	Chapter 7

TABLE 12.7 Optimization/troubleshooting guide 5: clarifier operation.

Indicator or observations	Probable cause	Check or monitor	Solutions	Reference
1.) Floating solids at surface, sometimes thick solids layer, possibly bulking solids. Higher effluent TSS or TP than normal.	1a.) Denitrification in clarifier causing rising sludge.	1a.) Check $NO_2 + NO_3$-N and DO concentrations entering and leaving clarifiers.	1a.) Increase RAS return rate and/or increase DO in clarifier influent.	
	1b.) RAS not being removed quickly enough.	1b.) Monitor blanket height and check RAS rate and detention time in clarifiers. Check orthophosphorus in RAS.	1b.) Verify that RAS is flowing normally from each clarifier. Reduce solids detention time in clarifiers (lower blanket height, increase RAS rate, and reduce number of clarifiers in service).	
	1c.) Excessive turbulence causing air bubbles in floc.	1c.) Observe turbulence and presence of bubbles on floc from mixed liquor effluent.	1c.) Reduce turbulence in aeration tank immediately upstream of clarifiers and reduce turbulence between aeration tank and clarifier inlet.	
	1d.) Filamentous organisms in mixed liquor causing bulking sludge.	1d.) Check settleability and SVI.	1d.) Minimize formation of filamentous organisms by adjusting conditions in process basin. See Chapter 7.	Chapter 7
2.) Solids overflowing clarifier weir. Higher effluent TSS and BOD than normal.	2a.) RAS not being removed quickly enough.	2a.) Monitor blanket height and check RAS rate.	2a.) See 1b in this guide.	
	2b.) Unequal flow distribution.	2b.) Check flow to each clarifier.	2b.) Adjust flow-splitting devices to provide equal flow to same-sized clarifiers or provide equal loadings to different-sized clarifiers.	
	2c.) Hydraulic overloading.	2c.) Check surface overflow rates at peak flow.	2c.) If available, place another clarifier in service. Reduce surges to the clarifiers by controlling pumping stations in collection system and recycle streams from solids handling to equalize flows.	
	2d.) Turbulence from collection rake.	2d.) Observe solids carry-over when rake passes by.	2d.) If possible, reduce rake travel speed.	

(continued)

TABLE 12.7 Optimization/troubleshooting guide 5: clarifier operation *(continued)*.

Indicator or observations	Probable cause	Check or monitor	Solutions	Reference
3.) Pin floc. Higher effluent TSS than normal.	3a.) Excessive turbulence upstream of clarifiers.	3a.) Observe turbulence upstream and check DO in last aeration zone.	3a.) Reduce aeration in aeration zone upstream of clarifiers. Adjust flow-splitting devices or water levels to reduce turbulence upstream of clarifiers.	
	3b.) SRT too long.	3b.) Calculate SRT (check MLSS and WAS solids concentration and WAS flowrate).	3b.) Reduce SRT by increasing sludge wasting. Limit wasting increase to 10% higher for 2 times SRT to minimize unstable operation. See OTG 4.	OTG 4
	3c.) Short-circuiting in clarifier.	3c.) Look for areas where floc carryover is heaviest. Observe water level along effluent weir.	3c.) Level weirs. If possible, add or adjust baffling to reduce excessive velocities and density currents.	
	3d.) Upset resulting from loading or colloidal solids in sidestreams from solids handling.	3d.) Measure TSS and turbidity in sidestreams from solids handling. Check DO profile through process.	3d.) Look at plant operations to see if sidestreams from solids handling are causing periodically high loads and adjust operations to reduce colloidal solids and even out BOD and TKN loadings to process. See Chapter 10.	Chapter 10 OTG 8
	3e.) Toxicity.	3e.) See OTG 8.	3e.) See OTG 8.	
4.) Thin RAS. Higher effluent TP than normal.	4a.) Plugging of sludge withdrawal.	4a.) Check RAS concentration, observe sludge withdrawal equipment.	4a.) Backflush RAS collection system.	
	4b.) RAS return rate too high.	4b.) Check RAS concentration.	4b.) Reduce RAS return rate.	
5.) Turbid effluent. Higher effluent TSS or TKN than normal.	5.) Toxic or acid constituents in wastewater.	5.) See OTG 8.	5.) See OTG 8.	OTG 8

TABLE 12.8 Optimization/troubleshooting guide 6: internal recycle.

Indicator or observations	Probable cause	Check or monitor	Solutions	Reference
1.) If removing nitrogen through internal recycle, high $NO_2 + NO_3$-N in effluent upstream of clarifiers. Higher effluent $NO_2 + NO_3$-N than normal.	1a.) Internal recycle flowrate is insufficient.	1a1.) Verify that internal recycle pumps are operating.	1a1.) If pumps are not operating, fix the problem and restore internal recycle flow.	
		1a2.) Compare internal recycle flowrate versus design rate. Look for clog in valve, line, or pump.	1a2.) If pumps are operating and flowrate is lower than normal, unclog internal recycle system and restore internal recycle flow.	
	1b.) Internal recycle is returning too much DO.	1b.) Check DO concentration in internal recycle.	1b.) Reduce DO concentration in internal recycle. See OTG 2 or 3, item 3a and 3b.	OTG 2 or 3
	1c.) Insufficient BOD for denitrification in anoxic zones.	1c.) See OTG 1, item 7.	1c.) See OTG 1, item 7.	OTG 1
2.) DO in anoxic zone. Higher effluent $NO_2 + NO_3$-N than normal.	2.) Internal recycle rate is too high for the loading to the process.	2.) Calculate internal recycle flowrate needed to achieve nitrogen removal and compare to actual rate.	2.) Lower internal recycle flowrate.	
3.) Low DO in aeration zone with aeration system at maximum; average or high NH_3-N and low $NO_2 + NO_3$-N in effluent upstream of clarifiers.	3.) Oxygen-limited condition in aeration zone is causing denitrification to occur simultaneously in aeration zone.	3.) Check profiles of DO, NH_3-N, and $NO_2 + NO_3$-N through process.	3.) For short-term solution, consider discontinuing internal recycle flow if aeration zone is large enough to accomplish simultaneous nitrification/denitrification. Increase aeration in aeration zone to reduce effluent NH_3-N, raise DO in aeration zone, and allow resumption of internal recycle.	

TABLE 12.9 Optimization/troubleshooting guide 7: pH/alkalinity.

Indicator or observations	Probable cause	Check or monitor	Solutions	Reference
1.) pH is average in influent, but there is low pH in process effluent. Higher effluent TKN than normal.		1.) Check pH profile through process and effluent NH_3-N.		
	1a.) SRT is too high.	1a.) Check MLSS, WAS flowrate, and WAS concentration and calculate SRT.	1a.) Decrease SRT by reducing MLSS through increased wasting if nitrification is occurring adequately or if nitrification is not desired.	
	1b.) Insufficient alkalinity to maintain pH.	1b.) Check effluent alkalinity.	1b1.) Add alkalinity through chemical addition. See Chapter 8.	Chapter 8
			1b2.) Consider implementing anoxic zone if none exists to recover alkalinity through denitrification. The anoxic zone can be created by shutting off aeration in the zone and ensuring good mixing through other means. Internal recycle may be used to maximize alkalinity recovery. See Chapter 3.	Chapter 3
2.) Low pH or alkalinity in influent to BNR process.	2a.) Low pH or alkalinity entering plant.	2a.) See 5 in OTG 1.	2a.) See 5 in OTG 1.	OTG 1
	2b.) Sidestreams in plant or acidic discharge in plant.	2b.) Check pH and alkalinity in sidestreams.	2b.) Look at plant operations to see if sidestreams are causing low pH or alkalinity and adjust operations to even out pH and alkalinity entering process. If acid is used in plant, consider neutralizing before returning to plant. See Chapter 10.	Chapter 10
	2c.) Insufficient alkalinity.	2c.) See 1b.	2c.) See 1b.	

TABLE 12.10 Optimization/troubleshooting guide 8: toxicity.

Indicator or observations	Probable cause	Check or monitor	Solutions	Reference
1.) Light color and/or high DO in aeration zones. Higher effluent TSS, TKN, or TP than normal.	1.) Toxic load inhibited nitrification, leading to overaeration.	1a.) Check pH profile through process, DO in aeration zones, NH_3-N and NO_2 + NO_3-N through process.	1a.) Maintain as much biomass in areas unaffected by the toxic load through rerouting of RAS, reduction in aeration or mixing, or isolation of basins to preserve biomass for reseeding.	Chapters 3 and 4
		1b.) Check raw wastewater samples and return streams from solids handling for toxic components.	1b.) Identify and eliminate the source of toxic load.	Chapters 3 and 4
2.) Raw or chemical odor in process, pin floc in clarifier effluent or cloudy effluent. Higher effluent TSS or TKN than normal.	2.) Toxic load has affected the biomass.	2a.) Take sample and refrigerate for later analysis; see 1a. Look for dead (inactive) microorganisms under the microscope to confirm.	2a.) See 1a in this guide.	Chapter 3
		2b.) See 1b.	2b.) See 1b in this guide.	
3.) For attached-growth process, excessive sloughing with little or no biomass on media. Higher effluent TSS or TKN than normal.	3.) Toxic load has affected the biomass.	3.) See lb. and 2a.	3.) See 1b in this guide. If recirculation is being used, turn off recirculation until toxic load has passed; restart recirculation to dilute any toxic residue after the load has passed.	Chapter 3

TABLE 12.11 Optimization/troubleshooting guide 9: sudden loss of chemical phosphorus removal.

Indicator or observations	Probable cause	Check or monitor	Solutions	Reference
1.) Pump failed.	1a.) Power failure.	1a.) Verify that the pump has power.	1a.) Provide power.	
	1b.) Mechanical failure.	1b.) Check the manufacturer's troubleshooting information.	1b.) Perform system checks and conduct maintenance as directed by the manufacturer.	
2.) Chemical feed piping plugged.	2.) Chemical precipitates have formed in the piping, restricting flow.	2.) Verify that chemical is reaching the application point and potentially break the piping in search of the suspected restriction.	2.) If the chemical flow is restricted, the operations staff should attempt to remove the restriction or restricted piping. Muriatic acid has reportedly been successful in dissolving the chemical buildup. Contact your chemical distributor for additional advice. Consider discontinuing carrier water that may be promoting the precipitate formation within the pipe.	
3.) Slug loading.	3.) Additional loading from an unknown source.	3a.) Monitor key industrial contributors.	3a.) Work with industries to get their loadings under control and/or arrange appropriate compensation for treatment. Apply penalties where appropriate.	OTG 1 Chapter 5
		3b.) Check sidestream contributions.	3b.) Sidestreams should be treated carefully. Contingencies may include adding chemical to the sidestreams, equalizing the flow, and recycling sidestreams during periods of low loadings.	Chapters 5 and 10
		3c.) Consider growth of the service area.	3c.) Increase dose rates, if possible. Evaluate treatment capacity and expand where warranted.	Chapter 8

(continued)

TABLE 12.11 Optimization/troubleshooting guide 9: sudden loss of chemical phosphorus removal *(continued)*.

Indicator or observations	Probable cause	Check or monitor	Solutions	Reference
4.) Loss of solids.	4.) Poor activated sludge treatment.	4.) Monitor effluent suspended solids.	4.) Improve effluent quality. See OTGs 1, 4, and 5 above. Note that because the effluent suspended solids will contain phosphorus, it will be considerably more difficult to meet a phosphorus effluent limit as the effluent suspended solids get higher.	OTG 1, 4, or 5 Chapter 5
5.) Analytical error.	5.) Human error.	5.) Laboratory quality assurance/quality control.	5.) If laboratory data are in error, the data should be excluded from the data set for process control purposes and should be noted as required if used for regulatory reporting.	Chapter 11

TABLE 12.12 Optimization/troubleshooting guide 10: gradual loss of chemical phosphorus removal.

Indicator or observations	Probable cause	Check or monitor	Solutions	Reference
1.) Additional loadings.	1a.) Industrial contributions.	1a.) Monitor key industrial contributors.	1a.) Work with industries to get their loadings under control and/or arrange appropriate compensation for treatment.	OTG 1
	1b.) Sidestream contributions.	1b.) Check sidestream contributions.	1b.) Sidestreams should be treated carefully. Contingencies may include adding chemical to the sidestreams, equalizing the flow, and recycling sidestreams during periods of low loadings.	Chapters 5 and 10
2.) Change in chemical strength.	2.) Inconsistent or low-grade chemical.	2.) Monitor chemical strength.	2a.) Increase the dose rate when using weaker chemicals; decrease the dose rate when using stronger chemicals.	Chapter 8
			2b.) Require more consistent product from supplier.	Chapter 8

polishing. Recovery following replacement of the failed mixers was quick, because the biological community had not changed significantly.

Reporting software allows the operators to quickly view data that they have collected, allowing the operators to monitor process changes against effluent quality. Many of the parameters they are concerned with have been described above, but because of the additional flexibility, a few additional parameters are monitored. The internal recycle allows biological solids to be returned to the first stage of the BPR tanks, where it comes into contact with the influent. This recycle rate is monitored through the supervisory control and data acquisition (SCADA) system and evaluated against a comparison of the mixed liquor concentrations in the select stages of the process. This allows the operators an objective method of monitoring the effect the recycle rate has on the process.

Shown in Table 12.13 are the control parameters that the Shawano plant uses to track the performance of their BPR system. Much of the data described in this table are either required by the regulatory agency or collected automatically through the SCADA system. The additional parameters were selected with the operations staff to provide the information that they would like to have available while operating the system. The additional effort is limited to a few hours per week.

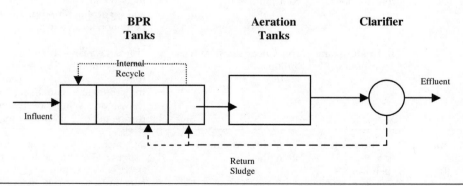

FIGURE 12.11 Version of the UCT process at Wolf Treatment Plant, Shawano, Wisconsin.

The Wolf plant has consistently met the effluent phosphorus limit through BPR, except during three mechanical failures. During these times, a chemical phosphorus removal system was used to allow them to meet their permit limits. The operators use the following fundamentals for successfully handling an upset:

- Baseline data collection is valuable to assist in recovery,
- Process flexibility allows operator control,
- Regular effluent monitoring allows timely adjustments, and
- Operator experience and training pays dividends.

CITY OF STEVENS POINT, WISCONSIN. Biological phosphorus removal at Stevens Point, Wisconsin, was started in October 1997 because of changed effluent phosphorus limits proposed by the Wisconsin Department of Natural Resources. The Stevens Point plant is a 15 140-m^3/d (4-mgd) average and 45 420-m^3/d (12-mgd) peak treatment plant using primary clarification, activated sludge biological treatment, anaerobic sludge digestion, and thickened liquid sludge disposal. Iron salt addition is provided for backup of the BPR system. The alternative limit at Stevens Point is 1.4 mg/L. Bench-scale testing was completed before the design of the phosphorus removal system, which has a modification to the A/O process that denitrifies the RAS. As indicated in Figure 12.12, during the first three months after startup, the plant was not consistently meeting the 1.4-mg/L limit, averaging between 2.5 and 3.25 mg/L in the effluent.

The plant was optimized by recommending process changes, such as increasing the primary sludge blanket to promote additional VFA and soluble BOD to pass through to the anaerobic tankage, increasing the MLSS (they were running approximately 600 mg/L), and limiting the peak recycle loads from the sludge decant tank. These efforts by plant staff have resulted in the plant consistently producing an effluent phosphorus concentration of approximately 0.5 mg/L, without the need for chemical addition (Stinson and Larson, 2003).

CITY OF DODGEVILLE, WISCONSIN. Biological phosphorus removal at Dodgeville, Wisconsin, was started in late 1999. The Dodgeville plant is a 3407-m^3/d (0.9-mgd) average, 13 630-m^3/d (3.6-mgd) peak flow oxidation ditch treatment plant. The modifications for BPR (Figure 12.13) were cost-effective, as they were retrofit into the existing primary clarifiers that were originally planned to be abandoned.

TABLE 12.13 Control parameters of the Wolf Treatment Plant, Shawano, Wisconsin.

Parameter	Purpose
Total phosphorus	Monitor loadings and treatment efficiencies.
Orthophosphorus	Check real-time effluent quality.
Orthophosphorus	Quick verification of phosphorus release and subsequent uptake through the activated sludge system.
Nitrate	Quick reference to nitrate concentrations in RAS and subsequent removal in BPR tanks.
ORP	Provides real-time indication of changes in the biological activity (aerobic to anoxic, anoxic to anaerobic).
Settleable solids	Settleable solids are run on samples from the anoxic and anaerobic sides of the BPR system when biophosphorus recirculation pump is running; used to evaluate relative solids concentrations.
RAS ratio (RAS-flow-to-influent-flow ratio)	Calculated to monitor changes in the relative amount of RAS pumped to allow interpretation of whether the RAS rate is affecting BPR performance.
Biophosphorus recirculation pump speed	Automatically collected through SCADA to allow observation of possible relationships between recycle rate and BPR performance.
BPR suspended solids ratio	Calculated from settleable solids numbers from above to indicate if biophosphorus recirculation pump speed needs adjustment.
Ferric chloride use	Manually recorded to track daily chemical use.
Fe:P ratio	Calculated to help determine how dependent the system is on chemical polishing when both chemical and biological removal systems are in operation.
MLVSS concentration	Also tracked for other activated sludge monitoring. The MLVSS is monitored with the BPR data to allow comparison with BPR efficiencies.

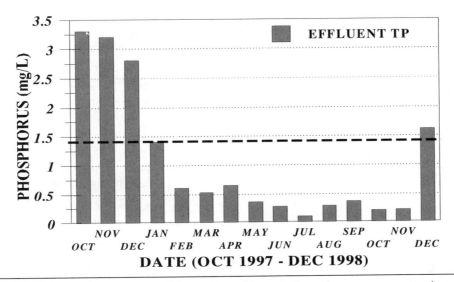

FIGURE 12.12 Stevens Point, Wisconsin, effluent phosphorus concentrations during initial startup.

The BPR system can be operated in various configurations and detention times to maximize operational flexibility. The plant was not consistently meeting the effluent phosphorus concentrations after startup with BPR alone. The plant staff and their engineer made several process adjustments and enhancements, such as lowering the DO concentration in the aerobic portion, decreasing the sludge age, and decreasing the RAS rate. The process modifications maximized the effectiveness of the system and decreased the levels of phosphorus to below their 1.4 mg/L alternative phosphorus limit (Stinson and Larson, 2003).

A profile of phosphorus through the system did not demonstrate a release of phosphorus in the anaerobic zones. It appeared that nitrate was not significantly reduced in the RAS denitrification tanks. However, the soluble BOD levels were decreasing through the BPR tanks. This suggests that nitrates were "bleeding through" to the BPR tanks and that the system was operating in an anoxic rather than

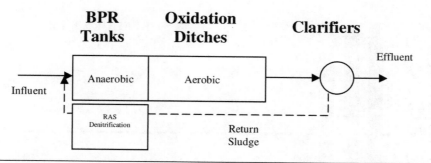

FIGURE 12.13 City of Dodgeville, Wisconsin, A/O process with RAS denitrification.

an anaerobic environment. The effluent phosphorus concentration also seemed to trend closely with DO concentration in the oxidation ditches.

After the first few months of operation, it was apparent that the RAS rates were higher than they needed to be. Using sludge quality settling tests similar to those described in Manual of Practice 11 (WEF, 1996) and evaluation of past data, it was decided to optimize the RAS rate. Reducing the RAS rate was not that simple because of the 200% of plant forward flow RAS design requirement in the State of Wisconsin for an oxidation ditch. For this reason, the two smaller clarifiers that had been kept and rehabilitated from the old facility were removed from service; the two small clarifiers are maintained in a state of readiness in case of a high-flow event. This action allowed a greater degree of turndown for the new larger clarifier. The lower RAS rate effectively increased the detention time for nitrate removal and reduced the amount of nitrates to be removed in the RAS denitrification zone.

The operators had spent the initial months of operating the facility in search of finding an optimal MLSS concentration. During this time, the MLSS rose to as high as 6000 mg/L before the plant staff reduced the concentration in an attempt to optimize settleability. The operators were concerned that the lower MLSS concentrations would not provide the effluent quality that the higher concentrations did. With this review, it was decided to depart from the constant MLSS approach to solids management and strive to achieve a 20-day SRT. This resulted in a lower MLSS concentration with excellent settleability and effluent quality.

Dissolved oxygen control was eliminated during design at the request of the city, a decision that was reversed following initial months of operation. This change occurred because the operations staff had tried to minimize the aeration rate on a couple of occasions, but found it to be difficult and unsuccessful. On one occasion, nitrification was inhibited as a result of these attempts. The operations staff then borrowed a data logging DO meter and proved that the DO varied greatly throughout the day. In addition to BPR process improvements, the plant staff also identified potential energy savings. The DO concentrations increased significantly over nighttime hours when loadings were low. This supported the theory that low loadings were resulting in excessive nitrate and DO levels in the RAS for a portion of the day.

Table 12.14 shows the difference in operation and performance when comparing October 2000 and March 2002. Note that October 2000 was the beginning of the optimization effort, as described above, and MLSS concentrations were already being brought back down, although not through SRT control methods.

EASTERN WATER RECLAMATION FACILITY, ORANGE COUNTY, FLORIDA.

The Eastern Water Reclamation Facility (EWRF) in Orange County, Florida, uses the modified (five-stage) Bardenpho process to meet advanced wastewater treatment standards of 5 mg/L BOD, 5 mg/L TSS, 3.0 mg/L TN, and 1.0 mg/L TP annual average. In addition to reclaimed water reuse, this facility discharges to created wetlands that discharge to natural wetlands, considered a surface water dis-

TABLE 12.14 Effect of operational adjustments on effluent phosphorus levels.

Date	RAS monthly average (m³/d)	RAS monthly average (mgd)	MLSS monthly average (mg/L)	DO daily range (mg/L)	Effluent phosphorus monthly average (mg/L)
October 2000	2105	0.556	3718	0.5 to 5	3.67
March 2002	662	0.175	3177	0.5 to 0.9	0.33

charge. There are no seasonal limits, and weekly average limits are 9.6 mg/L BOD, 9.6 mg/L TSS, 6.0 mg/L TN, and 2.4 mg/L TP.

The EWRF is rated for a 71 920-m^3/d (19.0-mgd) capacity, and is receiving approximately 53 000 to 54 900 m^3/d (14 to 14.5 mgd) influent flow. As shown in Figure 12.14, there are two separate trains providing treatment: the phases I and II train has four parallel process basins that share six secondary clarifiers and has a permitted capacity of 34 070 m^3/d (9.0 mgd); the phase III train has two larger parallel process basins and three larger clarifiers and has a permitted capacity of 37 850 m^3/d (10.0 mgd).

Operating staff regularly monitors combined clarifier effluent and individual clarifiers through grab samples taken the same time each day, coinciding with peak flow. The EWRF has consistently met effluent standards, and is now over 75% of permitted capacity.

In January 2004, EWRF experienced a slight rise in effluent turbidity, to 1.2 NTU; a small drop in effluent pH, from 7.5 to 7.4; and a small loss in settleability. Although there were no visual indications of upset, there was a slightly lower DO in the aeration zones than typical, and the laboratory data indicated a rise in effluent TN, primarily as NH_3-N from the phases I and II train. In February, the laboratory data

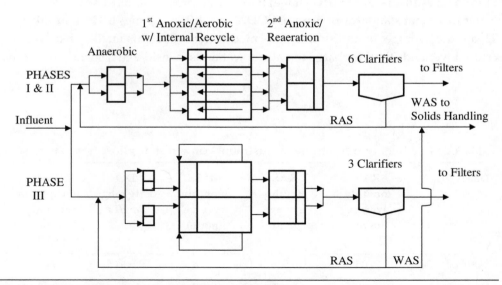

FIGURE 12.14 Schematic flow diagram of EWRF.

showed that the NH_3-N concentration exiting the phases I and II process basins had increased to a range of 4 to 8 mg/L. Also, there was a slightly sour smell from the RAS leaving the clarifiers in phases I and II. Testing with a portable TSS meter on MLSS and VSS with confirmation through laboratory results, field DO measurements, and laboratory NH_3-N tests were used over the next few weeks to evaluate the plant operation, diagnose problems, and take corrective actions. Plant staff made the following corrective actions in February and March:

- Shifted 2840 m^3/d (0.75 mgd) of influent wastewater from phases I and II to phase III, because the increased NH_3-N concentration exiting phases I and II and the lower DO concentrations in the aerobic zones indicated that aeration was insufficient for the organic loading;
- Increased aeration in aerobic zones by optimizing the submergence of surface aerators and adding supplemental floating aerators in two of the four aeration basins in phases I and II;
- Increased aeration in reaeration zones of phases I and II by cleaning grit plugging existing coarse-bubble diffusers and, where these diffusers could not be cleaned, installed temporary system using fine-bubble diffusers;
- Increased RAS flowrate for all clarifiers, from 75 to 100% of influent flow, to prevent the RAS from spending too much time in the clarifiers and becoming sour; and
- Increased wasting to reduce the MLSS in both trains to increase denitrification.

As shown in Figure 12.15, the phases I and II performance improved with lowered NH_3-N, lowered NO_3-N, and subsequently lower TN; and the phase III performance improved with lowered NO_3-N. For the months of March and April, combined effluent TN was reported to range from 1.8 to 2.2 mg/L, and effluent turbidity ranged between 0.6 to 0.8 nephelometric turbidity units (NTU).

WASTEWATER TREATMENT PLANT, STAMFORD, CONNECTICUT. The Stamford (Connecticut) Wastewater Treatment Facility (WWTF) is a 75 710-m^3/d (20-mgd) secondary activated sludge treatment plant that has been operating since 1976. In 1994, the plant treated an average daily flow of approximately 64 400 m^3/d (17 mgd) and removed 96% of the influent BOD and 91% of the influent TSS. Primary effluent enters two parallel trains of four aeration tanks in series. The Stamford

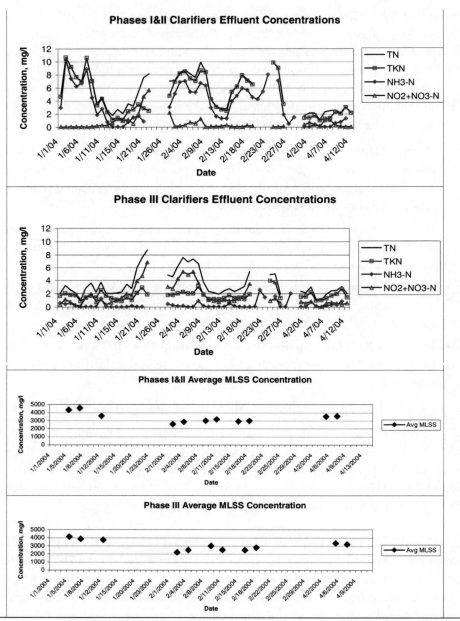

FIGURE 12.15 Clarifier effluent quality and MLSS in the two trains at EWRF.

WWTF was operated in both nitrification and denitrification modes with minor operational adjustments.

In nitrification mode, the SRT was increased from 2 to 4 days to 6 to 8 days, and the RAS flow was increased from 33% of the influent flow to 40%. Over 97% of the influent NH_3-N concentration was removed and/or converted by the treatment process. Aerators were run in the same mode as the original secondary treatment (i.e., aerators on high speed in the first and second tanks of each train and low speed in the remaining tanks).

In denitrification mode, the SRT was increased to 8 to 10 days, and the aerators in the first tank in each train were run on low speed to create an anoxic zone, as shown in Figure 12.16. The RAS flow was maintained between 30 and 50% of influent flow. There is no internal recycle, and no changes in plant equipment or structures. The primary effluent entering the process has the characteristics listed in Table 12.15.

During March through August 1990, the Stamford WWTF achieved nitrogen removal efficiencies ranging from 65 to 83%, with average effluent concentrations of 6 mg/L BOD, 10 mg/L TSS, and 7.3 mg/L TN. The average effluent nitrogen data between March and August 1990 is shown in Table 12.16.

During July and August 1990, increased *Nocardia* growth was experienced but was no worse than previous years when the plant was not operating in the denitrification mode. The *Nocardia* growth was controlled by low-level (9 mg/L) chlorination of the RAS for 72 to 96 hours. Chlorination at this level did not significantly affect nitrification or denitrification.

Nitrification and denitrification rate studies are performed monthly, and special test programs assist the plant staff in determining where the denitrification is occurring and how best to optimize nitrogen removal. The findings at Stamford were particularly useful, because they showed that secondary treatment plants may be able to reduce the amount of nitrogen being discharged to the environmentally sensitive Long Island Sound, with no capital investment and only minor process changes.

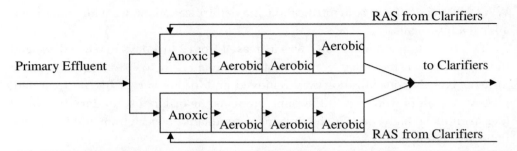

FIGURE 12.16 Stamford, Connecticut, process for nitrogen removal.

TABLE 12.15 Primary effluent characteristics.

Parameter	Concentration, mg/L (unless otherwise noted)
Alkalinity	130
BOD_5	116
TSS	100
TN	26
TKN	22
NH_3-N	15
NO_3-N	4
pH	7 standard units

TABLE 12.16 Average effluent nitrogen data, March to August 1990.

Month	TN, mg/L	TKN, mg/L	NH_3-N, mg/L	NO_3-N, mg/L
March	8.4	3.3	1.4	5.1
April	7.5	3.8	2.1	3.7
May	6.4	2.3	0.9	4.1
June	7.2	3.3	1.5	3.9
July	5.9	2.9	0.9	3
August	8.6	3.7	0.4	4.9

REFERENCES

Abu-ghararah Z. H.; Randall, C. W. (1991) *The Effect of Organic Compounds on Biological Phosphorus Removal. Water Sci. Technol.*, **23**, 585–94.

American Public Health Association; American Water Works Association; Water Environment Federation (1998) *Standard Methods for the Examination of Water and Wastewater*, 20th ed.; American Public Health Association: Washington, D.C.

Benisch, M.; Baur, R.; Neethling, J. B. (2004) Decision Tree for Troubleshooting Biological Phosphorus Removal. *Proceedings of the 77th Annual Water Environment Federation Technical Exhibition and Conference* [CD-ROM], New Orleans, Louisiana, Oct 2–6; Water Environment Federation: Alexandria, Virginia.

Ekama, G. A.; Marais, G. v. R.; Siebritz, I. P.; Pitman, A. R.; Keay, G. F. P.; Buchan, L.; Gerber, A.; Smollen, M. (1984) *Theory, Design and Operation of Nutrient Removal Activated Sludge Process.* Water Research Commission: Pretoria, South Africa.

Guidelines Establishing Test Procedures for the Analysis of Pollutants (2005) *Code of Federal Regulations*, Part 136, Title 40.

Mamais, D.; Jenkins, D.; Pitt, P. (1993) A Rapid Physical-Chemical Method for the Determination of Readily Biodegradable Soluble COD in Municipal Wastewater. *Water Res.*, **27** (1), 195–97.

Metcalf and Eddy (2003) *Wastewater Engineering: Treatment and Reuse*, 4th ed.; G. Tchobanoglous, F. L. Burton, H. D. Stensel (Eds.); McGraw-Hill: New York.

Park, J. K.; Wang, J.; Novotny, G. (1997) *Wastewater Characterization for Evaluation of Biological Phosphorus Removal*, Wisconsin Department of Natural Resources Research Report 174, PUBL-SS-574-97; Wisconsin Department of Natural Resources: Madison.

Park, J. K.; Whang, L. M.; Novotny, G. (1999) *Biological Phosphorus Removal Potential Test*, Wisconsin Department of Natural Resources Research Report 179, PUBL-SS-579-99; Wisconsin Department of Natural Resources: Madison.

Randall, C. W.; Barnard, J. L.; Stensel, H. D. (1992) *Design and Retrofit of Wastewater Treatment Plants for Biological Nutrient Removal.* Technomic Publishing: Lancaster, Pennsylvania.

Stinson, T. W.; Larson, T. A. (2003) Biological Phosphorus Removal—Optimizing System Performance. *Water Environ. Technol.*, **15**, 7.

Tillman, G. M. (1996) *Wastewater Treatment, Troubleshooting, and Problem Solving.* Lewis Publishers: Boca Raton, Florida.

Water Environment Federation (1996) *Operation of Municipal Wastewater Treatment Plants*, 5th ed.; Manual of Practice No. 11; Water Environment Federation: Alexandria, Virginia.

Water Environment Research Foundation (2003) *Methods for Wastewater Characterization in Activated Sludge Modeling.* Water Environment Research Foundation: Alexandria, Virginia.

Water Pollution Control Federation (1983) *Nutrient Control;* Manual of Practice No. FD-7; Water Pollution Control Federation: Washington, D.C.

Chapter 13

Instrumentation and Automated Process Control

Introduction	505	Accuracy and Repeatability	511
Online Analyzers	506	Maintenance	511
General Considerations	506	Zullig Technology	511
Meters Reproducibility and Accuracy	506	Fluorescent and Luminescent Dissolved Oxygen	512
Instrument Maintenance	507	*pH Measurement*	512
Specific Analyzers	508	Principles of Operation	513
Basic Instruments	508	Accuracy and Repeatability	514
Total Suspended Solids Meters	509	In-Tank Or Open-Channel Installation	514
Measuring Method	509	Flow-Through Installation	514
Accuracy and Repeatability	509	Maintenance Requirements	515
Installation	509	*Oxidation–Reduction Potential*	515
Maintenance Requirements	510	Introduction	515
Application	510	Principle of Operation	515
Dissolved Oxygen Measurement	510	Accuracy and Repeatability	515
Introduction	510		
Membrane Technology	510		

Installation	516
Maintenance Requirements	516
Application	516
Advanced Instruments	516
Ammonia and Ammonium	517
Measurement	517
Accuracy and Repeatability	518
Nitrate and Nitrite	518
Principle of Operation	518
Accuracy and Repeatability	519
Phosphorus and Orthophosphate	519
Orthophosphate	519
Total Phosphorus	519
Installation of Ammonia and Nutrient Analyzers	520
Maintenance Requirements	520
Applications	521
Process Parameters For Optimization and Automatic Control	521
General Considerations	521
Selecting Optimum Set Points	523
Basic Automatic Control	523
Excess Sludge Flow Control	523
Improved Calculation Methods	526
Selecting Sludge Age Target	526
Maintaining Optimum Sludge Age	527
Case Studies	528
Oxnard Trickling Filter Solids Contact Activated Sludge System	528
Toronto Main Wastewater Treatment Plant	528
Santa Clara/San Jose Water Pollution Control Plant	529
Dissolved Oxygen Control For Biological Nutrient Removal Plants	532
The Need For Good Dissolved Oxygen Control	532
Challenges	533
Control Strategies	534
Control of Chemical Addition	536
Advanced Control	537
Ammonia Control	537
Control of Denitrification	537
Respirometry	538
Intermittent Aeration	539
Sequencing Batch Reactors	540
COST Model For Control Strategy Development	541
Supervisory Control and Data Acquisition System Requirements	541
General Considerations	541
Supervisory Control and Data Acquisition Functions	541
Continuous Process Control	542

Programmable Logic Controller and Logic Program	542	Real-Time Trending	545
		Data Integrity	546
		Data Distribution	546
Programmable Logic Controller Programming Software	543	*Distributed Alarming*	546
		Historical Collection	547
		Historian Software	547
Data Acquisition	543	Historical Trend Charts	547
Input/Output Software	543	Historical Metric Charts	548
Data Highways and Ethernet Communications	543	*Information Systems*	548
		Reporting	548
		Automated Alarm Notification	548
Redundancy	544		
Network Mapping and Monitoring	544	Thin Client Software	549
		Server Emulation Sessions	549
Supervisory Control	544		
Supervisory Control and Data Acquisition Engine (Core)	544	*Security*	549
		Comprehensive Supervisory Control and Data Acquisition System Summary	550
Supervisory Control and Data Acquisition Database	545	References	551
Graphics	545		

INTRODUCTION

Biological nutrient removal (BNR) processes often require more sophistication in operation than conventional processes, and capital and operational costs of a BNR plant are much higher than of a regular activated sludge plant. It is a well-known fact that operation of BNR systems can significantly benefit from using online analyzers for both monitoring and automatic control. Experience also has proven that process parameters optimization and automation can reduce cost and improve reliability and operation of BNR plants.

This chapter provides an overview of online analyzers, process parameters optimization, automatic control, and requirements for control systems. For more detailed information, readers are encouraged to use both literature and website references that are included in the text. These references are provided to ease reader's access to detail information describing a particular technology, and they do not represent endorsement of the products described in some of the references.

ONLINE ANALYZERS

GENERAL CONSIDERATIONS. *Meters Reproducibility and Accuracy.* The most important consideration for the instrument used for process control is their reproducibility. For a control system, it is more important to use a reliable, low-maintenance meter that provides reproducible results than to have a very accurate meter that requires significant maintenance. This is especially true for instruments measuring operational parameters, such as dissolved oxygen (DO), oxidation-reduction potential (ORP), total suspended solids (TSS) of mixed liquor, and others.

The most obvious example for this is ORP measurement for airflow control. The ORP measurement reflects presence of so many different ionic species present in the wastewater that it makes no sense to try to measure to the nearest millivolt, and most control schemes using ORP use order-of-magnitude changes or sudden shifts in the rate of change rather than absolute values to initiate control. Similar consideration is applied to other control schemes, such as DO, nutrients, sludge age control, and others. For example, for good DO control, 10% accuracy of the measured DO value is typically adequate if the meter has good (within 0.1 mg/L) repeatability.

One important exception to this rule is pH control. Inaccuracies of a few tenths of a pH can cause serious problems with over- or under-dosing acids and alkalis because the pH reading is logarithmic.

Many plants, before purchasing instruments, test various brands of the same type of instruments. The critical issue during the testing is the standard that is used for judging meter performance. For example, use of one technology to check accuracy of a different technology proved to be a poor choice of standards selection. This statement is especially applicable to testing DO meters and nutrient meters. These meters have to be tested only against laboratory methods and not against portable meters that use different measurement technology than the meter tested. For example, comparison of reading of luminescent-based technology meters to reading of a meter manufactured that uses membrane-based technology may lead to wrong conclusions. At the same-time, when comparison is done to laboratory results, it is critical to perform laboratory testing in triplicates, because repeatability of laboratory tests generally has a lot of room for improvements.

A proper meter testing generally requires significant resources. That is why some users, instead of conducting their own testing, are using results of testing and surveys conducted by independent organizations. The most recent survey was conducted by the Water Environment Research Foundation (WERF) (Alexandria, Virginia) (Hill et al., 2002). Earlier review of DO monitoring was carried out by the Electric Power Research Institute (EPRI) (Palo Alto, California) (1996). The Instrumentation Test Association (ITA) (Henderson, Nevada) often sponsors not only side-by-side trials of different instruments, but also is establishing acceptable protocols for equipment testing (see ITA website for more details at http://www.instrument.org). In some cases, however, testing results at the individual plants are different than conclusions of these surveys and tests.

Instrument Maintenance. In the past, many instruments that were originally designed for clean water or other pure process fluids were unsuccessfully adapted for wastewater use. In recent years, more instruments have been specifically designed for wastewater applications. These instruments tend to be more reliable, although, in the harsh biofouling environment of wastewater treatment, these sensors still require routine maintenance.

If the primary sensor does not provide consistent and reasonably accurate readings, then the control system that uses the signal generated by the sensor cannot possibly function properly. Facilities with the best instrumentation and advanced control capabilities are invariably those where care and attention is paid to instrument maintenance. In plants that have inadequate instrument maintenance programs, the entire control system becomes nonfunctional and useless.

When considering instrumentation and control for any facility, it is important to be realistic about the amount of time and effort required to maintain the primary sensors. If there are insufficient resources to sustain the required manufacturer maintenance program for the instruments, then either these resources must be provided or an instrument with less maintenance requirements needs to be selected. Experience shows that presence of unreliable or unmaintained online instrumentation at the plant generates a negative attitude to instrumentation and control in general for years to come.

It cannot be overstressed how important it is to keep instruments properly cleaned and calibrated. The best starting point for setting up a good maintenance program is the instrument manual provided by the instrument supplier. As a bare minimum, the instruments must be cleaned and maintained as per the supplier's recommendations. One approach to overcoming labor-intensive maintenance is to provide automatic cleaning. Many instruments include automatic cleaning or can be retrofitted to provide this option. Some instruments even include automatic calibration. These instruments are generally more expensive; however, additional costs are typically offset by reduced maintenance time. Automated calibration systems typically have the added benefit of providing outputs that give an indication that a calibration succeeded. This information greatly increases the confidence level in the outputs from the instruments.

It is also important to provide special attention to the meters' installation. For example, the sample preparation system is often the weakness of many nutrient analyzers. The sample pump or filtration system that blocks up or fails to protect the analyzer become the least reliable part of the system. Attention should be paid to providing the proper design of a sample preparation system and making sure that the system is maintained regularly. Use of membrane filters to prepare samples can provide a high degree of analyzer protection; however, a filter assembly has to be either equipped with self-cleaning mechanism or users need to allocate time to clean or replace filters regularly. Pipelines used for transferring unfiltered sample should not be less than 25 mm (1 in.) in diameter; in addition, tight bends and any device that could restrict flow should be avoided.

Finally, it is important always to remember that, because of a variety of reasons, reliability of online analyzers rarely exceeds 90%. In other words, every analyzer will fail one day, and it is not a matter of whether it will happen, but when. However, if an automatic control system using this instrument, in addition to providing regular control during the time of proper performance of the meter, can automatically determine meters failure or significant inaccuracy and switch to a fail-safe mode, then less than 100% reliability of the meters does not represent significant problems. Unfortunately, development of these features is quite cumbersome, and most control systems do not have the described capabilities.

SPECIFIC ANALYZERS

BASIC INSTRUMENTS. Basic instruments are the instruments that are critical for BNR plant operation. They are, generally, fairly inexpensive, have been on the market for years, and have a less demanding maintenance program. Total suspended solids and DO analyzers described in this section are a bare minimum of analyzers that any plant needs (Metcalf and Eddy, 2003), especially a BNR plant. For a certain type of BNR plants, use of ORP and pH analyzers for both process monitoring and automatic control can also be beneficial.

Total Suspended Solids Meters. *MEASURING METHOD.* Optical techniques for measuring suspended solids are based on the scattering of a beam from a near infrared source by particles suspended in the process fluid. Two major technologies used are the backscattered at 90° (http://www.wtw-inc.com) and forward-scattered (http://www.insiteig.com) beams. There are attempts to improve accuracy and repeatability by increasing the number of beams. For example, one manufacturer uses dual 90° and 140° beams technology (http://www.hach.com), and another uses as many as six channels of multiangle measurement (http://www.na.water.danfoss.com). Depending on the application, an increased number of beams may or may not pay off.

The most important feature for any technology is color compensation or color-independent analysis. Blackish highly concentrated sludge is more difficult to measure than light-colored sludge with lower concentration. Multichannel analyzers may provide some advantages for these applications.

Another important feature is a method for sensor cleaning. The following automatic cleaning methods are available: water or air purging (http://www.insiteig.com), ultrasonic (http://www.wtw.com), and wiper cleaning (http://www.hach.com). Some manufacturers (http://www.cerlic.com and http://www.na.water.danfoss.com) claim that their products are not susceptible to active biofouling and, as a result, do not require a self-cleaning system. According to this manufacturers' information, infrequent manual cleaning is still required. It is always advisable, however, to consider self-cleaning systems, even for these analyzers.

ACCURACY AND REPEATABILITY. The accuracy of a suspended solids analyzer is typically 2% of reading or ±0.2 g/L, and several ranges of operation are available. The repeatability of solids analyzers is typically ±1% of reading or ±0.1 g/L, whichever is greater.

INSTALLATION. Suspended solids analyzers are typically installed as submersible probes with different devices to attach to concrete walls or handrails; and they can be installed as inline or insertion probes with devices for safe removal from a pressurized pipe.

MAINTENANCE REQUIREMENTS. Suspended solids analyzers with effective cleaning devices require little operator intervention. Calibration requirements depend on the changes in particle-size distribution of the measured stream. Experience shows that once a week is a reasonable calibration frequency. Calibration needs to be based on laboratory analysis of the same grab sample processed in triplicate.

APPLICATION. Suspended solids analyzers are used for wasted sludge flow control, in particular, sludge age and MLSS control.

Dissolved Oxygen Measurement. *INTRODUCTION.* Dissolved oxygen is defined as the measure of water quality, indicating free oxygen dissolved in water. The quantity of DO in water is typically expressed in parts per million or milligrams per liter.

There are generally three types of DO measuring technology: membrane, Zuligg, and florescent and luminescent.

MEMBRANE TECHNOLOGY. The operational theory of a membrane sensor is that oxygen in the wastewater diffuses through the membrane into the electrolyte. The concentration of gases always tends to equalize on both sides of the membrane. When the concentration is not equal, gas molecules migrate to the membrane side that has a lower concentration. When the membrane is functioning, the DO concentration in the electrolyte in the measurement cell approximately equals the DO concentration of the wastewater contacting the opposite side of the membrane. The diffusion process is extremely critical. The DO must be allowed to migrate freely through the membrane for the sensor to function properly.

Most membrane sensor designs use the following three basic elements:

(1) Electrodes. The electrodes provide the necessary reaction site for reduction of oxygen molecules and generation of electrons.
(2) Membrane. The gas-permeable membrane is designed to keep the electrolyte around the electrodes, while allowing only DO to diffuse into the measurement cell.
(3) Electrolyte. The electrolyte facilitates DO migration and provides an electrical path to complete the current loop. It also removes metal oxides (a byproduct of the reaction) from the electrodes, so that their metal surfaces are clean to react. The electrolyte must be periodically replenished to ensure that the electrodes remain clean.

More about this technology can be found on the following websites: http://www.analyticaltechnology.com, http://www.wtw.com, www.abb.com, and others.

The following sections present general comments on membrane-type sensors.

ACCURACY AND REPEATABILITY. Accuracy and repeatability of membrane-type sensors will vary by manufacturer, for example:

- Accuracy—0.10% of span
- Sensitivity—0.05% of span

MAINTENANCE. Membranes will need to be replaced regularly-generally, once per quarter or more often, as needed by process conditions. The sensor membrane or sensor cartridge is removed, and a new one is installed. The new sensor should be placed in the water and allowed to operate for at least 12 hours to polarize the electrodes for galvanic sensor. Polarographic electrodes do not need polarizing. Membranes are subject to biofouling and puncture. Automatic cleaning device need to be considered for the installations where automatic control of DO is practiced.

ZULLIG TECHNOLOGY. Zullig (ZULLIG, Ltd., Switzerland) DO sensor is a non-membrane, galvanic sensor, and the electrodes are two independently spring-loaded concentric rings, which are insulated from each other. The open electrodes in the Zullig sensor are protected from exposure to air bubbles and suspended solids in the process solution by a sample chamber, in which fresh sample is pumped to the electrodes through an oscillating chamber. This chamber also ensures that sufficient sampling occurs in wastewater with low flowrates. A rotating diamond grindstone continuously polishes the electrode surfaces in the Zullig sensor. This reduces cleaning and eliminates replacement membranes and replenishment of the electrolyte solution. Because of the high level of reliability and low maintenance requirements, the meters using this technology tend to be more expensive to purchase and to rebuild than any other described technology. More information describing this technology can be found at http://www.emersonprocess.com.

The accuracy of Zullig technology is ±0.2 ppm (mg/L) at less than 0.46 m/s (1.5 ft/sec). Accuracy degrades to ±0.3 ppm (mg/L) for DO levels greater than 5 ppm when flow is less than 0.4 m/s (1.5 ft/sec).

For maintenance, the sensor must be kept reasonably clean to maintain measurement accuracy. The time period between cleanings (days, weeks, etc.) is affected by the characteristics of the process solution and can only be determined by operating experience. Usually, the meters require once per six months cleaning interval. However, a sensor operating in wastewater that contains oil and/or grease may require more frequent cleaning. General Recommendation: At first, clean the sensor every 1 to 2 weeks until operating experience can determine the optimum time between cleanings that provides acceptable measurement results.

FLUORESCENT AND LUMINESCENT DISSOLVED OXYGEN. The newest technology being used for DO measurement and control is luminescent technology. A sensor is coated with a luminescent material. Blue light from a light-emitting diode (LED) strikes the luminescent chemical on the sensor. The luminescent chemical instantly becomes excited, and then, as the excited chemical relaxes, it releases red light. The red light is detected by a photo diode. The time it takes for the chemical to return to a relaxed state is measured; the higher the oxygen concentration, the fewer red lights given off by the sensor. The oxygen concentration is proportional to the time it takes for the luminescent material to return to a relaxed state. Unlike electrochemical DO probe technologies, the luminescent DO sensor does not consume oxygen. Similar to Zullig technology, there is no membrane to puncture, tear, or replace. There are no electrodes and no electrolyte to consume. The only maintenance requirement is sensor-cap replacement once a year.

Calibration is not required, as the instrument internally calibrates itself to a red LED of known intensity. These qualities make the sensor a very accurate and low-maintenance device. The only drawbacks are an increased response time (over a minute) and interference by fluorescent materials that rarely are present in the wastewater. More information regarding this technology can be found on http://www.hach.com and http://www.insiteig.com.

The accuracy should be <1 ppm ±0.1 ppm; >1 ppm ±0.2 ppm; repeatability of 0.05 ppm; and resolution of DO at 0.01 ppm or 0.01 mg/L or 0.01% saturation. Sensitivity should be ±0.05% of the span.

Generally, DO sensors require minimal maintenance, but that depends, to some degree, on the sensor design. Typically, the sensor can be hand wiped as needed. Weekly cleaning frequency sometimes is adequate; however, automatic cleaning using air or water blasts should be considered for DO control applications.

pH Measurement. In wastewater treatment, pH sensors are used to monitor plant conditions and biological treatment process conditions and control acid or base additions for pH adjustment. Regulatory agencies require measurement of plant influent and effluent pH to extrapolate overall plant conditions. It may also be necessary to monitor the pH of specific industrial discharges to give advance warning of possible toxic conditions.

While the activated sludge and most other biological processes can tolerate a pH variance of 5 to 9, some, such as anaerobic digestion, are pH-sensitive. Normal monitoring of plant influent and primary effluent or MLSS (if applicable) is sufficient to detect impending toxic conditions. The anaerobic digestion process requires a pH value in the range 6.6 to 7.6 and fails below 6.2. Because of this sensitivity, it is important to monitor the pH of anaerobic digester liquor. However, because of sensor fouling, continuous monitoring of digester pH is not recommended. Periodic sampling is preferred.

An online pH monitor can provide feedback for control of other processes requiring pH adjustment. For example, pH adjustment may be required to neutralize low-pH industrial wastes, enhance phosphorus removal by alum addition, or adjust pH to optimum ranges for nitrification/denitrification.

PRINCIPLES OF OPERATION. The heart of the sensor is the glass membrane. An electrical potential, varying with pH, is generated across the membrane. The difference between this potential and a reference electrode is measured and amplified by an electronic signal conditioner. The complete electric circuit includes the glass electrode wire, glass membrane, process fluid, reference electrode fill solution, and reference electrode wire.

A pH electrode assembly, or sensor, as it is sometimes called, consists of two primary parts.

(1) Measuring electrode. The measuring electrode is sometimes called the *glass electrode*, and is also referred to as a *membrane* or *active electrode*.
(2) Reference electrode. The reference electrode is also referred to as a *standard electrode*.

Just as a complete circuit requires the two half-cell potentials of a battery, so does a pH sensor. The mathematical expression for this is the following:

$$E = E_m - E_r \tag{13.1}$$

Where

E_m = electrode potential of the measuring electrode, and
E_r = electrode potential of the reference electrode.

The reference electrode is designed so that potential Er is constant with the pH and other chemical characteristics of the process fluid. The asymmetric potential varies from sensor to sensor, according to design preferences among manufacturers. It also changes as the sensor ages. For this reason, pH sensors must be periodically standardized against buffer solutions of known pH. Most commercial pH sensors also include automatic temperature compensation.

The measuring and reference electrodes can be in one of two forms: (1) two physically separate electrodes, known as an *electrode pair*; or (2) the electrodes can be joined together in a single glass body assembly, known as a *combination electrode*. The combination pH electrode is the most widely applied.

ACCURACY AND REPEATABILITY. Manufacturer claims for pH meter accuracy range from ± 0.02 to ± 0.2 pH units. This represents the combined accuracy of the electrodes and the signal conditioner or transmitter. Without temperature compensation, an additional error of 0.002 pH units per 1°C difference from the calibration temperature can be expected. The repeatability of pH meter measurements varies by manufacturer from 0.02 to 0.04 pH units. Stability (drift) is an important performance parameter that indicates how often meters must be recalibrated. Manufacturer claims for stability vary from 0.002 to 0.2 pH units drift per week. With flow-through probe mounts, the velocity of the sample can cause a shift (0.2 to 0.3 pH) in measured values.

Methods of reporting performance specifications vary among manufacturers. Adjustment of the method of reporting performance specifications to equal units of measure shows that there is large variance in the accuracy and stability claimed by different manufacturers. Typically, good pH meters achieve the following performance standards in wastewater treatment plants (WWTPs): accuracy, ±0.1 pH units; repeatability, ±0.03 pH units; and stability, ±0.02 pH units per week.

IN-TANK OR OPEN-CHANNEL INSTALLATION. This type of installation allows the sensor or pipe assembly to be lifted clear of the tank. A submersion probe installed in a well-mixed zone will provide a representative sample of the process. If the probe is installed in an open channel, it should be located in a free-flowing zone. The electrode assembly and support-pipe installation should be designed to inhibit collection of debris.

FLOW-THROUGH INSTALLATION. Where pH control is not the objective of the measurement, a pH analyzer can be installed as part of a sample system with other online analytical instruments. Locating the pH meter with other high-maintenance instruments permits easier service. Buffer solutions needed for standardization can conveniently be stored with other analytical instrument reagents.

Although dedicated flow-through pH sensors are available, most systems use a submersible-style pH sensor installed in a simple flow cell. Bypass and shutoff valves provide for instrument removal and service. Locating the pH analyzer near the probe mounting assembly makes for easier standardization. A work surface for setting containers of buffer solution is also helpful. A sample valve can be installed next to the sensor to collect a sample for conformance checks.

MAINTENANCE REQUIREMENTS. Frequent cleaning of the glass pH electrode may be required, depending on the nature of the process and the location of the probe in the treatment train. The time period between cleanings (days, weeks, etc.) is affected by the characteristics of the process solution and can only be determined by operating experience. Automatic cleaning systems (using air or water jets) are available to reduce cleaning frequency in difficult applications.

Oxidation–Reduction Potential. INTRODUCTION. As the term ORP implies, this measurement is used to determine the oxidizing or reducing activity present in a solution. The readout of the sensor is a voltage, where a solution containing a strong oxidizing agent, such as chlorine, has a positive ORP millivolt value (ability to accept electrons); and a solution containing a strong reducing agent, such as sodium bisulfate, has a negative ORP millivolt value (ability to furnish electrons). A solution that is neither oxidizing nor reducing has an ORP value of zero.

PRINCIPLE OF OPERATION. The ORP measurement is comprised of two half-cell, or electrode, potentials. One half-cell is the measuring electrode constructed with a platinum or gold tip. Gold must be used with solutions containing copper, lead, or zinc, because these elements will not work with platinum.

The other half-cell is the reference electrode. This is typically the same reference electrode used in pH instruments, namely a silver–silver chloride reference. As with pH, the two half-cell potentials are required to complete a circuit. The mathematical equation for ORP measurement is exactly the same as for pH measurement.

ACCURACY AND REPEATABILITY. The ORP is a nonspecific measurement; the measured potential is reflective of a combination of the effects of all the dissolved species in the medium. Users should thus be careful not to "over-interpret" ORP data, unless specific information about the site is known.

For most applications, the ability to achieve repeatable millivolt values is more important than the actual concentrations of the oxidants or reductants. As ORP is a surrogate for other parameters, such as chlorine, the absolute accuracy of the reading is less important than the repeatability.

Temperature has an integral effect on the electrodes, such that a change in temperature will change the voltage output value of the electrode measurement system. As this change cannot be easily compensated, ORP instruments typically do not incorporate temperature sensors. Rather, the operator must make adjustments to how the ORP reading is interpreted according to temperature changes.

Also, a change in the hydrogen ion (H^+) concentration will change the value of the ORP reading. Therefore, in ORP measurement applications, it is imperative that the pH of the process be tightly controlled to assure correct readings.

INSTALLATION. Care must be taken to mount the ORP sensor such that the electrode end is pointed down, +/-15 degrees from horizontal, to prevent an air bubble in the reference side from causing an air gap between the fill solution and the junction. An appropriate spot within a line or in a tank must be chosen to make sure that the sensor is not in a dead zone, where it is unable to see changes in the process.

Proper inline mixing or tank agitation may be necessary to ensure that the sensor is measuring a representative sample. Where the sensor is mounted in a sidestream, the sample line length should be as short as possible to minimize response time and keep the sensor "in sync" with the process.

MAINTENANCE REQUIREMENTS. Frequent cleaning of the sensors noble metal electrode may be required, depending on the nature of the process. The time period between cleanings (days, weeks, etc.) is affected by the characteristics of the process solution and can only be determined by operating experience.

Automatic cleaning systems (using air or water jets) are available to reduce cleaning frequency in difficult applications.

APPLICATION. The ORP is generally helpful for control of BNR processes implemented as sequencing batch reactors (SBRs) and intermittent aeration processes.

ADVANCED INSTRUMENTS. Generally, ammonia and nutrient analyzers are considered more advanced instruments than instruments described earlier, as a result of more sophisticated technology used for these measurements. At this time, ammonia and nutrient analyzers are considerably more expensive to purchase, install, and maintain than the described above meters. However, as a result of the complexity and high operating cost of BNR plants, ammonia and nutrient analyzers can be cost-effective tools for BNR process monitoring and control.

As a result of similarity in technologies often used for various nutrient analyzers, maintenance and installation requirements for all nutrient analyzers are described at the end of this section.

Ammonia and Ammonium. MEASUREMENT. The two principal methods for measuring ammonia with an online instrument are ion selective electrode and colorimetry.

- Ion selective electrode (ISE). Similar in principal to a pH electrode, ammonium ISEs can measure over a wide range and have a fast response time. They are, however, subject to interferences and operate over a limited pH range, so a gas-selective electrode (GSE) is more commonly used in wastewater applications (see below). Another ion selective measurement method includes addition of sodium hydroxide to the sample to raise the pH above 11 and drive all the ammonium to dissolved ammonia. An ammonia GSE is effectively a pH electrode behind a gas-permeable membrane. The ammonia gas is allowed to pass across the GSE membrane and re-dissolves in an electrolyte (typically ammonium chloride [NH_4Cl]) surrounding a pH electrode. The ammonia gas decreases the pH of the electrolyte. The analyzer then converts this change in pH to an ammonia reading. Gas-selective electrodes are much more selective than standard ammonium ISEs. More information regarding ISE technology can be found on the following web sites: http://www.wtw.com, http://www.isco.com, and http://www.myratek.com

- Colorimetry. There are several colorimetric methods for measuring ammonia, including the following:

    (1) Indophenol blue method. See http://www.na.water.danfoss.vom.
    (2) Monochloramine-F method. This method adds hypochlorite to the sample to convert all of the ammonia present to monochloramine. A second reagent containing an indicator is then added to the sample, and the resulting color development is measured using a colorimeter. Another option is to use UV technology to measure UV signature of conversion (see http://www.chemscan.com).

(3) A third colorimetric method works in a similar way to the GSE described above. Sodium hydroxide is added to the sample to raise the pH above 11. The ammonia gas is then stripped from the sample and re-dissolved in a liquid pH indicator solution. The change in color of the indicator is measured using a colorimeter and converted to an ammonia reading. (see http://www.wtw.com and http://www.abb.com).

ACCURACY AND REPEATABILITY. Most manufacturers claim a lower detection limit of 0.1 mg/L ammonia-nitrogen (NH_3-N) or ammonium-nitrogen (NH_4-N). The range of the instrument depends on the measurement method. The ISE and GSE instruments typically have the widest range, measuring as high as 1000 mg/L. Colorimetric analyzers can also measure up to very high levels, but require different reagents to extend the range In practice, this is not a great disadvantage, as ammonia levels typically do not vary widely for a single application.

Manufacturers claim accuracy of between 3 and 5% of reading, which is more than adequate for most wastewater applications.

Nitrate and Nitrite. PRINCIPLE OF OPERATION. Online measurement of nitrate in wastewater is most commonly achieved using UV absorbance (http://www.hach.com and http://www.wtw.com). Nitrate in water absorbs UV light up to 240 nm. Online instruments direct UV light through the sample and measure the amount of light that passes through to a detector. The absorbance is calculated and converted to a nitrate value. Light of a different wavelength is also passed through the sample to allow for compensation for interference resulting from turbidity. Because of solids interference with the measurements, a filtration system is typically required. In addition, there is also chemical oxygen demand (COD) interference. Some manufacturers are offering multiwavelength measurements to avoid interference problems (http://www.chemscan.com). Another solution to the turbidity interference problem is membrane filtration of the sample (http://www.na.water.danfoss.com).

In practice, UV absorbance instruments also measure nitrite (NO_2), as a result of the similar UV absorbance characteristics of the two substances. The result is provided in milligrams per liter of NO_x-N (where $NO_x = NO_3 + NO_2$). In wastewater plants, the concentration of NO_2 is generally insignificant, and the NO_x value is widely accepted for plant control purposes.

Another method of measuring nitrate uses ISE (http://www.myratek.com) and colorimetric (http://www.wtw.com and http://www.abb.com) technologies. However, these technologies are applied less frequently, because of the convenience of the direct UV absorbance method.

ACCURACY AND REPEATABILITY. The UV nitrate probes typically provide a lower detection limit of 0.1 to 0.2 mg/L NO_x-N. The measurement range may extend as high as 50 mg/L NO_x-N, although this may require a longer path length. Claimed accuracy varies from 2 to 5% of the reading, and repeatability is in the same range.

Phosphorus and Orthophosphate. Most discharge permits in the United States for phosphorus are written in terms of total phosphorus (TP). Online TP instruments are available; however, they are relatively expensive and require a high level of maintenance. As a result, orthophosphate (PO_4) instruments are more widely accepted, particularly for process monitoring and control, but, in many cases, also for effluent monitoring.

Note that an analyzer measuring 3 mg of orthophosphate is measuring the equivalent of roughly 1 mg of phosphorus. Thus, an orthophosphate analyzer measuring at 0.3 mg/L PO_4 is equivalent to a TP analyzer measuring at 0.1 mg/L phosphorus. Expressing PO_4 values as PO_4-P avoids this potential source of confusion.

ORTHOPHOSPHATE. All online orthophosphate instruments use colorimetry. Two colorimetric methods are commonly used.

(1) Molybdovanadate method, also referred to as the *yellow method*; and
(2) Ascorbic acid method, also referred to as the *blue method* (or *molybdenum blue method*).

In practice, the yellow method is most widely applied because of the relative simplicity of the instrument and related reagents. Where very low detection levels are required (<0.1 mg/L PO_4-P), the blue method is used.

Most orthophosphate analyzers using the yellow method are capable of measuring over a range from 0.1 mg/L to as high as 50 mg/L PO_4-P. Reported accuracy and repeatability are typically approximately 2%.

TOTAL PHOSPHORUS. Because of their inherent complexity (and corresponding expense), online TP analyzers are generally only used to monitor final effluent in wastewater treatment. Indeed, many plants prefer to use the simpler orthophosphate analyzers as a surrogate parameter on final effluent.

Online TP instruments first convert polyphosphate and organic phosphorus compounds to orthophosphate. Orthophosphate is then measured colorimetrically (typically using the molybdenum blue method). The conversion step requires high pressures and temperatures, and it is this step that makes online TP instruments relatively expensive and maintenance-intensive.

Total phosphorus analyzers are available with measuring ranges from as low as 0.01 mg/L to as high as 5 mg/L TP. Accuracy and repeatability are typically 2% or less.

Installation of Ammonia and Nutrient Analyzers. Analyzers come in two distinct styles: free-standing analyzers and submersible analyzers. The submersible analyzers can be installed directly to a channel or aeration tank, thereby providing faster response time and alleviating the need to transport sample to the analyzer by pump. Most of these analyzers also do not require a stand-alone sample preparation system. The disadvantages of this style of analyzer are the weight (a hoist may be required to facilitate removal for maintenance) and the inconvenience of having to remove the analyzer for reagent replacement and maintenance. This task may become overwhelming if analyzers require frequent maintenance.

Free-standing analyzers are generally wall-mounted and sometimes require a shelter. Sample must be delivered to the analyzer by a pump and often is required to be filtered. The design of sample pumping, piping, and a sample filtration system must have high degree of sophistication, and this system often costs more than an analyzer itself. Unfortunately, both design engineers and manufacturers do not always appreciate the importance of proper design of a sample delivery and preparation system, and, as a result, the entire analyzer assembly, costing thousands of dollars, becomes unusable and worthless. It is recommended to check an installation sites list, which is typically provided by an analyzer manufacturer, to learn more details about successful designs of a sample delivery and preparation system.

MAINTENANCE REQUIREMENTS. Most manufacturers provide a variety of features to reduce maintenance, such as automatic cleaning and calibration and the use of peristaltic pumps and pinch valves (so the sample stays inside tubing and does not contact any moving parts). Nevertheless, ammonia and nutrient analyzers are sophisticated instruments that generally require extensive routine maintenance.

The following maintenance steps and frequencies are common to most online analyzers:

(1) Routine replacement of reagents occurs typically monthly or quarterly, but can vary according to the measurement, calibration, and cleaning intervals chosen;

(2) Cleaning or replacement of electrodes (for ISE and GSE models) on a monthly basis;
(3) Replacement of tubing every three to six months; and
(4) Regular inspection (and cleaning if necessary) of the sample filtration system.

APPLICATIONS. Online nutrient are used in the following applications:

(1) Influent monitoring for feedforward control of chemical dosing and mass-load balancing.
(2) Mass-load equalization.
(3) Aeration control.
(4) Recirculation control.
(5) At exit of anaerobic zone to ensure adequate release of phosphorus from phosphate-accumulating organisms; can also be used to limit excessive phosphate release by controlling recirculation (containing nitrate) to the anaerobic zone.
(6) At exit of aeration basin to ensure adequate uptake of phosphorus.

PROCESS PARAMETERS FOR OPTIMIZATION AND AUTOMATIC CONTROL

GENERAL CONSIDERATIONS. If there were no changes in external conditions, such as flowrate and ammonia variations, process control would be a simple task. However, the ratio of maximum to minimum oxygen demand, consisting of demand for both biochemical oxygen demand (BOD) and ammonia removal, vary typically from 1.5:1 to 3:1 between the peak hours and off-peak hours. For small treatment plants, this ratio is sometimes as high as 10:1. This ratio for BNR plants is larger than for conventional plants as a result of higher variability of influent nitrogen concentration than influent BOD. Equalization of nitrogen load over 24 hours can substantially improve operation of the BNR plant. Fortunately, such equalization for a BNR plant is easier than for a conventional plant because approximately 30% of ammonia is recycled within the plant. A small storage tank for recycled flow coming from dewatering facility can significantly reduce variation of influent ammonia load. Filling the tank during peak hours and empting the tank during off-peak hours can be done automatically using a simple timer. Results of such operation at the City of San Jose (California) are shown on Figure 13.1. A more sophisticated control scheme, using an ammonia analyzer and a controller that maintains constant ammonia load by changing flow from a recycled water storage tank, can also be implemented. Similarly, a control loop could be implemented for equalization of influent ammonia load using the influent equalization tank.

However, even with the best efforts, it is impossible to completely eliminate nitrogen and COD load variation completely. Thus, to maintain optimum process performance, it is necessary to compensate for changing ammonia and COD load by changing airflow supply, waste flow, internal recycling flow, and other control parameters. Implementation of control schemes discussed in the previous chapters is the first step in improving performance of BNR plants. For example, DO control, in addi-

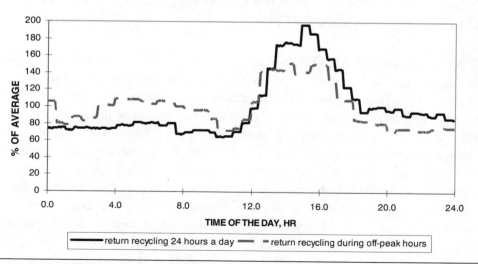

FIGURE 13.1 Influent ammonia concentration with and without ammonia equalization.

tion to decreasing energy costs, also reduces nitrate concentration by reducing inhibition of the dentitrification process caused by presence of oxygen in the anoxic zone. By maintaining optimum DO concentrations, it is possible to avoid both construction of special DO reduction zones between aerobic and anoxic zones and to decrease usage of methanol for denitrification.

SELECTING OPTIMUM SET POINTS. Selecting the optimum set point for each control loops, however, is a more cumbersome task for BNR plants than for conventional plants. For instance, an increase of solids retention time (SRT) may cause frequent problems for BNR plants, such as foam; and reduction of SRT may cause chlorine disinfection problems resulting from the presence of nitrite in the BNR effluent. By selecting and automatically maintaining optimum SRT according to Ekster and Rodríguez-Roda (2003), a 38 000 m^3/d (10-mgd) BNR plant can save between $10,000 and $25,000 per year. Table 13.1 contains considerations that must be taken into account when selecting the set point for control loops discussed in previous chapters.

BASIC AUTOMATIC CONTROL. *Excess Sludge Flow Control.* Optimum performance of activated sludge systems can be achieved only when sludge age is maintained at an optimum value. Below this value, a system can experience low DO filamentous bulking, poor ammonia removal and nitrite breakthrough, dispersed growth of biomass, and an overloaded thickening facility. At a sludge age above the optimum value, a system may face a low food-to-microorganism (F/M) filamentous bulking and foaming, increased oxygen demand, and increased clarifier loading.

Wasting mixed liquor is the simplest method to implement sludge age control, because it does not require measurement of solids concentration. In addition, such wasting can help, sometimes, to control foam (Parker et al., 2003). However, this simplicity comes at high price. As a result of a low solids concentration, the volume of excess sludge is several times larger than the volume of sludge to be wasted from return sludge line. As a result, the pumping cost of the wasted sludge and recycled water from the thickening facility is also much higher. The pipes, pumps, and thickening facilities need to be oversized to accommodate an increased flow. That is why wastage from the return sludge line alone or in a combination of wasting foam from mixed liquor is more widespread, At the same time, sludge age control for this popular wasting method is somewhat more challenging than for previously mentioned wasting methods.

TABLE 13.1 Considerations to be taken into account during selection of set points.

Process	Parameter	Potential problems is that value is Lower than optimum	Higher than optimum
Primary treatment	Sludge depth	• Reducton of VFA production and, as a result, reduction of influent denitrification potential. • Reduction of efficiency of subsequent sludge treatment processes as a result of excess water in thin sludge discharged from primary clarifiers.	• Increased effluent TSS, soluble and particulate BOD, and hydrogen sulfide (H_2S) generation. • Phosphate release in case of co-thickening of primary and BNR wasted sludge. • Increased floating solids resulting from gasification. • Increased mechanical stress on sludge and scum collection mechanism.
BNR process	SRT	• Chlorination problems resulting from the presence of nitrite in the BNR effluent. • Inadequate removal of phosphate, ammonia, and nitrate. • Deterioration of waste sludge thickening as a result of thin sludge discharged from secondary clarifiers.	• Low F/M foaming and bulking. • Increased clarifier solids loading. • Phosphate release. • Increased energy demand to sustain endogenous respiration.
BNR process	DO concentration	• Chlorination problems resulting from the presence of nitrite in the BNR effluent. • Inadequate removal of ammonia. • Foaming problems caused by *Microthrix*. • Gasification in the clarifier resulting from the presence of nitrogen (N_2).	• Inhibition of denitrification. • High energy cost resulting from excessive airflow supply. • Breakup of floc and, as a result, increased effluent TSS.
BNR process	Internal recycle	• Reduction of amount of nitrate to be denitrified.	• Decreased denitrification rate resulting from increased oxygen concentration in the anoxic compartment. • Increased pumping cost.

(continued)

TABLE 13.1 Considerations to be taken into account during selection of set points *(continued)*.

Process	Parameter	Potential problems is that value is Lower than optimum	Higher than optimum
BNR process	Return flowrate from the clarifier	• Improved denitrification in the clarifier. • Increased effluent TSS resulting from (a) increased sludge depth in clarifier and (b) solids floating resulting from denitrification if nitrate concentration is above 6 mg/L.	• Decreased denitrification rate in the clarifier. • Improved denitrification in the first anoxic compartment. • Increased pumping cost. • Deterioration of waste sludge thickening as a result of thin sludge discharged from secondary clarifiers.
BNR process	Methanol feed	Incomplete denitrification.	• Increased cost of methanol, aeration, and sludge processing. • Increased clarifier mass loading.

The traditional formula for calculating sludge age using the method of wasting of returned sludge is as follows:

$$\text{SRT} = V \times \text{MLSS} = (\text{WASSS} \times \text{waste flow}) \qquad (13.2)$$

Where

SRT = solids retention time,
V = volume of the aeration basins,
MLSS = mixed liquor suspended solids, and
WASSS = waste activated sludge suspended solids.

Most plants calculate waste flow based on this formula. However, there are several problems with using this formula for day-to-day operations. The first is the formula's susceptibility to the inaccuracy of TSS measurement, which, on average, is approximately 10%. The second problem is related to the fact that the results of grab sampling could be as much as 50% different than the daily average. Realizing these shortcomings in the formula, many operators use "one-time change of waste flow should not exceed 10%" as the general rule.

The third problem with the formula is that SRT is equal to sludge age only under static conditions. Under dynamic conditions, SRT does not reflect sludge age. For example, if the SRT is 10 days and the waste flow is reduced by a factor of two, the formula dictates that the sludge age immediately equals 20 days. However, it is clear that, even if wastage is stopped completely, the sludge age cannot be increased by 10 days earlier than in 10 days. If new wastage is calculated using the above-mentioned formula, the new sludge age target will be reached only after a period of time equal to approximately 20 days, and not instantaneously, as the formula suggests.

Another general rule states that a period of up to three sludge ages is required to see the results of a change in sludge age. This guideline likely considers that it takes a very long time to reach a new target sludge age if waste flow is calculated using the above formula. As can be seen, both general rules are related more to the deficiency of the formula than to the nature of the activated sludge process.

IMPROVED CALCULATION METHODS. Using several assumptions made two decades ago, Vaccari (1983) proposed an elaborate method for calculating sludge age. Sludge age that is calculated based on Vaccari's formula is called *dynamic SRT*. Dynamic SRT calculations take into account historical data and, as a result, better reflect sludge age.

One of the best methods for calculating sludge age under dynamic conditions is to calculate the mean solids retention time (MSRT), which is the average of SRT values. However, the challenge with using this method is determining the averaging time period. An improperly selected averaging time period can eliminate the advantages of calculating the MSRT. While there are several ways to determine the averaging time period for each operating condition, a method using mathematical modeling of activated sludge systems is one of the best (http://www.srtcontrol.com/SRTMaster/paper.shtml).

There are two obvious advantages to using a properly calculated MSRT or dynamic SRT instead of the traditional formula. First, the desired sludge age will be reached much earlier; and second, these calculations are considerably less susceptible to measurement inaccuracies, especially if frequency of sampling will be increased.

SELECTING SLUDGE AGE TARGET. After the method for calculating sludge age has been chosen, the next challenge is to correlate sludge age to activated sludge performance (such as effluent quality, sludge settleability, and operating costs) and find the optimum sludge age. Unfortunately, some practitioners pay little attention to the correlation between sludge age and operating costs, considering only effluent quality and operational convenience when selecting the desired sludge age.

The effect of sludge age on the operating cost of BNR plants is quite substantial. Ekster (2004) showed that the same effluent ammonia and nitrate concentrations can be achieved by different combinations of such operating parameters as sludge age, DO, and internal recycle flows, as in Table 13.1. The same table indicates that an increase in sludge age generally leads to lower BNR energy costs as a result of the reduced energy demand for aeration and internal recycling. However, an increased sludge age often increases foaming and effluent TSS concentration, and even may cause clarifier overload. When selecting the optimum parameters, it is important to count not only the energy cost, but also the cost of chemicals used to control these problems, the cost of foaming handling. As a result, the longest sludge age may not be the optimum one.

MAINTAINING OPTIMUM SLUDGE AGE. After the optimum sludge age is found, it should be maintained precisely. Variation of sludge age by just one day may cause an increase of operation cost by as much as $2500/mgd/yr (Ekster, 2004). An automatic waste control system can significantly improve the precision of sludge age control. It can provide constant accurate control over solids inventory, reduce the time required for sampling and sample processing, and eliminate human errors associated with wastage calculations and lag time. In addition, the automatic waste control can stabilize sludge load to the sludge processing facilities. A control system (see Figure 13.2) consists of two suspended solids meters; a controller (computer); and a waste flowmeter, with a valve for adjustment. Information from the suspended solid meters is sent to the controller; the controller compares the operational criteria, such as simple SRT or MSRT or mixed liquor TSS, with the target value, calculates the necessary adjustment of the waste flow (Q_w), and sends a signal to the control valve on the waste activated sludge line (Ekster, 2002).

There are several challenges that must be overcome to implement the control scheme described above.

(1) Poor reliability of TSS meters;
(2) Sludge age control may lead to excessive variability of the mass loading on thickening facility and, as a result, drastic deterioration of the sludge thickening process; and
(3) Potential overload of clarifiers and thickening facilities.

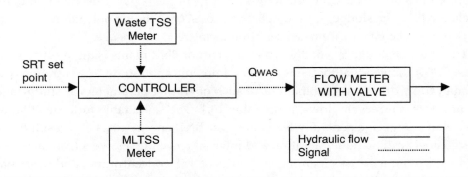

FIGURE 13.2 Automatic waste control system schematic.

To overcome these problems, use of a combination of pattern recognition algorithms and an International Association of Water Quality (London) activated sludge model was suggested (Ekster, 2002). Commercially available software that uses these principles is available (http://www.srtcontrol.com).

Case Studies. *OXNARD TRICKLING FILTER SOLIDS CONTACT ACTIVATED SLUDGE SYSTEM.* Operators at a 121 000-m^3/d (32-mgd) trickling filter activated sludge plant in Oxnard, California, observed considerable improvements in sludge settleability (Figure 13.3), including a 25% reduction in effluent TSS after implementation of sludge age optimization and automatic control program. In addition, oxygen demand was decreased by 4%, and polymer use by 25%. Operators at the Oxnard plant have realized that optimizing sludge age is a continuous improvement process and, as a result, work continuously to adjust the sludge age target based on quarterly performance, water temperature, and other factors.

TORONTO MAIN WASTEWATER TREATMENT PLANT. At the Toronto Main WWTP (606 000 m^3/d [160 mgd]), after installation of the sludge age control system, significant reduction of TSS and waste mass variability were observed (Table 13.2 and Figure 13.4).

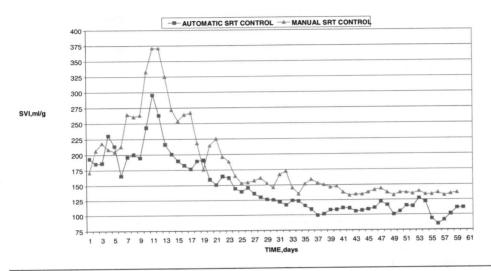

FIGURE 13.3 Oxnard trickling filter–solids contact activated sludge system improvements in sludge settleability.

Variation of return activated sludge suspended solids (RASSS) and mixed liquor suspended solids (MLSS) concentrations were reduced by approximately a factor of 3, while the variation of mass waste stream was reduced by a factor of 7.

Significant reduction of variation in the waste stream mass improved the operation of the thickening facility, and the earlier practice of bypassing caused by facility overload became unnecessary. Significant (up to 50%) reduction of polymer usage was also observed.

SANTA CLARA/SAN JOSE WATER POLLUTION CONTROL PLANT. Before implementation of automatic sludge age control, the 632 000-m^3/d (167-mgd) Santa Clara/San Jose Water Pollution Control Plant has experienced significant foaming problems. Following the installation of automated sludge wasting, no incidents of nocardioform foaming have occurred.

It is suggested (Ekster and Jenkins, 1999) that the reason for the absence of nocardioform foaming is that sludge age control with automatic sludge wasting allows

TABLE 3.2 Average percent deviation from 24-hour moving average.

	Manual control	Automatic control
Mixed liquor TSS	12	4
Return sludge TSS	14	5
Mass of waste load	23	3

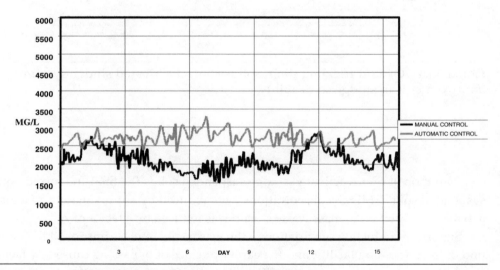

FIGURE 13.4 Effect of automatic control on MLSS.

much less variation in discrete calculated SRT values than manually controlled wasting based on laboratory TSS analyses does. This point is illustrated by SRT data in Figure 13.5.

During manually controlled sludge wasting based on daily laboratory TSS analyses, there are many incidents of very long SRT calculated based on the traditional formula. These may have been caused by errors in analyses or operation because of

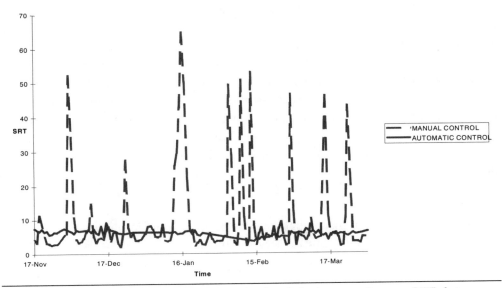

FIGURE 13.5 The effect of different methods of sludge wasting on SRT data.

the relatively infrequent data collection and waste flow adjustments. Of course, sludge age change did not exceed 1 day per day. But even a change in 1 day/day represents 15 to 20% instantaneous increase in sludge age and several-fold decreases in waste flow. A drastic increase in sludge age makes nocardiforms more competitive, and they start to grow in aeration tanks. These organisms concentrate in the foam layer and are retained in the activated sludge system by the surface-foam-trapping features of the aeration basin, even when the sludge age is lowered. With continuous TSS measurements and automatic sludge wasting control, there were no incidents of drastic decrease of waste flow and no nocardioform growth. It is interesting to note that, with automatic sludge age control, it was possible to operate in a nocardioform foam-free condition with a longer average sludge age (6.3 days) than with manual activated sludge wasting based on daily laboratory TSS analyses (5.1 days).

These case studies have shown that proper optimization and automation of sludge age is a cost-effective way to improve plant performance and enhance the bottom line.

Dissolved Oxygen Control for Biological Nutrient Removal Plants. THE NEED FOR GOOD DISSOLVED OXYGEN CONTROL. It is widely accepted and understood that good DO control is a desirable feature for all activated sludge systems. Without good DO control, there is a danger that the plant may suffer one or both extreme DO conditions. For any activated sludge plant, a very low DO concentration inhibits treatment effectiveness and possibly promotes filament growth. Unnecessarily high DO concentrations waste energy. Conservative estimates of energy savings through the use of automated DO control are in the range 5 to 10% of total power costs (Metcalf and Eddy, 2003), while more optimistic estimates put the potential savings over 30% for some facilities (EPRI, 1996). No matter which number is selected, the potential energy savings are substantial.

In addition to the described above considerations, BNR plants have several specific process features that make good DO control even more important. Based on the Monod equation. ammonia removal is proportional to the following ratio:

$$DO/(DO + K) \tag{13.3}$$

where K = coefficient.

The value of K changes depending on number of factors, such as SRT and temperature. However, for a particular facility, it can be considered constant. If K is not known, IAWQ recommends using a default value of 1 mg/L. By maintaining constant a DO concentration, the constant nitrification rate can be maintained.

At the same time, an increased DO in the recycle streams of a BNR plant inhibits both denitrification and phosphorus removal. This inhibition is proportional to the following ratio:

$$K_1/(DO + K_1) \tag{13.4}$$

where K_1 = coefficient.

Similarly to the K coefficient, the value of K_1 changes depending on number of factors, such as SRT and temperature. During low-flow periods, an increased DO concentration causes both energy waste and jeopardizes compliance with nitrate and phosphorus limits.

In completely mixed systems, it is possible to obtain effective DO control by simply controlling the DO at one location and adjusting blower output to the diffuser grid. However, in plug-flow systems, each pass has a different oxygen demand, and the air demand in the last pass is often lower than the air required for mixing. Therefore, the air flowrate required for each pass is different, and a common header system should ideally need to be equipped with multiple control valves to regulate the air flow; otherwise, not only nitrification process may suffer, but high DO in recycled flow may inhibit denitrification. In a multicompartment plug-flow system, it is necessary to provide at least the same number of DO control loops as there are number of aerobic compartments.

Automated control using online DO measurement offers the best option for providing good, accurate DO control in BNR plants; however, there are a number of challenges that must be overcome, and these are addressed in the following sections.

CHALLENGES. Despite the fact that automation of oxygen supply system is a well-known method of energy savings, it is estimated that as few as 10% of treatment plants in the United States use proper automatic DO control. Three main factors contributing to this are the following:

(1) Malfunction of primary sensors and valves, especially DO sensors. Most control systems are not able to detect malfunctions of DO meters. When these systems act using erroneous data, efficiency of treatment is severely jeopardized.
(2) Poor stability of control system resulting from control loops interaction. Finding the proper tuning parameters of the control algorithm is a difficult task because of the similarity between dynamic characteristics of multiple valves and blowers. Poor stability causes valve "hunting" or "chattering" and causes damage to the blower and excessive wear of valves.
(3) Improper place for DO sensors installation. For example, DO control meters at BNR plants need to be installed at one-quarter of the compartment length from the end of each nitrification compartment.

The following two approaches should be considered to overcome the problems of DO sensor malfunction:

(1) Improved DO sensor maintenance. To improve the quality of the DO signal and increase confidence that the sensor is giving sensible readings, it is imperative that the DO probe be properly maintained. All probes must be cleaned and calibrated on a regular basis. A DO sensor located in a highly loaded aeration tank (such as the first zone of a plug-flow system) operating at higher temperatures will require more frequent attention to ensure that it gives good readings than a sensor located in a low-loaded, low-temperature

aeration tank. It is beneficial to use automated cleaning and calibration to provide better sensor maintenance (see Marx et al. [1998] for a discussion of the merits of this approach).

(2) Fault detection. A major problem with incorporating the DO sensor is that most DO sensors and do not provide fault detection, beyond the loss of signal indication. Some sensors can provide an output to indicate when the probe was last cleaned or calibrated, and this can be used as part of routine manual cleaning and calibration. A more sophisticated and reliable approach to detecting erroneous DO readings is to use automatic calibration and checking. Currently, this feature is only available in one DO sensor developed by Dr. John Watts in the 1980s and 1990s and is widely used in the United Kingdom. Its features include automatic cleaning every four hours, automatic in situ, calibration every 24 hours, and sophisticated statistical analysis of the curve generated during calibration. Other facilities have added their own custom-made systems to simple DO sensors to provide feedback on sensor readings using automated or semiautomated calibrations. The most advanced approach to detecting faults in DO sensor readings and other monitors is to use some kind of fault detection algorithm in the control logic to detect when the readings are inconsistent with other measurements on the plant. This approach has been tried on a number of facilities; however, as a result of its complexity this method is not widespread. At the same time, some software manufacturers embrace this approach because of a dramatic increase in system reliability. Ekster and Wang (2005), for example, reported that testing and use of this commercially available software DOmaster™ (Ekster and Associates, Inc., Fremont) at the full-scale facility showed good results.

CONTROL STRATEGIES. The following are descriptions of the traditional approaches:

(1) Direct control. The DO signal is used to control the aeration device directly (blower or surface aerator).
(2) On/off control. A variation on the direct control approach, where aeration devices are shut off when the DO reaches an upper set point, and then switched on again when it reaches a lower set point, or after a set time interval.

(3) Cascade control. A DO signal is used in each basin, in conjunction with a control valve to adjust the air flow to the basin, being fed from a common header. A second control loop is set up between a pressure transducer on the main header and the blower system. The blowers act to maintain the header pressure near the desired set point, as the individual control valves feeding each basin open and close.

Direct control DO is the simplest control approach and can provide very good, tight control. However, this approach can only be used in a limited number of cases, where each aeration basin has its own dedicated aeration source and DO sensor. The following plants are most suited to using direct control: plants with mechanical aerators and oxidation ditches; and plants with a small number of completely mixed aeration basins, with blowers dedicated to each basin.

On/off control is most applicable where mixing and aeration are separate actions (e.g., pure oxygen plants and some oxidation ditches). Using this approach for other plant types is not widely accepted because of concerns over solids settling when the aerator is switched off. However, this control strategy has been used at many BNR European plants with good success. Intermittent aeration can produce very low effluent nitrates.

Cascade control was identified by the WERF report (Hill et al., 2002) as the most common approach for DO control. For multiple-tank and plug-flow systems especially, it is the most logical control approach. Providing a common header for multiple tanks and compartments within each tank enables the plant to be designed with the most cost-effective number and size of blowers, providing air throughout the plant. The blowers are used to maintain the header pressure using feedback from a pressure transducer, and this control loop can be set up to provide a rapid response to changes header pressure. The pressure set point is either based on operator's experience or calculated automatically, based on the requirement to maintain one air control valve within the air distribution system at 75 to 90% opening. The airflow through individual basin feed pipes is regulated by a control valve, which is controlled by a DO sensor. A flow meter can be installed on the feed pipes and included in the cascade loop to provide more precise control.

As mentioned above, one of the main difficulties in using a cascade control is the interaction between different elements in the system. For example, when one aeration basin experiences a decrease in DO concentration, an air control valve starts to open and pressure starts dropping. If the header pressure controller does not increase blower output quickly, then the air flow to other basins will be reduced. When this happens, the header pressure drops more noticeably and the blower control system may respond quickly, raising the header pressure and increasing the flow to the one basin we considered at first. Now the airflow in the basin is too high, and the valve may close down again, causing the other elements in the control system to have to respond again. In this simple example, it can be seen that all of the elements in the system are interconnected and influence each another. An extreme sensitivity of air flow to a change of air pressure and hydraulic flow and nonlinearity of each of cascaded control loops are the main challenges in designing and tuning of DO control systems.

Recently, considerable work has been done with IAWQ and other computer models to help determine the best parameters for different control schemes, including simple proportional-integral-derivative (PID) loops and more advanced approaches using techniques such as time series and regression analyses and a multicascade, nonlinear, proportional-integral- derivative Olsson and Newell, 1999). Using a mathematical model of an aeration system, Ekster (2004) theoretically proved that a traditional pressure-based control system has stability problems. Ekster and Wang (2005) also showed that DOmaster that uses an IAWQ activated sludge model and dynamically updated database instead of a pressure-based blower control algorithm was able to provide precise and reliable DO control at a 120 000-m^3/d (32-mgd) facility with no oscillation of blower and valves outputs. The previous traditional DO control system that was based on pressure based control and PID logic failed to provide satisfactory performance.

Development and tuning of cascade DO control systems generally is not trivial task and requires specialized expertise that may not be readily available at some treatment plants. This expertise can be obtained by using either specialized consulting, system integration firms, or by purchasing commercially available turn-key DO control software and hardware that is provided by some blower manufacturers (http://www.turblex.com) and by manufacturers of process control systems and software (http://www.srtcontrol.com).

Control of Chemical Addition. The final area that can be considered for basic control, and arguably the simplest to implement, is the automated control of chemical addition. The first level of control for chemical addition is to base the chemical dose rate on the flow (i.e., to flow-pace the addition). In addition to this, the chemical feed can be regulated based on a water analyzer or probe.

The most widely implemented automated chemical addition control system is for chlorination, where the addition of chlorine is controlled to maintain the desired chlorine residual. The main feature of this control system that makes it relatively straight forward to implement, at least in principle, is that the addition of chlorine has a direct effect on the chlorine residual, enabling a simple feedback loop to be used to regulate the chlorination. Other examples of chemical feed systems amenable to automated control include ferric and alum feed control for chemical phosphorus removal, methanol feed control for denitrification (using either nitrate or ORP measurements), and alkalinity measurement and control.

ADVANCED CONTROL

AMMONIA CONTROL. The influent ammonia concentration changes over 24 hours as a result to achieve the same ammonia concentration 24 hours a day downstream of an aerobic compartment, increasing the removal rate during the peak time and reducing it during the off-peak time is desirable. This can be achieved by changing the DO set point instead of maintaining it at a constant level. The challenges of such control, however, are slow dynamic of ammonia change and reliability and cost of ammonia online analyzers. For such control, either a PID cascade control scheme (Ingildsen et al., 2001) or rule-based control (Krause et al., 2001) can be used. Cascade control using the Monod equation instead of PID was used by Liu et al. (2003). Another method of ammonia control is changing air flow proportionally to the change in mixed liquor ammonia concentration (Pelliter et al., 1999). A proportionality coefficient is changed based on the rules developed for a particular facility. This method lacks the precision of cascade control schemes that use DO sensors. All discussed control schemes require significant development efforts and may not be practical for some facilities, especially if ammonia concentration at the end of aerobic compartment does not need to be constant. A simpler control method is to turn on and off the air supply to a so-called swinging compartment (i.e., compartment that can be used for either nitrification or denitrification processes) based on the ammonia concentration (Samuelsson and Carlsson, 2001).

CONTROL OF DENITRIFICATION. At the BNR plant that includes internal mixed liquor recycle (MLR) from the aerobic zone to the anoxic zone, denitrification is controlled by MLR flowrate and, sometimes, chemical addition of methanol or other sources of COD. Increasing the MLR increases the extent of denitrification, but only up to a certain point. If the MLR is increased too high, it can cause excessive DO to be returned to the anoxic zone, inhibiting denitrification. For practical purposes, an MLR up to four to six times the influent flow is typically sufficient to provide improved denitrification.

The MLR flow can control denitrification in proportion to the influent or based on nitrate concentration in the anoxic compartment using either proportional algorithm (Yuan et al., 2001) or fuzzy logic (Serralta et al., 2001). Implementation of these control loops can significantly reduce power usage for recirculation and improve nitrogen removal. These control schemes are fairly easy to implement; however, nitrate analyzer reliability, initial cost, and maintenance requirements need be factored in before making a decision. Oxidation-reduction potential probes can be used as a cheaper alternative for this control approach; however, the control range is dependent on the presence of ionic species, other than nitrate, in the mixed liquor, making it difficult to define the desired operating set points, and they still have issues with cleaning and maintenance.

For chemical addition, the traditional control approach is to provide a flow-proportional dose system. More advanced control uses the measurement of nitrate and controlling chemical (generally methanol) addition, using either a simple proportional control algorithm (Devisscher et al., 2001) or fuzzy logic (Marsilli-Libelli et al., 2001). Another advanced chemical addition control scheme is using a cascade PID algorithm and two nitrate analyzers (Cho, Chang, et al., 2001). One analyzer is installed in the mixed liquor channel and another one located in the anoxic compartment. The one located in the mixed liquor channel is used to provide signals to the primary PID controller that calculates nitrate concentration target in the anoxic compartment. The secondary PID controller maintains this target, using information obtained from nitrate analyzer installed in the anoxic compartment. Difficulties in maintaining the two analyzers and tuning the controllers are drawbacks of this advanced control method.

Both ammonia and nitrate control schemes need to take into account challenges of operating ammonia and nitrate analyzers. Even the most advanced control method can cause significant operational problems if it does not use fault-safe logic. Development of such logic sometimes requires three to four times more effort than development of the control algorithm itself.

RESPIROMETRY. Respirometry has been widely used to characterize waste and as a method for determining kinetics (Chandran and Smets, 2001; Spanjers et al., 2002). Development of respirometric control has been made possible by the use of the online respirometer. This instrument has been used to conduct surveys of wastewater treatment facilities, to develop diurnal load profiles, and to optimize the operation of many plants. Several control approaches have been suggested, and they are dependent on the type of activated sludge plant that is to be controlled. Oxidation ditches, completely mixed tanks, SBRs, and plug-flow reactors are all suitable candidates for respirometric control. (Barnard et al., 2003).

Respirometric control was shown to be the most effective approach for improving nitrogen removal in an oxidation ditch in Beemster, Holland (Draaijer et al., 1997). In this application, an in situ respirometer was used to measure the oxygen load present in the ditch and simply control the number of aerators in operation to match the demand.

Respirometric control is an obvious choice for SBRs, and some promising investigations have been carried out in this field (i.e., Cohen, et al., 2003; Yoong et al., 2000). The simplest control using respirometric control, proposed by Shaw and Watts (2002), is to discontinue the aeration once the respirometric measurement indicates that the biomass is endogenous and allow an idle phase that makes use of endogenous denitrification to improve overall nitrogen removal (Shaw and Watts, 2002).

Plug-flow reactors provide the greatest challenge for implementation of respirometric control, although it was one of the first activated sludge processes considered for advanced control by the foremost proponent of respirometric control, Dr. John Watts (Watts and Garber, 1993). Watts and Garber proposed that anoxic/aerobic swing zones could be used with respirometric measurements to adjust the overall aerobic volume to match the influent load. Rapid respirogram generation is required to facilitate dynamic control; or, alternatively, respirometry can be used to build a picture of typical load variations, and this can be used to adjust anoxic/aerobic volumes.

No matter which activated sludge process is used, respirometric control can be used to improve the performance of the plant. There is great potential to provide better nitrogen removal; to increase nitrifier viability, by reducing the amount of time that the biomass is endogenous; and to reduce energy costs, by matching aeration to the nitrogenous oxygen demand in the plant.

INTERMITTENT AERATION. Intermittent aeration in a continuous-flow system (as opposed to SBRs) has been proposed as an effective way to achieve nitrification and denitrification in a single reactor. In this process, the biomass in the main reactor (or reactors) is subjected to a cycling of aerobic, and then anoxic, conditions, as the aeration system is switched on and off. Because the reactor repeatedly changes from an aerobic state to an anoxic state and back again, an obvious measurement choice for control purposes is ORP. Several examples of ORP control systems exist, including those described by Caulet et al. (1997), Charpentier et al. (1998), and Mori et al. (1997). In these systems, ORP is used to determine when denitrification is complete under anoxic conditions and when nitrification is complete under aerobic conditions. Carucci et al. (1997) also recognized the potential for using ORP to control intermittent aeration, but contended that pH was a more suitable control parameter, based on their experiments at the Cisterna di Latina WWTP.

The other main approach to controlling intermittent aeration is to simply use DO and timers to achieve the required treatment. In their publication "Introduction of Japanese Advanced Environmental Equipment" (2001), the Japan Society of Industrial Machinery Manufacturers describes a simple control scheme that runs the aeration system to a very low DO (0.2 mg/L) for a set period to provide mixing, but no nitrification, and then increases the DO to 2.5 mg/L to provide fully aerobic conditions for nitrification.

SEQUENCING BATCH REACTORS. As previously mentioned in the respirometric control section of this chapter, SBRs offer the simplest application to control by adjusting the duration of the different treatment phases. Several examples of control of pilot-scale SBRs exist in the literature; though there are few instances where control has been applied to full-scale plants. This is, in part, a result of difficulties with synchronizing multiple basin timings, particularly for SBR basins that share a single blower system. In pilot-scale tests, this is not generally a concern.

Cho, Sung, and Lee (2001) tested ORP, DO and pH control options in a pilot SBR and concluded that ORP provided the best total nitrogen removal. Similarly, Andreottola et al. (2001) found that ORP provided the best control. In their control scheme, they used an algorithm to detect the characteristic nitrate "knee" that appears in the ORP, at the point where nitrate has been reduced to zero, to determine when denitrification was complete. Neither of these examples is based on a full-scale application. Demoulin et al. (1997) describes a control system for the cyclical activated sludge technology system at the Großarl WWTP, Austria. In their system, they use ORP to adjust the DO set points in the aerobic treatment steps, though it is not clear if this process is fully automated.

COST MODEL FOR CONTROL STRATEGY DEVELOPMENT. Much of the current development of control algorithms uses process models. To standardize the way in which models are used to develop control approaches, in 2001, the benchmark group of the European Co-operation in the field of Scientific and Technical Research (COST) Action 624 published guidelines for several benchmark models that can be used to test control strategies (Copp, 2001). Many of the control algorithms developed and presented in many European papers have been developed using computer-based simulations in accordance with the COST guidelines. This has enabled control algorithms to be developed without the expense and difficulties of using real instrumentation and control equipment.

SUPERVISORY CONTROL AND DATA ACQUISITION SYSTEM REQUIREMENTS

GENERAL CONSIDERATIONS. In essence, the purpose of a comprehensive supervisory control and data acquisition (SCADA) system is to provide timely information to various operational and management personnel so that they can perform their job function properly and efficiently. It is because of this that a comprehensive SCADA system becomes the central engine for processing all types of essential information for a variety of job functions. Until recently, proprietary distributed control systems were recommended for installations at WWTPs. Because of their extremely high cost and proprietary nature, more and more facilities are implementing various SCADA architectures. This section describes essential components of a SCADA system. Because of specific computer terms used in this section, the glossary of terms is provided at the beginning of the section.

SUPERVISORY CONTROL AND DATA ACQUISITION FUNCTIONS. The comprehensive SCADA system encompasses seven multiple levels of process control, data manipulation, and management systems. The following are these seven core functions:

(1) Continuous process control;
(2) Process data acquisition;
(3) Supervisory control;
(4) Distributed alarming;

(5) Historical collection, display, and analysis;
(6) Information systems; and
(7) Security.

Finally, a comprehensive SCADA architecture should be designed with future expansion (scalability) as a major consideration in mind. Effort should be made to adopt industry standards that avoid proprietary software and proprietary configurations as much as possible. The following is a more detailed list of these architecture standards and functions.

This function includes I/O device communication between field devices, such as instrumentation, starters, switches and motors, and PLCs. Continuous control includes features discussed in the following sections.

Continuous Process Control. PROGRAMMABLE LOGIC CONTROLLER AND LOGIC PROGRAM.

- Provides continuous process control logic (i.e., simple to complex calculations, high-speed logic, process alarm initiation, life safety logic, process interlocks, and other custom commands and controls).
- Communicates with field devices and instrumentation via drivers, including Ethernet, and other protocols.
- Compliance to IEC 61131-3 PLC program logic standards. The body for adopting and maintaining these standards is "PLC OPEN" (WWW.PLCOPEN.ORG).
- Communicates with HMI server system via digital nonproprietary communication protocol called OPC (Microsoft-based object linking and embedding for process control). The body for adopting and maintaining these standards is "OPC Foundation" (www.opcfoundation.org).
- May also communicate with HMI server via TCP/IP protocols such as Ethernet or Ethernet/IP. This is a very popular communication protocol because it is readily available, cost effective, and easy to implement. However, caution must be taken if the Ethernet communication path is shared with other data because this may affect data throughput.
- Finally, may also communicate with HMI server over proprietary protocols such as ControlNet by Allen-Bradley. This communication protocol supports high-speed deterministic data transmission plus additional integrated PLC/HMI features.

PROGRAMMABLE LOGIC CONTROLLER PROGRAMMING SOFTWARE.

(1) Used for developing PLC logic,
(2) Troubleshooting PLC program, and
(3) IEC1131-3 compliant programming languages, including the following: ladder logic, function block diagramming, sequential function charts, instruction list, and structured text.

Data Acquisition. This second function establishes communication between the controller (specialized simple computer designed only for process control) and the HMI server software. It is generally performed by an I/O driver (i.e., communication software), which converts the controller data into a format required by the SCADA software. The data travels over a "data highway" or local area network. Data acquisition features include those described in the following sections.

INPUT/OUTPUT SOFTWARE.

(1) Sits between PLC and HMI server software.
(2) Software generally resides in HMI server.
(3) Passes bidirectional data between PLC and HMI server.
(4) Monitors communication between PLC and HMI server.
(5) Two types of communication software called drivers: proprietary I/O drivers and open standard OPC I/O drivers.
(6) The OPC drivers can communication with other OPC-compliant software applications, including the following: historical archiving (historical collection and display applications), thin client applications (browsers), and alarm notification applications (e-mail, cell phone, etc.).

DATA HIGHWAYS AND ETHERNET COMMUNICATIONS.

(1) Proprietary data highways: PLC controller to remote I/O and PLC controller to SCADA server.
(2) TCP/IP (Ethernet) networks:
—PLC controller to remote I/O,
—PLC controller to SCADA server,
—SCADA server to SCADA client,
—SCADA server to thin clients, and
—OPC servers to OPC clients.

(3) Other methods of communication: COM/DCOM and serial.

REDUNDANCY.
- Supports primary and backup data highways,
- Supports primary and backup TCP/IP Ethernet networks as well as proprietary networks such as ControlNet,
- Automatic fail-over, and
- Automatic recovery.

NETWORK MAPPING AND MONITORING.
- Specialized stand-alone software that monitors network configuration,
- Monitoring of network active connections,
- Alarm generation of network errors, and
- Network log.

Supervisory Control. This third function is one of the key characteristics of the comprehensive SCADA system. It allows the operator to enter set points; start, pause, and stop major process functions; provide data to remote SCADA clients, and monitor overall communication health. Supervisory control also includes real-time trending for process monitoring. Supervisory control features include those described in the following sections.

SUPERVISORY CONTROL AND DATA ACQUISITION ENGINE (CORE).
- Establishes and maintains sessions between SCADA nodes;
- Maintains alarm queues;
- Maintains network queues;
- Minimizes network traffic between clients;
- Provides communication monitoring, alarming, and alarm distribution;
- Establishes ODBC connections to compliant software (relational databases); and
- Establishes OPC connections to compliant software (I/O drivers, etc.).

SUPERVISORY CONTROL AND DATA ACQUISITION DATABASE.

- Process database obtains real-time data from I/O drivers;
- Allows operator to send setpoints and supervisory commands to the process;
- Process database obtains data from relational databases;
- Provides data distribution to remote SCADA clients;
- Allows distribution of data to other third-party applications;
- Accepts alarming set points;
- Provides advanced alarm-handling features;
- Provides security protection for process outputs and modifications;
- Provides scaling, signal conditioning, engineering units, etc.; and
- Allows additional text information related to data.

GRAPHICS.

- Provides operators a real-time "window" into the process;
- Commonly used for data monitoring;
- Provides mechanism to enter process and alarm set points;
- Interface for alarm acknowledgement;
- Interface for real-time and historical charts;
- Graphical re-creation of process piping, tanks, pumps, valves, etc.;
- Supports online configuration and troubleshooting; and
- Optimized for reduced network traffic.

REAL-TIME TRENDING.

- Display data changes in real-time (20 times per second or higher).
- Prebuilt trends and custom trends.
- Select multiple pens (data points).
- Selectable time scale: start time, end time, duration, etc.
- Selectable mode: average, high, low, sample, etc.

- Adjustable range: engineering units, limits, zoom, etc.
- Operator may build or modify custom charts on demand.

DATA INTEGRITY.

- Monitor data integrity between HMI server and I/O devices,
- Monitor data integrity between I/O devices and field devices (instrumentation),
- Minimize network traffic,
- Monitor sessions between HMI clients, and
- Optimize communications with I/O drivers.

DATA DISTRIBUTION.

- Provide data to other software applications,
- Maintain communication with remote SCADA clients,
- Communication with relational databases (ODBC), and
- Communication with historical archiving software.

Distributed Alarming. This function is the fourth key characteristic of the comprehensive SCADA system. It provides a comprehensive alarming information distribution to all SCADA clients. This is done in a manner to minimize network traffic, while keeping all clients informed of current alarm conditions. Distributed alarming features include the following:

- Alarm set point adjustment (low, low low, high, high high, rate of change, etc.).
- Alarm detection.
- Alarm distribution (dissemination of alarm information over local area network [LAN]).
- Alarm acknowledgement.
- Alarm filtering (filter by priority, area or process, node, time, date, etc.).
- Alarm text messaging.
- Alarm logging.
- Alarm security (for alarm acknowledgement and set point adjustment).

Historical Collection. Display and Analysis. Stand-alone historical collection software is a desirable feature for a comprehensive SCADA system. In the event that the SCADA system is offline, the historical collection may continue to collect and archive process data to ensure that no data is lost if it is communicating with an OPC data server. Many historians are easily configured to provide the data to the SCADA displays via an OPC connection.

HISTORIAN SOFTWARE.

(1) Stand-alone software may continue to collect if SCADA offline.
(2) Collects real-time data directly from the following:
 —OPC servers and
 —HMI SCADA servers.
(3) Optimized data collection.
(4) Advanced compression algorithms.
(5) Automatic archiving.
(6) Mathematical functions.
(7) Built-in report generation.
(8) Built-in troubleshooting tools.
(9) Report generation.
(10) Proprietary third-party drivers and ODBC connectors.

HISTORICAL TREND CHARTS.

- Typically time-series chart format.
- Displays historical data from archive files.
- Displayed on HMI server screens.
- Compatible with ActiveX control charts and browsers such as MS Explorer.
- Prebuilt and custom trends for users.
- Select multiple pens (data points).
- Selectable time scale: start time, end time, duration, etc.
- Selectable mode: average, high, low, sample, etc.
- Adjustable range: engineering units, limits, zoom, etc.

HISTORICAL METRIC CHARTS.

- The bar chart is the most common format (others are also available).
- Custom analysis tools for determining equipment optimization include the following components:
  - Overall equipment effectiveness, including equipment downtime;
  - Production efficiency;
  - Equipment maintenance monitoring;
  - Key performance indicators;
  - Process optimization; and
  - Statistical analysis tools.

Information Systems. This sixth function is another key characteristic of a comprehensive SCADA system. It provides timely and easy to access information to all types of personnel in the format, which is required for their job function. This information is typically available on their desktop, without the need for expensive FAT client software. Information systems features include the following.

REPORTING.

(1) Custom and predetermined report formats.
(2) Provide real-time process data in report format.
(3) Provide historical data reports.
(4) Security reports.
(5) Set point change reports.
(6) Reports generated in multiple formats, including the following:
  - Microsoft Excel,
  - Browser viewable with HTML,
  - Microsoft Word or WordPerfect, and
  - Browser viewable from Java applets or ActiveX controls.

AUTOMATED ALARM NOTIFICATION.

(1) Alarm conditions automatically alerted to the following:
  - Cell phone,
  - Pager,

—Blackberry,
—E-mail,
—Voicemail,
—Palm pilot, and
—Personal digital assistants (PDAs) and pocket personal computers (PCs).
(2) Critical process information available to many of the devices listed above.

THIN CLIENT SOFTWARE.

(1) Leverages use of Microsoft Internet Information Services.
(2) Generally restricted to LAN.
(3) Allows thin client connections to view process data from any browser.
(4) Supports the following functions:
—View real-time process data and trends,
—Automatic report generation and viewing,
—Historical data charts and analysis,
—Key performance indicators,
—Process optimization, and
—Statistical analysis.

SERVER EMULATION SESSIONS.

(1) Emulates server software from another computer on LAN.
(2) Allows remote session established with SCADA server.
(3) Eliminates need for thick client licenses.
(4) Reduces costly client computer configuration and maintenance.
(5) Multiple connection options available:
—Microsoft terminal services and
—Citrix services (Citrix, Fort Lauderdale, Florida).

Security. Security is the seventh feature that encompasses all applications related to the comprehensive SCADA system. It protects from unauthorized entry and also serves to only enable features and functions based on the individual's user's rights. Security features include the following:

(1) Several levels of protection, including the following:
—Individual user-based security system,
—Group-based security system (i.e., operators),
—Application-based security,

—Process screen-based security, and

—Process value- (database tag-) based security.

(2) Protection against unauthorized operation.
(3) Custom security modifications supported.
(4) Compatible with Microsoft Windows 2000 and 2003 security.
(5) Level C2 security-compliant (a high level of security per a standard implemented by the U.S. Department of Defense).
(6) Security audit logs, including the following:

—Logs all process changes by user and

—Automatic timeout and logout.

COMPREHENSIVE SUPERVISORY CONTROL AND DATA ACQUISITION SYSTEM SUMMARY. In summary, the installation of a comprehensive SCADA system should be thoroughly planned and designed for each of the following components:

- Continuous process control;
- Process data acquisition;
- Supervisory control;
- Distributed alarming;
- Historical collection, display, and analysis;
- Information systems; and
- Security.

Standards for installation and programming should be determined before installation begins. Methods of installation and programming execution should be determined with requirements of detailed documentation provided by the system integrator performed concurrently as installation progresses. Finally, custom training with a detailed operations and maintenance manual is an essential part of the comprehensive SCADA system installation.

REFERENCES

Andreottola, G.; Foladori, P.; Ragazzi, M. (2001) On-Line Control of a SBR System for Nitrogen Removal from Industrial Wastewater. *Water Sci. Technol.*, **43** (3), 93–100.

Barnard, J. L.; Shaw, A.; Watts, J. B. (2003) Optimized Control of Nitrifying Suspended Growth Systems Using Respirometry. *Proceedings of the Ozwater Convention and Exhibition, AWA 20th Convention*, Perth, Australia; Australian Water Association: Sydney.

Carucci, A.; Rolle, E.; Smurra, P. (1997) Experiences of On-Line Control at a Wastewater Treatment Plant for Nitrogen Removal. *IAWQ Specialist Group on Instrumentation, Control and Automation 7th International Workshop*, Brighton, United Kingdom, July; International Association on Water Quality: London.

Caulet, P.; Lefevre, F.; Bujon, B.; Reau, P.; Philippe, J. P.; Audic, J. M. (1997) Automated Aeration Management in Waste Water Treatment: Interest of the Application to Serial Basins Configuration. *IAWQ Specialist Group on Instrumentation, Control and Automation 7th International Workshop*, Brighton, United Kingdom, July; International Association on Water Quality: London.

Chandran, K.; Smets, B. F. (2001) Estimating Biomass Yield Coefficients for Autotrophic Ammonia and Nitrite Oxidation from Batch Respirograms. *Water Res.*, **35** (13), 3041–3282.

Charpentier, J.; Martin, G.; Wacheux, H.; Gilles, P. (1998) ORP Regulation and Activated Sludge: 15 Years of Experience. *Water Sci. Technol.*, **38** (3), 197–208.

Cho, B. C.; Chang, C. N.; Liaw, S. L.; Huang, P. T. (2001a) The Feasible Sequential Control Strategy of Treating High Strength Organic Nitrogen Wastewater with Sequencing Batch Biofilm Reactor. *Water Sci. Technol.*, **43** (3), 115–122.

Cho, J.-H.; Sung, S. W.; Lee, I.-B. (2001b) Cascade Control Strategy for External Carbon Dosage in Predenitrification Process. *Proceedings of the First IWA Conference on Instrumentation, Control and Automation*, Malmo, Sweden, June 3–7; International Water Association: London, p 67–75.

Cohen, A.; Hegg, D.; de Michele, M.; Song, Q.; Kasabov, N. (2003) An Intelligent Controller for Automated Operation of Sequencing Batch Reactors. *Water Sci. Technol.*, **47** (12), 57–63.

Copp, J. (2001) *The COST Simulation Benchmark-Description and Simulator Manual.* COST (European Cooperation in the field of Scientific and Technical Research): Brussels, Belgium (available online at http://www.ensic.inpl-nancy.fr/COSTWWTP/Pdf/Simulator_manual.pdf).

Demoulin, G.; Goronszy, M. C.; Wutscher, K.; Forsthuber, E. (1997) Co-Current Nitrification/Denitrification and Biological P-Removal in Cyclic Activated Sludge Plants by Redox Controlled Cycle Operation. *Water Sci. Technol.,* **35** (1), 215.

Devisscher, M.; Bogaert, H.; Bixio, D.; Van de, V. J.; Thoeye, C. (2001) Feasibility of Automatic Chemical Dosage Control—A Full Scale Evaluation. *Proceedings of the First IWA Conference on Instrumentation, Control and Automation,* Malmo, Sweden, June 3–7; International Water Association: London, p 623–631.

Draaijer, H.; Buunen, A. H. M.; van Dijk, J. W. (1997) Full-Scale Respirometric Control of an Oxidation Ditch. *IAWQ Specialist Group on Instrumentation, Control and Automation 7th International Workshop,* Brighton, United Kingdom, July; International Association on Water Quality: London.

Ekster, A.; Jenkins, D. (1999) Nocardioform Foam Control in a BNR Plant. *Proceedings of the 72nd Annual Water Environment Federation Technical Exposition and Conference* [CD-ROM], New Orleans, Louisiana, Oct 9–13; Water Environment Federation: Alexandria, Virginia.

Ekster, A.; Rodríguez-Roda, I. (2003) The Effect of Sludge Age on the Operation Cost of Activated Sludge System. *Proceedings of the 76th Annual Water Environment Federation Technical Exhibition and Conference* [CD-ROM], Los Angeles, Oct 11–15; Water Environment Federation: Alexandria, Virginia.

Ekster, A. (2004) Golden Age. *Water Environ. Technol.,* **16** (5), 62–64.

Ekster, A (2004) Development of Universal DO Control Software. Energy Innovation Small Grant Program Report; California Energy Commission: Sacramento, California.

Ekster, A.; Wang, J. (2005) Reliable DO Control is Available. *Water Environ. Technol.,* **17** (2), 41–43.

Electric Power Research Institute (1996) Review of Dissolved Oxygen Monitoring Equipment. Electric Power Research Institute (EPRI). Community Environment Center Report CR-106373, April; Electric Power Research Institute: Palo Alto, California.

Hill, R. D.; Manross, R. C.; Davidson, E. V.; Palmer, T. M.; Ross, M. C.; Nutt, S. G. (2002) *Sensing and Control Systems: A Review of Municipal and Industrial Experiences.* WERF Report 99-WWF-4; Water Environment Research Foundation: Alexandria, Virginia.

Ingildsen, P.; Jeppsson, U.; Olsson, G. (2001) Dissolved Oxygen Controller Based on On-Line Measurement of Ammonia Combining Feed-Forward and Feedback. *Proceedings of the First IWA Conference on Instrumentation, Control and Automation,* Malmo, Sweden, June 3–7; International Water Association: London, p 631–639.

Japan Society of Industrial Machinery Manufacturers (2001) Introduction of Japanese Advanced Environmental Equipment. Japan Society of Industrial Machinery Manufacturers (JSIM): Tokyo, Japan (available online at http://nett21.gec.jp/JSIM_DATA/index.html).

Krause, K.; Bocker, K.; Londong, J. (2001) Simulation of a Nitrification Concept Considering Influent Ammonium Load. *Proceedings of the First IWA Conference on Instrumentation, Control and Automation,* Malmo, Sweden, June 3–7; International Water Association: London, p 555–563.

Liu, W.; Lee, G.; Goodley, J. (2003) Using On-Line Ammonia and Nitrate Instruments to Control Modified Ludzack-Ettinger (MLE) Process. *Proceedings of the 76th Annual Water Environment Federation Technical Exhibition and Conference* [CD-ROM], Los Angeles, California, Oct 11–15; Water Environment Federation: Alexandria, Virginia.

Marsilli-Libelli, S.; Giunti, L. (2001) Fuzzy Predictive Control for Organic Carbon Dosing in Denitrification. *Proceedings of the First IWA Conference on Instrumentation, Control and Automation,* Malmo, Sweden, June 3–7; International Water Association: London, p 35–43.

Marx, J. J.; Rieth, M. G.; Baker, G. (1998) Dissolved Oxygen Meters for Activated Sludge Monitoring and Control. *Proceedings of the 71st Annual Water Environment Federation Technical Exposition and Conference,* Orlando, Florida, Oct 3–7; Water Environment Federation: Alexandria, Virginia; 6, p 35–45.

Metcalf and Eddy, Inc. (2003) *Wastewater Engineering: Treatment and Reuse*, 4th ed.; G. Tchobanoglous, F. L. Burton, H. D. Stensel (Eds.); McGraw-Hill: New York.

Mori, Y.; Sasaki, K.; Yamamoto, Y.; Tsumura, K.; Ouchi, S. (1997) Countermeasures for Hydraulic Load Variation in Intermittently Aerated 2-Tank Activated Sludge Process for Simultaneous Removal of Nitrogen and Phosphorus. *IAWQ Specialist Group on Instrumentation, Control and Automation 7th International Workshop*, Brighton, United Kingdom, July; International Association on Water Quality: London.

Olsson, G.; Newell, B. (1999) *Wastewater Treatment Systems.* International Water Association: London.

Parker, D.; Geary, S.; Jones, G.; McIntyre, L.; Oppenheim, S.; Pedregon, V.; Pope, R.; Richards, T.; Voigt, C.; Volpe, G.; Willis, J.; Witzgall, R. (2003) Making Classifying Selectors Work for Foam Elimination in the Activated Sludge Process. *Water Environ. Res.*, **75**, 83.

Pelliter, et al. (1999) Using On-Line Ammonia Measurement for Activated Sludge Control. *Proceedings of the 72nd Annual Water Environment Federation Technical Exposition and Conference* [CD-ROM], New Orleans, Louisiana, Oct 9–13; Water Environment Federation: Alexandria, Virginia.

Samuelsson, P.; Carlsson, B. (2001) Control of Aeration Volume in an Activated Sludge for Nutrient Removal. *Proceedings of the First IWA Conference on Instrumentation, Control and Automation*, Malmo, Sweden, June 3–7; International Water Association: London, p 43–51.

Serralta, J.; Ribes, J.; Seco, A.; Ferrer, J. (2001) A Supervisory Control System for Optimizing Nitrogen Removal and Aeration Consumption in Wastewater Treatment Plants. *Proceedings of the First IWA Conference on Instrumentation, Control and Automation*, Malmo, Sweden, June 3–7; International Water Association: London, p 429–437.

Shaw, A.; Watts, J. (2002) The Use of Respirometry for the Control of Sequencing Batch Reactors-Principles and Practical Application. *Proceedings of the 75th Annual Water Environment Federation Technical Exhibition and Conference* [CD-ROM], Chicago, Illinois, Sep 28–Oct 2; Water Environment Federation: Alexandria, Virginia.

Spanjers, H.; Patry, G. G.; Keesman, K. J. (2002) Respirometry-Based On-Line Model Parameter Estimation at a Full-Scale WWTP. *Water Sci. Technol.*, **45** (4–5), 335–343.

Vaccari, D. A. (1983) Calculation of Mean Cell Residence Time for Unsteady-State Activated Sludge Systems. *Biotechnol. Bioeng.*, **27**, 695.

Watts, J.; Garber, W. (1993) On-Line Respirometry: A Powerful Tool for ASP Operation and Design. *6th IAWQ Specialized Conference, Sensors in Waste Water Technology*, Hamilton, Canada, June 17–23; International Water Association: London.

Yoong, E. T.; Lant, P. A.; Greenfield, P. F. (2000) In Situ Respirometry in an SBR Treating Wastewater with High Phenol Concentrations. *Water Res.*, **34** (1), 239–245.

Yuan, Z.; Oehmen, A. M.; Ingeldsen, I. (2001) Control of Nitrate Recirculation Flow in Predenitrification Systems. *Proceedings of the First IWA Conference on Instrumentation, Control and Automation*, Malmo, Sweden, June 3–7; International Water Association: London, p 27–35.

Glossary

Acclimation The dynamic response of a system to the addition or deletion of a substance until equilibrium is reached; adjustment to a change in the environment. Typically used to describe the response of microorganisms to a change in environment.

Accuracy The absolute nearness to the truth. In physical measurements, it is the degree of agreement between the quantity measured and the actual quantity. Accuracy should not be confused with "precision", which denotes the reproducibility of the measurement.

Acid (1) A substance that tends to lose a proton. (2) A substance that dissolves in water with the formation of hydrogen ions. (3) A substance containing hydrogen, which may be replaced by metals to form salts.

Activated sludge Solids produced by the growth of organisms in the aeration tank in the presence of dissolved oxygen. Also also called *biomass*.

Activated sludge loading The kilograms (pounds) of biochemical oxygen demand applied per unit volume of aeration capacity or per kilogram (pound) of mixed liquor suspended solids per day.

Activated sludge process A biological wastewater treatment process that converts organic materials to carbon dioxide, water, and energy for new growth. In this process, nonsettleable (suspended, dissolved, and colloidal solids) organic materials are converted to a settleable product using aerobic and facultative microorganisms.

ActiveX ActiveX is a programming code that defines Microsoft's (Microsoft Corporation, Redmond, Washington) interaction between Web servers, clients, add-ins, and Microsoft Office applications.

Advanced waste treatment Any physical, chemical, or biological treatment process used to accomplish a degree of treatment greater than that achieved by secondary treatment (85% removal of biochemical oxygen demand and suspended solids).

Aeration A process that produces contact between air and a liquid by one or more of the following methods: (a) spraying the liquid in the air, (b) bubbling air through the liquid, and (c) agitating the liquid to promote surface absorption of air.

Aeration period (1) The theoretical time, typically expressed in hours, during which mixed liquor is aerated in a biological reactor while undergoing activated sludge treatment. It is equal to the volume of the tank divided by the combined volumetric rate of wastewater and return sludge flow. (2) The theoretical time during which water is subjected to aeration.

Aeration tank A biological reactor in which wastewater or other liquid is aerated.

Aerator A device that brings air and a liquid into intimate contact. See *Diffuser*.

Aerobic Requiring free or dissolved oxygen in an aqueous environment. Nitrification and biochemical oxygen demand removal requires an aerobic environment.

Aerobic bacteria Bacteria that require free elemental oxygen to sustain life and metabolize substrate.

Agglomeration Coalescence of dispersed suspended matter into larger flocs or particles.

Air diffuser Devices of varied design that transfer oxygen from air into a liquid.

Air diffusion The transfer of air into a liquid through an oxygen-transfer device.

Air stripping A technique for removal of volatile substances from a solution; uses the principles of Henry's law to transfer volatile pollutants from a solution of high concentration to an air stream of lower concentration. The process typically is designed so that the solution containing the volatile pollutant contacts large volumes of air. The method is used to remove ammonia in advanced waste treatment. Ammonia can be removed from wastewater by air stripping.

Algae Photosynthetic microscopic plants that contain chlorophyll that float or are suspended in water. They may also be attached to structures, rocks, etc. In high concentrations, algae may deplete dissolved oxygen in receiving waters.

Alkaline The condition of water, wastewater, or soil that contains a sufficient amount of alkali substances to raise the pH above 7.0.

Alkalinity The capacity of water to neutralize acids; a property imparted by carbonates; bicarbonates; hydroxides; and, occasionally, borates, silicates, and phosphates. It is expressed in milligrams of equivalent calcium carbonate per liter (mg/L $CaCO_3$).

Alum, aluminum sulfate [$Al_2(SO_4)_3 \cdot 18H_2O$] Used as a coagulant in filtration. Dissolved in water, it hydrolyses into $Al(OH)_2$ and sulfuric acid (H_2SO_4). To precipitate the hydroxide, as needed for coagulation, the water must be alkaline.

Ambient Generally refers to the prevailing environmental condition in a given area at a given time.

Ammonia, ammonium (NH_3, NH_4^+) Urea and proteins are degraded into dissolved ammonia and ammonium ion in raw wastewaters. Typically raw wastewater contains 15 to 50 mg/L of NH_3.

Ammonia-nitrogen The quantity of elemental nitrogen present in the form of ammonia (NH_3).

Ammonification Bacterial decomposition of organic nitrogen to ammonia.

Amoeba A group of simple protozoans, some of which produce diseases, such as dysentery, in humans.

Anaerobic (1) A condition in which free and dissolved oxygen is unavailable. (2) Requiring or not destroyed by the absence of air or free oxygen. Biological phosphorus removal requires an anaerobic environment.

Anaerobic bacteria Bacteria that grow only in the absence of free and dissolved oxygen.

Anion A negatively charged ion attracted to the anode under the influence of electrical potential.

Anoxic Condition in which oxygen is available in the combined form only (e.g., NO_2^-, NO_3^-); there is no free oxygen. Anoxic zones in biological reactors are used for denitrification.

Appurtenances Machinery, appliances, or auxiliary structures attached to a main structure enabling it to function but not considered an integral part of it.

Automatic sampling Collecting of samples of prescribed volume over a defined time period by an apparatus designed to operate remotely without direct manual control. See also *Composite sample*.

Autotrophic organisms Organisms, including nitrifying bacteria and algae, that use carbon dioxide as a source of carbon for cell synthesis.

Average daily flow The total flow past a point over a period of time divided by the number of days in that period (million gallons per day, cubic meters per day, liters per second, etc.).

Bacteria A group of universally distributed, rigid, essentially unicellular microscopic organisms lacking chlorophyll. They perform a variety of biological treatment processes, including biological oxidation, sludge digestion, nitrification, and denitrification.

Beggiatoa A filamentous organism whose growth is stimulated by hydrogen sulfide (H_2S).

Bicarbonate alkalinity Alkalinity caused by bicarbonate ions.

Bioassay (1) An assay method using a change in biological activity as a qualitative or quantitative means of analyzing a material's response to biological treatment. (2) A method of determining the toxic effects of industrial wastes and other wastewaters by using viable organisms; exposure of fish to various levels of a chemical under controlled conditions to determine safe and toxic levels of that chemical.

Biochemical (1) Pertaining to chemical change resulting from biological action. (2) A chemical compound resulting from fermentation. (3) Pertaining to the chemistry of plant and animal life.

Biochemical oxidation The microbial conversion of organic materials to carbon dioxide, water, and energy in an aerobic environment.

Biochemical oxygen demand (BOD) A measure of the quantity of oxygen used in the biochemical oxidation of organic matter in a specified time, at a specific temperature, and under specified conditions. Typically, five days at 20°C for wastewater monitoring and process control.

Biochemical oxygen demand load The BOD concentration, typically expressed in kilograms (pounds) per unit of time, of wastewater entering a waste treatment system or body of water.

Biodegradation The destruction of organic materials by microorganisms, soils, natural bodies of water, or wastewater treatment systems.

Biological denitrification The transformation of nitrate-nitrogen to inert nitrogen gas by microorganisms in an anoxic environment in the presence of an electron donor to drive the reaction.

Biological nutrient removal The reduction in concentration of nitrogen or phosphorus compounds using microorganisms.

Biological oxidation The process by which living organisms convert organic matter to carbon dioxide, water, and energy for new cell growth in the presence of oxygen. Also, the process by which ammonia-nitrogen is converted to nitrate-nitrogen.

Biomass The mass of living organisms contained in a biological treatment process.

Breakpoint chlorination Addition of chlorine to water or wastewater until the chlorine demand has been satisfied, with further addition resulting in a residual that is directly proportional to the amount added beyond the breakpoint.

Brush aerator A surface aerator that rotates about a horizontal shaft with metal blades attached to it; commonly used in oxidation ditches.

Buffer A substance that resists a change in pH.

Bulking Inability of activated sludge solids to separate from liquid under quiescent conditions (clarification); may be associated with the growth of filamentous organisms, low DO, or high solids loading rates. Bulking sludge typically has a sludge volume index >150 mL/g.

Calibration (1) The determination, checking, or rectifying of the graduation of any instrument giving quantitative measurements. (2) The process of taking measurements or of making observations to establish the relationship between two quantities.

Carbon (C) (1) A chemical element essential for growth. (2) A solid material used for adsorption of pollutants.

Carbonaceous biochemical oxygen demand (CBOD) A quantitative measure of the amount of dissolved oxygen required for the biological oxidation of carbon containing compounds in a sample. See *Biochemical oxygen demand*.

Cation A positively charged ion attracted to the cathode under the influence of electrical potential.

Centrate Liquid removed by a centrifuge; typically contains high concentrations of suspended, nonsettling solids and nitrogen.

Chemical dose A specific quantity of chemical applied to a specific quantity of fluid for a specific purpose. For example, chlorine added to final effluent for disinfection or polymer added to sludge for conditioning.

Chemical equilibrium The condition that exists when there is no net transfer of mass or energy between the components of a system. This is the condition in a reversible chemical reaction for which the rate of the forward reaction equals the rate of the reverse reaction.

Chemical equivalent The weight (in grams) of a substance that combines with or displaces 1 g of hydrogen. It is found by dividing the formula weight by its valence.

Chemical feeder A device for dispensing a chemical at a predetermined rate for the treatment of water or wastewater. The change in feed rate occurs by manually or automatically adjusting the flowrate, while maintaining the same concentration. Feeders are designed for solids, liquids, or gases.

Chemical oxidation The oxidation of compounds in wastewater or water by chemical means. Typical oxidants include ozone, chlorine, and potassium permanganate.

Chemical oxygen demand (COD) A quantitative measure of the amount of oxygen required for the chemical oxidation of carbonaceous (organic) material in wastewater using inorganic dichromate or permanganate salts as oxidants in a two hour test.

Chemical precipitation (1) Formation of particulates by the addition of chemicals. (2) The process of removing phosphorus by the addition of lime or alum to form insoluble compounds. Phosphorus is typically removed from wastewater using chemical precipitation processes.

Chemical reaction A transformation of one or more chemical species into other species, resulting in the evolution of heat or gas, color formation, or precipitation. It may be initiated by a physical process, such as heating, by the addition of a chemical reagent, or it may occur spontaneously.

Chemical treatment Any treatment process involving the addition of chemicals to obtain a desired result, such as precipitation, coagulation, flocculation, sludge conditioning, disinfection, or odor control.

Chlorination The application of chlorine or chlorine compounds to water or wastewater, generally for the purpose of disinfection, but frequently for chemical oxidation and odor control.

Ciliated protozoa Protozoans with cilia (hairlike appendages) that assist in movement; common in trickling filters and healthy activated sludge. Free-swimming ciliates are present in the bulk liquid; stalked ciliates are commonly attached to solids matter in the liquid.

Coagulant A simple electrolyte, typically an inorganic salt containing a multivalent cation of iron, aluminum, or calcium (i.e., $FeCl_3$, $FeCl_2$, $Al_2[SO_4]_3$, and CaO). Also, an inorganic acid or base that induces coagulation of suspended solids. See also *Flocculant*.

Coagulant or flocculant aid An insoluble particulate used to enhance solid-liquid separation by providing nucleating sites or acting as a weighting agent or sorbent; also used colloquially to describe the action of flocculants in water treatment.

Coagulation The conversion of colloidal (<0.001 mm) or dispersed (0.001- to 0.1-mm) particles into small visible coagulated particles (0.1 to 1 mm) by the addition of a coagulant, compressing the electrical double layer surrounding each suspended particle, decreasing the magnitude of repulsive electrostatic interactions between particles, thereby destabilizing the particle. See also *Flocculation*.

Cocci Sphere shaped bacteria.

Coefficient A numerical quantity, determined by experimental or analytical methods, interposed in a formula that expresses the relationship between two or more variables to include the effect of special conditions or correct a theoretical relationship to one found by experiment or actual practice.

Coliform group bacteria A group of bacteria predominantly inhabiting the intestines of man or animal, but also occasionally found elsewhere. It includes all aerobic and facultative anaerobic, Gram-negative, non-spore forming, rod shaped bacteria that ferment lactose with the production of gas. Also included are all bacteria that produce a dark, purplish green metallic sheen by the membrane filter technique used for coliform identification. The two groups are not always identified, but they are generally of equal sanitary significance.

Colloids Finely divided solids (smaller than 0.002 mm and larger than 0.000 001 mm) that will not settle but may be removed by coagulation, biochemical action, or membrane filtration; they are intermediate between true solutions and suspensions.

Colony A discrete clump of microorganisms on a surface as opposed to dispersed growth throughout a liquid culture medium.

Complete mix Activated sludge process where wastewater is rapidly and evenly distributed throughout the biological reactor.

Component object model (COM) A software architecture that allows interactions among different software.

Composite sample A combination of individual samples of water or wastewater taken at preselected intervals to minimize the effect of the variability of the individual sample. Individual samples may be of equal volume or may be proportional to the flow at time of sampling.

Concentration (1) The amount of a given substance dissolved in a discrete unit volume of solution or applied to a unit weight of solid. (2) The process of increasing the dissolved solids per unit volume of solution, typically by evaporation of the liquid. (3) The process of increasing the suspended solids per unit volume of sludge by sedimentation or dewatering.

Contact stabilization   Modification of the activated sludge process involving a short period of contact between wastewater and sludge for rapid removal of soluble biochemical oxygen demand by adsorption, followed by a longer period of aeration in a separate tank, where sludge is oxidized and new sludge is synthesized.

Conventional aeration   Process design configuration, whereby the aeration tank organic loading is higher at the influent end than at the effluent end. Flow passes through a serpentine tank system, typically side-by-side, before passing on to the secondary clarifier. Also called plug flow.

Correlation   (1) A mutual relationship or connection. (2) The degree of relative correspondence, as between two sets of data.

Declining growth phase   Period of time between the log-growth phase and the endogenous phase, during which the amount of food is in short supply, leading to ever-slowing bacterial growth rates.

Decomposition of wastewater   (1) The breakdown of organic matter in wastewater by bacterial action, either aerobic or anaerobic. (2) Chemical or biological transformation of the organic or inorganic materials contained in wastewater.

Defoamer   A material having low compatibility with foam and a low surface tension. Defoamers are used to control, prevent, or destroy various types of foam, the most widely used being silicone defoamers. A droplet of silicone defoamer contacting a bubble of foam will cause the bubble to undergo a local and drastic reduction in film strength, thereby breaking the film. Unchanged, the defoamer continues to contact other bubbles, thus breaking up the foam. A valuable property of most defoamers is their effectiveness in extremely low concentration. In addition to silicones, defoamers for special purposes are based on polyamides, vegetable oils, and stearic acid.

Denitrification   The biological reduction of nitrate-nitrogen to nitrogen gas in an anoxic environment; also, removal of total nitrogen from a system. See also *Nitrification*.

Design criteria   (1) Engineering guidelines specifying construction details and materials. (2) Objectives, results, or limits that must be met by a facility, structure, or process in performance of its intended functions.

Design flow   Engineering guidelines that typically specify the amount of influent flow that can be expected on a daily basis over the course of a year. Other design flows can be set for monthly or peak flows.

Design loadings   Flowrates and constituent concentrations that determine the design of a process unit or facility necessary for proper operation.

Detention time The period of time that a water or wastewater flow is retained in a basin, tank, or reservoir for storage or completion of physical, chemical, or biological reaction. See also *Retention time*.

Detergent (1) Any of a group of synthetic, organic, liquid, or water soluble cleaning agents that are inactivated by hard water and have wetting and emulsifying properties but, unlike soap, are not prepared from fats and oils. (2) A substance that reduces the surface tension of water.

Diffused aeration Injection of air under pressure through submerged porous plates, perforated pipes, or other devices to form small air bubbles, from which oxygen is transferred to the liquid as the bubbles rise to the water surface.

Diffused air Small air bubbles formed below the surface of a liquid to transfer oxygen to the liquid.

Diffuser A porous plate, tube, or other device through which air is forced and divided into minute bubbles for diffusion in liquids. In the activated sludge process, it is a device for dissolving air into mixed liquor. It is also used to mix chemicals, such as chlorine, through perforated holes.

Dissolved oxygen oxygen dissolved in liquid, typically expressed in milligrams per liter or percent saturation.

Dissolved solids Solids in solution that cannot be removed by filtration; for example, NaCl and other salts that must be removed by evaporation.

Distributed client/server A communication architecture that exists between a server application and a remote (distributed) client application. Typically, the client makes a request for specific data from the server application. Upon successfully receiving the information requested from the server, the client returns an error checking response to the server application.

Distributed component object model (DCOM) A protocol that enables software components to communicate directly over a network in a reliable, secure, and efficient manner. Previously called "Network OLE", DCOM is designed for use across multiple network transports, including Internet protocols, such as http.

Diurnal (1) Occurring during a 24 hour period; diurnal variation. (2) Occurring during the day (as opposed to night). (3) In tidal hydraulics, having a period or cycle of approximately one tidal day.

Driver A piece of software that enables a computer to communicate with a remote input/output device (printer, scanner, programmable logic controller, or remote transmitting unit).

Dry weather flow (1) The flow of wastewater in a combined sewer during dry weather. Such flow consists primarily of wastewater, with no stormwater included. (2) The flow of water in a stream during dry weather, generally contributed entirely by groundwater.

E. coli See *Escherichia coli*.

Efficiency The relative results obtained in any operation in relation to the energy or effort required to achieve such results. It is the ratio of the total output to the total input, expressed as a percentage: ([In–Out]/In) × 100.

Effluent Wastewater or other liquid, partially or completely treated or in its natural state, flowing out of a reservoir, basin, treatment plant, or industrial treatment plant, or part thereof.

Effluent quality The physical, biological, and chemical characteristics of a wastewater or other liquid flowing out of a basin, reservoir, pipe, or treatment plant.

Endogenous decay The loss of biomass resulting from the autooxidation of organisms in the biological process.

Endogenous respiration Autooxidation by organisms in biological processes.

Enhanced biological phosphorus removal (EBPR) Relies on the selection and proliferation of a microbial population capable of storing orthophosphate in excessive of their biological growth requirements.

Enterococci A group of cocci that normally inhabit the intestines of humans and animals. Incorrectly used interchangeably with fecal *Streptococci*.

Enzyme A catalyst produced by living cells. All enzymes are proteins, but not all proteins are enzymes.

Equalization In wastewater systems, the storage and controlled release of wastewater to treatment processes at a controlled rate determined by the capacity of the processes or at a rate proportional to the flow in the receiving stream; used to smooth out variations in temperature, composition, and flow.

Equalizing basin A holding basin, in which variations in flow and composition of a liquid are averaged. Such basins are used to provide a flow of reasonably uniform volume and composition to a treatment unit. Also called *balancing reservoir*.

Equilibrium A condition of balance, in which the rate of formation and the rate of consumption or degradation of various constituents are equal. See also *Chemical equilibrium*.

Escherichia coli (E. coli) One of the species of bacteria in the fecal coliform group. It is found in large numbers in the gastrointestinal tract and feces of warm blooded animals and humans. Its presence is considered indicative of fresh fecal contamination, and it is used as an indicator organism for the presence of less easily detected pathogenic bacteria.

Eutrophication Nutrient enrichment of a lake or other waterbody, typically characterized by increased growth of planktonic algae and rooted plants. It can be accelerated by wastewater discharge and polluted runoff.

Extended aeration A modification of the activated sludge process using long aeration periods to promote aerobic digestion of the biological mass by endogenous respiration. The process includes stabilization of organic matter under aerobic conditions and disposal of the gaseous end products into the air. Effluent contains finely divided suspended matter and soluble matter.

Extended aeration process A modification of the activated sludge process. See *Extended aeration*.

Facultative Having the ability to live under different conditions; for example, with or without free oxygen.

Facultative bacteria Bacteria that can grow and metabolize in the presence and in the absence of dissolved oxygen.

Fecal coliform Aerobic and facultative, Gram-negative, non-spore forming, rod shaped bacteria capable of growth at 44.5°C (112°F) and associated with fecal matter of warm blooded animals.

Ferric chloride ($FeCl_3$) A soluble iron salt often used as a sludge conditioner to enhance precipitation or bind up sulfur compounds in wastewater treatment. See also *Coagulant*.

Ferric sulfate [$Fe_2(SO_4)_3$] A water soluble iron salt formed by reaction of ferric hydroxide and sulfuric acid or by reaction of iron and hot concentrated sulfuric acid; also obtainable in solution by reaction of chlorine and ferrous sulfate; used in conjunction with lime as a sludge conditioner to enhance precipitation.

Ferrous chloride ($FeCl_2$) A soluble iron salt used as a sludge conditioner to enhance precipitation or bind up sulfur. See also *Coagulant*.

Ferrous sulfate ($FeSO_4 \cdot 7H_2O$) A water soluble iron salt, sometimes called copperas; used in conjunction with lime as a sludge conditioner to enhance precipitation.

Filamentous growth Intertwined, threadlike biological growths characteristic of some species of bacteria, fungi, and algae. Such growths reduce sludge settleability and dewaterability.

Filamentous organisms Bacterial, fungal, and algal species that grow in threadlike colonies, resulting in a biological mass that will not settle and may interfere with drainage through a filter.

Final effluent The effluent from the final treatment unit of a wastewater treatment plant.

Final sedimentation The separation of solids from wastewater in the last settling tank of a treatment plant.

Five-day biochemical oxygen demand (BOD$_5$) A standard test to assess wastewater pollution resulting from organic substances, measuring the oxygen used under controlled conditions of temperature (20°C) and time (five days).

Floc Collections of smaller particles agglomerated into larger, more easily settleable particles through chemical, physical, or biological treatment. See also *Flocculation*.

Flocculant Water soluble organic polyelectrolytes that are used alone or in conjunction with inorganic coagulants, such as aluminum or iron salts, to agglomerate the solids present to form large, dense floc particles that settle rapidly.

Flocculating tank A tank used for the formation of floc by the gentle agitation of liquid suspensions, with or without the aid of chemicals.

Flocculation In water and wastewater treatment, the agglomeration of colloidal and finely divided suspended matter after coagulation by gentle stirring by either mechanical or hydraulic means. For biological wastewater treatment in which coagulation is not used, agglomeration may be accomplished biologically.

Flocculation agent A coagulating substance that, when added to water, forms a flocculent precipitate that will entrain suspended matter and expedite sedimentation; examples are alum, ferrous sulfate, and lime.

Flow equalization Transient storage of wastewater for release to a sewer system or wastewater treatment plant at a controlled rate to provide a reasonably uniform flow for treatment.

Flowrate The volume or mass of a gas, liquid, or solid material that passes through a cross-section of conduit in a given time; measured in such units as kilograms per hour (kg/h), cubic meters per second (m^3/s), liters per day (L/d), or gallons per day (gpd).

Foam (1) A collection of minute bubbles formed on the surface of a liquid by agitation, fermentation, etc. (2) The frothy substance composed of an aggregation of bubbles on the surface of liquids and created by violent agitation or by the admission of air bubbles to liquid containing surface active materials, solid particles, or both. Also called froth. Certain microorganisms, such as *Nocardia*, will cause significant foaming problems, especially at the long solid retention times required for nutrient removal.

Food-to-microorganism ratio (F/M) In the biological wastewater process, the loading rate expressed as kilograms of biochemical oxygen demand or five-day chemical oxygen demand per kilogram of mixed liquor volatile suspended solids per day (kg BOD_5 or COD/kg MLVSS·d).

Free oxygen Elemental oxygen (O_2).

Freeboard The vertical distance between the normal maximum level of the surface of the liquid in a conduit, reservoir, tank, or canal and the top of the sides of an open conduit or the top of a dam or levee, which is provided so that waves and other movements of the liquid will not overflow the confining structure.

Free-swimming ciliate Mobile, one-celled organisms using cilia (hairlike projections) for movement.

Fungi Small non-chlorophyll bearing plants that lack roots, stems, or leaves; occur (among other places) in water, wastewater, or wastewater effluent; and grow best in the absence of light. Their decomposition may cause disagreeable tastes and odors in water; in some wastewater treatment processes, they are helpful, and, in others, they are detrimental.

Grab sample A sample taken at a given place and time. It may be representative of the flow. See also *Composite sample*.

Grease and oil In wastewater, a group of substances, including fats, waxes, free fatty acids, calcium and magnesium soaps, mineral oils, and certain other nonfatty materials; water insoluble organic compounds of plant and animal origins or industrial wastes that can be removed by natural flotation skimming.

Hardness A characteristic of water imparted primarily by salts of calcium and magnesium, such as bicarbonates, carbonates, sulfates, chlorides, and nitrates, that causes curdling and increased consumption of soap; deposition of scale in boilers; damage in some industrial processes; and, sometimes, objectionable taste. It may be determined by a standard laboratory titration procedure or computed from the amounts of calcium and magnesium expressed as equivalent calcium carbonate.

Heavy metals Metals that can be precipitated by hydrogen sulfide in acid solution, for example, lead, silver, gold, mercury, bismuth, and copper.

High-purity oxygen A modification of the activated sludge process using relatively pure oxygen and covered aeration tanks in a conventional flow arrangement.

High-rate aeration A modification of the activated sludge process, whereby the mixed liquor suspended solids loadings are kept high, allowing high food-to-microorganism ratios and shorter detention times.

Human machine interface (HMI)   Also referred to as a graphical user interface or man machine interface, this is a process that displays graphics and allows operators to interface with the control system in graphic form. It may contain trends, alarm summaries, pictures, or animation.

Human machine interface server (HMI server)   A computer server that communicates directly with remoted input/output devices such as programmable logic controllers and maintains a central database. The HMI server also communicates with third-party applications such as data loggers, historians, reporting, management information systems, and other data repositories such as structured query language servers and Oracle.

Hydrated lime   Limestone that has been "burned" and treated with water under controlled conditions until the calcium oxide portion has been converted to calcium hydroxide.

Hydraulic loading   The amount of water applied to a given treatment process, typically expressed as volume per unit time, or volume per unit time per unit surface area.

Hydrogen ion concentration   The concentration of hydrogen ions in moles per liter of solution. Commonly expressed as the pH value, which is the logarithm of the reciprocal of the hydrogen ion concentration. See also *pH*.

Hypochlorite   Calcium, sodium, or lithium hypochlorite.

Industrial wastewater   Wastewater derived from industrial sources or processes.

Influent   Water, wastewater, or other liquid flowing into a reservoir, basin, treatment plant, or treatment process. See also *Effluent*.

Inorganic   All combinations of elements that do not include organic carbon.

Inorganic matter   Mineral type compounds that are generally nonvolatile, not combustible, and not biodegradable. Most inorganic type compounds or reactions are ionic in nature; therefore, rapid reactions are characteristic.

Instrumentation   Use of technology to control, monitor, or analyze physical, chemical, or biological parameters.

Ion   A charged atom, molecule, or radical that affects the transport of electricity through an electrolyte or, to a certain extent, through a gas. An atom or molecule that has lost or gained one or more electrons.

Ion exchange   (1) A chemical process involving reversible interchange of ions between a liquid and a solid, but no radical change in structure of the solid. (2) A chemical process in which ions from two different molecules are exchanged. (3) The reversible transfer or sorption of ions from a liquid to a solid phase by replacement with other ions from the solid to the liquid.

Jar test A laboratory procedure for evaluating coagulation, flocculation, and sedimentation processes in a series of parallel comparisons.

Kinetics The study of the rates at which changes occur in chemical, physical, and biological treatment processes.

Kjeldahl nitrogen The combined amount of organic and ammonia nitrogen. Also called *total Kjeldahl nitrogen*.

Lag-growth phase The initial period following bacterial introduction, during which the population grows slowly as bacteria acclimate to the new environment.

Lethal concentration The concentration of a test material that causes death of a specified percentage of a population, typically expressed as the median or 50% level (L50).

Lime Any of a family of chemicals consisting essentially of calcium hydroxide made from limestone (calcite) composed almost wholly of calcium carbonate or a mixture of calcium and magnesium carbonate; used to increase pH to promote precipitation reactions or for lime stabilization to kill pathogenic organisms.

Log-growth phase Initial stage of bacterial growth, during which there is an ample food supply, causing bacteria to grow at their maximum rate.

Mean (1) The arithmetic average of a group of data. (2) The statistical average (50% point) determined by probability analysis.

Mean cell residence time (MCRT) The average time that a given unit of cell mass stays in the activated sludge aeration tank. It is typically calculated as the total mixed liquor suspended solids in the aeration tank divided by the combination of solids in the effluent and solids wasted. Mean cell residence time is equivalent to solids retention time.

Mechanical aeration (1) The mixing, by mechanical means, of wastewater and activated sludge in the aeration tank of the activated sludge process to bring fresh surfaces of liquid into contact with the atmosphere. (2) The introduction of atmospheric oxygen to a liquid by the mechanical action of paddle, paddle wheel, spray, or turbine mechanisms.

Metabolism (1) The biochemical processes in which food is used and wastes, formed by living organisms. (2) All biochemical reactions involved in cell synthesis and growth.

Microbial activity The activities of microorganisms resulting in chemical or physical changes.

Microbial film A gelatinous film of microbial growth attached to or spanning the interstices of a support medium. Also called *biomass*.

Microorganisms Very small organisms, either plant or animal, invisible or barely visible to the naked eye. Examples are algae, bacteria, fungi, protozoa, and viruses.

Microscopic Very small, typically between 0.5 and 100 mm, and visible only by magnification with an optical microscope.

Microscopic examination (1) The examination of water to determine the presence and amounts of plant and animal life, such as bacteria, algae, diatoms, protozoa, and crustacea. (2) The examination of water to determine the presence of microscopic solids. (3) The examination of microbiota in process water, such as the mixed liquor in an activated sludge plant.

Milligrams per liter (mg/L) A measure of concentration equal to and replacing parts per million in the case of dilute concentrations.

Million gallons per day (mgd) A measure of flow equal to 1.547 cfs, 681 gpm, or 3785 m^3/d.

Mixed liquor A mixture of raw or settled wastewater and activated sludge contained in an aeration tank in the activated sludge process. See also *Mixed liquor suspended solids*.

Mixed liquor suspended solids (MLSS) The concentration of suspended solids in activated sludge mixed liquor, expressed in milligrams per liter. Commonly used in connection with activated sludge aeration units.

Mixed liquor volatile suspended solids (MLVSS) That fraction of the suspended solids in activated sludge mixed liquor that can be driven off by combustion at 550°C (1022°F); it indicates the concentration of microorganisms available for biological oxidation.

Mole (1) Molecular weight of a substance, typically expressed in grams. (2) A device to clear sewers and pipelines. (3) A massive harbor work, with a core of earth or stone, extending from shore into deep water. It serves as a breakwater, a berthing facility, or a combination of the two.

Monitoring (1) Routine observation, sampling, and testing of designated locations or parameters to determine the efficiency of treatment or compliance with standards or requirements. (2) The procedure or operation of locating and measuring radioactive contamination by means of survey instruments that can detect and measure, as dose rate, ionizing radiations.

Monod equation A mathematical expression first used by Monod in describing the relationship between the microbial growth rate and concentration of growth limiting substrate.

Moving average Trend analysis tool for determining patterns or changes in treatment process, for example, a seven-day moving average would be the sum of the datum points for seven days divided by seven.

National Pollutant Discharge Elimination System (NPDES) A permit that is the basis for the monthly monitoring reports required by most states in the United States.

Nematode Member of the phylum (Nematoda) of elongated cylindrical worms parasitic in animals or plants or free-living in soil or water.

Network A system of computers interconnected by telephone wires or other means to share information. Also called *net*.

Nitrate (NO_3) An oxygenated form of nitrogen.

Nitrate recycle The recycle flow from the end of the aerobic zone of a biological reactor to the anoxic zone for denitrification. Nitrate recycle is typically 2 to $4Q$.

Nitrification The biological oxidation of ammonia-nitrogen to nitrite-nitrogen and nitrate-nitrogen.

Nitrifying bacteria Bacteria capable of oxidizing nitrogenous material.

Nitrite (NO_2) An intermediate oxygenated form of nitrogen.

Nitrogen (N) An essential nutrient that is often present in wastewater as ammonia, nitrate, nitrite, and organic nitrogen. The concentrations of each form and the sum (total nitrogen) are expressed as milligrams per liter elemental nitrogen. Also present in some groundwater as nitrate and in some polluted groundwater in other forms. See also *Nutrient*.

Nitrogen cycle A graphical presentation of the conservation of matter in nature showing the chemical transformation of nitrogen through various stages of decomposition and assimilation. The various chemical forms of nitrogen as it moves among living and nonliving matter are used to illustrate general biological principles that are applicable to wastewater and sludge treatment.

Nitrogen removal The removal of nitrogen from wastewater through physical, chemical, or biological processes, or by some combination of these.

Nitrogenous oxygen demand (NOD) A quantitative measure of the amount of oxygen required for the biological oxidation of nitrogenous material, such as ammonia-nitrogen and organic nitrogen, in wastewater; typically measured after the carbonaceous oxygen demand has been satisfied. See also *Biochemical oxygen demand*, *Nitrification*, and *Second stage biochemical oxygen demand*.

Nitrosomonas A genus of bacteria that oxidize ammonia to nitrate.

Nocardia Irregularly bent, short filamentous organisms in the biological process that produce a very stable, brown foam.

Nutrient Any substance that is assimilated by organisms and promotes growth; generally applied to nitrogen and phosphorus in wastewater, but also to other essential and trace elements.

Object linking and embedding (OLE) for process control (OPC) A relatively new technology designed to communicate between Windows-based applications, such as human machine interface (HMI) and others, and process control hardware, such as programmable logic controllers or even other HMI servers.

Open database connectivity (ODBC) A standard database access method developed by Microsoft Corporation. Essentially, ODBC is a communication driver that passes data between a database, such as a structured query language server, and an application, such as a supervisory control and data acquisition program. For this to work, both programs must be ODBC-compliant.

Open systems In open systems, no single manufacturer controls the specifications for the architecture. The specifications are in the public domain, and developers can legally write to them. Open systems are crucial for interoperability. Open systems must use nonproprietary programming languages, such as Visual Basic.

Operating system (OS) The software on your computer that controls the basic operation of the machine. The operating system performs tasks such as recognizing keyboard input, sending output to the monitor, keeping track of files and directories on the disk, and controlling other connected devices, such as disk drives and printers.

Organic Refers to volatile, combustible, and sometimes biodegradable chemical compounds containing carbon atoms (carbonaceous) bonded together with other elements. The principal groups of organic substances found in wastewater are proteins, carbohydrates, and fats and oils. See also Inorganic.

Organic loading The amount of organic material, typically measured as five-day biochemical oxygen demand, applied to a given treatment process; expressed as weight per unit time per unit surface area or per unit weight.

Organic nitrogen Nitrogen chemically bound in organic molecules, such as proteins, amines, and amino acids.

Orthophosphate (1) A salt that contains phosphorus as PO_4^{3-}. (2) A product of hydrolysis of condensed (polymeric) phosphates. (3) A nutrient required for plant and animal growth. See also *Nutrient* and *Phosphorus removal*.

Overflow rate One of the criteria in the design of settling tanks for treatment plants; expressed as the settling velocity of particles that are removed in an ideal basin if they enter at the surface. It is expressed as a volume of flow per unit water surface area.

Oxidant A chemical substance capable of promoting oxidation, for example, O_2, O_3, and Cl_2. See also *Oxidation* and *Reduction*.

Oxidation (1) A chemical or biological reaction in which the oxidation number (valence) of an element increases because of the loss of one or more electrons by that element. Oxidation of an element is accompanied by simultaneous reduction of the other reactant. See also *Reduction*. (2) The conversion or organic materials to simpler, more stable forms with the release of energy. This may be accomplished by chemical or biological means. (3) The addition of oxygen to a compound.

Oxidation ditch A biological wastewater treatment facility that uses an oval channel with a rotor placed across it to provide aeration and circulation. The screened wastewater in the ditch is aerated by the rotor and circulated at approximately 0.3 m/s (1 to 2 ft/sec). See also *Secondary treatment*.

Oxidation process Any method of wastewater treatment for the oxidation of the putrescible organic matter.

Oxidation-reduction potential (ORP) The potential required to transfer electrons from the oxidant to the reductant and used as a qualitative measure of the state of oxidation in wastewater treatment systems.

Oxidized wastewater Wastewater in which the organic matter has been stabilized.

Oxygen (O) A chemical element necessary for aerobic processes. Typically found as O_2 and used in biological oxidation. It constitutes approximately 20% of the atmosphere.

Oxygen consumed A measure of the oxygen-consuming capability of inorganic and organic matter present in water or wastewater. See also *Chemical oxygen demand*.

Oxygen deficiency (1) The additional quantity of oxygen required to satisfy the oxygen requirement in a given liquid; typically expressed in milligrams per liter. (2) Lack of oxygen.

Oxygen transfer (1) Exchange of oxygen between a gaseous and a liquid phase. (2) The amount of oxygen absorbed by a liquid compared to the amount fed to the liquid through an aeration or oxygenation device; typically expressed as percent.

Oxygen uptake rate The oxygen used during biochemical oxidation, typically expressed as milligrams oxygen per liter per hour in the activated sludge process.

Oxygen use (1) The portion of oxygen effectively used to support aerobic treatment processes. (2) The oxygen used to support combustion in the degradation of sludge by incineration or wet air oxidation.

Oxygenation capacity In treatment processes, a measure of the ability of an aerator to supply oxygen to a liquid.

Peak flow The maximum quantity of influent flow that occurs over a relatively short period of time.

pH A measure of the hydrogen ion concentration in a solution, expressed as the logarithm (base ten) of the reciprocal of the hydrogen ion concentration in gram moles per liter. On the pH scale (0 to 14), a value of 7 at 25°C (77°F) represents a neutral condition. Decreasing values indicate increasing hydrogen ion concentration (acidity); increasing values indicate decreasing hydrogen ion concentration (alkalinity).

Phenolic compounds Hydroxyl derivatives of benzene. The simplest phenolic compound is hydroxyl benzene (C_6H_5OH).

Phosphate A salt or ester of phosphoric acid. See also *Orthophosphate* and *Phosphorus*.

Phosphorus An essential chemical element and nutrient for all life forms. Occurs in orthophosphate, pyrophosphate, tripolyphosphate, and organic phosphate forms. Each of these forms and their sum (total phosphorus) is expressed as milligrams per liter elemental phosphorus. See also *Nutrient*.

Phosphorus removal The reduction in phosphorus concentration by either a biological process or precipitation of soluble phosphorus by coagulation and subsequent flocculation and sedimentation.

Physical-chemical treatment Treatment of wastewater by unit processes other than those based on microbiological activity. Unit processes commonly included are precipitation with coagulants, flocculation with or without chemical flocculants, filtration, adsorption, chemical oxidation, air stripping, ion exchange, reverse osmosis, and several others. Chemical precipitation of phosphorus is a physical-chemical process.

Phytoplankton Plankton consisting of plants, such as algae.

Phosphorus-accumulating organisms (PAOs) Those organisms that exist in wastewater capable of storing orthophosphate in excessive of their biological growth requirements.

Pin floc Small floc particles that settle poorly.

Plug flow Flow in which fluid particles are discharged from a tank or pipe in the same order in which they entered it. The particles retain their discrete identities and remain in the tank for a time equal to the theoretical detention time.

Polyelectrolytes Complex polymeric compounds, typically composed of synthetic macromolecules that form charged species (ions) in solution; water soluble polyelectrolytes are used as flocculants; insoluble polyelectrolytes are used as ion-exchange resins. See also *Polymers*.

Polymers Synthetic organic compounds with high molecular weights and composed of repeating chemical units (monomers); they may be polyelectrolytes, such as water soluble flocculants or water insoluble ion exchange resins, or insoluble uncharged materials, such as those used for plastic or plastic lined pipe and plastic trickling filter media.

Population dynamics The ever-changing numbers of microscopic organisms within the activated sludge process.

Precipitate (1) To condense and cause to fall as precipitation, as water vapor condenses and falls as rain. (2) The separation from solution as a precipitate. (3) The substance that is precipitated.

Primary treatment (1) The first major treatment in a wastewater treatment facility, used for the purpose of sedimentation. (2) The removal of a substantial amount of suspended matter, but little or no colloidal and dissolved matter. (3) Wastewater treatment processes typically consisting of clarification with or without chemical treatment to accomplish solid-liquid separation.

Programmable logic controller (PLC) A highly reliable, special-purpose computer processor based input/output (I/O) device used in industrial monitoring and control applications. The PLCs can have proprietary programming and networking protocols or have open system programming, such as IEC-61131-3 and Ethernet networking. The PLCs also have special-purpose digital and analog I/O ports.

Protocol A standard procedure for regulating data transmission between computers.

Protozoa Small one celled animals, including amoebae, ciliates, and flagellates.

Quicklime A calcined material, the major part of which is calcium oxide, or calcium oxide in natural association with a lesser amount of magnesium oxide. It is capable of combining with water, that is, being slaked.

Rate (1) The speed at which a chemical reaction occurs. (2) Flow volume per unit time. See also *Kinetics*.

Reaction rate The rate at which a chemical or biological reaction progresses. See also *Kinetics* and *Rate*.

Reactor The container, vessel, or tank in which a chemical or biological reaction is carried out.

Receiving water A river, lake, ocean, or other watercourse into which wastewater or treated effluent is discharged.

Recirculation (1) In the wastewater field, the return of all or a portion of the effluent in a trickling filter to maintain a uniform high rate through the filter. Return of a portion of the effluent to maintain minimum flow is sometimes called *recycling*. (2) The return of effluent to the incoming flow. (3) The return of the effluent from a process, factory, or operation to the incoming flow to reduce the water intake. The incoming flow is called makeup water.

Recycle (1) To return water after some type of treatment for further use; generally implies a closed system. (2) To recover useful values from segregated solid waste.

Reduce The opposite of oxidize. The action of a substance to decrease the positive valence of an ion.

Reduction The addition of electrons to a chemical entity decreasing its valence. See also *Oxidation*.

Removal efficiency A measure of the effectiveness of a process in removing a constituent, such as biochemical oxygen demand or total suspended solids. Removal efficiency is calculated by subtracting the effluent value from the influent value and dividing it by the influent value. Multiply the answer by 100 to convert to a percentage.

Respiration Intake of oxygen and discharge of carbon dioxide as a result of biological oxidation.

Retention time The theoretical time required to displace the contents of a tank or unit at a given rate of discharge (volume divided by the rate of discharge). Also called *detention time*.

Return activated sludge (RAS) Settled activated sludge returned to mix with incoming raw or primary settled wastewater. Also called *returned sludge*.

Rotifer Minute, multicellular aquatic animals with rotating cilia on the head and forked tails. Rotifers help stimulate microfloral activity and decomposition, enhance oxygen penetration, and recycle mineral nutrients.

Secondary effluent (1) The liquid portion of wastewater leaving secondary treatment. (2) An effluent that, with some exceptions, contains not more than 30 mg/L each (on a 30 day average basis) of five-day biochemical oxygen demand and suspended solids.

Secondary (biological) treatment Biological wastewater treatment, particularly activated sludge treatment, which generally produces an effluent with less than 30 mg/L of biochemical oxygen demand and total suspended solids.

Second stage biochemical oxygen demand That part of the oxygen demand associated with the biochemical oxidation of nitrogenous material. As the term implies, the oxidation of the nitrogenous materials typically does not start until a portion of the carbonaceous material has been oxidized during the first stage.

Seed sludge In biological treatment, the inoculation of the unit process with biologically active sludge, resulting in acceleration of the initial stage of the process.

Selector A zone in a biological treatment process with specific environmental conditions that allow for the growth or lack of growth of particular organisms.

Settleometer A 2-L or larger beaker used to conduct the settleability test.

Settling velocity Velocity at which subsidence and deposition of settleable suspended solids in wastewater will occur.

Sludge blanket Accumulation of sludge hydrodynamically suspended within an enclosed body of water or wastewater.

Sludge volume index (SVI) The ratio of the volume (in milliliters) of sludge settled from a 1000 mL sample in 30 minutes to the concentration of mixed liquor (in milligrams per liter multiplied by 1000.

Soda ash A common name for commercial sodium carbonate (Na_2CO_3).

Sodium carbonate (Na_2CO_3) A salt used in water treatment to increase the alkalinity or pH of water or to neutralize acidity. Also called *soda ash*.

Sodium hydroxide (NaOH) A strong caustic chemical used in treatment processes to neutralize acidity, increase alkalinity, or raise the pH value. Also known as *caustic soda, sodium hydrate, lye,* and *white caustic*.

Sodium hypochlorite (NaOCl) A water solution of sodium hydroxide and chlorine, in which sodium hypochlorite is the essential ingredient.

Sodium metabisulfite ($Na_2S_2O_5$) A cream-colored powder used to conserve chlorine residual; 1.34 parts of $Na_2S_2O_5$ will consume 1 part of chlorine residual.

Solids inventory Amount of sludge in the treatment system, typically expressed as kilogram (tons). Inventory of plant solids should be tracked through the use of mass-balance set of calculations.

Solids loading Amount of solids applied to a treatment process per unit time per unit volume.

Solids retention time (SRT) The average time of retention of suspended solids in a biological waste treatment system, equal to the total weight of suspended solids leaving the system, per unit time.

Specific oxygen uptake rate (SOUR) Measures the microbial activity in a biological system expressed in milligrams oxygen per hour per gram volatile suspended solids. Also called *respiration rate*.

Stalked ciliates Small, one-celled organisms possessing cilia (hairlike projections used for feeding) that are not motile. They develop at lower prey densities, long solids retention times, and low food-to-microorganism ratios.

Standard Methods (1) An assembly of analytical techniques and descriptions commonly accepted in water and wastewater treatment (*Standard Methods for the Examination of Water and Wastewater*) published jointly by the American Public Health Association, the American Water Works Association, and the Water Environment Federation. (2) Validated methods published by professional organizations and agencies covering specific fields or procedures. These include, among others, the American Public Health Association, American Public Works Association, American Society of Civil Engineers, American Society of Mechanical Engineers, American Society for Testing and Materials, American Water Works Association, U.S. Bureau of Standards, U.S. Standards Institute (formerly American Standards Association), U.S. Public Health Service, Water Environment Federation, and U.S. Environmental Protection Agency.

Step aeration A procedure for adding increments of settled wastewater along the line of flow in the aeration tanks of an activated sludge plant. Also called *step feed*.

Stoichiometric Pertaining to or involving substances that are in the exact proportions required for a given reaction.

Straggler floc Large (6-mm or larger) floc particles that have poor settling characteristics.

Substrate (1) Substances used by organisms in liquid suspension. (2) The liquor in which activated sludge or other matter is kept in suspension.

Suctoreans Ciliates that are stalked in the adult stage and have rigid tentacles to catch prey.

Supernatant (1) The liquid remaining above a sediment or precipitate after sedimentation. (2) The most liquid stratum in a sludge digester.

Surface overflow rate A design criterion used for sizing clarifiers; typically expressed as the flow volume per unit amount of clarifier space (cubic meters per square meters per second [gallons per day per square foot]).

Surfactant A surface active agent, such as alkyl benzene sulfonate or linear alkylbenzene sulfonate, that concentrates at interfaces, forms micelles, increases solution, lowers surface tension, increases adsorption, and may decrease flocculation.

Suspended solids (1) Insoluble solids that either float on the surface of, or are in suspension in, water, wastewater, or other liquids. (2) Solid organic or inorganic particles (colloidal, dispersed, coagulated, or flocculated) physically held in suspension by agitation or flow. (3) The quantity of material removed from wastewater in a laboratory test, as prescribed in *Standard Methods* and referred to as nonfilterable residue.

Synergism Interaction between two entities producing an effect greater than a simple additive one.

Thick client application A software application that is loaded on a server computer and provides information that can only communicate with a remote client that has additional software loaded. Thick client implies that additional software <u>is required</u> to be loaded or configured on the remote client computer.

Thin client application A software application that is loaded on a server computer and provides information or data that can be viewed or manipulated from a browser on a remote client. Thin client implies that additional software <u>is not required</u> to be loaded or configured on the remote client computer.

Titration The determination of a constituent in a known volume of solution by the measured addition of a solution of known strength to completion of the reaction as signaled by observation of an endpoint.

Total organic carbon (TOC) The amount of carbon bound in organic compounds in a sample. Because all organic compounds have carbon as the common element, total organic carbon measurements provide a fundamental means of assessing the degree of organic pollution.

Total oxygen demand (TOD) A quantitative measure of all oxidizable material in a sample water or wastewater as determined instrumentally by measuring the depletion of oxygen after high temperature combustion. See also *Chemical oxygen demand* and *Total organic carbon*.

Total solids (TS) The sum of dissolved and suspended solid constituents in water or wastewater.

Total suspended solids (TSS) The amount of insoluble solids floating and in suspension in the wastewater. Also referred to as *total nonfilterable residue*.

Toxicant A substance that kills or injures an organism through chemical, physical, or biological action; examples include cyanides, pesticides, and heavy metals.

Toxicity The adverse effect that a biologically active substance has, at some concentration, on a living entity.

Trace nutrients Substances vital to bacterial growth. Trace nutrients are defined in this text as nitrogen, phosphorus, and iron.

Transmission control protocol/Internet protocol (TCP/IP) A protocol for communication between computers, used as a standard for transmitting data over networks and as the basis for standard Internet protocols.

Ultimate biochemical oxygen demand (BOD_u) (1) Commonly, the total quantity of oxygen required to completely satisfy the first stage BOD. (2) More strictly, the quantity of oxygen required to completely satisfy both the first and second stage BODs.

Upflow Term used to describe treatment units in which flow enters at the bottom and exits at the top.

Valence An integer representing the number of hydrogen atoms with which one atom of an element (or one radical) can combine (negative valence) or the number of hydrogen atoms the atom or radical can displace (positive valence).

Virus The smallest (10 to 300 μm in diameter) life form capable of producing infection and diseases in humans and animals.

Volatile acids Fatty acids containing six or fewer carbon atoms. They are soluble in water and can be steam distilled at atmospheric pressure. They have pungent odors and are often produced during anaerobic decomposition.

Volatile solids (VS) Materials, generally organic, that can be driven off from a sample by heating, typically to 550°C (1022°F); nonvolatile inorganic solids (ash) remain.

Volatile suspended solids (VSS) That fraction of suspended solids, including organic matter and volatile inorganic salts, that will ignite and burn when placed in an electric muffle furnace at 550°C (1022°F) for 60 minutes.

Volumetric Pertaining to measurement by volume.

Washout Condition whereby excessive influent flows (typically at peak flow conditions) cause the solids in the aeration basins and/or clarifiers to be carried over into downstream processes or discharged to the receiving stream.

Waste activated sludge (WAS) Solids removed from the activated sludge process to prevent an excessive buildup in the system.

Wastewater The spent or used water of a community or industry containing dissolved and suspended matter.

Index

A

Accumulation of growth, 97
Accuracy and repeatability,
    ammonia/ammonium analyzers, 518
    dissolved oxygen meter, 511
    oxidation–reduction potential sensor, 515
    pH meter, 514
    total suspended solids meter, 509
    UV nitrate probes, 519
Acetic acid, 273
Acetogenesis, 318
Acid fermentation, 141
Acidogenesis, 317
Activated primary fermentation tanks, 322
Activated sludge BNR processes, 259
Advanced control, 537
Aeration
    efficiency, 372
    intermittent, 539
    optimization/troubleshooting guide, 475
    sidestream, 366
    zones, 185
Aeration device, selection of, 187
Air distribution system, 96
Alarm notification, SCADA, 548
Alkalinity, 24, 42, 283, 434
    considerations, 282
    feed sources, 371
    measurement, 284
    optimization/troubleshooting guide, 486
    supplementation, 277
Alpha factor, 199
Alternate carbon sources, denitrification, 266
Aluminum compound chemical addition, 297
Ammonia
    analyzers, 520
    control, 537
    measurement, 517
    stripping, 374
    toxicity, 12
Ammonia-nitrogen, analysis, 401

Ammonia-nitrogen, monitoring, 440
Ammonification, 35
Ammonium, measurement, 517
Ammonium, utilization, 38
Anaerobic ammonia oxidation process, 373
Anaerobic/anoxic/oxic configurations, 130
Anaerobic/oxic configurations, 130
Analytical methods,
    nitrogen, 401
    phosphorus, 412
    short-chain volatile fatty acids, 415
Analyzers, 508, 520
Analyzers, phosphorus/orthophosphate, 519
ANAMOX process, 373
Annual average, 30
Ascorbic acid method, orthophosphate, 519
Ascorbic acid method, phosphorus, 413
Assimilation, 35
Attached growth systems, 55, 76, 102
Automated control, dissolved oxygen, 532
Automated process control, 503
Automatic control, process parameters, 521
Automatic control, set points, 523

B

BABE process, 376
Bardenpho process, 82
Belt filter presses, struvite, 384
Bioaugmentation batch enhanced process, 376
Biodenipho configuration, 137
Biological aerated filters, 67, 257, 263
Biological chemical nitrogen removal, 139
Biological chemical phosphorus removal, 139
Biological phosphorus removal
    chemical environment, 464
    data analysis/interpretation, 464
    performance indicators, 465
    potential test, 454
    solids retention time, 465
Biological selectors, 242
Biomass growth, 38, 71
Biomass inventory, optimization/troubleshooting guide, 481
Black foam, 247
Brown foam, 247

C

Calcium hydroxide, 281
Carbon augmentation, 74
Carbon supplementation, 257
Carbonaceous materials, 13
Cascade control, 535
Case studies, City of Bowie Wastewater Ttreatment Plant, 216
    Clean Water Services, 148
    Dodgeville, Wisconsin, 491
    Goldsboro Water Reclamation Facility, 217
    Greenville Utilities Commission Wastewater Treatment Plant, 221
    Havelock Wastewater Treatment Plant, 268
    Kalispell, Montana, 338
    Kelowna, Canada, 337
    Lethbridge Wastewater Treatment Plant, 146
    Long Creek Wastewater Treatment Plant, 269
    McAlpine Creek Wastewater Management Facility, 150
    McDowell Creek Wastewater Treatment Plant, 274

New York City Department of Environmental Protection, 376
North Cary Water Reclamation Facility, 219
Northwest Cobb Water Reclamation Facility, 306
Orange County, Florida, 495
Oxnard, California, 528
Potsdam Wastewater Treatment Plant, 217
Prague, Czech Republic, 374
Santa Clara/San Jose, California, 529
Shawano, Wisconsin, 468
South Cary Water Reclamation Facility, 218
South Cary, North Carolina, 341
Stamford, Connecticut, 497
Stevens Point, Wisconsin, 491
struvite control, 395
Toronto, Canada, 528
Traverse City Regional Wastewater Treatment Plant, 150
Virginia Initiative Plant, 222
Wilson Hominy Creek Wastewater Management Facility, 220
Centrate nitrification, 368
Centrate storage and conveyance systems, struvite, 384
Channel type systems, 171
Charts, SCADA historical, 547
Chemical addition, 241, 256, 536
Chemical environment,
    biological phosphorus removal, 464
    denitrification, 463
    nitrification, 462
Chemical feed control, 256, 306
Chemical feed system design, 309
Chemical oxygen demand, 186, 445

Chemical phosphorus removal, 189
Chemical polishing, 144
Chlorination, 241
Chlorine mass dose, 242
Chromotropic acid method, nitrate analysis, 406
City of Bowie Wastewater Treatment Plant, 216
Clarifier operation, optimization/troubleshooting guide, 483
Clarifier selection, 187
Clean Water Services, 148
Cleaning loops, struvite, 389
Coatings, 332
Colorimetric method,
    phosphorus, 413
    nitrite analysis, 404
    nitrogen analysis, 401
Colorimetry,
    ammonia/ammonium measurement, 517
    nitrate measurement, 518
    orthophosphate, 519
Combined denitrification systems, 78
Combined nitrification systems, 78
Combined nitrogen and phosphorus removal, 162, 192
Combined sidestream treatment processes, 374
Complete-mix fermenter, 326
Composite sampling, 428
Compressive biological model, 229
Continuous process control, 542
Control,
    advanced, 537
    ammonia, 537
    cascade, 535
    chemical addition, 536
    denitrification, 537

direct, 534
dissolved oxygen, 532
intermittent aeration, 539
on/off, 534
respirometric, 538
return activated sludge fermentation, 335
SCADA, 541
sequencing batch reactors, 540
Corrosion, 332
COST model, control approaches, 541
Counter current aeration, 87
Coupled systems, 67
Covers, fermenter, 331
Cyclical nutrient removal systems, 173
Cyclically aerated activated sludge, 86

D

Dark brown foam, 247
Dark tan foam, 247
Data,
    acquisition, 543
    analysis and interpretation, 462
    highways, 543
    model input, 230
Decision tree, optimization/troubleshooting, 467
Deflocculation, 203
Denitrification, 68
    chemical environment, 464
    control, 537
    data analysis/interpretation, 463
    filter, 76
    kinetics, 71
    performance indicators, 464

solids retention time, 464
test, 452
Dewatering filtrate, 275
Dewatering, 355
Diffused aeration, optimization/troubleshooting guide, 475
Diffusers, access to, 94
Digested sludge transfer pump stations, struvite, 386
Digester decant boxes, struvite, 385
Digestion methods, phosphorus analysis, 412
Dilution water, struvite, 389
Direct control, 534
Dissolved air flotation, 188
Dissolved oxygen
    adequate levels, 94
    concentration, 42, 182
    control, 532
    measurement, 510
    monitoring, 437
    sensors, 533, 534
    set point, 524
Distributed alarming, SCADA, 546
Dodgeville, Wisconsin, 491
Dose curve, 292, 299
Dynamic solids retention time, 526

E

Effluent concentration control, 266
Effluent filtration, 144
Effluent permit requirements, 29
Effluent requirements, 192
Electrode screening method, nitrate analysis, 405

Enhanced biological phosphorus removal (EBPR), 106, 108, 113, 129
Equalization, ammonia, 522
Equalization, sidestream, 366
Equipment selection, struvite, 389
Equipment, fermentation, 334
Ethernet communications, 543
Eutrophication, 12
Excess sludge flow control, 523

F

Facility design, struvite control, 391
Fault detection, dissolved oxygen sensors, 534
Feed-forward control, 265
Fermentate pumping, 330
Fermentation, 313
    primary sludge, 319
    return activated sludge, 319, 333
    sidestream equipment, 334
Fermented sludge, pumping, 330
Fermenters,
    covers, 331
    complete-mix, 326
    configurations, 322
    return activated sludge configuration, 333
    single-stage static, 327
    two-stage, 328
    unified fermentation and thickening, 329
Ferric compounds, 290
Ferrous compounds, 290
Filamentous bulking, 236, 239
Filamentous foaming, 247
Filamentous organisms, 234
Filtrate pump stations, struvite, 385
Final clarification, 354

Five-stage Bardenpho process, 165
Flow measurement, 332
Flow sheets for combined nutrient removal, 162
Flow variations, 25, 43
Flow-composite sampling, 430
Flow-paced control, 265
Flow-through pH meter, 514
Fluorescent technology, dissolved oxygen measurement, 512
Foam, 98, 214, 233, 243, 372
Froth, 243

G

Gas-selective electrodes, 517
Glycogen involvement, 113
Glycogen-accumulating organisms, 127
Goldsboro Water Reclamation Facility, 217
Grab sampling, 424
Greenville Utilities Commission Wastewater Treatment Plant, 221
Grinders, 330
Groundwater, 13
Growth, 97

H

Havelock Wastewater Treatment Plant, 268
Headspace monitoring, 333
High percentage nitrogen and phosphate removal, 195
High purity oxygen activated sludge, 284
High-activity ammonium removal, 372
Historian software, SCADA, 547
Historical collection, data, 547

Hybrid systems, 87, 138
Hydraulic retention time, 117
Hydroblasting, struvite, 389
Hydrolysis, 35, 317

I

In situ sampling, 424
Indophenol blue method, 517
Influent amenability to BNR, 351
Influent carbon augmentation, 140
Influent composition, 113
Influent solids concentration, 370
Information systems, SCADA, 548
Inhibition, 43
Input/output software, 543
Installation, analyzers, 520
    dissolved oxygen sensors, 533
    oxidation–reduction potential sensor, 516
    pH meter, 514
    total suspended solids meter, 509
Instrumentation, 332, 499
Instruments, advanced, 516
Instruments, maintenance, 507
In-tank ph meter, 514
Integrated fixed film activated sludge, 87
Interaction of nutrients in biological nutrient removal plants, 178
Interference, nitrogen sampling, 410
Intermittent aeration, 539
Internal recycle, 127
    rates, 212
    optimization/troubleshooting guide, 485
    set point, 524
    International Water Association, 119, 228
Interval sampling, 427, 429

Ion chromatography,
    nitrate analysis, 407
    nitrite analysis, 407
    nitrogen analysis, 401
    phosphorus, 413
Ion selective electrodes,
    ammonia/ammonium measurement, 517
    nitrate measurement, 518
    nitrogen analysis, 401
Iron compound chemical addition, 290

J

Jar testing, 292
Johannesburg configuration, 135
Johannesburg process, 168

K

Kalispell, Montana, 338
Kelowna, Canada, 337
Kinetic rate, 92
Kjeldahl method, organic nitrogen analysis, 409

L

Laboratory analyses, 399
Lagoon flushing, struvite, 391
Lethbridge Wastewater Treatment Plant, 146
Level measurement, 333
Lime addition, 303
Load variations, 25, 43, 351
Loading, optimization/troubleshooting guide, 469
Logic program, 542

Long Creek Wastewater Treatment Plant, 269
Loss of solids, 96
Loss of sponges, 95
Luminescent technology, dissolved oxygen measurement, 512

M

Magnesium hydroxide, 282, 305
Magnetic treatment, 389
Maintenance,
    ammonia analyzers, 520
    dissolved oxygen meter, 511, 533
    nutrient analyzers, 520
    oxidation-reduction potential sensor, 516
    pH meter, 515
    total suspended solids meter, 510
Management of return flows, 28
McAlpine Creek Wastewater Management Facility, 150
McDowell Creek Wastewater Treatment Plant, 274
Mean cell residence time, 352
Mean solids retention time, calculating, 526
Measurement,
    ammonia/ammonium, 517
    dissolved oxygen, 510
    nitrate and nitrite, 518
    oxidation-reduction potential, 515
    pH, 512
    phosphorus/orthophosphate, 519
    total suspended solids, 509

Mechanical aeration, optimization/troubleshooting guide, 478
Mechanistic models, 228
Media breakage, 93, 97
Media location, 92
Media mixing, 97
Membrane bioreactor, 98
Membrane technology, dissolved oxygen measurement, 510
Meters,
    dissolved oxygen, 510
    pH, 512
    reproducibility and accuracy, 506
    total suspended solids, 509
Methanogenesis, 318
Methanol,
    addition, 257, 263
    feed control, 265
    feed set point, 525
    storage, 258
Metric charts, historical, 548
Microbiological activity, monitoring, 460
Microscopic examination, 234
Mixed liquor suspended solids, monitoring, 430
Mixed liquor volatile suspended solids, monitoring, 430
Mixers, 331, 334
Mixing, 94
Mixing, optimization/troubleshooting guide, 475
Models, description, 228
Modified Bardenpho configuration, 132
Modified Johannesburg process, 169

Modified Ludzack-Ettinger process, 79
Modified University of Cape Town configuration, 134
Modified University of Cape Town Process, 168
Molybdovanadate method, orthophosphate, 519
Monitoring,
    alkalinity, 434
    ammonia-nitrogen, 440
    biological phosphorus removal potential, 454
    chemical oxygen demand, 445
    denitrification test, 452
    dissolved oxygen, 437
    microbiological activity, 460
    mixed liquor, 430
    network, 544
    nitrate-nitrogen, 442
    nitrification test, 451
    nitrite-nitrogen, 441
    orthophosphorus, 444
    oxidation-reduction potential, 438
    pH, 433
    return sludge, 430
    settleability, 432
    sludge volume index, 432
    soluble biochemical oxygen demand, 450
    temperature, 436
    total Kjeldahl nitrogen, 440
    total phosphorus, 443
    volatile fatty acids, 447
    waste sludge, 430
Monochloramine-F method, 517
Monthly average, 29
Moving bed biofilter reactor, 77

N

Network mapping/monitoring, 544
New York City Department of Environmental Protection, 376
Nitrate in groundwater, 13
Nitrate measurement, 518
Nitrate utilization, 71
Nitrate-nitrogen analysis, 404
Nitrate-nitrogen, monitoring, 442
Nitrification, 37
    kinetics, 38
    test, 451
    chemical environment, 462
    data analysis/interpretation, 462
    performance indicators, 463
    solids retention time, 463
Nitrifier growth rate, 37
Nitrite measurement, 518
Nitrite-nitrogen analysis, 404
Nitrite-nitrogen, monitoring, 441
Nitrogen analysis, 400
Nitrogen cycle, 16
Nitrogen removal, sidestream, 367
Nocardia, 247
Nonfilamentous bulking, 243
North Cary Water Reclamation Facility, 219
Northwest Cobb Water Reclamation Facility, 306
Nutrient analyzers, 520
Nutrient balance, 240
Nutrient release, 353
Nutrient sources, 8

O

Odor control, 331, 335

Odors, 94
On/off control, 534
Online analyzers, 506, 519 – 521
Open-channel pH meter, 514
Operation,
    nitrate/nitrite analyzers, 518
    oxidation–reduction potential measurement, 515
    pH meter, 513
Optimization,
    chemical dosages, 275
    guides, 465
    process parameters, 521
    techniques, 419
Orange County, Florida, 495
Orange Water and Sewer Authority process, 171
Organic nitrogen analysis, 408
Orthophosphate, 288, 306, 413, 519
Orthophosphorus, monitoring, 444
Oxidation ditches, 86, 136
Oxidation–reduction potential, 332, 438, 515
Oxnard, California, 528
Oxygen, effects, 182

P

Performance indicators,
    biological phosphorus removal, 465
    denitrification, 464
    nitrification, 463
Performance optimization, 523
Permit requirements, 29
pH, 24, 42, 184, 203, 241, 512
    meters, 333
    monitoring, 433
    optimization/troubleshooting guide, 486
Phoredox process, 166, 206
Phosphate accumulating organisms, 106
Phosphate removal but no nitrification, 192
Phosphate removal with nitrification but no denitrification, 193
Phosphate removal with nitrification only in summer, 194
Phosphate stripper, 138
Phosphate-accumulating organisms, 169, 178, 259, 271
Phosphorus
    analysis, 411
    measurement, 519
    precipitation, 285, 288, 305
    precipitating agents, struvite, 388
    recovery, struvite crystallization, 391
    removal, optimization/troubleshooting guide, 488
Phostrip configuration, 138
Pipe fittings, struvite, 385
Pipe lining, struvite, 389
Plant recycles, 129
Plastic media, 90, 96
Polymer, 305
Potsdam Wastewater Treatment Plant, 217
Prague, Czech Republic, 374
Pre-fermentation, 140
Primary clarification, 354
Primary clarifiers, 189
Primary nutrient release, 353
Primary sludge fermentation, 319, 322, 329
Primary sludge pumping, 330
Process control parameters, 521
Process design, struvite control, 392
Process evaluation, 422

Process fundamentals, 37
Programmable logic controller (PLC), 542
Protective coatings, 332
Pumping,
    fermentate, 330
    fermented sludge, 330
    primary sludge, 330

Q

Quicklime, 281

R

Reactor configuration, recycle nitrification, 372
Receiving waters, 12
Recycle flows, 27, 127
Recycle loads, eliminating/minimizing, 360
Recycle loads, estimating, 357
Recycle nitrification and denitrification, 372
Recycle nitrification, 370
Recycle stream, equalization, 366
Recycle stream, operational issues, 367
Redundancy, data, 544
Reporting, SCADA, 548
Respirometry, 538
Restoring alkalinity, 198
Retrofitting for nutrient removal, 208
Return activated sludge, 212
Return activated sludge fermentation, 319, 333
Return activated sludge, monitoring, 430
Return flowrate, set point, 525
Return flows, 28
Return stream management, 215
Rope type media, 88, 92
Rotating biological contactors, 62, 92

Rotifers, 235

S

Sampling, 424
    locations, 422, 428
    nitrogen analysis, 400
    phosphorus analysis, 411
    plan, 422
    sample handling, 423
    sidestream fermentation parameters, 336
    techniques, 422
Santa Clara/San Jose, California, 529
SCADA, 541, 544
    alarm notification, 548
    data distribution, 546
    data integrity, 546
    database, 545
    distributed alarming, 546
    engine (core), 544
    graphics, 545
    historian software, 547
    historical collection, 547
    historical trend charts, 547
    information systems, 548
    real-time trending, 545
    reporting, 548
    security, 549
    server emulation, 549
    thin client software, 549
Screen clogging, 94, 97
Screens, 330
Scum removal, 331
Seasonal permit, 30
Secondary clarification, 100
Secondary clarifiers, 190

Secondary nutrient release, 354
Security, SCADA, 549
Sedimentation tanks, activated primary, 322
Separate RAS regeneration, 376
Separate sludge systems, 53
Separate stage denitrification, 75
Sequencing batch reactors, 84, 173, 540
Server emulation, SCADA, 549
Set points, automatic control, 523
Settleability, monitoring, 432
SHARON process, 373
Shawano, Wisconsin, 468
Short-chain volatile fatty acids, analysis, 415
Sidesteams
    treatment, 367
    fermentation equipment, 334
    loads, 358
    equalization, 366
    management and treatment, 362, 365
Simulators, 228, 230
Simultaneous nitrification and denitrification, 163, 172
Single reactor systems, 372
Single sludge systems, 52
Single-stage static fermenter, 327
Sinking sponges, 95
Slaking, 281
Sludge age, automatic control, 525, 527
Sludge
    bulking problems, 233, 243
    collector drives, 329
    density meters, 333
    depth set point, 524
    fermentation, 313
    grinders, 330
    processing, 143

    production, 353
    settleability, 199
    struvite, 385
    transfer lines, 385
Sludge volume index, 241, 432
Sodium bicarbonate, 282
Sodium carbonate, 282
Sodium hydroxide, 279
Software, thin client, 549
Solids handling and processing, 349
Solids recycle pumps, 334
Solids removal, sidestream, 366
Solids retention time, 117
    biological phosphorus removal, 465
    denitrification, 464
    nitrification, 463
    set point, 524
Solids separation, 143
Solids, 21
Solids, loading variations, 351
Soluble biochemical oxygen demand, monitoring, 450
Source of phosphorus, 9
Sources of nitrogen, 8
South Cary Water Reclamation Facility, 218
South Cary, North Carolina, 341
Sponge media, 90, 94
Stabilization, 355
Stamford, Connecticut, 497
Stand-alone sidestream treatment, 368
Startup procedures, 96
Stevens Point, Wisconsin, 491
Stiff white foam, 247
Stoichiometry, 37, 70
Storage, nitrogen sampling, 400, 409
Storage, phosphorus sampling, 411

Storm flows, 213
Struvite, 371
    chemistry, 377
    control alternatives, 385
    crystallization, phosphorus recovery, 391
    formation, 352, 376, 382
Supervisory control and data acquisition (SCADA), 541, 544
Suspended growth systems, 43, 75, 78, 100, 130

T

Taking tank out of service, 96, 98
Technology based permits, 29
Temperature, 23, 122
Temperature, effects, 183
Temperature, monitoring, 436
Tertiary denitrification processes, 263
Tertiary filters, 190
Thickening, 354
Thin client software, SCADA, 549
Time-composite sampling, 429
Titrimetric methods, nitrogen analysis, 401
Toronto, Canada, 528
Total Kjeldahl nitrogen, 186, 440
Total phosphorus, 443, 519
Total suspended solids meters, 509
Toxicity, optimization/troubleshooting guide, 487
Traverse City Regional Wastewater Treatment Plant, 150
Trend charts, historical, 547
Trickling filter, 57
Troubleshooting, 202, 231, 243, 248, 419
Troubleshooting guides, 465
    aeration, 475, 478
    alkalinity, 486
    biomass inventory, 481
    chemical phosphorus removal, 488
    clarifier operation, 483
    diffused aeration, 475
    internal recycle, 485
    loadings, 469
    mechanical aeration, 478
    mixing, 475, 478
    mixing, 473
    pH, 486
    toxicity, 487
Two-stage fermenter/thickener, 328

U

Ultrasonic treatment, struvite, 389
Unified fermentation and thickening fermenter, 329
University of Cape Town (UCT) process, 121, 133, 166
UV absorbance, nitrate/nitrite measurement, 518

V

Vaccari method, sludge age, 526
Vanadomolydophosphoric acid colorimetric method, 413
Very dark foam, 247
Virginia Initiative Process (VIP) configuration, 135, 167, 222
Viscous bulking, 238
Volatile fatty acids, 108, 169, 178, 283
    addition, 140
    monitoring, 447

supplementation, 271

W

Washout, 119
Waste activated sludge, monitoring, 430
Waste sludge, 215
Wastewater characteristics, 13, 35
Wastewater temperature, 41
Water quality based permit, 30
Westbank process, 169
Wilson Hominy Creek Wastewater Management Facility, 220
Worms, 93, 97
Wuhrmann process, 79

Z

Zullig technology, dissolved oxygen measurement, 511